Springer Collected Works in Mathematics

For further volumes:
http://www.springer.com/series/11104

Chern at 3, with Grandmother

Hamburg, 1935

Shiing-Shen Chern

Selected Papers II

Reprint of the 1989 Edition

Springer

Shiing-Shen Chern
(1911 Jiaxing, China –
2004 Tianjin, China)
The University of Chicago
Chicago, IL, USA

ISSN 2194-9875
ISBN 978-1-4614-8976-4 (Softcover)
978-0-387-96816-2 (Hardcover)
DOI 10.1007/978-1-4614-9343-3
Springer New York Heidelberg Dordrecht London

Library of Congress Control Number: 2012954381

Mathematics Subject Classification (2010): 01A75, 53-02

Printed on acid-free paper

Springer is part of Springer Science+Business Media (www.springer.com)

Contents

*Numbers in brackets refer to the Bibliography of the Publications of S.S. Chern (see pages xxiii–xxx).

Curriculum Vitae of Shiing-shen Chern

I. Born October 26, 1911 in Kashing, Chekiang Province, China.

B.Sc., Nankai University, Tientsin, China, 1930.

M.Sc., Tsinghua University, Peiping, China, 1934.

D.Sc., University of Hamburg, Germany, 1936.

China Foundation Postdoctoral Fellow at the Sorbonne, Paris, France, 1936–37.

Professor of Mathematics, Tsinghua University and Southwest Associated University, Kunming, China, 1937–43.

Member, Institute for Advanced Study, Princeton, USA, 1943–45.

Acting Director, Institute of Mathematics, Academia Sinica, Nanking, China, 1946–48.

Professor of Mathematics, University of Chicago, 1949–60.

Professor of Mathematics, University of California at Berkeley, 1960–79; Professor Emeritus, 1979 to present.

Director, Mathematical Sciences Research Institute, Berkeley, 1981–84; Director Emeritus, 1984 to present.

Became US citizen 1961.

Director, Nankai Institute of Mathematics, Tianjin, China, 1984 to present.

II. Visiting Professor or Member: Harvard University 1952, Eidgenossische Technische Hochschule, Zurich 1953, Massachusetts Institute of Technology 1957, Institute for Advanced Study 1964, University of California at Los Angeles 1966, Institut des Hautes Etudes Scientifiques, Paris 1967, Instituto de Matematica Pura e Aplicada, Rio de Janeiro, Brazil 1970, University of Warwick, Coventry, England 1972, Rockefeller University, New York 1973, Eidgenossische Technische Hochschule 1981, Max Planck Institut für Mathematik, Bonn, Germany 1982.

III. Member, Academia Sinica, 1948 to present.

Honorary Member, Indian Mathematical Society, 1950 to present.

Member, National Academy of Sciences, 1961 to present.

Vice-President, American Mathematical Society, 1962–64.

Fellow, American Academy of Arts and Sciences, 1963 to present.

Corresponding Member, Brazilian Academy of Sciences, 1971 to present.

Associate Founding Fellow, Third World Academy of Sciences, 1983 to present.

Foreign Member, Royal Society of London, 1985 to present.

Honorary Member, London Mathematical Society, 1986 to present.

Corresponding Member, Academia Peloritana, Messina, Sicily, 1986 to present.

Honorary Life Member, New York Academy of Sciences, 1987 to present.

Foreign Member, Academia dei Lincei, Rome, 1988 to present.

IV. LL.D. (hon), The Chinese University of Hongkong, 1969.

D.Sc. (hon), University of Chicago, 1969.

D.Sc. (hon), University of Hamburg, 1971.

Dr. Math. Eidgenossische Technische Hochschule Zurich, 1982.

D.Sc. (hon), SUNY at Stony Brook, 1985.

Hon Doctor, Nankai University, 1985.

V. Honorary Professor: Beijing University 1978, Nankai University 1978, Institute of Systems Science, Academy of Sciences 1980, Chinan University, Canton 1980, Graduate School, Academy of Sciences 1984, Nanjing University 1985, East China Normal University 1985, Chinese University of Science and Technology 1985, Beijing Normal University 1985, Chekiang University 1985, Hangchow University 1986, Fudan University 1986, Shanghai University of Technology 1986, Tianjin University 1987, Tohoku University, Japan 1987.

VI. Chauvenet Prize, Mathematical Association of America 1970.

National Medal of Science 1975.

Alexander von Humboldt Award, Germany 1982.

Steele Prize, American Mathematical Society 1983.

Wolf Prize, Israel 1983–84.

A Summary of My Scientific Life and Works*

By Shiing-shen Chern

I was born on October 26, 1911 in Kashing, Chekiang Province, China. My high school mathematics texts were the then popular books *Algebra* and *Higher Algebra* by Hall and Knight, and *Geometry* and *Trigonometry* by Wentworth and Smith, all in English. Training was strict and I did a large number of the exercises in the books. In 1926 I enrolled as a freshman in Nankai University, Tientsin, China. It was clear that I should study science, but my disinclination with experiments dictated that I should major in mathematics. The Mathematics Department at Nankai was a one-man department whose Professor, Dr. Li-Fu Chiang, received his Ph.D. from Harvard with Julian Coolidge. Mathematics was at a primitive state in China in the late 1920s. Although there were universities in the modern sense, few offered a course on complex function theory and linear algebra was virtually unknown. I was fortunate to be in a strong class of students and such courses were made available to me, as well as courses on non-Euclidean geometry and circle and sphere geometry, using books by Coolidge.

The period around 1930, when I graduated from Nankai University, saw great progress in Chinese science. Many students of science returned from studies abroad. At the center of this development was Tsing Hua University of Peking (then called Peiping), founded through the return of the Boxer's Indemnity by the U.S. I was an assistant at Tsing Hua in 1930–1931 and was a graduate student from 1931–1934. My teacher was Professor Dan Sun, a former student of E.P. Lane at Chicago. Therefore, I began my mathematical career by writing papers on projective differential geometry.

In 1934 I was awarded a fellowship to study abroad. I went to Hamburg, Germany, because Professor W. Blaschke lectured in Peking in 1933 on the geometry of webs and I was attracted by the subject. I arrived at Hamburg in the fall of 1934 when Kähler's book *Einführung in die Theorie der Systeme von Differentialgleichungen* was published and he gave a seminar based on it. In a less than two-year stay in Hamburg I worked in more depth on the Cartan–Kähler theory than any other topic. I received my D.Sc. in February 1936.

The completion of the degree fulfilled my obligation to the fellowship. It was natural to look forward to a carefree postdoctoral year in Paris with the master himself, Elie Cartan. It turned out to be a year of hard work. In 1936–1937 in Paris I learned moving frames, the method of equivalence, more Cartan–Kähler theory, and, most importantly, the mathematical language and the way of thinking of Cartan. Even now I frequently find Cartan easier to follow than some of his expositors.

* Originally written 1978; updated and revised 1988.

I returned to China in the summer of 1937 to become Professor of Mathematics at Tsing Hua University. I crossed the Atlantic on the S.S. Queen Elizabeth and, after a month long tour of the United States, I crossed the Pacific on the S.S. Empress of Canada. The Sino-Japanese war broke out while I was on board and I never reached Peking.

During the war Tsing Hua University moved to Kunming in Southwest China and became a part of Southwest Associated University. Mathematically it was a period of isolation. I taught courses on advanced topics (such as conformal differential geometry, Lie groups, etc.) and had good students.

In 1943 I became a member of the Institute for Advanced Study; both Veblen and Weyl were aware of my work. During the period 1943–1945 I learned algebraic topology and fiber bundles and did my work on characteristic classes, among other things. The war ended in 1945 and I decided to return to China. Postwar transportation difficulties delayed my trip so that I did not arrive in Shanghai until March 1946. I was called to organize a new institute of Mathematics of the Academic Sinica in Nanking. The work lasted only for about two years. On December 31, 1948 I left Shanghai for the United States, again on an invitation of the Institute for Advanced Study. (See Weil's article in Volume I. Before leaving China I was offered a position at the Tata Institute in Bombay, then at a planning stage, which I was not able to accept. The offer must have come on the initiative of D.D. Kosambi, the first professor of mathematics at Tata, who knew well my work on path geometry.) I spent the winter term of 1949 at the Institute. During 1949–1960 I was a Professor at the University of Chicago.

In 1960 I moved to Berkeley where I became Professor Emeritus in 1979. Together with C.C. Moore and I.M. Singer I submitted a proposal to the National Science Foundation for a Mathematical Institute in Berkeley. It was granted and I became the Director of the Mathematical Sciences Research Institute in 1981–84. After my retirement I started a mathematical institute at my alma mater, Nankai University, Tianjin, China. I am hoping that my last retirement will come soon.

In the following I will try to give a summary of my mathematical works.

1. Projective Differential Geometry

Einstein's general relativity provided the great impetus to the study of Riemannian geometry and its generalizations. Before that, geometry was dominated by Felix Klein's Erlangen Program announced in 1871, which assigns to a space a group of transformations which is to play the fundamental role. Thus the Euclidean space has the group of rigid motions and the projective space has the group of projective collineations, etc. Along the lines to classical curve and surface theory in the tradition of Serret–Frenet, Euler, Monge, and Gauss, projective differential geometry was founded by E.J. Wilczynski and G. Fubini and E. Cech. Its main problem is to find a complete system of local invariants of a submanifold under the projective group and interpret them geometrically through osculation by simpler geometrical figures. The main difficulty lies in that the projective group is relatively large and invariants can only be reached through a high order of osculation. Moreover, the group of isotropy is non-compact, a fact which excludes many beautiful geometrical properties.

In my first papers [1], [2] I avoided the first difficulty by studying more complicated figures. The papers are nothing more than exercises, but the philosophy behind them found an echo in the recent works of P.A. Griffiths on webs, Abel's theorem, and their applications to algebraic geometry. For example, instead of studying an algebraic curve of degree d in the plane, one can study the configuration consisting of d points on each line of the plane, its points of intersection with the curve. One gets in this way d arcs in correspondence. Paper [1] studies two arcs in correspondence.

My next paper [3] concerns projective line geometry, now a forgotten subject. A line complex is, in modern terminology, a hypersurface in the Plüker-Grassmann manifold of all lines in the three-dimensional projective space. While the consideration of tangent spheres of a surface leads to the fundamental notions of lines of curvature and principal curvatures and that of the tangent quadrics of a projective surface leads to the quadrics of Darboux and Lie, the use of quadratic line complexes in the study of general line complexes was initiated in this paper.

Several years later I returned to projective differential geometry by introducing new invariants of contact of a pair of curves in a projective space of n dimensions, and also of surfaces [17], [19]. They include as a special case, the invariant of Mehmke–Smith, which plays a role in some questions on singularities in several complex variables. Generally speaking, the study of diffeomorphism invariants of a jet at a singularity has recently attracted wide attention (H. Whitney, R. Thom). The projective invariants, studied extensively by Italian differential geometers, should enter into the more refined questions.

The Laplace transforms of a conjugate net was a favorite topic in the theory of transformations of surfaces. It is a beautiful geometric construction which leads to a transformation of linear homogeneous hyperbolic partial differential equations of the second order in two variables. In [24], [35] a generalization was given to a class of submanifolds of any dimension. This generalization could be related to the recent search of high-dimensional solitons and their Bäcklund transformations.

From projective spaces it is natural to pass to spaces with paths where the straight lines are replaced by the integral curves of a system of ordinary differential equations of the second order, an idea which could be traced back to Hermann Weyl. Such spaces are said to be projectively connected or to have a projective connection. Projective relativity (O. Veblen, J.A. Schouten) aims at singling out the projectively connected spaces whose paths are to be identified with the trajectories of a free particle in a unified field theory. They are defined by a system of "field equations." A new system of field equations was proposed in [111].

From the mathematical viewpoint, projectively connected spaces are of intrinsic interest. Relating projective spaces and general projectively connected spaces is the imbedding problem. Given a submanifold M in a projective space, an induced projective connection can be defined on M by taking a field of linear subspaces transversal to the tangent spaces of M and projecting neighboring tangent spaces from them. In [7] I proved an analogue of the Schläfli–Janet–Cartan imbedding theorem for Riemannian spaces of which the following is a special case: A real analytic normal (in the sense of Cartan) projective connection on a space of dimension n can be locally induced by an imbedding in a projective space of dimension $n(n+1)/2 + [n/2]$. The dimension needed is thus generally higher than in Schläfi's case.

The fundamental theorem on projective connections is the theorem associating a unique normal projective connection to a system of paths. I announced in [23] that the same is true when there is in a space of dimension n a family of k-dimensional submanifolds depending on $(k+1)(n-k)$ parameters and satisfying a completely integrable system of differential equations. The case $k = 1$ is classical and the case $k = n - 1$ was the main conclusion of M. Hachtroudi's Paris thesis. My derivation was long and was never published. A geometrical treatment was later given by C.T. Yen (*Annali di Matematica* 1953).

In the Princeton approach to non-Riemannian geometry led by Veblen and T.Y. Thomas, a main tool is the use of normal coordinates relative to which the normal extensions of tensors are defined. Normal coordinates in the projective geometry of paths can be given different definitions; their existence is generally not easy to establish. In [8] I showed that Thomas's normal coordinates are in general different from the normal coordinates defined naturally from Cartan's concept of a projective connection.

In my recent joint works with Griffiths on webs [112] we came across a theorem characterizing a flat normal projective connection as one with ∞^2 totally geodesic hypersurfaces suitably distributed; the classical theorem needs ∞^n totally geodesic hypersurfaces, n being the dimension of the space.

In concluding this section, I wish to say that I believe that projective differential geometry will be of increasing importance. In several complex variables and in the transcendental theory of algebraic varieties the importance of the Kähler metric cannot be over-emphasized. On the other hand, projective properties are in the holomorphic category. They will appear when the problems involve, directly or indirectly, the linear subspaces or their generalizations.

2. Euclidean Differential Geometry

Before the nineteen-forties, a mathematical student was usually introduced to differential geometry through a course on curves and surfaces in Euclidean space, known in European universities as "applications of the infinitesimal calculus to geometry". I was particularly fascinated by Blaschke's book for its emphasis on global problems. I was, however, able to do some work only after I began to treat surface theory by moving frames. In [29] I observed that Hilbert's proof of the rigidity of the sphere gives the more general theorem that a closed strictly convex surface in E^3 (= three-dimensional Euclidean space) is a sphere if one principal curvature is a monotone decreasing function of the other.

More generally, a natural area of investigation in Euclidean differential geometry is concerned with the W-hypersurfaces, where there is a functional relation between the principal curvatures. If a hypersurface is closed and strictly convex, its Gauss map into the unit hypersphere is one-to-one and we can identify functions on the hypersurface with those on the unit hypersphere. Let σ_r, $1 \leq r \leq n$, be the rth elementary symmetric function of the reciprocals of the principal curvatures of a convex hypersurface in

E^{n+1}. In [68] I proved that if, for a certain r, the σ_r functions of two closed strictly convex hypersurfaces Σ, Σ^* in E^{n+1} agree as functions on the unit hypersphere, then Σ and Σ^* differ by a translation. The condition means geometrically that σ_r are the same at points of Σ, Σ^* at which the normals are parallel. In [69], Hano, Hsiung, and I proved a similar uniqueness theorem by replacing the conditions by $\sigma_r \leq \sigma_r^*$, $\sigma_{r+1} \geq \sigma_{r+1}^*$ for a certain r. The proofs depend on the establishment of some integral formulas.

In [81] I considered hypersurfaces with boundary in the Euclidean space and found upper bounds on their size if certain curvature conditions are satisfied. This generalized some work of E. Heinz and S. Bernstein for surfaces in E^3.

Again using integral formulas, Hsiung and I proved in [77] that a volume-preserving diffeomorphism of two k-dimensional compact submanifolds in E^n is an isometry if a certain additional condition is satisfied.

In [62] and [66] Lashof and I studied the total curvature of a compact immersed submanifold in E^n. The total curvature is defined as the measure of the image of the unit normal bundle on the unit hypersphere of E^n under the Gauss map. (Observe that independent of the dimension of a submanifold the unit normal bundle has dimension $n - 1$, which is the dimension of the unit hypersphere of E^n.) The total curvature was considered by J. Milnor following his work on that of a knot. Generalizing the classical theorems of Fenchel for the total curvature of a closed space curve, Lashof and I proved that the total curvature of a compact immersed submanifold in E^n, when properly normalized, has a universal lower bound and that it is reached when and only when the submanifold is a convex hypersurface. As a corollary, it is proved that a closed surface of non-negative Gaussian curvature in E^3 is convex, generalizing a classical theorem of Hadamard. In this work, a lemma on the local behavior of a hypersurface with degenerate second fundamental form plays a fundamental role. Total curvature and tight immersion have received many interesting developments in recent years (Kuiper, Banchoff, Pohl, and Chen).

Among these is Banchoff's introduction of the notion of a taut immersion, which means that the distance function of a point of the submanifold from any point in space has the smallest number of critical points. This is a stronger property than tight immersion. In [143] Tom Cecil and I proved that tautness is invariant under the Lie group of sphere transformations (=group formed by all contact transformations carrying spheres to spheres). We also introduced some basic notions of the differential geometry in Lie sphere geometry, such as the Legendre map and the Dupin submanifold.

It was Bonnet who studied isometric deformations of surfaces in E^3 preserving the mean curvature. The problem leads to a complicated over-determined system of partial differential equations which has been studied by many authors. In [133] I showed that these are either surfaces of constant mean curvature or form an exceptional family, depending on 6 constants, which consists of W-surfaces. In the analytical treatment the connection form of the unit tangent bundle is heavily used.

Given an oriented (two-dimensional) surface in E^4, its Gauss map has as image the Grassmann manifold of all oriented planes through a point. The latter is homeomorphic to $S^2 \times S^2$. As a result the map defines a pair of integers. Spanier and I [47] proved that if the surface is imbedded, these two integers are equal when the spaces are properly oriented.

In [50] Kuiper and I introduced two integers to an immersed manifold in E^n: the indices of nullity and of relative nullity. Inequalities are established between them and the dimension and codimension of a compact submanifold in E^n. The origin of this work was a theorem of Tompkins that there is no closed surface in E^3 whose Gaussian curvature is identically zero.

The smoothness requirements of various theorems in surface theory have been thoroughly investigated by P. Hartman and A. Wintner in a long series of papers. In [55] we studied the critical case for the isothermic coordinates, namely, the minimum conditions so that the metric in the isothermic coordinates has the same smoothness.

Finally I wish to mention a result on complex space-forms. In his thesis, Brian Smyth determined the complete Einstein hypersurfaces in a Kählerian manifold of constant holomorphic sectional curvature by using the classification of symmetric Hermitian spaces. The result turns out to be a local one. The problem leads to an over-determined differential system and I showed in [87] that the theorem follows from a careful study of the integrability conditions. The hypersurfaces in question are either totally geodesic or are hyperspheres.

Euclidean differential geometry is comparable to elementary number theory in its beauty of simplicity. Unlike the latter more remains to be discovered.

3. Geometrical Structures and Their Intrinsic Connections

A Riemannian structure is governed by its Levi-Civita connection, and a path structure by its normal projective connection. A fundamental problem of local differential geometry is to associate to a structure a connection which describes all the properties. An effective way of doing this is by Elie Cartan's method of equivalence. In the years 1937–1943 when I was isolated in the interior of China I carried out the program in many cases:

The geometry of the equation of the second order

$$y'' = F(x, y, y'), \qquad y' = dy/dx, \qquad y'' = d^2y/dx^2$$

in the (x, y)-plane was studied by A. Tresse. Tresse's results were formulated in terms of the Lie theory; it would be more geometrical to say that a normal projective connection can be defined in the space of line elements (x, y, y'). I studied the equation of the third order

$$y''' = F(x, y, y', y'')$$

under the group of contact transformations in the plane and showed that in an important case a conformal connection can be defined intrinsically [6], [13]. I also defined affine connections from structures arising from webs [9] (cf. §8).

Local differential geometrical structures are defined either by differential systems or by metrics, the two typical cases being projective geometry and Euclidean geometry. When the paths are the integral curves of a system of ordinary differential equations, the allowable parameter change has an important bearing on the resulting geometry. D.D. Kosambi considered a system of differential equations of the second order with an allowable affine transformation of parameters and attached to the structure an affine connection. I proved in [10] the result by the method of equivalence and went on in [11] to solve the corresponding problem when the paths are defined by a system of differential equations of higher order.

Geometrically it is more natural that a family of submanifolds is given with unrestricted parametrization, i.e., the parameters are allowed arbitrary (smooth) changes. Generalizing Tresses's problem to n dimensions, the given data should be $\infty^{2(n-1)}$ curves satisfying a differential system such that through any point and tangent to any direction at the point there is exactly one such curve. With these curves taking the place of the straight lines, a generalized projective geometry, i.e., a normal projective connection, can be defined. As mentioned in §1, I extended this result to the case when there is given $\infty^{(k+1)(n-k)}$ k-dimensional submanifolds satisfying a differential system. In the same vein I defined in [20] a Weyl connection, giving ∞^2 surfaces in $\mathbb{R}^3$ as "isotropic surfaces." This was extended to n dimensions in [21], but the details of the n-dimensional case were never published.

In [22], [42] I studied the connections to be attached to a Finsler metric and showed that there is more than one natural choice.

In 1972 Moser found a local normal form of a non-degenerate real hypersurface in C_2 and asked me to identify his invariants with those of Elie Cartan. Years before I had extended Cartan's work to a real hypersurface in C_{n+1}. I have not published the results, partly because a paper of Tanaka on the same subject appeared in the meantime, although Tanaka made an assumption on the hypersurface (which he removed in a later paper). In [105] Moser and I gave both the normal form of a non-degenerate real hypersurface in C_{n+1} and its intrinsic connection as a CR-manifold and identified the two sets of invariants. When the hypersurface is real analytic, I defined in [107] a projective connection. The latter does not give all the invariants, but has the advantage that its invariants are in the holomorphic category.

All these are special cases of a G-structure. Some G-structures, such as the complex structures, admit an infinite pseudo-group of transformations. In [54] I gave an introduction to G-structures, including the notion of a torsion form and an exposition of Cartan's theory of infinite continuous pseudo-groups. A more complete account of G-structures was given in [83].

In [61] I observed that the Hodge harmonic theory is valid for a torsionless G-structure, with $G \subset O(n)$; the Hodge decomposition can then be generalized to the decomposition of a harmonic form into irreducible summands under the action of G. This viewpoint also gives a better understanding of Hodge's results.

Among mathematical disciplines the area of geometry is not so well defined. Perhaps the notion of a G-structure is of sufficient scope to fulfill the current requirements for the mainstream of geometry.

4. Integral Geometry

I went to Hamburg in 1934 when Blaschke, in his usual style, started a series of papers entitled "Integral Geometry". Although I have a keen interest in the subject, my works on it have been scattered.

I observed that integral geometry in the tradition of Crofton deals with two homogeneous spaces with the same group. Call the group G. If the homogeneous spaces are realized as coset spaces G/H and G/K, H and K being subgroups of G, two cosets aH and bK, $a, b \in G$, are called incident if they have an element in common. With this notion of incidence, Crofton's formula was established in a very general context [14], [16], [18]. This notion of incidence was appreciated by Weil and found useful in later works of Helgason and Tits.

My other work on integral geometry concerns the kinematic density of Poincaré. With Chih-Ta Yen I gave a proof of the fundamental kinematic formula in E^n [15], [48].

In his formula for the volume of a tube, Weyl introduced a number of scalar invariants of an imbedded manifold in E^n, half of which depend only on the induced metric. If M^p and M^q are closed imbedded manifolds of E^n, with M^p fixed and M^q moving, I proved in [84] a simple formula expressing the integral of an invariant of the intersection $M^p \cap M^q$ over the kinematic measure. This complements the fundamental kinematic formula, which deals with hypersurfaces.

5. Characteristic Classes

My introduction to characteristic classes was through the Gauss–Bonnet formula, known to every student of surface theory. Long before 1943, when I gave an intrinsic proof of the n-dimensional Gauss–Bonnet formula [25, 30], I knew, by using orthonormal frames in surface theory, that the classical Gauss–Bonnet is but a global consequence of the Gauss formula which expresses the "theorema egregium." The algebraic aspect of the proof in [25] is the first instance of a construction later known as transgression, which is destined to play a fundamental role in the homology theory of fiber bundles, and in other problems.

The Gauss–Bonnet formula is concerned with the Euler–Poincaré characteristic. It was natural to look at corresponding results for the general Stiefel–Whitney characteristic classes, then newly introduced. I soon realized that the latter are essentially defined only mod two and relating them with curvature forms would be artificial. Technically its cause lies in the complicated homology structure of the orthogonal group, such as the presence of torsion. The Grassmann manifold and the Stiefel manifold over the complex numbers have no torsion, and the same is true of the unitary group. In [33] I introduced the characteristic classes of complex vector bundles and related them via the de Rham theorem, with the curvature forms of an Hermitian structure in the bundle. Actually this paper contains, through the explicit construction of differential forms, the essence of the homology structure of a principal bundle with the unitary group as structure group: transgression, characteristic classes, universal bundle, etc. These characteristic classes are defined for algebraic manifolds, but their definition, whether via an Hermitian structure or via the universal bundle, is not algebraic. In [51] I showed that by considering an associated bundle with the flag manifold as fibers the characteristic classes can be defined in terms of those of line-bundles. As a consequence the dual homology class of a characteristic class of an algebraic manifold contains a representative algebraic cycle.

The study of the homology structure of a fiber bundle through the use of a connection merges local properties into global properties and combines differential geometry with differential topology. The general case of a principal bundle with an arbitrary Lie group as structure group, of which my work above is a special case concerning the unitary group, was carried out by Weil in 1949 in an unpublished manuscript. Part of Weil's results was presented in [I, 1]. The main conclusion is the so-called Weil homomorphism which identifies the characteristic classes (through the curvature forms) with the invariant polynomials under the action of the adjoint group. This identification, whose importance should be immediately recognizable, has recently been found crucial in the heat equation proof of the Atiyah–Singer index theorem and in Bott's theorem on foliations.

Actually, the characteristic forms themselves, which represent the characteristic classes via the de Rham theorem, contain more information. The vanishing of the characteristic forms, not just their classes, (which only means that the forms are exact), leads to the secondary characteristic classes. These were studied with James Simons in [98, 103]. The secondary characteristic classes depend on the choice of the connection, but enjoy strong invariance properties under a change of the connection. They have been found to play a role in various problems, such as conformal immersions and the η-invariant defined by the spectrum of a compact Riemannian manifold. A duality theorem for characteristic forms was given in a joint paper with White [108].

In [39] I determined the mod two cohomology ring of the real Grassmann manifold. As a consequence it follows that the Stiefel–Whitney classes generate the mod two characteristic ring of a sphere bundle. The result plays a role in the estimation of the number of closed geodesics on a compact Riemannian manifold.

When the base manifold has a complex structure, its ring of complex-valued exterior differential forms has also a more refined structure. Forms have a bidegree and there are two exterior differentiations, one with respect to the complex structure and the other to its conjugate complex structure, denoted usually by ∂, $\bar{\partial}$ respectively. In [80, 92] Bott and I studied the forms of a holomorphic Hermitian vector bundle relative to the operator $i\partial\bar{\partial}$. This has applications to complex geometry, and in particular to the study of the zeroes of holomorphic sections, which contains as a particular case the classical theory of value distributions of meromorphic functions.

In [101] I gave an elementary proof (without sheaf cohomology) of Bott's theorem on characteristic numbers and the residues of a meromorphic vector field on a compact complex manifold. The proof is in the spirit of a transgression.

On a manifold it is necessary to use covariant differentiation; curvature measures its non-commutativity. Its combination as a characteristic form measures the non-triviality of the underlying bundle. This train of ideas is so simple and natural that its importance can hardly be exaggerated.

6. Holomorphic Mappings

The simplest case of a holomorphic mapping is $\mathbb{C} \to P_1$, where $\mathbb{C}$ is the complex line and P_1 is the complex projective line. In usual terminology $\mathbb{C}$ is called the Gaussian plane and P_1 the Riemann sphere; the mapping is known as a meromorphic function. The geometrical basis of the classical value distribution theory consists of two theorems, known as the first and second main theorems, which are but the Gauss–Bonnet theorem applied to the Hopf bundle and the canonical bundle of P_1 respectively. From these the Nevanlinna defect relation follows by calculus-type inequalities.

In [70] these viewpoints were made precise by the study of holomorphic mappings of a non-compact Riemann surface into a compact one. As a differential geometer I have naturally been interested in the theory of a family of meromorphic functions interpreted as a holomorphic curve in P_n, the complex projective space of n dimensions, as developed by Henri Cartan, H. and J. Weyl, and Ahlfors. A geometrical treatment was given in [99] for P_2; the corresponding results for P_n were worked out by H. Yamaguchi in an unpublished manuscript. An essential ingredient for the good distributional behavior of a non-compact holomorphic curve lies in the validity of Frenet-type formulas. Cowan, Vitter, and I considered holomorphic curves in any complex manifold M and showed that Frenet formulas will be valid only when M has very special properties, which are close to being of constant holomorphic sectional curvature [104].

It is natural to consider holomorphic mappings in higher dimensions, a broad subject of which much remains to be understood. In [75] I gave some general observations. Following some work of H. Levine, done with my supervision, I studied in [71] a holomorphic mapping $f: C_n \to P_n$ and proved that under some growth conditions the set $P_n - f(C_n)$ is of measure zero.

In [80] Bott and I reformulated the value distribution problem as one on the distribution of zeroes of the holomorphics sections of a holomorphic vector bundle. A preparatory algebraic problem consists of the study of complex transgression, i.e., transgression relative to the operation $i\partial\bar{\partial}$ [80], [92]. Characteristic classes are defined in a refined sense, which is of importance in applications to problems pertaining to the holomorphic category.

In [88] I proved a Schwarz lemma in high dimensions as a volume-decreasing property. With S.I. Goldberg [106] an analogous theorem was proved for a class of harmonic mappings of Riemannian manifolds.

In [90] I introduced with H. Levine and L. Nirenberg intrinsic pseudo-norms in the real cohomology vector spaces of a complex manifold M. The definition utilizes the pluri-subharmonic functions. The pseudo-norm becomes a norm when there are enough pluri-subharmonic functions in M. Under a holomorphic mapping the pseudo-norm is a non-increasing function.

Geometry occupies an important position in complex function theory. Its role in several complex variables will be even greater in the future.

7. Minimal Submanifolds

The Grassmann manifold $\tilde{G}_{2,n}$ of all the oriented planes through a point in E^n has a complex structure invariant under the action of $SO(n)$. On the other hand, an oriented surface in E^n has a complex structure through its induced Riemannian metric. The surface is minimal if and only if the Gauss map is anti-holomorphic [79]. This theorem was proved by Pinl for $n = 4$ and is clearly the starting point in relating minimal surfaces with complex function theory. One of the fundamental theorems on minimal surfaces is the Bernstein uniqueness theorem which says that a minimal surface $z = f(x, y)$ in E^3 defined for all x, y must be a plane. As generalized by Osserman, Bernstein's theorem is a consequence of the theorem that if a complete minimal surface is not a plane, its image on the unit sphere under the Gauss map is dense. In [79] this was generalized to a density theorem on the image under the Gauss map of a complete minimal surface in E^n, which is not a plane. More refined density theorems were established in [86], a joint paper with Osserman.

The geometry of minimal surfaces on a sphere S^n takes a different course, because one naturally starts by studying the compact ones. As the codimension is arbitrary, it is necessary to consider higher osculating spaces. This was carried out in [96], in which it is shown that their dimensions are successive even integers. In fact, this is true for a minimal surface in any space form. According to E. Calabi, minimal two-spheres in S^n can be enumerated, because of the simple conformal behavior of the two-sphere. In [94] I gave a simple construction of all the minimal two-spheres in S^4. This work was continued by Lucas Barbosa for minimal 2-spheres in S^n.

In [91], following the celebrated work of Simons on closed minimal submanifolds (of dimension k) on the unit sphere S^n, do Carmo, Kobayashi, and I considered the square of the length of the second fundamental form, to be denoted by $\|h\|^2$. It follows from Simons' inequality that if $\|h\|^2 \leq k(n-k)/(2n-2k-1)$, then $\|h\|^2 = 0$ or $k(n-k)/(2n-2k-1)$. We determined all the closed minimal submanifolds of S^n with $\|h\|^2 = k(n-k)/(2n-2k-1)$.

Another space whose closed minimal surfaces can be described more concretely is the complex projective space, or more generally, the complex Grassmann manifold. The solutions find applications in theoretical physics and are known as the σ-models. Mathematically, such harmonic maps have been studied by J. Eells and his coworkers. Wolfson and I, in [131], [135], [142], applied the method of moving frames, thus giving a different and more geometrical treatment of the subject.

In [125] Osserman and I studied the question of intrinsic characterization of the induced metric on a minimal submanifold in an Euclidean space, generalizing the famous theorem of Ricci. We gave various results, but the problem is not completely solved.

In [114] I studied affine minimal hypersurfaces, where the integral to be minimized involves second partial derivatives. I computed the first variation and showed that its vanishing is characterized by the vanishing of the affine mean curvature.

Affine minimal surfaces play a role in Bäcklund-type transformations, which remains to be better understood. In fact, Terng and I proved in [116] that if the focal surfaces of a W-congruence have parallel affine normals at corresponding points, then both are affine minimal surfaces.

One of the objectives of the geometry of minimal submanifolds is to provide examples. This is of importance, because general theorems are frequently derived through comparison with them.

8. Webs

The papers [4], [5] constitute my thesis in Hamburg. In [4] I gave a sharp upper bound for the rank of a d-web of hypersurfaces in $\mathbb{R}^n$. Paper [5] was a natural outgrowth in a web-geometry atmosphere of my efforts to understand the Cartan–

Kähler theory. I defined an affine connection, and from which a complete system of invariants, of a three web of r-dimensional submanifolds in $\mathbb{R}^{2r}$, and gave geometrical interpretations of some of these invariants.

It was an irony of fate that I should return to the subject forty years later. When Phillip Griffiths visited Berkeley in 1975–1976, he was interested in web geometry as a generalization of the geometry of projective varieties. For instance, an algebraic curve of degree d in P_n is met by a generic hyperplane in d points; by duality this gives a d-web of hyperplanes in the dual space P_n^*. We were interested in generalizing Bol's linearization theorem to n dimensions. Our effort in [110], [112] to prove the generalization was not entirely successful. A supplementary condition has to be added and a more algebraic geometrical proof substituted. The correction appeared in [124]. The problem is, however, an interesting one and deserves further study.

In [113] we gave a sharp upper bound for the k-rank of a d-web of codimension k in $\mathbb{R}^{kn}$.

9. Exterior Differential Systems and Partial Differential Equations

As I mentioned at the beginning, I was interested in exterior differential systems. This is very natural, because problems in differential geometry often lead to such systems. The basic general notion is that of a system in involution. In [122], [141] Bryant, Griffiths, and I gave an exposition of the Cartan-Kähler theory and pfaffian systems in involution. This will be expanded to a book with the collaboration of R. Gardner and H. Goldschmidt.

In applications, there is no essential difference between exterior differential systems and partial differential equations, because one can be converted to the other. With large over-determined systems the former has the advantage owing to its compact notation and the fact that the forms often have geometrical meaning. In global problems the latter presents readily the differential operators, to which functional analysis can be applied. There is also a considerable gap between the general theory and the special cases. They compete for the soul of a mathematician working in this area. I think it will be advisable to keep an open mind.

An example is the evolution equations such as the sine-Gordon and KdV equations. It has been observed that these are the integrability conditions on a pseudo-spherical surface. After a preliminary observation in [127] this question is investigated in detail by K. Tenenblat in [139]. A simpler approach, using only the structure equations of SL(2, R), was made with Peng in [118], [128].

In [134] R. Hamilton and I studied three-dimensional compact oriented manifolds and their contact structures and admissible Riemannian metrics. On such a manifold the contact structure always exists, as does the metric. The structure defines a curvature first introduced by S. Webster. We showed that with the contact structure given, the contact form can be so modified as to make the Webster curvature positive, zero, or negative and that in the last case it can be made a negative constant. The problem is analogous to the Yamabe problem on the scalar curvature of an n-dimensional

Riemannian manifold, $n \geq 3$, under a conformal transformation. It leads to a second order subelliptic equation; the contact structure pushed the arguments through.

10.

I would not conclude this account without mentioning my wife's role in my life and work. Through war and peace and through bad and good times we have shared a life for almost fifty years, which is both simple and rich. If there is credit for my mathematical works, it will be hers as well as mine.

Bibliography of the Publications of S.S. Chern

Note: **Boldface** numbers at the end of each entry denote the volume in which the entry appears.

I. *Books and Monographs*

1. *Topics in Differential Geometry* (mimeographed), Institute for Advanced Study, Princeton (1951), 106 pp.
2. *Differentiable Manifolds* (mimeographed), University of Chicago, Chicago (1953), 166 pp.
3. *Complex Manifolds*
   a. University of Chicago, Chicago (1956), 195 pp.
   b. University of Recife, Recife, Brazil (1959), 181 pp.
   c. Russian translation, Moscow (1961), 239 pp.
4. *Studies in Global Geometry and Analysis* (Editor), Mathematical Association of America (1967), 200 pp.
5. *Complex Manifolds without Potential Theory*, van Nostrand (1968), 92 pp. Second edition, revised, Springer-Verlag (1979) 152 pp.
6. *Minimal Submanifolds in a Riemannian Manifold* (mimeographed), University of Kansas, Lawrence (1968), 55 pp.
7. (with Wei-huan Chen) *Differential Geometry Notes*, in Chinese, Beijing University Press (1983), 321 pp.
8. *Studies in Global Differential Geometry* (Editor), Mathematical Association of America (1988), 350 pp.

II. *Papers*

1932

[1] Pairs of plane curves with points in one-to-one correspondence. *Science Reports Nat. Tsing Hua Univ.* 1 (1932) 145–153. **(II)**

1935

*[2] Triads of rectilinear congruences with generators in correspondence. *Tohoku Math. J.* **40** (1935) 179–188.

[3] Associate quadratic complexes of a rectilinear congruence. *Tohoku Math. J.* **40** (1935) 293–316. **(II)**

[4] Abzählungen für Gewebe. *Abh. Math. Sem. Univ. Hamburg* **11** (1935) 163–170. **(I)**

1936

[5] Eine Invariantentheorie der Dreigewebe aus r-dimensionalen Mannigfaltigkeiten im R_{2r}. *Abh. Math. Sem. Univ. Hamburg* **11** (1936) 333–358. **(I)**

* Does not appear in these volumes.

1937

[6] Sur la géométrie d'une équation différentielle du troisième ordre. *C. R. Acad. Sci. Paris* **204** (1937) 1227–1229. **(II)**

[7] Sur la possibilitè de plonger un espace à connexion projective donné dans un espace projectif. *Bull. Sci. Math.* **61** (1937) 234–243. **(I)**

1938

[8] On projective normal coördinates. *Ann. of Math.* **39** (1938) 165–171. **(II)**

[9] On two affine connections. *J. Univ. Yunnan* **1** (1938) 1–18. **(II)**

1939

[10] Sur la géométrie d'un système d'équations différentielles du second ordre. *Bull. Sci. Math* **63** (1939) 206–212. **(II)**

1940

*[11] The geometry of higher path-spaces. *J. Chin. Math. Soc.* **2** (1940) 247–276.

[12] Sur les invariants intégraux en géométrie. *Science Reports Nat. Tsing Hua Univ.* **4** (1940) 85–95. **(II)**

[13] The geometry of the differential equation $y''' = F(x, y', y'')$. *Science Reports Nat. Tsing Hua Univ.* **4** (1940) 97–111. **(I)**

[14] Sur une généralisation d'une formule de Crofton. *C.R. Acad. Sci. Paris* **210** (1940) 757–758. **(II)**

[15] (with C.T. Yen) Sulla formula principale cinematica dello spazio ad n dimensioni. *Boll. Un. Mat. Ital.* **2** (1940) 434–437. **(II)**

*[16] Generalization of a formula of Crofton. *Wuhan Univ. J. Sci.* **7** (1940) 1–16.

1941

[17] Sur les invariants de contact en géométrie projective différentielle. *Acta Pontif. Acad. Sci.* **5** (1941) 123–140. **(II)**

1942

[18] On integral geometry in Klein spaces. *Ann. of Math.* **43** (1942) 178–189. **(I)**

[19] On the invariants of contact of curves in a projective space of N dimensions and their geometrical interpretation. *Acad. Sinica Sci. Record* **1** (1942) 11–15. **(I)**

[20] The geometry of isotropic surfaces. *Ann. of Math.* **43** (1942) 545–559. **(II)**

[21] On a Weyl geometry defined from an $(n-1)$-parameter family of hypersurfaces in a space of n dimensions. *Acad. Sinica Sci. Record* **1** (1942) 7–10. **(I)**

1943

[22] On the Euclidean connections in a Finsler space. *Proc. Nat. Acad. Sci. USA*, **29** (1943) 33–37. **(II)**

[23] A generalization of the projective geometry of linear spaces. *Proc. Nat. Acad. Sci. USA*, **29** (1943) 38–43. **(I)**

1944

[24] Laplace transforms of a class of higher dimensional varieties in a projective space of n dimensions. *Proc. Nat. Acad. Sci. USA*, **30** (1944) 95–97. **(II)**

[25] A simple intrinsic proof of the Gauss–Bonnet formula for closed Riemannian manifolds. *Ann of Math.* **45** (1944) 747–752. **(I)**

[26] Integral formulas for the characteristic classes of sphere bundles. *Proc. Nat. Acad. Sci. USA* **30** (1944) 269–273. **(II)**

*[27] On a theorem of algebra and its geometrical application. *J. Indian Math. Soc.* **8** (1944) 29–36.

BIBLIOGRAPHY OF THE PUBLICATIONS OF S.S. CHERN

1945

*[28] On Grassmann and differential rings and their relations to the theory of multiple integrals. *Sankhya* **7** (1945) 2–8.

[29] Some new characterizations of the Euclidean sphere. *Duke Math. J.* **12** (1945) 279–290. **(II)**

[30] On the curvature integra in a Riemannian manifold. *Ann. of Math.* **46** (1945) 674–684. **(I)**

*[31] On Riemannian manifolds of four dimensions. *Bull. Amer. Math. Soc.* **51** (1945) 964–971.

1946

[32] Some new viewpoints in the differential geometry in the large. *Bull. Amer. Math. Soc.* **52** (1946) 1–30. **(II)**

[33] Characteristic classes of Hermitian manifolds. *Ann. of Math.* **47** (1946) 85–121. (I)

1947

[34] (with H.C. Wang). Differential geometry in symplectic space I. *Science Report Nat. Tsing Hua Univ.* **4** (1947) 453–477. **(II)**

[35] Sur une classe remarquable de variétés dans l'espace projectif à *N* dimensions. *Science Reports Nat. Tsing Hua Univ.* **4** (1947) 328–336. **(I)**

*[36] On the characteristic classes of Riemannian manifolds. *Proc. Nat. Acad. Sci USA*, **33** (1947) 78–82.

*[37] Note of affinely connected manifolds. *Bull. Amer. Math. Soc.* **53** (1947) 820–823; correction ibid **54** (1948) 985–986.

*[38] On the characteristic ring of a differentiable manifold. *Acad. Sinica. Sci. Record* **2** (1947) 1–5.

1948

[39] On the multiplication in the characteristic ring of a sphere bundle. *Ann. of Math.* **49** (1948) 362–372. **(I)**

[40] Note on projective differential line geometry. *Acad. Sinica Sci. Record* **2** (1948) 137–139. **(II)**

*[41] (with Y.L. Jou) On the orientability of differentiable manifolds. *Science Reports Nat. Tsing Hua Univ.* **5** (1948) 13–17.

[42] Local equivalence and Euclidean connections in Finsler spaces. *Science Reports Nat. Tsing Hua Univ.* **5** (1948) 95–121. **(II)**

1949

[43] (with Y.F. Sun). The imbedding theorem for fibre bundles. *Trans. Amer. Math. Soc* **67** (1949) 286–303. **(II)**

*[44] (with S.T. Hu) Parallelisability of principal fibre bundles. *Trans. Amer. Math. Soc.* **67** (1949) 304–309.

1950

[45] (with E. Spanier). The homology structure of sphere bundles. *Proc. Nat. Acad. Sci. USA*, **36** (1950) 248–255. **(II)**

[46] Differential geometry of fiber bundles. *Proc. Int. Congr. Math.* (1950) **II** 397–411. **(II)**

1951

[47] (with E. Spanier). A theorem on orientable surfaces in four-dimensional space. *Comm. Math. Helv.* **25** (1951) 205–209. **(I)**

1952

[48] On the kinematic formula in the Euclidean space of *N* dimensions. *Amer. J. Math* **74** (1952) 227–236. **(II)**

[49] (with C. Chevalley). Elie Cartan and his mathematical work. *Bull. Amer. Math. Soc.* **58** (1952) 217–250. **(II)**

[50] (with N.H. Kuiper) Some theorems on the isometric imbedding of compact Riemann manifolds in Euclidean space. *Ann. of Math.* **56** (1952) 422–430. **(II)**

1953

[51] On the characteristic classes of complex sphere bundles and algebraic varieties. *Amer. J. of Math.*, **75** (1953) 565–597. **(I)**

*[52] Some formulas in the theory of surfaces. *Boletin de la Sociedad Matematica Mexicana*, **10** (1953) 30–40.

[53] Relations between Riemannian and Hermitian geometries. *Duke Math. J.*, **20** (1953) 575–587. **(II)**

1954

[54] Pseudo-groupes continus infinis *Colloque de Geom. Diff.* Strasbourg (1954) 119–136. **(I)**

[55] (with P. Hartman and A. Wintner) On isothermic coordinates. *Comm. Math. Helv.* **28** (1954) 301–309. **(I)**

1955

[56] La géométrie des sous-variétés d'un espace euclidien à plusieurs dimensions. *l'Ens. Math.*, **40** (1955) 26–46. **(II)**

[57] An elementary proof of the existence of isothermal parameters on a surface. *Proc. Amer. Math. Soc.*, **6** (1955) 771–782. **(II)**

[58] On special *W*-surfaces. *Proc. Amer. Math. Soc.*, **6** (1955) 783–786. **(II)**

[59] On curvature and characteristic classes of a Riemann manifold. *Abh. Math. Sem. Univ. Hamburg* **20** (1955) 117–126. **(II)**

1956

*[60] Topology and differential geometry of complex manifolds. *Bull. Amer. Math. Soc.*, **62** (1956) 102–117.

1957

[61] On a generalization of Kähler geometry. *Lefschetz jubilee volume.* Princeton Univ. Press (1957) 103–121. **(I)**

[62] (with R. Lashof) On the total curvature of immersed manifolds. *Amer. J. of Math.* **79** (1957) 306–318. **(I)**

[63] (with F. Hirzebruch and J-P. Serre) On the index of a fibered manifold. *Proc. Amer. Math. Soc.*, **8** (1957) 587–596. **(I)**

[64] A proof of the uniqueness of Minkowski's problem for convex surfaces. *Amer. J. of Math.*, **79** (1957) 949–950. **(II)**

1958

*[65] Geometry of submanifolds in complex projective space. *Symposium International de Topologia Algebraica* (1958) 87–96.

[66] (with R.K. Lashof) On the total curvature of immersed manifolds, II. *Michigan Math. J.* **5** (1958) 5–12. **(II)**

[67] Differential geometry and integral geometry. *Proc. Int. Congr. Math. Edinburgh* (1958) 441–449. **(II)**

1959

[68] Integral formulas for hypersurfaces in Euclidean space and their applications to uniqueness theorems. *J. of Math. and Mech.* **8** (1959) 947–956. **(I)**

1960

[69] (with J. Hano and C.C. Hsiung) A uniqueness theorem on closed convex hypersurfaces in Euclidean space. *J. of Math. and Mech.* **9** (1960) 85–88. **(I)**

[70] Complex analytic mappings of Riemann surfaces I. *Amer. J. of Math.* **82** (1960) 323–337. **(I)**

[71] The integrated form of the first main theorem for complex analytic mappings in several complex variables. *Ann. of Math.* **71** (1960) 536–551. **(I)**

*[72] Geometrical structures on manifolds. *Amer. Math. Soc. Pub.* (1960) 1–31.

*[73] La géométrie des hypersurfaces dans l'espace euclidean. *Seminaire Bourbaki*, **193** (1959–1960).

*[74] Sur les métriques Riemanniens compatibles avec une reduction du groupe structural. *Séminaire Ehresmann*, January 1960.

1961

[75] Holomorphic mappings of complex manifolds. *L'Ens. Math.* **7** (1961) 179–187. **(I)**

1962

*[76] Geometry of quadratic differential form. *J. of SIAM* **10** (1962) 751–755.

1963

[77] (with C.C. Hsiung) On the isometry of compact submanifolds in Euclidean space. *Math. Annalen* **149** (1963) 278–285. **(II)**

[78] Pseudo-Riemannian geometry and Gauss–Bonnet formula. *Academia Brasileira de Ciencias* **35** (1963) 17–26. **(I)**

1965

[79] Minimal surfaces in an Euclidean space of N dimensions. *Differential and Combinatorial Topology*, Princeton Univ. Press (1965) 187–198. **(I)**

[80] (with R. Bott) Hermitian vector bundles and the equidistribution of the zeroes of their holomorphic sections. *Acta. Math.* **114** (1965) 71–112. **(II)**

[81] On the curvatures of a piece of hypersurface in Euclidean space. *Abh. Math. Sem. Univ. Hamburg* **29** (1965) 77–91. **(III)**

[82] On the differential geometry of a piece of submanifold in Euclidean space. *Proc. of U.S.–Japan Seminar in Diff. Geom.* (1965) 17–21. **(III)**

1966

[83] Geometry of G-structures. *Bull. Amer. Math. Soc.* **72** (1966) 167–219. **(III)**

*[84] On the kinematic formula in integral geometry. *J. of Math. and Mech.* **16** (1966) 101–118. **(III)**

*[85] Geometrical structures on manifolds and submanifolds. *Some Recent Advances in Basic Sciences*, Yeshiva Univ. Press (1966) 127–135.

1967

[86] (with R. Osserman) Complete minimal surfaces in Euclidean n-space. *J. de l'Analyse Math.* **19** (1967) 15–34. **(III)**

[87] Einstein hypersurfaces in a Kählerian manifold of constant holomorphic curvature. *J. Diff. Geom.* **1** (1967) 21–31. **(III)**

1968

[88] On holomorphic mappings of Hermitian manifolds of the same dimension. *Proc. Symp. Pure Math.* **11**. Entire Functions and Related Parts of Analysis (1968) 157–170. **(I)**

1969

[89] Simple proofs of two theorems on minimal surfaces. *L'Ens. Math.* **15** (1969) 53–61. **(I)**

1970

[90] (with H. Levine and L. Nirenberg) Intrinsic norms on a complex manifold. *Global analysis*, Princeton Univ. Press (1970) 119–139. **(I)**

[91] (with M. do Carmo and S. Kobayashi) Minimal submanifolds of a sphere with second fundamental form of constant length. *Functional Analysis and Related Fields*, Springer-Verlag (1970) 59–75. **(I)**

[92] (with R. Bott) Some formulas related to complex transgression. *Essays on Topology and Related Topics*, Springer-Verlag, (1970) 48–57. **(I)**

*[93] Holomorphic curves and minimal surfaces. *Carolina Conference Proceedings* (1970) 28 pp.

[94] On minimal spheres in the four-sphere, Studies and Essays Presented to Y. W. Chen, Taiwan, (1970) 137–150. **(I)**

[95] Differential geometry: Its past and its future. *Actes Congrès Intern. Math.* (1970) **1**, 41–53. **(III)**

[96] On the minimal immersions of the two-sphere in a space of constant curvature. *Problems in Analysis*, Princeton Univ. Press, (1970) 27–40. **(III)**

1971

[97] Brief survey of minimal submanifolds. *Differentialgeometrie im Grossen.* W. Klingenberg (ed.), **4** (1971) 43–60. **(III)**

*[98] (with J. Simons) Some cohomology classes in principal fibre bundles and their application to Riemannian geometry. *Proc. Nat. Acad. Sci. USA*, **68** (1971) 791–794.

1972

[99] Holomorphic curves in the plane. *Diff. Geom., in honor of K. Yano*, (1972) 73–94. **(III)**

*[100] Geometry of characteristic classes. *Proc. 13th Biennial Sem. Canadian Math. Congress*, (1972) 1–40. Also pub. in Russian translation.

1973

[101] Meromorphic vector fields and characteristic numbers. *Scripta Math.* **29** (1973) 243–251. **(I)**

*[102] The mathematical works of Wilhelm Blaschke. *Abh. Math. Sem. Univ. Hamburg* **39** (1973) 1–9.

1974

[103] (with. J. Simons) Characteristic forms and geometrical invariants. *Ann. of Math.* **99** (1974) 48–69. **(I)**

[104] (with M. Cowen, A. Vitter III) Frenet frames along holomorphic curves. *Proc. of Conf. on Value Distribution Theory*, Tulane Univ. (1974) 191–203. **(III)**

[105] (with J. Moser) Real hypersurfaces in complex manifolds. *Acta. Math.* **133** (1974) 219–271. **(III)**

1975

[106] (with S.I. Goldberg) On the volume decreasing property of a class of real harmonic mappings. *Amer. J. of Math.* **97** (1975) 133–147. **(III)**

[107] On the projective structure of a real hypersurface in C_{n+1}. *Math. Scand.* **36** (1975) 74–82. **(III)**

1976

[108] (with J. White) Duality properties of characteristic forms. *Inv. Math.* **35** (1976) 285–297. **(III)**

1977

*[109] Circle bundles. *Geometry and topology, III*. Latin Amer. School of Math, Lecture Notes in Math. Springer-Verlag, **597** (1977) 114–131.

*[110] (with P.A. Griffiths) Linearization of webs of codimension one and maximum rank. *Proc. Int. Symp. on Algebraic Geometry, Kyoto* (1977) 85–91.

1978

[111] On projective connections and projective relativity. *Science of Matter*, dedicated to Ta-you Wu, (1978) 225–232. **(III)**

[112] (with P.A. Griffiths) Abel's theorem and webs. *Jber. d. Dt. Math. Verein.* **80** (1978) 13–110. **(III)**

[113] (with P.A. Griffiths) An inequality for the rank of a web and webs of maximum rank. *Annali Sc. Norm. Super.–Pisa, Serie IV*, **5** (1978) 539–557. (III)

[114] Affine minimal hypersurfaces. *Minimal Submanifolds and Geodesics.* Kaigai Publications, Ltd. (1978) 1–14. **(III)**

1979

*[115] Herglotz's work on geometry. *Ges. Schriften Gustav Herglotz*, Göttingen (1979) xx–xxi.

[116] (with C.L. Terng) An analogue of Bäcklund's theorem in affine geometry. *Rocky Mountain J. Math.* **10** (1979) 105–124. **(III)**

[117] From triangles to manifolds. *Amer. Math. Monthly* **86** (1979) 339–349. **(III)**

[118] (with C.K. Peng) Lie groups and KdV equations. *Manuscripta Math.* **28** (1979) 207–217. **(III)**

1980

[119] General relativity and differential geometry. *Some Strangeness in the Proportion: A Centennial Symp. to Celebrate the Achievements of Albert Einstein, Harry Woolf* (ed.), Addison-Wesley Publ. (1980) 271–287. **(III)**

[120] (with W.M. Boothby and S.P. Wang) The mathematical work of H.C. Wang. *Bull. Inst. of Math,* **8** (1980) xiii–xxiv. **(III)**

*[121] Geometry and physics. *Math. Medley*, Singapore, **8** (1980) 1–6.

*[122] (with R. Bryant and P.A. Griffiths) Exterior differential systems. *Proc. of 1980 Beijing DD-Symposium,* (1980) 219–338.

1981

[123] Geometrical interpretation of the sinh-Gordon equation. *Annales Polonici Mathematici* **39** (1981) 63–69. **(IV)**

[124] (with P.A. Griffiths) Corrections and addenda to our paper: "Abel's theorem and webs." *Jber. d. Dt. Math.–Verein.* **83** (1981) 78–83. **(III)**

[125] (with R. Osserman) Remarks on the Riemannian metric of a minimal submanifold. *Geometry Symposium Utrecht 1980*, Lecture Notes in Math. Springer-Verlag **894** (1981) 49–90. **(IV)**

[126] (with J. Wolfson) A simple proof of Frobenius theorem. *Manifolds and Lie Groups, Papers in Honor of Y. Matsushima.* Birkhäuser (1981) 67–69. **(IV)**

[127] (with K. Tenenblat) Foliations on a surface of constant curvature and modified Korteweg–de Vries equations. *J. Diff. Geom.* **16** (1981) 347–349. **(IV)**

[128] (with C.K. Peng) On the Bäcklund transformations of KdV equations and modified KdV equations. *J. of China Univ. of Sci. and Tech.*, **11** (1981) 1–6. **(IV)**

1982

[129] Web geometry. *Proc. Symp. in Pure Math.* **39** (1983) 3–10. **(IV)**

[130] Projective geometry, contact transformations, and *CR*-structures. *Archiv der Math.* **38** (1982) 1–5. **(IV)**

1983

[131] (with J. Wolfson) Minimal surfaces by moving frames. *Amer. J. Math.* **105** (1983) 59–83. **(IV)**

[132] On surfaces of constant mean curvature in a three-dimensional space of constant curvature. *Geometric Dynamics*, Springer Lecture Notes **1007** (1983) 104–108. **(IV)**

1984

[133] Deformation of surfaces preserving principal curvatures, *Differential Geometry and Complex Analysis*, Volume in Memory of H. Rauch, Springer-Verlag (1984) 155–163. **(IV)**

1985

[134] (with R. Hamilton) On Riemannian metrics adapted to three-dimensional contact manifolds. *Arbeitstagung Bonn 1984* Springer Lecture Notes **1111** (1985) 279–308. **(IV)**

[135] (with J. Wolfson) Harmonic maps of S^2 into a complex Grassmann manifold. *Proc. Nat. Acad. Sci.* USA **82** (1985) 2217–2219. **(IV)**

[136] Moving frames, *Soc. Math. de France*, Astérisque, (1985) 67–77. **(IV)**

*[137] Wilhelm Blaschke and web geometry, Wilhelm Blaschke—Gesammelte Werke. **5**, Thales Verlag, (1985) 25–27.

*[138] The mathematical works of Wilhelm Blaschke—an update. Thales Verlag, (1985), 21–23.

1986

[139] (with K. Tenenblat) Pseudospherical surfaces and evolution equations. *Studies in Applied Math.* MIT **74** (1986) 55–83. **(IV)**

[140] On a conformal invariant of three-dimensional manifolds. *Aspects of Mathematics and Its Applications* Elsevier Science Publishers B.V. (1986) 245–252. **(IV)**

*[141] (with P.A. Griffiths) Pfaffian systems in involution. *Proceedings of 1982 Changchun Symposium on Differential Geometry and Differential Equations*, Science Press, China, (1986) 233–256.

1987

[142] (with J. Wolfson) Harmonic maps of the two-sphere into a complex Grassmann manifold II. *Ann. of Math.* **125** (1987) 301–335. **(IV)**

[143] (with T. Cecil) Tautness and Lie Sphere geometry *Math. Annalen*, Volume Dedicated to F. Hirzebruch **278** (1987) 381–399. **(IV)**

1988

[144] Vector bundles with a connection. *Studies in Global Differential Geometry*, MAA, no. 27 (1988), 1–26.

1989

[145] (with T. Cecil) Dupin submanifolds in Lie sphere geometry, to appear in *Differential Geometry and Topology*, Springer Lecture Notes 1989.

[146] Historical remarks on Gauss–Bonnet, to appear in Moser Volume, Academic Press.

[147] An introduction to Dupin submanifolds, to appear in Do Carmo Volume.

陳省身数学
論文選集

Pairs of Plane Curves with Points in One-to-One Correspondence

By S. S. CHERN (陳省身)
Department of Mathematics

Introduction

In a series of papers published in the Transactions of the American Mathematical Society, Wilczynski first gave a systematic treatment of projective differential geometry based upon a system of differential equations. It is the purpose of this paper to apply his theory to a pair of plane curves whose points are in one-to-one correspondence.

Consider the system of differential equations:

$$(1)\qquad \begin{aligned} y'' &= a_{11}y + a_{12}z + b_1 y', \\ z' &= a_{21}y + a_{22}z + b_2 y', \end{aligned}$$

where the a's and b's are analytic functions of x in a certain neighborhood at consideration and the primes indicate differentiations with respect to x. A fundamental system of solutions $y_1, y_2, y_3, z_1, z_2, z_3$ of the equations (1) gives the parametric equations of two curves C_y and C_z. The change of fundamental systems corresponds to the projective transformation of the pair of curves. It is from this fact that equations (1) can be taken as a basis for the study of the projective differential theory of a pair of plane curves. We call x a parameter and we say that two points P_y and P_z are in correspondence when they correspond to the same parametric value of x.

We shall first reduce the system of equations to a certain canonical form, determine the complete system of invariants and covariants, and give their geometric interpretations. Then we expand y, z in power series and study the properties of osculants, of which those having contact with both curves are especially of interest.

1. Canonical Form and the Complete System of Invariants and Covariants

If the transformation of dependent variables

$$(2)\qquad y = \lambda(x)\bar{y}, \qquad z = \mu(x)\bar{z},$$

which leaves the curves invariant is applied to the system (1), the corresponding coefficients $\bar{b}_1, \bar{a}_{22}$ of the resulting system of equations are given by

$$(3)\qquad \bar{b}_1 = b_1 - 2\frac{\lambda'}{\lambda}, \qquad \bar{a}_{22} = a_{22} - \frac{\mu'}{\mu}.$$

If λ and μ are so chosen that

$$\lambda' - \frac{b_1}{2}\lambda = 0, \qquad \mu' - a_{22}\mu = 0,$$

i.e.,

(4) $$\lambda = \text{const. } e^{\int \frac{b_1}{2}\, dx}, \qquad \mu = \text{const. } e^{\int a_{22}\, dx},$$

the resulting system assumes the form

(5) $$\begin{aligned} y'' &= A_{11}y + A_{12}z, \\ z' &= A_{21}y + B_2 y', \end{aligned}$$

where A_{11}, A_{12}, A_{21}, B_2 are functions of a_{11}, a_{12}, a_{21}, a_{22}, b_1, b_2. The system of equations (5) is called the semi-canonical form of the system (1). The most general transformation of dependent variables leaving the semi-canonical form invariant is one for which λ and μ are constants.

Suppose that the system has been reduced to its semi-canonical form. The transformation

(6) $$\bar{x} = \xi(x), \qquad y = \lambda(x)\bar{y}, \qquad z = \mu(x)\bar{z}$$

converts it into

(7) $$\begin{aligned} \frac{d^2\bar{y}}{d\bar{x}^2} &= \bar{A}_{11}\bar{y} + \bar{A}_{12}\bar{z} + \bar{B}_1\frac{d\bar{y}}{d\bar{x}}, \\ \frac{d\bar{z}}{d\bar{x}} &= \bar{A}_{21}\bar{y} + \bar{A}_{22}\bar{z} + \bar{B}_2\frac{d\bar{y}}{d\bar{x}}, \end{aligned}$$

where

(8) $$\begin{aligned} \bar{B}_1 &= \frac{1}{\lambda\xi'^2}(-2\lambda'\xi' - \lambda\xi''), \\ \bar{A}_{21} &= \frac{1}{\mu\xi'}(A_{21}\lambda + B_2\lambda'), \\ \bar{A}_{22} &= -\frac{\mu'}{\mu\xi'}. \end{aligned}$$

Hence the system of differential equations is again in its semi-canonical form if

$$\lambda = \frac{c}{\xi'}, \qquad \mu = \bar{c},$$

where c and $\bar{c}$ are constants. The most general transformation of the dependent and independent variables which leaves the semi-canonical form invariant is, therefore,

(9) $$\bar{x} = \xi(x), \qquad y = \frac{c\bar{y}}{\xi'}, \qquad z = \bar{c}\bar{z}.$$

By actually applying this transformation, the coefficient $\bar{A}_{21}$ is found to be

(10) $$\bar{A}_{21} = \frac{c}{\bar{c}\xi'^{3/2}}\left(A_{21} - \frac{1}{2}B_2\frac{\xi''}{\xi'}\right).$$

If $B_2 \neq 0$, i.e., if the point P_y does not lie on the tangent to C_z at P_z for all values of x, we can make $\bar{A}_{21} = 0$ by choosing ξ such that

$$B_2\xi'' - 2A_{21}\xi' = 0. \tag{11}$$

It is possible, therefore, to reduce the system (1) into the canonical form

$$\begin{aligned} y'' &= \alpha_{11}y + \alpha_{12}z, \\ z' &= \beta_2 y'. \end{aligned} \tag{12}$$

The most general transformation of the dependent and independent variables which leaves the canonical form invariant is of the form

$$\bar{x} = ax + b, \qquad y = c\bar{y}, \qquad z = \bar{c}\bar{z}, \tag{13}$$

where $a, b, c, \bar{c}$ are constants.

It is evident that $\alpha_{11}, \alpha_{12}, \beta_2$ form a complete system of invariants and y, z, y' form a complete system of covariants of (12).

2. Adjoint System

The line coordinates of the tangents l_y and l_z to C_y and C_z at the corresponding points P_y and P_z are respectively given by

$$u = (y, y'), \qquad v = (z, z'), \tag{14}$$

which form a fundamental system of solutions of a new system of differential equations, called the adjoint of (1). Put $\eta = (y, z)$. Differentiating (14) and making use of (1), we have

$$\begin{aligned} u' &= a_{12}\eta + b_1 u, \\ v' &= \left(\frac{a_{21}^2}{b_2} + \frac{a_{21}b_2'}{b_2} + a_{21}b_1 - a_{11}b_2 - a_{21}'\right)\eta + \left(\frac{a_{21}}{b_2} + \frac{b_2'}{b_2} + a_{22} + b_1\right)v, \\ u'' &= \left(-\frac{a_{12}a_{21}}{b_2} + b_1 a_{12} + a_{12}a_{22} + a_{12}'\right)\eta + (a_{12}b_2 + b_1^2 + b_1')u - \frac{a_{12}}{b_2}v. \end{aligned} \tag{15}$$

Eliminating η, we obtain the adjoint system:

$$\begin{aligned} u'' &= \left(\frac{a_{21}b_1}{b_2} - \frac{a_{12}'b_1}{a_{12}} + a_{12}b_2 - a_{22}b_1 + b_1'\right)u - \frac{a_{12}}{b_2}v \\ &\quad + \left(-\frac{a_{21}}{b_2} + \frac{a_{12}'}{a_{12}} + b_1 + a_{22}\right)u', \\ v' &= \left(\frac{a_{21}^3 b_1}{-a_{12}b_2} - \frac{a_{21}b_1b_2'}{a_{12}b_2} - \frac{a_{21}b_1^3}{a_{12}} + \frac{a_{11}b_1b_2}{a_{12}} + \frac{a_{21}'b_1}{a_{12}}\right)u \\ &\quad + \left(\frac{a_{21}}{b_2} + \frac{b_2'}{b_2} + b_1 + a_{22}\right)v \\ &\quad + \left(\frac{a_{21}^3}{a_{12}b_2} + \frac{a_{21}b_2'}{a_{12}b_2} + \frac{a_{21}b_1}{a_{12}} - \frac{a_{11}b_2}{a_{12}} - \frac{a_{21}'}{a_{12}}\right)u'. \end{aligned} \tag{16}$$

The relation between the system (1) and its adjoint (16) is a reciprocal one.

If the system of differential equations of the curves C_y and C_z has been reduced to its canonical form (12), the adjoint system is

$$(17)\qquad \begin{aligned} u'' &= \alpha_{12}\beta_2 u - \frac{\alpha_{12}}{\beta_2} v + \frac{\alpha'_{12}}{\alpha_{12}} u', \\ v' &= \frac{\beta'_2}{\beta_2} v - \frac{\alpha_{11}\beta_2}{\alpha_{12}} u'. \end{aligned}$$

The system (12) and its adjoint system (17) are equivalent if and only if there exists a transformation of the form

$$y = c\bar{y}, \qquad z = \bar{c}\bar{z},$$

which leaves the canonical form (12) invariant and carries (12) to (17). By actually carrying out the transformation, we get the conditions

$$(18)\qquad \alpha'_{12} = 0, \quad \beta'_2 = 0, \quad \alpha_{11}^2 - \alpha_{12}^2\beta_2^2 = 0.$$

Therefore a necessary and sufficient condition that (1) be equivalent to its adjoint is that $\alpha_{11}, \alpha_{12}, \beta_2$ are constants connected by the relation

$$(18')\qquad \alpha_{11}^2 - \alpha_{12}^2\beta_2^2 = 0.$$

Let us now proceed to find the geometrical interpretation of the conditions (18). By hypothesis, we have

$$\sum_{i=1}^{3} u_i y_i = 0, \qquad \sum_{i=1}^{3} v_i z_i = 0.$$

If the differential equations for u and v are equivalent to those for y and z, we have

$$u_i = \sum_{k=1}^{3} c_{ik} y_k, \qquad v_i = \sum_{k=1}^{3} c_{ik} z_k \qquad (i = 1, 2, 3),$$

where c_{ik} are constants such that $|c_{ik}| \neq 0$. Therefore, the system (1) is equivalent to its adjoint system if and only if the pair of curves C_y and C_z is a pair of superposed conics.

3. Differential Equations of C_y, C_z, and $C_{y'}$

The differential equations of C_y, C_z, and the covariant curve $C_{y'}$ are easily found. Differentiating (12), we get

$$(19)\qquad y''' = \alpha_{11} y' + \alpha_{12} z' + \alpha'_{11} y + \alpha'_{12} z.$$

The elimination of z and z' from this equation and the equations (12) gives

$$(20)\qquad y''' - \frac{\alpha'_{12}}{\alpha_{12}} y'' - (\alpha_{11} + \alpha_{12}\beta_2) y' - \frac{1}{\alpha_{12}}(\alpha'_{11}\alpha_{12} - \alpha_{11}\alpha'_{12}) y = 0,$$

which is the differential equation of the curve C_y. In like manner, the differential equation of the curve C_z is

$$
\begin{aligned}
(21)\qquad & z''' - \left(2\frac{\beta_2'}{\beta_2} + \frac{\alpha_{11}'}{\alpha_{11}}\right)z'' + \left(2\frac{\beta_2'^2}{\beta_2^2} - \frac{\beta_2''}{\beta_2} + \frac{\beta_2'\alpha_{11}'}{\beta_2\alpha_{11}} - \alpha_{11} - \alpha_{12}\beta_2\right)z' \\
& + \frac{\beta_2}{\alpha_{11}}(\alpha_{11}'\alpha_{12} - \alpha_{11}\alpha_{12}')z = 0.
\end{aligned}
$$

The conditions for C_y and C_z to be conics are respectively*

$$
(22)\quad 27\alpha_{11}' - 36\alpha_{11}\frac{\alpha_{12}'}{\alpha_{12}} - 9\alpha_{12}'\beta_2 - 27\alpha_{12}\beta_2' + 9\frac{\alpha_{12}'''}{\alpha_{12}} - 45\frac{\alpha_{12}'\alpha_{12}''}{\alpha_{12}^2} + 40\frac{\alpha_{12}'^3}{\alpha_{12}^3} = 0
$$

and

$$
\begin{aligned}
(23)\qquad & 9\frac{\beta_2'''}{\beta_2} - 45\frac{\beta_2'\beta_2''}{\beta_2^2} + 40\frac{\beta_2'^3}{\beta_2^3} - 9\frac{\beta_2''\alpha_{11}'}{\beta_2\alpha_{11}} + 15\frac{\beta_2'^2\alpha_{11}'}{\beta_2^2\alpha_{11}} + 9\frac{\beta_2'\alpha_{11}''}{\beta_2\alpha_{11}} - 15\frac{\beta_2'\alpha_{11}'^2}{\beta_2\alpha_{11}^2} \\
& - 36\alpha_{11}\frac{\beta_2'}{\beta_2} - 9\frac{\alpha_{11}'''}{\alpha_{11}} + 45\frac{\alpha_{11}'\alpha_{11}''}{\alpha_{11}^2} - 40\frac{\alpha_{11}'^3}{\alpha_{11}^3} + 36\beta_2\alpha_{12}\frac{\alpha_{11}'}{\alpha_{11}} \\
& + 9\alpha_{11}' - 9\alpha_{12}\beta_2' - 27\beta_2\alpha_{12}' = 0.
\end{aligned}
$$

It follows from (22) and (23) that if the invariants $\alpha_{11}, \alpha_{12}, \beta_2$ are constants, the curves C_y and C_z are conics.

Differenting (19), we have

$$
y^{(4)} = (\alpha_{11} + \alpha_{12}\beta_2)y'' + (2\alpha_{11}' + 2\alpha_{12}'\beta_2 + \alpha_{12}\beta_2')y' + \alpha_{11}''y + \alpha_{12}''z.
$$

If we put $y' = \tau$ and eliminate y and z from this equation and (12), (19), we get the differential equation of $C_{y'}$

$$
\begin{aligned}
& (\alpha_{11}\alpha_{12}' - \alpha_{12}\alpha_{11}')\tau''' - (\alpha_{11}\alpha_{12}'' - \alpha_{12}\alpha_{11}'')\tau'' + [(\alpha_{11}'\alpha_{12}'' - \alpha_{12}'\alpha_{11}'') \\
(24)\qquad & - (\alpha_{11} + \alpha_{12}\beta_2)(\alpha_{11}\alpha_{12}' - \alpha_{12}\alpha_{11}')]\tau' + [(\alpha_{11} + \alpha_{12}\beta_2)(\alpha_{11}\alpha_{12}'' - \alpha_{12}\alpha_{11}'') \\
& - (2\alpha_{11}' + 2\alpha_{12}'\beta_2 + \alpha_{12}\beta_2')(\alpha_{11}\alpha_{12}' - \alpha_{12}\alpha_{11}')]\tau = 0,
\end{aligned}
$$

which is the locus of the points of intersection of corresponding tangents to the two curves C_y and C_z. The locus is a straight line if and only if $\alpha_{11}/\alpha_{12} = \text{const}$.

4. Limiting Point and Limiting Line

Let l_{yz} be the line joining the corresponding points P_y and P_z on C_y and C_z and l_{YZ} the line joining the neighboring points P_Y and P_Z corresponding to the parametric value $x + \Delta x$. Then, by means of (12),

$$
Y = y + y'\Delta x + \cdots,
$$

$$
Z = z + \beta_2 y'\Delta x + \cdots.
$$

*Cf. E. J. Wilczynski, Projective Differential Geometry of Curves and Ruled Surfaces, p. 61.

The equations of the lines l_{YZ} and l_{yz} are respectively

$$|X \quad y \quad z| = 0$$

and

$$|X \quad y + y'\Delta x \quad z + \beta_2 y'\Delta x| + \cdots = 0.$$

The point of intersection of these lines is given by $\lambda y + \mu z$, where $\lambda : \mu$ satisfies the relation

$$\lambda |yy'z| + \beta_2 \mu |yy'z| + \cdots = 0.$$

When Δx approaches zero, we have $\lambda : \mu = -\beta_2$. The point $\rho = -\beta_2 y + z$ is called the limiting point on l_{yz}. Differentiating ρ successively, we have

$$\begin{aligned} \rho' &= -\beta_2' y, \\ \rho'' &= -\beta_2' y' - \beta_2'' y, \\ \rho''' &= -2\beta_2'' y' - (\beta_2'\alpha_{11} + \beta_2''')y - \beta_2'\alpha_{12} z. \end{aligned}$$

Eliminating y', y, z, we obtain the differential equation of the curve enveloped by l_{yz}

$$\beta_2'^3 \rho''' - 2\beta_2'\beta_2''\rho'' + (2\beta_2''^2 - \beta_2'\beta_2''' - \beta_2'^3\alpha_{11} - \beta_2'^3\alpha_{12}\beta_2)\rho' + \beta_2'^3\alpha_{12}\rho = 0. \tag{25}$$

If β_2 is a constant, we have $\rho' = 0$ so that $C\rho$ is a point. Hence $\beta_2 = \text{const.}$ is a necessary and sufficient condition that the lines l_{yz} joining the corresponding points P_y and P_z on C_y and C_z are concurrent. In other words, the correspondence of the two curves is established by a perspectivity with $P\rho$ as a center.

Dually, the tangent to $C_{y'}$ at y' is defined as the limiting line of the point y', which is given by

$$|X \quad y' \quad y''| = 0.$$

Making use of (12), this can be written

$$|X\, y\, y'| + \frac{\alpha_{12}}{\alpha_{11}} |X\, z\, y'| = 0.$$

5. Conics of Contact

Let us take the points P_y, P_z, $P_{y'}$ as the vertices of the local triangle of reference at P_y, P_z and choose the unit point of the coordinate system such that an expression of the form

$$y_1 y + y_2 z + y_3 y'$$

will represent a point whose coordinates are y_1, y_2, y_3. Then the coordinates of the point P_Y on C_y in the neighborhood of P_y are given by

$$\begin{aligned} y_1 &= 1 + \tfrac{1}{2}\alpha_{11}\Delta x^3 + \tfrac{1}{6}\alpha_{11}'\Delta x^3 + \tfrac{1}{24}(\alpha_{11}'' + \alpha_{11}^2 + \alpha_{11}\alpha_{12}\beta)\Delta x^4 + \cdots, \\ y_2 &= \tfrac{1}{2}\alpha_{12}\Delta x^2 + \tfrac{1}{6}\alpha_{12}'\Delta x^3 + \tfrac{1}{24}(\alpha_{12}'' + \alpha_{11}\alpha_{12} + \alpha_{12}^2\beta_2)\Delta x^4 + \cdots, \\ y_3 &= \Delta x + \tfrac{1}{6}(\alpha_{11} + \alpha_{12}\beta_2)\Delta x^3 + \tfrac{1}{24}(2\alpha_{11}' + 2\alpha_{12}'\beta_2 + \alpha_{12}\beta_2')\Delta x^4 + \cdots, \end{aligned} \tag{26}$$

and the coordinates of the point P_Z on C_z in the neighborhood of P_z are given by

$$
(27)\quad
\begin{aligned}
z_1 &= \tfrac{1}{2}\alpha_{11}\beta_2\Delta x^2 + \tfrac{1}{6}(\alpha'_{11}\beta_2 + 2\alpha_{11}\beta'_2)\Delta x^3 \\
&\quad + \tfrac{1}{24}(\alpha''_{11}\beta_2 + 3\alpha'_{11}\beta'_2 + 3\alpha_{11}\beta''_2 + \alpha_{11}^2\beta_2 + \alpha_{11}\alpha_{12}\beta_2^2)\Delta x^4 + \cdots, \\
z_2 &= 1 + \tfrac{1}{2}\alpha_{12}\beta_2\Delta x^2 + \tfrac{1}{6}(\alpha'_{12}\beta_2 + 2\alpha_{12}\beta'_2)\Delta x^3 \\
&\quad + \tfrac{1}{24}(\alpha''_{12}\beta_2 + 3\alpha'_{12}\beta'_2 + 3\alpha_{12}\beta''_2 + \alpha_{11}\alpha_{12}\beta_2 + \alpha_{12}^2\beta_2^2)\Delta x^4 + \cdots, \\
z_3 &= \beta_2\Delta x + \tfrac{1}{2}\beta'_2\Delta x^2 + \tfrac{1}{6}(\beta''_2 + \alpha_{11}\beta_2 + \alpha_{12}\beta_2^2)\Delta x^3 \\
&\quad + \tfrac{1}{24}(\beta'''_2 + 2\alpha'_{11}\beta_2 + 3\alpha_{11}\beta'_2 + 2\alpha'_{12}\beta_2^2 + 4\alpha_{12}\beta_2\beta'_2)\Delta x^4 + \cdots.
\end{aligned}
$$

Let us denote by $C\lambda\mu$ the conic having λ-point contact with C_y at P_y and μ-point contact with C_z at P_z, for which $\lambda + \mu = 5$. We have six covariant conics associated with C_y, C_z at every pair of corresponding points, namely, C_{50}, C_{41}, C_{32}, C_{23}, C_{14}, C_{05}. By imposing the conditions of contact, we find the equations of the six conics

$$
(28)\quad
\begin{aligned}
&C_{50}\colon && \Theta x_2^2 + \alpha_{12}x_3^2 + \frac{2}{3}\frac{\alpha'_{13}}{\alpha_{12}}x_2x_3 - 2x_1x_2 = 0, \\
&C_{41}\colon && \alpha_{12}x_3^2 + \frac{2}{3}\frac{\alpha'_{12}}{\alpha_{12}}x_2x_3 - 2x_1x_2 = 0, \\
&C_{32}\colon && \alpha_{12}x_3^2 - 2x_1x_2 = 0, \\
&C_{23}\colon && \alpha_{11}x_3^2 - 2\beta_2x_1x_2 = 0, \\
&C_{14}\colon && \alpha_{11}x_3^2 + 2\Phi x_1x_3 - 2\beta_2x_1x_2 = 0, \\
&C_{50}\colon && \Delta x_1^2 + \alpha_{11}x_3^2 + 2\Phi x_1x_3 - 2\beta_2x_1x_2 = 0,
\end{aligned}
$$

where

$$
(29)\quad
\begin{aligned}
\Theta &= -\frac{4}{9}\frac{\alpha_{12}'^2}{\alpha_{12}^3} + \frac{1}{3}\frac{\alpha''_{12}}{\alpha_{12}^2} + \frac{\alpha_{11}}{\alpha_{12}} - \beta_2, \\
\Phi &= \frac{1}{3}\left(\frac{\alpha'_{11}}{\alpha_{11}} - \frac{\beta'_2}{\beta_2}\right), \\
\Delta &= \frac{1}{3}\frac{\alpha''_{11}}{\alpha_{11}^2} - \frac{4}{9}\frac{\alpha_{11}'^2}{\alpha_{11}^2} - \frac{1}{9}\frac{\alpha'_{11}\beta'_2}{\alpha_{11}^2\beta_2} - \frac{1}{3}\frac{\beta''_2}{\alpha_{11}\beta_2} + \frac{5}{9}\frac{\beta_2'^2}{\alpha_{11}\beta_2^2} + \frac{\alpha_{12}\beta_2}{\alpha_{11}} - 1.
\end{aligned}
$$

It is seen that no three of the conics can belong to the same pencil unless two of them coincide, i.e., unless there exist conics like C_{51}, C_{42}, C_{33}, C_{24}, C_{15}. The conditions are respectively given by

$$
(30)\quad \Theta = 0, \qquad \alpha'_{12} = 0, \qquad \alpha_{11} - \alpha_{12}\beta_2 = 0, \qquad \Phi = 0, \qquad \Delta = 0.
$$

It is easily shown that there exists a curve of the 12-th order, the polars of whose points with respect to the six conics envelop a conic. This curve consists of a curve of the 9-th order and the lines $x_1 = 0, x_2 = 0, x_3 = 0$. In fact, two of the six polars coincide when the point lies on any of these lines.

6. Contact Nodal Cubics

Let us now proceed to find the pencil of cubics having four-point contact with C_y at P_y and with C_z at P_z.

By means of (26) and (27) we find that each of the cubic curves

$$\Omega_1(x) \equiv \frac{\alpha'_{11}\beta_2 - \alpha_{11}\beta'_2}{\alpha_{11}\beta_2^2}\left(x_1^2x_2 - \frac{1}{3}\frac{\alpha'_{12}}{\alpha_{12}}x_1x_2x_3 - \frac{1}{2}\alpha_{12}x_1x_3^2\right)$$

$$+\frac{\alpha'_{12}}{\alpha_{12}}\left(x_1x_2^2 - \frac{1}{2}\frac{\alpha_{11}}{\beta_3}x_2x_3^2\right) = 0, \tag{31}$$

$$\Omega_2(x) \equiv 3\beta_3^2(\alpha_{11} - \alpha_{12}\beta_2)\left(-x_1x_2^2 + \frac{1}{2}\frac{\alpha_{11}}{\beta_2}x_2x_3^2\right)$$

$$+ (\alpha'_{11}\beta_2 - \alpha_{11}\beta'_2)(x_1x_2x_3 - \tfrac{1}{2}\alpha_{12}x_3^3) = 0$$

is a cubic of the required pencil. Consequently, the equation

$$\lambda\Omega_1(x) - \mu\Omega_2(x) = 0 \tag{32}$$

represents the pencil which we wish to find. Among these there is a cubic having a node at P_y which is given by $\lambda = 0$. Its nodal tangents are

$$x_2 = 0 \quad \text{and} \quad 3\beta_2^2(\alpha_{11} - \alpha_{12}\beta_2)x_2 - (\alpha'_{11}\beta_2 - \alpha_{11}\beta'_2)x_3 = 0. \tag{33}$$

The former is the tangent to C_y at P_y while the latter is a covariant line which we shall denote by l_c.

Similarly, the cubic of the pencil having a node at P_z is given by

$$\frac{\lambda}{\mu} = \frac{3\alpha_{12}\beta_2^2}{\alpha'_{12}}(\alpha_{11} - \alpha_{12}\beta_2).$$

Its nodal tangents are

$$x_1 = 0 \quad \text{and} \quad 3(\alpha_{11} - \alpha_{12}\beta_2)x_1 + \beta_2\alpha'_{12}x_3 = 0. \tag{34}$$

Let us denote the second line by $l_{c'}$. It is obvious that the lines l_c and $l_{c'}$ coincide with l_{yz} if and only if

$$\alpha_{11} - \alpha_{12}\beta_2 = 0,$$

provided that $\alpha'_{12} \neq 0$, $\alpha'_{11}\beta_2 - \alpha_{11}\beta'_2 \neq 0$, i.e., if and only if the points P_y, P_z, ρ, and the point of intersection of the limiting line and l_{yz} form a harmonic range.

In conclusion, the author wishes to express his gratefulness to Professor Dan Sun for encouragement and suggestions.

February 15, 1932.

Reset from
Science Reports Nat. Tsing Hua Univ.
1 (1932) 145–153

Associate Quadratic Complexes of a Rectilinear Congruence

By SHIING-SHEN CHERN, Peiping, China

Introduction

In his Brussels paper([1]) Wilczynski has laid the foundation to the study of the projective differential geometry of rectilinear congruences by a method which is peculiarly his own. We shall summarize here the essentials of his results.

Let the congruence be formed by the lines l_{yz} joining the corresponding points of the surfaces S_y and S_z defined by the equations

$$(1)\qquad \begin{cases} y_i = y_i(u, v), \\ z_i = z_i(u, v), \end{cases} \qquad i = 1, 2, 3, 4,$$

where the corresponding points are given by the same set of parameters u, v. The homogeneous point coordinates y_i and z_i may be considered as solutions of a system of partial differential equations which, by means of allowable transformations of the dependent and independent variables, can be reduced to the cononical form

$$(2)\qquad \begin{cases} y_v = mz, \qquad z_u = ny, \\ y_{uu} = ay + bz + cy_u + dz_v, \\ z_{vv} = a'y + b'z + c'y_u + d'z_v, \end{cases}$$

where $m, n, a, b, c, d, a', b', c', d'$ are functions of u, v satisfying a set of integrability conditions of the form

$$(3)\qquad \begin{cases} c = f_u, \quad d' = f_v, \quad b = -d_v - df_v, \quad a' = -c'_u - c'f_u, \\ mn - c'd = f_{uv}, \\ m_{uu} + d_{vv} + df_{vv} + d_v f_v - f_u m_u = ma + db', \\ n_{vv} + c'_{uu} + c'f_{uu} + c'_u f_u - f_v n_v = c'a + nb', \\ 2m_u n + mn_u = a_v + f_u mn + a'd, \\ m_v n + 2mn_v = b'_u + f_v mn + bc', \end{cases}$$

where $f(u, v)$ is a new function introduced. In fact, when a system of differential equations of the form (2) is given whose coefficients satisfy the integrability conditions

([1]) E. J. Wiczynski, Sur la théorie générale des congruences, Classe des Sciences de l'Académie Royale de Belgique, Mémoires in 4°, 2. série, t. III (1911). This paper will hereafter be cited as the Brussels paper.

(3), the surfaces S_y, S_z, and the associated congruence Γ_{yz} will be defined up to a projective transformation. This is the reason that system (2) can be taken as a basis for the study of the projective differential geometry of rectilinear congruences.

The most general transformation in dependent and independent variables which leaves the canonical form (2) unaltered is of the form

$$y = k(u)\bar{y}, \qquad z = l(v)\bar{z}, \qquad \bar{u} = U(u), \qquad \bar{v} = V(v), \tag{4}$$

where k, l, U, V are arbitrary functions in the respective arguments. The canonical form (2) is geometrically characterized by the fact that surfaces S_y and S_z are the focal sheets of the congruence and that the line l_{yz} describes the developables of the congruence when u or v alone varies.

Waelsch has proved([1]) the important theorem that there exists a linear complex to which belong all the central radii through P_y (or P_z) and all these in the neighborhood of l_{yz}. Put

$$\begin{cases} \rho = y_u - \dfrac{m_u}{m} y, \\ \sigma = z_v - \dfrac{n_y}{n} z. \end{cases} \tag{5}$$

Then, with reference to the local tetrahedron $yz\rho\sigma$ in the neighborhood of l_{yz} and with a unit point properly chosen, the equations of these linear complexes in the usual Plückerian line coordinates are respectively

$$\begin{cases} mw_{13} + c'w_{42} = 0, \\ dw_{13} + nw_{42} = 0. \end{cases} \tag{6}$$

We shall call the former the y-associate linear complex and the latter the z-associate linear complex. The two associate linear complexes coincide and become the osculating linear complex at l_{yz} if and only if

$$W \equiv mn - c'd = 0, \tag{7}$$

which is a characterization of the well known W-congruences.

The generalization of Waelsch's associate linear complexes seems to be a promising way to enrich the projective theory of rectilinear congruences. It is for this purpose that the present work is carried out.

1. Definition and Equations of the Associate Quadratic Complexes

Let us confine our attention to the neighborhood of the line l_{yz} and take the points P_y, P_z, P_ρ, P_σ as the vertices of the local tetrahedron of reference. The unit point will be chosen such that an expression of the form

$$x_1 y + x_2 z + x_3 \rho + x_4 \sigma$$

([1]) E. Waelsch, Zur Infinitesimalgeometrie der Strahlenkongruenzen und Flächen, Sitzungsberichte der Wiener Akademie der Wissenschaften, Bd. 100 (1891), pp. 167–172.

represents a point whose local coordinates are (x_1, x_2, x_3, x_4). For simplicity in writing, we shall change the notation of the Plückerian line coordinates in putting

$$p_1 = w_{12}, \quad p_2 = w_{13}, \quad p_3 = w_{14}, \quad p_4 = w_{34}, \quad p_5 = w_{42}, \quad p_6 = w_{23}, \tag{8}$$

so that they are connected by the identity

$$\Omega(p) \equiv p_1 p_4 + p_2 p_5 + p_3 p_6 = 0. \tag{9}$$

Let P_Y, P_Z, P_P, P_Σ be the points corresponding respectively to $P_y, P_z, P_\rho, P_\sigma$ for the set of parameters $u + \Delta u, v + \Delta v$. Then we may write

$$\begin{cases} Y = y_1 y + y_2 z + y_3 \rho + y_4 \sigma, \\ Z = z_1 y + z_2 z + z_3 \rho + z_4 \sigma, \\ P = \rho_1 y + \rho_2 z + \rho_3 \rho + \rho_4 \sigma, \\ \Sigma = \sigma_1 y + \sigma_2 z + \sigma_3 \rho + \sigma_4 \sigma, \end{cases} \tag{10}$$

where, up to terms of the third order,

$$\begin{cases} y_1 = 1 + \dfrac{m_u}{m}\Delta u + \dfrac{1}{2}\left(a + c\dfrac{m_u}{m}\right)\Delta u^2 + mn\Delta u\Delta v \\ \qquad + \dfrac{1}{6}\left(a_u + c_u\dfrac{m_u}{m} + dn_v + ac + bn + a\dfrac{m_u}{m} + c^2\dfrac{m_u}{m}\right)\Delta u^3 \\ \qquad + \dfrac{1}{2}(3m_u n + mn_u)\Delta u^2\Delta v + \dfrac{1}{2}(m_v n + mn_v)\Delta u\Delta v^2 \\ \qquad + \dfrac{1}{6}\left(a' + c'\dfrac{m_u}{m}\right)\Delta v^3 + \cdots, \\ y_2 = m\Delta v + \dfrac{1}{2}\left(b + d\dfrac{n_v}{n}\right)\Delta u^2 + m_u\Delta u\Delta v + \dfrac{1}{2}\left(m_v + m\dfrac{n_v}{n}\right)\Delta v^2 \\ \qquad + \dfrac{1}{6}\left(b_u + d_u\dfrac{n_v}{n} + cd\dfrac{n_v}{n} + bc + dmn\right)\Delta u^3 + \dfrac{1}{2}m_{uu}\Delta u^2\Delta v \\ \qquad + \dfrac{1}{2}\left(m_{uv} + \dfrac{m_u n_v}{n} + m^2 n\right)\Delta u\Delta v^2 \\ \qquad + \dfrac{1}{6}\left(m_{vv} + 2m_v\dfrac{n_v}{n} + b'm + d'm\dfrac{n_v}{n}\right)\Delta v^3 + \cdots, \\ y_3 = \Delta u + \dfrac{1}{2}c\Delta u^2 + \dfrac{1}{6}(c_u + a + c^2)\Delta u^3 + \dfrac{1}{2}mn\Delta u^2\Delta v + \dfrac{1}{6}mc'\Delta v^3 + \cdots, \\ y_4 = \dfrac{1}{2}d\Delta u^2 + \dfrac{1}{2}m\Delta v^2 + \dfrac{1}{6}(d_u + cd)\Delta u^3 + \dfrac{1}{2}m_u\Delta u\Delta v^2 \\ \qquad + \dfrac{1}{6}(2m_v + md')\Delta v^3 + \cdots, \end{cases} \tag{11}$$

and, up to terms of the second order,

$$
(12)\quad \begin{cases}
\rho_1 = \left(a + c\dfrac{m_u}{m} - \dfrac{m_{uu}}{m}\right)\Delta u + \left(mn - \dfrac{m_{uv}}{m} + \dfrac{m_u m_v}{m^2}\right)\Delta v \\
\quad + \dfrac{1}{2}\left(a_u + c_u\dfrac{m_u}{m} - \dfrac{m_{uuu}}{m} + bn + dn_v + ac + c^2\dfrac{m_u}{m} - c\dfrac{m_u^2}{m^2} + \dfrac{m_u m_{uu}}{m^2}\right)\Delta u^2 \\
\quad + \left(2m_u n + mn_u - \dfrac{m_{uuv}}{m} + \dfrac{m_u m_{uv}}{m^2} + \dfrac{m_v m_{uu}}{m^2} - \dfrac{m_u^2 m_v}{m^3}\right)\Delta u \Delta v \\
\quad + \dfrac{1}{2}\left(mn_v + nm_v - \dfrac{m_{uvv}}{m} + 2\dfrac{m_v m_{uv}}{m^2} + \dfrac{m_u m_{vv}}{m^2} - 2\dfrac{m_u m_v^2}{m^3}\right)\Delta v^2 + \cdots, \\
\rho_2 = \left(b + d\dfrac{n_v}{n}\right)\Delta u + \dfrac{1}{2}\left(b_u + d_u\dfrac{n_v}{n} + dmn + bc + cd\dfrac{n_v}{n} - b\dfrac{m_u}{m}\right. \\
\quad \left. - d\dfrac{m_u n_v}{mn}\right)\Delta u^2 + \dfrac{1}{2}\left(m^2 n - m_{uv} + \dfrac{m_u m_v}{m}\right)\Delta v^2 + \cdots, \\
\rho_3 = 1 + \left(c - \dfrac{m_u}{m}\right)\Delta u + \dfrac{1}{2}\left(c_u + a + c^2 - c\dfrac{m_u}{m} + 2\dfrac{m_u^2}{m^2} - 2\dfrac{m_{uu}}{m}\right)\Delta u^2 \\
\quad + \left(mn - \dfrac{m_{uv}}{m} + \dfrac{m_u m_v}{m^2}\right)\Delta u \Delta v + \cdots, \\
\rho_4 = d\Delta u + \dfrac{1}{2}\left(d_u + cd - d\dfrac{m_u}{m}\right)\Delta u^2 + \cdots.
\end{cases}
$$

The expansions for z_i and σ_i may be obtained from (11) and (12) by means of the mutual substitution

$$
(13)\qquad \begin{pmatrix} u, m, a, b, c, d, y, \rho, y_1, y_2, y_3, y_4, \rho_1, \rho_2, \rho_3, \rho_4 \\ v, n, b', a', d', c', z, \sigma, z_2, z_1, z_4, z_3, \sigma_2, \sigma_1, \sigma_4, \sigma_3 \end{pmatrix},
$$

in which each letter is to be replaced by the one above it or below it.

Suppose $\Phi = hZ + k\Sigma$ be a point on the line $l_{Z\Sigma}$. The central radii at l_{YZ} through P_Y are the lines joining P_Y to the points P_Φ on $l_{Z\Sigma}$. By the expansion (11) and the one obtained from (12) by the substitution (13), we get, for the expansions of the Plückerian line coordinates of the line $l_{Y\Phi}$.

$$
\begin{cases}
p_1 = h + \left[\dfrac{m_u}{m}h + \left(mn - \dfrac{n_{uv}}{n} + \dfrac{n_u n_v}{n^2}\right)k\right]\Delta u \\
\quad + \left[\dfrac{n_v}{n}h + \left(b' + d'\dfrac{n_v}{n} - \dfrac{n_{uv}}{u}\right)k\right]\Delta v \\
\quad + \dfrac{1}{2}\left[\left(a + c\dfrac{m_u}{m}\right)h + \left(mn_u + 3m_u n - 2\dfrac{m_u n_{uv}}{mn} + 2\dfrac{m_u n_u n_v}{mn^2}\right.\right. \\
\quad \left.\left. - \dfrac{n_{uuv}}{n} + 2\dfrac{m_u n_{uv}}{n^2} + \dfrac{n_{uu} n_v}{n^2} - 2\dfrac{n_u^2 n_v}{n^3}\right)k\right]\Delta u^2
\end{cases}
$$

$$
(14)\quad
\begin{cases}
\quad + \left[\left(mn + \dfrac{m_u n_v}{mn}\right)h + \left(2mn_v + m_v n - \dfrac{n_{uvv}}{n} + \dfrac{n_{uv}n_v}{n^2} + \dfrac{n_u n_{vv}}{n^2}\right.\right. \\
\qquad \left.\left. - \dfrac{u_u n_v^2}{n^3} + b'\dfrac{m_u}{m} + d'\dfrac{m_u n_v}{mn} - \dfrac{m_u n_{vv}}{mn}\right)k\right]\Delta u\Delta v \\
\quad + \dfrac{1}{2}\left[\left(b' + d'\dfrac{n_v}{n}\right)h + \left(b'_v + d'_v\dfrac{n_v}{n} - \dfrac{n_{vvv}}{n} + b'd' + d'^2\dfrac{n_v}{n}\right.\right. \\
\qquad \left.\left. - d'\dfrac{n_v^2}{n^2} + \dfrac{n_v n_{vv}}{n^2} - a'm - c'm_u\right)k\right]\Delta v^2 + \cdots, \\
p_2 = c'k\Delta v - \dfrac{1}{2}nh\Delta u^2 - a'k\Delta u\Delta v \\
\quad + \dfrac{1}{2}\left[c'h + \left(c'_v + c'd' - c'\dfrac{n_v}{n}\right)k\right]\Delta v^2 + \cdots, \\
p_3 = k + \dfrac{m_u}{m}k\Delta u + \left[h + \left(d' - \dfrac{n_v}{n}\right)k\right]\Delta v + \dfrac{1}{2}\left(a + c\dfrac{m_u}{m}\right)k\Delta u^2 \\
\quad + \left[\dfrac{m_u}{m}h + \left(2mn - \dfrac{m_u n_v}{mn} - \dfrac{n_{uv}}{n} + \dfrac{n_u n_v}{n^2} + d'\dfrac{m_u}{m}\right)k\right]\Delta u\Delta v \\
\quad + \dfrac{1}{2}\left[d'h + \left(d'_v + b' + d'^2 - d'\dfrac{n_v}{n} + 2\dfrac{n_v^2}{n^2} - 2\dfrac{n_{vv}}{n}\right)k\right]\Delta v^2 + \cdots, \\
p_4 = k\Delta u + \dfrac{1}{2}ck\Delta u^2 + \left[h + \left(d' - \dfrac{n_v}{n}\right)k\right]\Delta u\Delta v + \cdots, \\
p_5 = -mk\Delta v + \dfrac{1}{2}\left[dh - \left(b + d\dfrac{n_v}{n}\right)k\right]\Delta u^2 - m_u k\Delta u\Delta v \\
\quad + \dfrac{1}{2}\left[-mh + \left(-m_v + m\dfrac{n_v}{n} - 2md'\right)k\right]\Delta v^2 + \cdots, \\
p_6 = -h\Delta u + \dfrac{1}{2}\left[-ch - 2\left(mn - \dfrac{n_{uv}}{n} + \dfrac{n_u n_v}{n^2}\right)k\right]\Delta u^2 \\
\quad - \left[\dfrac{n_v}{n}h + \left(b' + d'\dfrac{n_v}{n} - \dfrac{n_{vv}}{n}\right)k\right]\Delta u\Delta v + c'mk\Delta v^2 + \cdots.
\end{cases}
$$

The power series (14) satisfy the equation of the y-associate linear complex

$$(6_1)\qquad mp_2 + c'p_5 = 0$$

identically in h, k up to terms of the first order in Δu, Δv inclusive. We shall define as the y-associate quadratic complex at l_{yz} the quadratic complex whose equation is satisfied by the power series (14) identically in h, k up to and including terms of the second order in Δu, Δv.

Let us assume that the focal sheets S_y and S_z of the congruence are neither curves nor developables so that

$$mnc'd \neq 0.$$

Let

(15) $$a_{11}p_1^2 + a_{12}p_1p_2 + a_{13}p_1p_3 + \cdots = 0$$

be the equation of the y-associate quadratic complex. By making use of the power series (14), we may determine the coefficients of the equation (15). The final results are contained in the

Theorem 1. *If the invariant*

(16_1) $$\vartheta_1 \equiv c'm_v - mc'_v + mc'd' \neq 0,$$

there is a four-parameter family of y-associate quadratic complexes at l_{yz}. These complexes have the equation

(17_1) $$\begin{aligned} p \equiv a_1\left(p_2^2 - \frac{c'^2}{m^2}p_5^2\right) &+ a_2\left[p_2p_3 + \frac{c'}{m}p_3p_5 + \frac{1}{2}\frac{c'}{m}\left(b + d\frac{n_v}{n}\right)p_4^2\right. \\ &\left. -\frac{1}{m}\left(a' + c'\frac{m_u}{m}\right)p_4p_5 - \frac{1}{2}\frac{W}{m}p_4p_6 + \frac{1}{2}\frac{\vartheta_1}{m^3}p_5^2\right] \\ &+ a_3\left(p_2p_4 + \frac{c'}{m}p_4p_5\right) + a_4\left(p_2p_5 + \frac{c'}{m}p_5^2\right) + a_5\left(p_2p_6 + \frac{c'}{m}p_5p_6\right) = 0, \end{aligned}$$

where a_1, a_2, a_3, a_4, a_5 are the parameters. All the complexes of the family have in common the central radii through P_y and the lines of the (2,2)-congruence

(18_1) $$\begin{aligned} &mp_2 + c'p_5 = 0, \\ &c'\left(b + a\frac{n_v}{n}\right)p_4^2 - 2\left(a' + c'\frac{m_u}{m}\right)p_4p_5 - Wp_4p_6 + \frac{\vartheta_1}{m^2}p_5^2 = 0. \end{aligned}$$

Similar definitions and results hold for the central radii through P_z. By means of (13) and the further mutual substitution

(13_a) $$\begin{pmatrix} p_1, & p_2, & p_3, & p_4, & p_5, & p_6 \\ -p_1, & -p_5, & -p_6, & -p_4, & -p_2, & -p_3 \end{pmatrix},$$

we get from (16_1), (17_1), (18_1) the

Theorem 2. *If the invariant*

(16_2) $$\vartheta_2 \equiv dn_u - nd_u + ncd \neq 0;$$

there is a four-parameter family of z-associate quadratic complexes at l_{yz}. These complexes have the equation

$$Q \equiv b_1\left(p_5^2 - \frac{d^2}{n^2}p_2^2\right) + b_2\left[-p_5p_6 - \frac{d}{n}p_2p_6 + \frac{1}{2}\frac{d}{n}\left(a' + c'\frac{m_u}{m}\right)p_4^2\right.$$

$$(17_2) \qquad \left. - \frac{1}{n}\left(b + d\frac{n_v}{n}\right)p_2p_4 + \frac{1}{2}\frac{W}{n}p_3p_4 + \frac{1}{2}\frac{\vartheta_2}{n_3}p_2^2\right]$$

$$+ b_3\left(p_4p_5 + \frac{d}{n}p_2p_4\right) + b_4\left(p_2p_5 + \frac{d}{n}p_2^2\right) + b_5\left(-p_3p_5 - \frac{d}{n}p_2p_3\right) = 0.$$

All these complexes have in common the central radii through P_z and the lines of the (2,2)-congruence

$$dp_2 + np_5 = 0,$$

$$(18_2) \qquad \frac{\vartheta_2}{n^2}p_2^2 + Wp_3p_4 - 2\left(b + d\frac{n_v}{n}\right)p_2p_4 + d\left(a' + c'\frac{m_u}{m}\right)p_4^2 = 0.$$

We shall call the congruences (18_1) and (18_2) the y-associate and the z-associate (2,2)-congruences at l_{yz} respectively. Immediately we have the

Corollary. *If $\vartheta_1\vartheta_2 \neq 0$, the y-associate and the z-associate (2,2)-congruences decompose into pairs of linear congruences when and only when the congruence Γ_{yz} is a W-congruence.*

The parametric net on the surfaces S_y and S_z are conjugate nets. Their rays at P_y and P_z are respectively the lines P_yP_σ and P_zP_ρ. Their axes, on the other hand, have respectively the equations([1])

$$(19) \qquad \begin{cases} x_3 = 0, & x_2 + \left(f_v + \dfrac{d_v}{d} - \dfrac{n_v}{n}\right)x_4 = 0, \\[2ex] x_4 = 0, & x_1 + \left(f_u + \dfrac{c_u'}{c'} - \dfrac{m_u}{m}\right)x_3 = 0. \end{cases}$$

It may be easily verified that *the y-associate and the z-associate guadratic complexes at l_{yz} always contain these four lines.*

2. ϑ_1- and ϑ_2-Congruences. Geometrical Interpretation of the Invariants ϑ_1 and ϑ_2

An examination of the derivation of the equations (17_1) and (17_2) discloses the fact that the invariants ϑ_1 and ϑ_2 enter so essentially in the discussion of the associate quadratic complexes as to alter the main feature of the problem. On account of their importance connected herewith, congruences for which $\begin{Bmatrix}\vartheta_1 = 0\\ \vartheta_2 = 0\end{Bmatrix}$([2]) will be called

([1]) E. J. Wilczynski, The general theory of congruences, Transactions of the American Mathematical Society, vol. 16 (1915), p. 314.

([2]) With such a notation either the words in the upper row or those in the lower row are true. But when two or more braces appear, the two rows should be read consistently.

$\begin{Bmatrix} \vartheta_1\text{-congruences} \\ \vartheta_2\text{-congruences} \end{Bmatrix}$. For such congruences the nature of the associate quadratic complexes is given by the

Theorem 3. *The y-associate quadratic complexes of a ϑ_1-congruences at l_{yz} form a five-parameter family and have the equation*

$$\begin{aligned} &a_1\left[p_1p_2 + \frac{c'}{m}p_1p_5 - \frac{1}{2}\frac{c'}{m}\left(b + d\frac{n_v}{n}\right)p_4p_6 + \frac{1}{m}\left(a' + c'\frac{m_u}{m}\right)p_5p_6\right. \\ &\left.+\frac{1}{2}\frac{W}{m}p_6^2\right] + a_2\left(p_2^2 - \frac{c'^2}{m^2}p_5^2\right) + a_3\left[p_2p_3 + \frac{c'}{m}p_3p_5 + \frac{1}{2}\frac{c'}{m}\left(b + d\frac{n_v}{n}\right)p_4^2\right. \\ &\left.-\frac{1}{m}\left(a' + c'\frac{m_u}{m}\right)p_4p_5 - \frac{1}{2}\frac{W}{m}p_4p_6\right] + a_4\left(p_2p_4 + \frac{c'}{m}p_4p_5\right) \\ &+ a_5\left(a_2p_5 + \frac{c'}{m}p_5^2\right) + a_6\left(p_2p_6 + \frac{c'}{m}p_5p_6\right) = 0, \end{aligned} \tag{20}$$

where $a_1, a_2, a_3, a_4, a_5, a_6$ are the parameters. All these complexes have in common the central radii through P_y and the lines of the linear congruence

$$mp_2 + c'p_5 = 0, \qquad c'\left(b + d\frac{n_v}{n}\right)p_4 - 2\left(a' + c'\frac{m_u}{m}\right)p_5 - Wp_6 = 0. \tag{21}$$

The directrices of the linear congruence (21) are the lines

$$c'\left(b + d\frac{n_v}{n}\right) : -2\left(a' + c'\frac{m_u}{m}\right) : -W : 0 : 0 : 0, \tag{22$_1$}$$

$$c'\left(b + d\frac{n_v}{n}\right) : 0 : -W : 0 : 2\frac{m}{c'}\left(a' + c'\frac{m_u}{m}\right) : 0. \tag{22$_2$}$$

The line (22_1) is a line joining the points P_y and

$$c'\left(b + d\frac{n_v}{n}\right)z - 2\left(a' + c'\frac{m_u}{m}\right)\rho - W\sigma,$$

and the line (22_2) joins the points

$$Wc'y + 2m\left(a' + c'\frac{m_u}{m}\right)z \quad\text{and}\quad c'\left(b + d\frac{n_v}{n}\right)z - W\sigma.$$

Similar results hold for the z-associate quadratic complexes, as may be obtained by making use of the substitutions (13) and (13_a).

We shall give a simple geometric interpretation of the condition $\vartheta_1 = 0$. The y-associate linear complex at the line corresponding to the parameters $u + \delta u$, $v + \delta v$ has the equation([1])

([1]) Brussel's paper, formula (72).

$$\begin{aligned} &mp_2 + c'p_5 + [(m_u - mc)\delta u + m_v\delta v]p_2 + c'_u\delta v + (c'_v - c'd')\delta v]p_5 \\ &\quad - W\delta u p_6 + \left[\left(c'b + c'd\frac{n_v}{n}\right)\delta u + (a'm + c'm_u)\delta v\right]p_4 = 0, \end{aligned} \tag{23}$$

referred to the tetrahedron $yz\rho\sigma$. It intersects the y-associate linear complex of l_{yz} in a linear congruence whose equations are

$$\begin{cases} \Omega_1 \equiv mp_2 + c'p_5 = 0, \\ \Omega_2 \equiv [(m_u - mc)\delta u + m_v\delta v]p_2 + \left[c'\left(b + d\frac{n_v}{n}\right)\delta u + m\left(a' + c'\frac{m_u}{m}\right)\delta v\right]p_4 \\ \qquad + [c'_u\delta u + (c'_v - c'd')\delta v]p_5 - W\delta u p_6 = 0. \end{cases} \tag{24}$$

This congruence belongs to the pencil of linear complexes

$$\Omega_2 + \lambda\Omega_1 = 0,$$

whose special linear complexes correspond to the parameters satisfying the equation

$$[(m_u - mc)\delta u + m_v\delta v + m\lambda][c'_u\delta u + (c'_v - c'd')\delta v + c'\lambda] = 0.$$

This equation has equal roots in λ if and only if

$$(mc'_u - m_u c' + mcc')\delta u - \vartheta_1\delta v = 0. \tag{25}$$

But the axes of the special linear complexes of the pencil are the directrices of the linear congruence (24). Hence we get the

Theorem 4. *The equation (25) defines a neighboring generator of l_{yz} such that the y-associate linear complexes at these two neighboring generators intersect in a linear congruence with coincident directrices.*

A geometric interpretation of the condition $\vartheta_1 = 0$ is then given by the

Corollary. *The linear congruence common to the y-associate linear complexes at l_{yz} and at a neighboring generator of l_{yz} for which v alone varies is a linear congruence with coincident directrices if and only if $\vartheta_1 = 0$.*

The discussion for the z-associate linear complexes and the interpretation of the condition $\vartheta_2 = 0$ are then easily carried out. A systematic study of the ϑ-congruences will be left for a later occasion.

3. Characteristics of the Associate Quadratic Complexes. Their Double Lines.

Throughout the following discussions except in the last section we shall assume that the congruence l_{yz} is neither a W-congruence nor a ϑ-congruence, so that

$$W\vartheta_1\vartheta_2 \not\equiv 0.$$

To Klein and Weiler[1] is due a complete classification of the quadratic line complexes by means of elementary divisors. The quadratic complexes are thus classified into 49 distinct species, each having a characteristic of its own. The present section is devoted to examine the characteristics of the complexes $P = 0$, and, since there are parameters at our disposal, to find the complexes of the family whose characteristics are of a more particular type. Our discussion is restricted to the y-associate quadratic complexes, as the study of the z-associate quadratic complexes is then merely a matter of translation of the letters.

Consider first the λ-matrix

$$
(26)\qquad \|p + \lambda\Omega\| = \left\| \begin{matrix}
0 & 0 & 0 & \lambda & 0 & 0 \\
0 & a_1 & \frac{1}{2}a_2 & \frac{1}{2}a_3 & \frac{1}{2}a_4 + \lambda & \frac{1}{2}a_5 \\
0 & \frac{1}{2}a_2 & 0 & 0 & \frac{1}{2}\frac{c'}{m}a_2 & \lambda \\
\lambda & \frac{1}{2}a_3 & 0 & \frac{1}{2}\frac{c'}{m}\left(b + d\frac{n_v}{n}\right)a_2 & -\frac{1}{2m}\left(a' + c'\frac{m_u}{m}\right)a_2 + \frac{1}{2}\frac{c'}{m}a_3 & -\frac{1}{4}\frac{W}{m}a_2 \\
0 & \frac{1}{2}a_4 + \lambda & \frac{1}{2}\frac{c'}{m}a_2 & -\frac{1}{2m}\left(a' + c\frac{m_u}{m}\right)a_2 + \frac{c'}{2m}a_3 & -\frac{c'^2}{m^2}a_1 + \frac{1}{2}\frac{\vartheta_1}{m^3}a_2 + \frac{c'}{m}a_4 & \frac{1}{2}\frac{c'}{m}a_5 \\
0 & \frac{1}{2}a_5 & \lambda & -\frac{1}{4}\frac{W}{m}a_2 & \frac{1}{2}\frac{c'}{m}a_5 & 0
\end{matrix} \right\|.
$$

We find, for the value of the determinant of the matrix,

$$
(26_1)\qquad \begin{aligned}
|p + \lambda\Omega| = -\lambda^3\Bigg[\lambda^3 + a_4\lambda^2 + \Bigg\{&\frac{c'^2}{m^2}a_1^2 - \frac{c'}{m}(a_1a_4 + a_2a_5) \\
&-\frac{1}{2}\frac{\vartheta_1}{m^3}a_1a_2 + \frac{1}{4}a_4^2\Bigg\}\lambda + \frac{1}{4}\frac{\vartheta_1}{m^3}a_2^2a_5\Bigg].
\end{aligned}
$$

Hence the characteristic of the complex $P = 0$ is in general of the form [3°111], where the notation 3° denotes the fact that the corresponding root of λ is 0.

As subcases of the characteristic [3°111], we may have characteristics of the form [3°(11)1], [3°21], [3°(21)], [3°3], [3°(111)]. It is easy to see that the other possible forms of the characteristic of the complex $P = 0$ are [4°11], [4°(11)], [4°2], [5°1], [6°]. But a close examination shows that the case [3°(111)] can not happen while the other cases can actually occur with one or more conditions imposed on the parameters.

It is well known that there are points in space, called the singular points of the complex $P = 0$, through which the complex cones of $P = 0$ degenerate into pairs of planes; the singular planes $P = 0$ are defined dually. A line belonging to the complex

(1) F. Klein, Über die Transformation der allgemeinen Gleichung zweiten Grades zwischen Linien-Koordinaten auf eine canonische Form, Diss. Bonn (1868) and Mathematische Annalen, Bd. 23.

A. Weiler, Über die verschiedenen Gattungen der Komplexe zweiten Grades, Mathematische Annalen, Bd. 7.

such that every point on and every plane through which is singular is called a double line of the complex. The coordinates of a double line must satisfy the conditions([1])

$$\frac{\partial p}{\partial p_i} + k\frac{\partial \Omega}{\partial p_i} = 0, \qquad i = 1, \ldots, 6, \tag{27}$$

where k is a factor of proportionality. The nature of the double lines of a complex, if there are any, is intimately connected with the characteristic of the complex. Since $\dfrac{\partial p}{\partial p_i}$ does not involve p_1, the complex $P = 0$ has a double line given by $k = 0$ and

$$p_1 = 1, \qquad p_2 = p_3 = p_4 = p_5 = p_6 = 0,$$

which is the line l_{yz}. In case $k \neq 0$, we have, since

$$\frac{\partial p}{\partial p_1} = 0,$$

the condition

$$p_4 = 0.$$

Therefore, the line l_{yz} is a double of the complex $P = 0$. Any other line of $P = 0$ must intersect the line l_{yz}.

We shall give the conditions for the different species of the complexes $P = 0$ and their double lines as follows:

1. The general case [3°111]. The complex has only one double line, the line l_{yz}.
2. The case [3°(11)1]. Here the parameters are connected by the conditions

$$\frac{\vartheta_1}{m^2 c'} a_2 = 4\frac{c'}{m} a_1 - 2a_4 = 2\frac{a_2 a_5}{a_1}, \tag{28}$$

so that there is a two-parameter family of such complexes. Each complex has three double lines, the line l_{yz} and two lines intersecting l_{yz}. The totality of the latter lines forms the linear congruence

$$p_4 = 0, \qquad 2m^2 c' p_2 + 2mc'^2 p_5 + \vartheta_1 p_6 = 0. \tag{29}$$

The congruence (29) is a congruence with the double directrix l_{yz}.

3. The case [3°21]. There is a three-parameter family of such complexes, the parameters being connected by the relation obtained by setting the discriminant of the cubic equation (26′) to zero. The double lines of these complexes are the line l_{yz} and the lines intersecting l_{yz}.

4. The case [3°(21)]. There is a one-parameter family of such complexes, the parameters being connected by the relations

$$a_2 = -\frac{8mc'^2}{\vartheta_1} a_1, \qquad a_4 = 6\frac{c'}{m} a_1, \qquad a_5 = \frac{1}{2}\frac{\vartheta}{m^2 c'} a_1. \tag{30}$$

([1]) C. M. Jessop, A Treatise on the Line Complex (Cambridge University Press, 1903), p. 199.

Each complex has the double line l_{yz} and another double line. The latter lines form a pencil in the plane

$$\vartheta_1 x_3 - 4mc'^2 x_4 = 0 \tag{31}$$

with the vertex $(-\vartheta_1, 4m^2c', 0, 0)$.

5. The case [3°3]. There is a two-parameter family of such complexes, the parameters being connected by the relations

$$\begin{aligned} &12\frac{c'^2}{m^2}a_1^2 - 12\frac{c'}{m}(a_1a_4 + a_2a_5) - 6\frac{\vartheta_1}{m^3}a_1a_2 - a_4^2 = 0,\\ &27\frac{\vartheta_1}{m^2}a_2^2a_5 - 4a_4^3 = 0. \end{aligned} \tag{32}$$

The double lines of these complexes are the line l_{yz} and the lines of the (3, 3)-congruence

$$p_4 = 0, \qquad 2\vartheta_1 p_3 p_6^2 - m^3 p_2^3 - 3mc^{2\prime} p_2^2 p_5 - 3mc'^2 p_2 p_5^2 - c'^3 p_5^3 = 0. \tag{33}$$

6. The case [4°11]. There is a three-parameter family of such complexes for which we have([1])

$$a_5 = 0. \tag{34}$$

Each complex has l_{yz} as its only double line.

7. The case [4°(11)]. There is a one-parameter family of such complexes, whose parameters are connected by the relations

$$\vartheta_1 a_2 + 2m^2c'a_4 = 0, \qquad a_1 = a_5 = 0. \tag{35}$$

The double lines of these complexes are l_{yz}, the lines of the pencil in the plane

$$\vartheta_1 x_3 - 2mc'^2 x_4 = 0 \tag{36}$$

with vertex at P_z, and the lines of the pencil in the plane $yz\rho$ with vertex at $(-\vartheta_1, 2m^2c', 0, 0)$.

8. The case [4°2]. There is a two-parameter family of such complexes, whose parameters are connected by the relations

$$a_5 = a_1(2mc'^2a_1 - 2m^2c'a_4 - \vartheta_1 a_2) = 0. \tag{37}$$

The double lines of these complexes are l_{yz}, the lines through P_σ, and the lines in the plane $y\rho\sigma$.

9. The case [5°1]. There is a two-parameter family of such complexes, whose parameters satisfy the relations

$$a_5 = \frac{c'^2}{m^2}a_1^2 - \frac{c'}{m}a_1a_4 - \frac{1}{2}\frac{\vartheta_1}{m^3}a_1a_2 + \frac{1}{4}a_4^2 = 0. \tag{38}$$

Each complex has only one double line l_{yz}.

([1]) We do not consider the case $a_2 = 0$, as the quadratic complex $P = 0$ degenerates into a pair of linear complexes.

10. The case [6°]. There is a one-parameter family of such complexes, for which

$$a_1(2mc'^2a_1 - \vartheta_1 a_2) = a_4 = a_5 = 0. \tag{39}$$

Each complex has three double lines coincident at l_{yz}.

4. Some Particular Associate Quadratic Complexes

Let the y-associate quadratic complex contain the central radii through P_z up to the first order inclusive. In the equation $P = 0$, we have then (supposing $W \neq 0$)

$$a_2 = a_5 = 0,$$

which is obtained by making use of the power series (14) with the substitutions (13) and (13_a). Equations (17_1) then becomes

$$\left(p_2 + \frac{c'}{m}p_5\right)\left[a_1\left(p_2 - \frac{c'}{m}p_5\right) + a_3 p_4 + a_4 p_5\right] = 0. \tag{40}$$

Hence such a complex degenerates into a pair of linear complexes, of which one is the y-associate linear complex and the other is a tangent linear complex.

Again, suppose the y-associate quadratic complex contains the line l_{yz} up to and including those of the third order. By the power series (11) and the one obtained from (11) by (13), we find, for the linear coordinates of a neighboring line of l_{yz},

$$\left\{\begin{aligned}
p_2 &= -\frac{1}{2}n\Delta u^2 + \frac{1}{2}c'\Delta v^2 - \frac{1}{3}\left(\frac{1}{2}n_u + nc\right)\Delta u^3 - \frac{1}{2}n_v\Delta u^2\Delta v \\
&\quad - \frac{1}{2}a'\Delta u\Delta v^2 + \frac{1}{6}(c'_v + c'd')\Delta v^3 + \cdots, \\
p_3 &= \Delta v + \frac{m_u}{m}\Delta u\Delta v + \frac{1}{2}d'\Delta v^2 - \frac{1}{3}nd\Delta u^3 + \frac{1}{2}\left(a + c\frac{m_u}{m}\right)\Delta u^2\Delta v \\
&\quad + \left(\frac{1}{2}d'\frac{m_u}{m} + mn\right)\Delta u\Delta v^2 + \frac{1}{6}(d'_v + b' + d'^2)\Delta v^3 + \cdots, \\
p_4 &= \Delta u\Delta v + \frac{1}{2}c\Delta u^2\Delta v + \frac{1}{2}d'\Delta u\Delta v^2 + \cdots, \\
p_5 &= \frac{1}{2}d\Delta u^2 - \frac{1}{2}m\Delta v^2 + \frac{1}{6}(d_u + cd)\Delta u^3 - \frac{1}{2}b\Delta u^2\Delta v \\
&\quad - \frac{1}{2}m_u\Delta u\Delta v^2 - \frac{1}{3}\left(\frac{1}{2}m_v + md'\right)\Delta v^3 + \cdots, \\
p_6 &= -\Delta u - \frac{1}{2}c\Delta u^2 - \frac{n_v}{n}\Delta u\Delta v - \frac{1}{6}(c_u + a + c^2)\Delta u^3 \\
&\quad - \left(\frac{1}{2}c\frac{n_v}{n} + mn\right)\Delta u^2\Delta v - \frac{1}{2}\left(b' + d'\frac{n_v}{n}\right)\Delta u\Delta v^2 + \frac{1}{3}mc'\Delta v + \cdots,
\end{aligned}\right. \tag{41}$$

where the expansion for p_1 is omitted since it is of no use in our subsequent discussion. The power series (41) satisfy the equation $P = 0$ up to and including terms of the third order in Δu, Δv if and only if

$$a_5 = 0.$$

Hence we have the

Theorem 5. *Every* $\left\{\begin{array}{l}\textit{y-associate}\\ \textit{z-associate}\end{array}\right\}$ *quadratic complex at* l_{yz} *which contains the line* l_{yz} *up to the third order has a characteristic of the form* $[4^\circ e_1 e_2]$ *(where* $e_1 + e_2 = 2$*), and conversely.*

By means of (41), it is also easy to verify the

Theorem 6. *For a non-W, non-ϑ-congruence whose focal sheets are neither curves nor developables, there can not exist a* $\left\{\begin{array}{l}\textit{y-associate}\\ \textit{z-associate}\end{array}\right\}$ *quadratic complex containing the line* l_{yz} *up to the fourth order inclusive.*

5. Singular Surface of an Associate Quadratic Complex

The singular surface of a quadratic complex is at the same time the locus of its singular points and the envelope of its singular planes. It is in general a surface of the fourth order and the fourth class. We proceed to find the equation of the singular surface of $P = 0$ in the local point coordinates.

Let (x_1, x_2, x_3, x_4) be a point in space. With X_1, X_2, X_3, X_4 as current coordinates, the equation of the complex cone through (x_1, x_2, x_3, x_4) is

$$p(x_i X_k - x_k X_i) = 0. \tag{42}$$

We shall introduce non-homogeneous coordinates by putting

$$x_4 = X_4 = 1.$$

Then

$$p_1 = x_1 X_2 - x_2 X_1, \qquad p_2 = x_1 X_3 - x_3 X_1, \qquad p_3 = x_1 - X_1,$$
$$p_4 = x_3 - X_3, \qquad p_5 = X_2 - x_2, \qquad p_6 = x_2 X_3 - x_3 X_2.$$

By the change of coordinates

$$\xi = X_1 - x_1, \qquad \eta = X_2 - x_2, \qquad \zeta = X_3 - x_3,$$

we get

$$p_1 = x_1 \eta - x_2 \xi, \qquad p_2 = x_1 \zeta - x_3 \xi, \qquad p_3 = -\xi,$$
$$p_4 = -\zeta, \qquad p_5 = \eta, \qquad p_6 = x_2 \zeta - x_3 \eta.$$

Hence, in terms of the coordinates ξ, η, ζ, the complex cone (42) has the equation

$$(a_1 x_3^2 + a_2 x_3)\xi^2 + \left(-\frac{c'}{m} a_5 x_3 - \frac{c'^2}{m^2} a_1 + \frac{1}{2}\frac{\vartheta_1}{m^3} a_2 + \frac{c'}{m} a_4\right)\eta^2$$

$$+ \left[a_1 x_1^2 - a_3 x_1 + a_5 x_1 x_2 + \frac{1}{2}\frac{W}{m} a_2 x_2 + \frac{1}{2}\frac{c'}{m}\left(b + d\frac{n_v}{n}\right) a_2\right]\zeta^2$$

$$(42') \qquad + \left[-a_5 x_1 x_3 + a_4 x_1 + \frac{c'}{m} a_5 x_2 - \frac{1}{2}\frac{W}{m} a_2 x_3 + \frac{1}{m}\left(a' + c'\frac{m_u}{m}\right) a_2\right.$$

$$\left. - \frac{c'}{m} a_3\right]\eta\zeta + (-2a_1 x_1 x_3 - a_5 x_2 x_3 - a_2 x_1 + a_3 x_3)\zeta\xi$$

$$+ \left(a_5 x_3^2 - a_4 x_3 - \frac{c'}{m} a_2\right)\xi\eta = 0.$$

The equation of the singular surface is then given by the condition that the complex cone (42) degenerates into a pair of planes. On returning to homogeneous coordinates, we get the equation of the singular surface in the form

$$(43) \qquad \begin{vmatrix} f_{11} & f_{12} & f_{13} \\ f_{21} & f_{22} & f_{23} \\ f_{31} & f_{32} & f_{33} \end{vmatrix} = 0,$$

where

$$(43') \qquad \begin{cases} f_{11} = a_1 x_3^2 + a_2 x_3 x_4, \\ f_{22} = -\dfrac{c'}{m} a_5 x_3 x_4 + \left(-\dfrac{c'^2}{m^2} a_1 + \dfrac{1}{2}\dfrac{\vartheta_1}{m^3} a_2 + \dfrac{c'}{m} a_4\right) x_4^2, \\ f_{33} = a_1 x_1^2 + a_5 x_1 x_2 - a_3 x_1 x_4 + \dfrac{1}{2}\dfrac{W}{m} a_2 x_2 x_4 + \dfrac{1}{2}\dfrac{c'}{m}\left(b + a\dfrac{n_v}{n}\right) a_2 x_4^2, \\ f_{12} = f_{21} = \dfrac{1}{2} a_5 x_3^2 - \dfrac{1}{2} a_4 x_3 x_4 - \dfrac{1}{2}\dfrac{c'}{m} a_2 x_4^2, \\ f_{13} = f_{31} = -a_1 x_1 x_3 - \dfrac{1}{2} a_2 x_1 x_4 - \dfrac{1}{2} a_5 x_2 x_3 + \dfrac{1}{2} a_3 x_3 x_4, \\ f_{23} = f_{32} = -\dfrac{1}{2} a_5 x_1 x_3 + \dfrac{1}{2} a_4 x_1 x_4 + \dfrac{1}{2}\dfrac{c'}{m} a_5 x_2 x_4 - \dfrac{1}{4}\dfrac{W}{m} a_2 x_3 x_4 \\ \qquad + \left[\dfrac{1}{2m}\left(a' + c'\dfrac{m_u}{m}\right) a_2 - \dfrac{1}{2}\dfrac{c'}{m} a_3\right] x_4^2. \end{cases}$$

The equation appears to be of the sixth degree in x_i. But it contains x_4^2 as an extraneous factor which must be discarded. The resulting equation is therefore of the fourth degree.

We want to find the singular points in the plane $yz\sigma$. For this purpose, put $x_3 = 0$ in (43) and simplify the resulting determinant. Removing the extraneous factor x_4^2, we get the equation

$$x_4^2\left[\vartheta_1 x_1^2 - 2c'(a'm + c'm_u)x_1x_4 + Wc'^2x_2x_4 + c'^3\left(b + d\frac{n_v}{n}\right)x_4^2\right] = 0.$$

This locus consists of the line l_{yz} counted twice and a conic. We have therefore the

Theorem 7. *The four-parameter family of singular surfaces of the complexes $P = 0$ cut the plane $yz\sigma$ in the line l_{yz} counted twice and the fixed conic*

$$(44_1)\qquad \vartheta_1 x_1^2 - 2c'(a'm + c'm_n)x_1x_4 + W_c'^2x_2x_4 + c'^3\left(b + d\frac{n_v}{n}\right)x_4^2 = 0.$$

The conic (44_1) is tangent to l_{yz} at P_z and meets the line $l_{z\sigma}$ at the further point

$$(45_1)\qquad \tau_1 = c'\left(b + d\frac{n_v}{n}\right)z - W\sigma.$$

Similarly, the locus of the singular points of the z-associate quadratic complexes in the plane $yz\rho$ is the line l_{yz} counted twice and conic

$$(44_2)\qquad \vartheta_1 x_2^2 - 2d(bn + dn_v)x_2x_3 + Wd^2x_1x_3 + d^3\left(a' + c'\frac{m_u}{m}\right)x_3^2 = 0,$$

which is tangent to l_{yz} at P_y and meets the line $l_{y\rho}$ at the further point

$$(45_2)\qquad \tau_2 = d\left(a' + c'\frac{m_u}{m}\right)y - W\rho.$$

We shall speak of the conic (44_1) as the y-associate conic and the conic (44_2) as the z-associate conic at l_{yz}.

The points τ_1 and τ_2 have another geometrical interpretation. Let us find the common lines of the y-associate and the z-associate (2,2)-congruences. Since $W \neq 0$, we have

$$p_2 = p_5 = 0$$

and

$$c'\left(b + d\frac{n_v}{n}\right)p_4^2 - Wp_4p_6 = 0,$$

$$d\left(a' + c'\frac{m_u}{m}\right)p_4^2 + Wp_3p_4 = 0.$$

Hence the common lines are either lines for which

$$p_2 = p_4 = p_5 = 0$$

or the line

$$p_1 : p_2 : p_3 : p_4 : p_5 : p_6$$
$$= \frac{c'd}{W}\left(a' + c'\frac{m_u}{m}\right)\left(b + d\frac{n_v}{n}\right) : 0 : -d\left(a' + c'\frac{m_u}{m}\right) : W : 0 : c'\left(b + d\frac{n_v}{n}\right).$$

This gives the

Theorem 8. *The y-associate and the z-associate* (2,2)*-congruences have in common the central radii through* P_y *and through* P_z *and the line* $\tau_1\tau_2$.

6. The Associate (2,2)-congruences and Allied Considerations

We proceed to find the focal surfaces of the associate (2,2)-congruences. Consider first the *y*-associated (2,2)-congruence (18_1). It is represented as the intersection of a linear complex (the *y*-associate linear complex) with a quadratic complex. Let (q_i) be one of its lines. The tangent linear complexes of the quadratic complex at (q_i) have the equation(1)

$$\mu q_4 p_1 + \mu q_5 p_2 + \mu q_6 p_3 + \left[\lambda\left\{2c'\left(b + d\frac{n_v}{n}\right)q_4 - 2\left(a' + c'\frac{m_u}{m}\right)q_5 - Wq_6\right\} + \mu q_1\right]p_4 + \left[\lambda\left\{-2\left(a' + c'\frac{m_u}{m}\right)q_4 + 2\frac{\vartheta_1}{m^2}q_5\right\} + \mu q_2\right]p_5 + \left[-\lambda W q_4 + \mu q_3\right]p_6 = 0, \tag{46}$$

where λ and μ are homogeneous parameters. Each of the complexes (46) is special. In fact, their axes form a pencil of lines through the point $\left(\text{putting } a' + c'\dfrac{m_u}{m} = \alpha', b + d\dfrac{n_v}{n} = \beta\right)$

$$x_1 : x_2 : x_3 : x_4 = -2\alpha' q_3 q_4 + 2\frac{\vartheta_1}{m^2} q_3 q_5 + W q_2 q_4 : 2c'\beta q_4^2 - 2\alpha' q_4 q_5 - W q_4 q_6 : -2\alpha' q_4^2 + 2\frac{\vartheta_1}{m^2} q_4 q_5 : -W q_4^2, \tag{47}$$

and lying in the plane

$$q_4 x_2 + q_5 x_3 + q_6 x_4 = 0. \tag{48}$$

Hence the normal correlation of the quadratic complex on *q* is a singular one, of which the point (47) and the plane (48) form the singular couple(2).

Also, the *y*-associate linear complex defines at *q* a normal correlation. The two normal correlations have in common two couples. They are found by taking the null plane of the point (47) and the null point of the plane (48) with respect to the null system of the *y*-associate linear complex. The inverse couples of these couples common to the two normal correlations are the focal couples of the *y*-associate (2,2)-congruence at (q_i),(3), in the sense that the line (q_i) is tangent to the focal sheets of the congruence at these focal couples. Thus we find that one focal sheet of the congruence is the locus

(1) Jessop, loc. cit., p. 88.
(2) G. Koenigs, La Géométrie Réglée et ses Applications, Paris Gauthier-Villars (1895), p. 19 and p. 72.
(3) Koenigs, loc. cit., p. 72.

of the point

$$(c'q_5, -mq_6, 0, mq_4)$$

and the another focal sheet is the envelope of the planes (48). Since q_i satisfy the equations (18_1), we have on eliminating q_i the

Theorem 9. *The lines of the y-associate (2,2)-congruence intersect the y-associate conic and are tangent to the quadric cone*

$$(49_1)\qquad u_1 = 0, \quad c'\beta u_2^2 - 2\alpha' u_2 u_3 - W u_2 u_4 + \frac{\vartheta_1}{m^2} u_3^2 = 0.$$

Likewise, the lines of the z-associate (2,2)-congruence intersect the z-associate conic and are tangent to the quadric cone

$$(49_2)\qquad u_2 = 0, \quad d\alpha' u_1^2 - 2\beta u_1 u_4 - W u_1 u_3 + \frac{\vartheta_2}{n^2} u_4^2 = 0.$$

It must be noticed that not every line intersecting the y-associate conic and tangent to the cone (49_1) belongs to the y-associate (2,2)-congruence, as the totality of such lines will form a (4,4)-congruence. But the (4,4)-congruence so formed does not degenerate into two (2,2)-congruences. In fact, its relation to the y-associate (2,2)-congruence may be explained in the following manner. To find the two lines of the y-associate (2,2)-congruence through a point, we take the quadric cone joining it to the points of the y-associate conic and the two planes through it tangent to the cone (49_1). The quadric cone and the two planes so constructed intersect in general in four lines, of which the two required are the lines lying in the null plane corresponding to the point in the null system of the y-associate linear complex. It is dual to find the lines of the y-associate (2,2)-congruence in a plane and similar discussion holds of course for the z-associate (2,2)-congruence.

The point equations of the cones (49_1) and (49_2) are respectively

$$(50_1)\qquad -\frac{W\vartheta_1}{m^2} x_2 x_4 + \frac{1}{4} W^2 x_3^2 - W\alpha' x_3 x_4 + \left(\alpha'^2 - \frac{\vartheta_1 c'\beta}{m^2}\right) x_4^2 = 0,$$

$$(50_2)\qquad -\frac{W\vartheta_2}{n^2} x_1 x_3 + \left(\beta_2 - \frac{\vartheta_2 d\alpha'}{n^2}\right) x_3^2 - W\beta x_3 x_4 + \frac{1}{4} W^2 x_4^2 = 0.$$

They intersect in the line l_{yz} and the twisted cubic

$$(51)\qquad \begin{cases} x_1 = \left(\dfrac{n^2\beta^2}{W\vartheta_2} - \dfrac{d\alpha'}{W}\right) t^2 - \dfrac{n^2\beta}{\vartheta_2} t + \dfrac{1}{4}\dfrac{Wn^2}{\vartheta_2}, \\[2ex] x_2 = \dfrac{1}{4}\dfrac{Wm^2}{\vartheta_1} t^3 - \dfrac{m^2 a'}{\vartheta_1} t^2 + \left(\dfrac{m^2\alpha'^2}{W\vartheta_1} - \dfrac{c'\beta}{W}\right) t, \\[2ex] x_3 = t^2, \\[1ex] x_4 = t. \end{cases}$$

The cubic (51) never degenerates. We shall call it the primary associate cubic of the congruence at l_{yz}.

Dually, the common tagent planes of the y-associate and the z-associate conics form a pencil of planes through l_{yz} and a cubic developable. The parametric equations of the cubic developable are

$$(52)\qquad \begin{cases} u_1 = t^2, \\ u_2 = t, \\ u_3 = \left(\dfrac{d\alpha'}{W} - \dfrac{n^2\beta^2}{W\vartheta_2}\right)t^2 - \dfrac{nd\beta}{\vartheta_2}t - \dfrac{1}{4}\dfrac{Wd^2}{\vartheta_2}, \\ u_4 = -\dfrac{1}{4}\dfrac{Wc'^2}{\vartheta_1}t^3 - \dfrac{mc'\alpha'}{\vartheta_1}t^2 + \left(\dfrac{c'\beta}{W} - \dfrac{m^2a'^2}{W\vartheta_1}\right)t. \end{cases}$$

Its cuspidal edge is a twisted cubic with the parametric equations

$$(53)\qquad \begin{cases} x_1 = \dfrac{c'^2}{\vartheta_1}\left(d\alpha' - \dfrac{n^2\beta^2}{\vartheta_2}\right)t^3 + \dfrac{3}{4}\dfrac{W^2c'^2d^2}{\vartheta_1\vartheta_2}t + \dfrac{Wmc'd^2\alpha'}{\vartheta_1\vartheta_2}, \\ x_2 = -\dfrac{Wnc'^2d\beta}{\vartheta_1\vartheta_2}t^3 - \dfrac{3}{4}\dfrac{W^2c'^2d^2}{\vartheta_1\vartheta_2}t^2 - \dfrac{d^2}{\vartheta_2}\left(c'\beta - \dfrac{m^2\alpha'^2}{\vartheta_1}\right), \\ x_3 = -\dfrac{Wc'^2}{\vartheta_1}t^3, \\ x_4 = \dfrac{Wd^2}{\vartheta_2}. \end{cases}$$

The cubic (53) will be called the secondary associate cubic of the congruence at l_{yz}.

A twisted cubic in space defines a linear complex, to which belong all the tangents of the cubic. By expressing the line coordinates of a tangent of the cubic as functions of the parameter t and eliminating t therefrom, we get as the equation of the linear complex of the primary associate cubic

$$(54)\qquad \begin{aligned} K_1 \equiv p_1 &+ \frac{m^2\alpha'}{\vartheta_1}p_2 + \frac{1}{W}\left(c'\beta - \frac{m^2\alpha'^2}{\vartheta_1}\right)p_3 \\ &+ \left\{\frac{1}{W^2}\left(d\alpha' - \frac{n^2\beta^2}{\vartheta_2}\right)\left(c'\beta - \frac{m^2\alpha'^2}{\vartheta_1}\right) - \frac{m^2n^2\alpha'\beta}{\vartheta_1\vartheta_2} + \frac{3}{16}\frac{W^2m^2n^2}{\vartheta_1\vartheta_2}\right\}p_4 \\ &+ \frac{n^2\beta}{\vartheta_2}p_5 - \frac{1}{W}\left(d\alpha' - \frac{n^2\beta^2}{\vartheta_2}\right)p_6 = 0; \end{aligned}$$

and as that of the secondary associate cubic

$$(55)\qquad \begin{aligned} K_2 \equiv p_1 &- \frac{nd\beta}{\vartheta_2}p_2 + \frac{1}{W}\left(c'\beta - \frac{m^2\alpha'^2}{\vartheta_1}\right)p_3 \\ &+ \left\{\frac{1}{W^2}\left(d\alpha' - \frac{n^2\beta^2}{\vartheta_2}\right)\left(c'\beta - \frac{m^2\alpha'^2}{\vartheta_1}\right) - \frac{mnc'd\alpha'\beta}{\vartheta_1\vartheta_2} + \frac{3}{16}\frac{W^2c'^2d^2}{\vartheta_1\vartheta_2}\right\}p_4 \\ &- \frac{mc'\alpha'}{\vartheta_2}p_5 - \frac{1}{W}\left(d\alpha' - \frac{n^2\beta^2}{\vartheta_2}\right)p_6 = 0. \end{aligned}$$

None of the complexes (54) and (55) can be special. *They are in involution if and only if*

$$\frac{mn\alpha'\beta}{\vartheta_1\vartheta_2}(mn+c'd)+\frac{m^3c'\alpha'^2}{\vartheta_1^2}+\frac{n^3d\beta^2}{\vartheta_2^2}-\frac{3}{16}\frac{W^2}{\vartheta_1\vartheta_2}(m^2n^2+c'^2d^2)=0. \tag{56}$$

The lines common to the complex $K_1=0$ and the associate linear complexes form a regulus, which consists of one family of generators of a quadric. The lines $l_{y\rho}$ and $l_{z\sigma}$ lie completely on this quadric and its equation is found to be

$$\begin{aligned} &x_1x_2+\frac{1}{W}\left(d\alpha'-\frac{n^2\beta^2}{\vartheta_2}\right)x_2x_3+\frac{1}{W}\left(c'\beta-\frac{m^2\alpha'^2}{\vartheta_1}\right)x_1x_4\\ &+\left\{\frac{1}{W^2}\left(d\alpha'-\frac{n^2\beta^2}{\vartheta_2}\right)\left(c'\beta-\frac{m^2\alpha'^2}{\vartheta_1}\right)-\frac{m^2n^2\alpha'\beta}{\vartheta_1\vartheta_2}+\frac{3}{16}\frac{W^2m^2n^2}{\vartheta_1\vartheta_2}\right\}x_3x_4=0. \end{aligned} \tag{57}$$

Similarly, the complex $K_2=0$ and the associate linear complexes define a quadric with the equation

$$\begin{aligned} &x_1x_2+\frac{1}{W}\left(d\alpha'-\frac{n^2\beta^2}{\vartheta_2}\right)x_2x_3+\frac{1}{W}\left(c'\beta-\frac{m^2\alpha'^2}{\vartheta_1}\right)x_1x_4\\ &+\left\{\frac{1}{W^2}\left(d\alpha'-\frac{n^2\beta^2}{\vartheta_2}\right)\left(c'\beta-\frac{m^2\alpha'^2}{\vartheta_1}\right)-\frac{mnc'd\alpha'\beta}{\vartheta_1\vartheta_2}+\frac{3}{16}\frac{W^2c'^2d^2}{\vartheta_1\vartheta_2}\right\}x_3x_4=0. \end{aligned} \tag{58}$$

The quadrics (57) and (58) have in common the lines $l_{y\rho}$, $l_{z\sigma}$, the line joining y to $-\frac{1}{W}\left(c'\beta-\frac{m^2\alpha'^2}{\vartheta_1}\right)z+\sigma$, and the line joining z to $-\frac{1}{W}\left(d\alpha'-\frac{n^2\beta^2}{\vartheta_1}\right)y+\rho$.

7. The Transformation of Laplace

Wilczynski has defined([1]) the congruence Γ_1 formed by the lines $l_{y\rho}$ as the first Laplace transform of the congruence Γ_{yz} and the congruence Γ_{-1} formed by the lines $l_{z\sigma}$ as the minus first Laplace transform of Γ_{yz}. He thus obtained a Laplace sequence of congruences extending in general in both directions, such that each congruence of the sequence is the first Laplace transform of its successor and is the minus first Laplace transform of its predecessor. We shall in this section examine the Laplace sequence for W-congruences and ϑ-congruences, thus giving up the assumption $W\vartheta_1\vartheta_2\not\equiv 0$ from now on.

The condition $\vartheta_1=0$ can be written as

$$\frac{m_v}{m}-\frac{c'_v}{c'}+f_v=0,$$

i.e.,

$$\frac{\partial}{\partial v}\left(e^f\frac{m}{c'}\right)=0. \tag{59_1}$$

Similarly, the condition $\vartheta_2=0$ can be written as

([1]) Brussels paper, § 10.

$$(59_2) \qquad \frac{\partial}{\partial u}\left(e^{\int}\frac{n}{d}\right) = 0.$$

By making use of the systems of differential equations for the congruences Γ_1 and Γ_{-1} and taking care of the integrability conditions (3) when necessary, the condition for Γ_1 to be a $\begin{Bmatrix}\vartheta_1\text{-congruence}\\ \vartheta_2\text{-congruence}\end{Bmatrix}$ is found to be

$$(60) \qquad \begin{Bmatrix} \dfrac{\partial}{\partial v}\{e^{\int}d^2[mn - (\log m)_{uv}]\} = 0, \\ \dfrac{\partial}{\partial u}\{e^{-\int}m^2[c'd - (\log d)_{uv}]\} = 0; \end{Bmatrix}$$

and the conditions for Γ_{-1} to be a $\begin{Bmatrix}\vartheta_1\text{-congruence}\\ \vartheta_2\text{-congruence}\end{Bmatrix}$ is

$$(61) \qquad \begin{Bmatrix} \dfrac{\partial}{\partial v}\{e^{-\int}n^2[c'd - (\log c')_{uv}]\} = 0, \\ \dfrac{\partial}{\partial u}\{e^{\int}c'^2[mn - (\log n)_{uv}]\} = 0. \end{Bmatrix}$$

From the conditions (59), (60), (61), we get easily the

Theorem 10. *If the minus first Laplace transform of a* $\begin{Bmatrix}W\vartheta_1\\ W\vartheta_2\end{Bmatrix}$*-congruence is a* $\begin{Bmatrix}\vartheta_1\\ \vartheta_2\end{Bmatrix}$*-congruence, then the first Laplace transform is also a* $\begin{Bmatrix}\vartheta_1\\ \vartheta_2\end{Bmatrix}$*-congruence.*

Wilczynski has proved that if any two consecutive congruences of a Laplace sequence are W-congruence, then every congruence of the Laplace sequence is a W-congruence([1]). Therefore, by Theorem 10, *if any two consecutive congruences of a Laplace sequence are* $\begin{Bmatrix}W\vartheta_1\\ W\vartheta_2\end{Bmatrix}$*-congruences, then every congruence of the sequence is a* $\begin{Bmatrix}W\vartheta_1\\ W\vartheta_2\end{Bmatrix}$*-congruence.*

Finally, let us consider the nature of a $W\vartheta_1\vartheta_2$-congruence. If $W = \vartheta_1 = \vartheta_2 = 0$, we can, by a transformation of the independent variables, make

$$W \equiv mn - c'd = 0,$$

$$e^{\int}\frac{m}{c'} = 1,$$

$$e^{\int}\frac{n}{d} = 1.$$

([1]) Brussels paper, § 10.

This gives

$$f = 0, \qquad m = c', \qquad n = d.$$

Hence a congruence for which $W \equiv \vartheta_1 \equiv \vartheta_2 \equiv 0$ *and whose focal sheets are neither degenerate nor developable necessarily belongs to a linear complex.*

Tsing Hua University
November 11, 1933.

Reset from
Tohoku Math. J. **40** (1935) 293–316

Sur la Géométrie d'une Équation Différentielle du Troisième Ordre

Note de M. SHIING-SHEN CHERN

Présentée par M. Elie Cartan

Donnons-nous une équation différentielle du troisième ordre

$$y''' = F(x, y, y', y''). \tag{1}$$

Nous nous proposons à étudier ses propriétés géométriques par rapport au groupe de transformations de contact du plan, en employant la méthode d'équivalence de M. E. Cartan ([1]). D'après cette méthode, nous introduisons quatre expressions de Pfaff covariantes, à savoir:

$$\begin{cases} \omega_1 = \alpha\{dy'' - F\,dx + \beta(dy - y'\,dx) + \gamma(dy' - y''\,dx)\}, \\ \omega_2 = \lambda(dy - y'\,dx), \\ \omega_3 = \mu\{dy' - y''\,dx + \nu(dy - y'\,dx)\}, \\ \omega_4 = u\{dx + v(dy - y'\,dx) + w(dy' - y''\,dx)\}, \end{cases} \tag{2}$$

où les α, β, γ, λ, μ, ν, u, v, w sont neuf variables auxiliaires nouvelles.

En appliquant la méthode, on trouve qu'il y a un invariant relatif qui joue un rôle important. Son expression est

$$I = -\frac{\partial F}{\partial y} - \frac{1}{3}\frac{\partial F}{\partial y'}\frac{\partial F}{\partial y''} - \frac{2}{27}\left(\frac{\partial F}{\partial y''}\right)^3 + \frac{1}{2}\frac{d}{dx}\frac{\partial F}{\partial y'} + \frac{1}{3}\frac{\partial F}{\partial y''}\frac{d}{dx}\frac{\partial F}{\partial y''} - \frac{1}{6}\frac{d^2}{dx^2}\frac{\partial F}{\partial y''},$$

en posant, Φ étant une fonction quelconque de x, y, y', y'',

$$\frac{d\Phi}{dx} = \frac{\partial \Phi}{\partial x} + y'\frac{\partial \Phi}{\partial y} + y''\frac{\partial \Phi}{\partial y'} + F\frac{\partial \Phi}{\partial y''}.$$

La condition $I = 0$ *est la condition nécessaire et suffisante pour que la condition de contact de deux courbes intégrales infiniment voisines de l'équation* (1) *soit donnée par une équation de Monge du second ordre.*

Dans le cas $I = 0$, l'espace généralisé qu'on définit a les équations de structure suivantes:

([1]) *Ann. Éc. Norm.*, 25, 1908, Chap. I, p. 57.

$$
(3)\quad
\begin{cases}
\omega_1' = [\varpi_1\omega_1] + [\varpi_2\omega_3], \\
\omega_2' = [\varpi_1\omega_2] + 2[\varpi_3\omega_2] - [\omega_3\omega_4], \\
\omega_3' = [\varpi_2\omega_2] + [\varpi_1\omega_3] + [\varpi_3\omega_3] - [\omega_1\omega_4], \\
\omega_4' = [\varpi_4\omega_2] + [\varpi_5\omega_3] + [\varpi_3\omega_4], \\
\varpi_1' = [\varpi_2\omega_4] + 2[\varpi_5\omega_1] + [\varpi_4\omega_3], \\
\varpi_2' = [\varpi_2\varpi_3] + [\varpi_4\omega_1] + [\varpi_6\omega_3], \\
\varpi_3' = -[\varpi_2\omega_4] - [\varpi_5\omega_1] + [\varpi_6\omega_2], \\
\varpi_4' = -[\varpi_1\varpi_4] - [\varpi_2\varpi_5] - [\varpi_3\varpi_4] + [\varpi_6\omega_4] + e[\omega_2\omega_3] \\
\qquad - c[\omega_3\omega_1] - 2a[\omega_1\omega_2], \\
\varpi_5' = -[\varpi_1\varpi_5] + [\varpi_4\omega_4] + a[\omega_2\omega_3] + b[\omega_3\omega_1] + c[\omega_1\omega_2], \\
\varpi_6' = -[\varpi_1\varpi_6] - [\varpi_2\varpi_4] - 2[\varpi_3\varpi_6] + f[\omega_2\omega_3] + a[\omega_3\omega_1] - e[\omega_1\omega_2].
\end{cases}
$$

C'est un espace généralisé dont les éléments sont les courbes intégrales de l'équation (1). Regardant ces courbes intégrales comme les points d'un espace à trois dimensions, *cet espace généralisé est un espace à connexion conforme normale.* Les cônes des directions isotropes de la connexion (3) sont donnés par la condition de contact de deux courbes intégrales infiniment voisines. Rappelons qu'en partant de cette famille de cônes on peut définir, d'une manière intrinsèque, une connexion conforme normale. Cette connexion est nécessairement la connexion (3).

Dans le cas général où $\mathrm{I} \neq 0$, on définit une géométrie généralisée dont les équations de structure sont

$$
(4)\quad
\begin{cases}
\omega_1' = [\varpi\omega_1] + [\omega_2\omega_4] + a[\omega_2\omega_3] + b[\omega_3\omega_4], \\
\omega_2' = [\varpi\omega_2] - [\omega_3\omega_4] + c[\omega_2\omega_3], \\
\omega_3' = [\varpi\omega_3] - [\omega_1\omega_4] + e[\omega_1\omega_2] + f[\omega_2\omega_3] + b[\omega_2\omega_4], \\
\omega_4' = g[\omega_1\omega_2] + h[\omega_1\omega_3] + k[\omega_2\omega_3] + l[\omega_2\omega_4], \\
\varpi' = m[\omega_1\omega_2] + n[\omega_1\omega_3] + p[\omega_1\omega_4] + q[\omega_2\omega_3] + r[\omega_2\omega_4] + s[\omega_3\omega_4],
\end{cases}
$$

avec les relations

$$p = e = -\tfrac{1}{2}c, \qquad r = a + \tfrac{1}{2}bc.$$

Cet espace généralisé est fondé sur les éléments de contact du second ordre, x, y, y', y'', et il a comme groupe fondamental un certain groupe à cinq paramètres.

Reset from
C. R. Acad. Sci. Paris **204** (1937) 1227–1229

On Projective Normal Coördinates

By SHIING-SHEN CHERN*
(Received August 11, 1937)

1. Introduction

The normal coördinates in a projectively connected space were first introduced by O. Veblen and J. M. Thomas.[1] In a purely geometrical way E. Cartan has also defined a system of projective normal coördinates.[2] In this paper we shall give an analytic characterization of the normal coördinates of Cartan, from which it results that the two definitions are not the same. Further, some necessary conditions for the spaces for which the two definitions agree are obtained. I want to express here my hearty thanks to Professor Cartan for his valuable suggestions and criticisms.

2. Analytic Characterization of the Normal Coördinates of Cartan

Consider a space of n dimensions with the coördinates $u^1, \ldots, u^n$. According to Cartan a projective connection is defined in the space if there is given at each point A a projective reference ("repère projectif") $AA_1 \ldots A_n$ and a law of infinitesimal displacement between the references

$$(1)\qquad \begin{cases} dA = \omega^1 A_1 + \cdots + \omega^n A_n, \\ dA_1 = \omega_1^0 A + \omega_1^1 A_1 + \cdots + \omega_1^n A_n, \\ \cdots \\ dA_n = \omega_n^0 A + \omega_n^1 A_1 + \cdots + \omega_n^n A_n, \end{cases}$$

where $\omega^i, \omega_i^0, \omega_i^j$ are differential forms in u^i, thus[3]

$$(2)\qquad \begin{cases} \omega^i = \Pi_k^i du^k, \qquad \omega_i^0 = \Pi_{ik}^0 du^k, \\ \omega_i^j = \Pi_{ik}^j du^k. \end{cases}$$

* Research fellow of the China Foundation for the Promotion of Culture and Education.

[1] O. Veblen and J. M. Thomas, "Projective normal coördinates for the geometry of paths," Proc. N. A. S., vol. 11 (1925), pp. 204–7. See also T. Y. Thomas, Differential Invariants of Generalized Spaces, pp. 91–96.

[2] E. Cartan, Leçons sur la théorie des espaces à connexion projective, Paris 1936, pp. 220–4. This book will be cited as "Cartan, Leçons."

[3] For all details the reader is referred to Cartan, Leçons, Part 2.

It is possible to change the projective reference $AA_1 \dots A_n$ about its origin A without changing the projective connection. For a given system of coördinates u^i there is at each point a uniquely determined reference, called the natural reference ("repère naturel"), for which we have

$$\Pi^i_k = \delta^i_k, \qquad \Pi^i_{ik} = 0. \tag{3}$$

For the sake of avoiding ambiguity we shall denote by Γ^j_{ik} the components Π^j_{ik} when the space is referred to the system of natural references. It may be verified that the components Γ^j_{ik} are the components of a projective connection in the sense of T. Y. Thomas.[4] This would mean that under a transformation of coördinates

$$u^i = u^i(\bar{u}^1, \dots, \bar{u}^n), \qquad \Delta = \left|\frac{\partial u^i}{\partial \bar{u}^k}\right| \neq 0, \tag{4}$$

they are transformed according to the equations

$$\bar{\Gamma}^l_{ik}\frac{\partial u^j}{\partial \bar{u}^l} = \frac{\partial^2 u^j}{\partial \bar{u}^i\, \partial \bar{u}^k} + \Gamma^j_{lm}\frac{\partial u^l}{\partial \bar{u}^i}\frac{\partial u^m}{\partial \bar{u}^k} - \frac{1}{n+1}\frac{\partial u^j}{\partial \bar{u}^k}\frac{\partial \log \Delta}{\partial \bar{u}^i} - \frac{1}{n+1}\frac{\partial u^j}{\partial \bar{u}^i}\frac{\partial \log \Delta}{\partial \bar{u}^k}, \tag{5}$$

which play a fundamental part in the theory of Thomas.

The coördinates u^i are said to be normal coördinates in the sense of Veblen and Thomas[5] if the following relations are satisfied

$$\Gamma^j_{ik}u^i u^k = 0. \tag{6}$$

The coördinates u^i remain normal when they undergo the transformation

$$u^i \to \frac{a^i_k u^k}{1 + a^0_k u^k}, \qquad |a^i_k| \neq 0, \tag{7}$$

where the a^0_k, a^i_k are arbitrary constants.

The definition of Cartan's normal coördinates is purely geometrical. The coördinates u^i are said to be normal in Cartan's sense at the point $O(u^i = 0)$ if the geodesics through O are given by the equations

$$u^i = a^i t \tag{8}$$

and if by developing the geodesics through O in the tangent space associated at O and by assigning the corresponding points the same coördinates u^i the coördinates u^i form a system of projective coördinates in the tangent space at O. The existence of Cartan's normal coördinates and the fact that they remain normal under the transformations (7) are geometrically evident. Analytically, in order that the coördinates u^i be normal it is necessary and sufficient that along the curves $u^i = a^i t$ (a^i being arbitrary constants) the projective acceleration of the geometrical point A be zero, i.e., there exists a factor $\lambda(u^1, \dots, u^n) \neq 0$ such that, for $u^i = a^i t$,

[4] T. Y. Thomas, "On the projective and equi-projective geometries of paths," Proc. N. A. S. vol. 11 (1925), pp. 199–203.

[5] Veblen and Thomas, loc. cit.

(9) $$\frac{d^2(\lambda A)}{dt^2} = 0.$$

But we have, on referring to the system of natural references.

$$\frac{dA}{dt} = \frac{du^i}{dt} A_i,$$

$$\frac{d^2 A}{dt^2} = \Pi^0_{ij} \frac{du^i}{dt}\frac{du^j}{dt} A + \left(\frac{d^2 u^i}{dt^2} + \Gamma^i_{jk}\frac{du^j}{dt}\frac{du^k}{dt}\right) A_i.$$

Hence the condition (9) is equivalent to the conditions

$$\frac{\partial^2 \lambda}{\partial u^i \partial u^j} a^i a^j + \lambda \Pi^0_{ij} a^i a^j = 0,$$

$$2a^i a^k \frac{\partial \lambda}{\partial u^k} + \lambda \Gamma^i_{jk} a^j a^k = 0.$$

As this is true for all values of a^i, we get

(10) $$\begin{cases} \dfrac{\partial^2 \lambda}{\partial u^i \partial u^j} u^i u^j + \lambda \Pi^0_{ij} u^i u^j = 0, \\ 2\dfrac{\partial \lambda}{\partial u^k} u^i u^k + \lambda \Gamma^i_{jk} u^j u^k = 0. \end{cases}$$

The necessary and sufficient conditions such that a solution $\lambda(u^1, \ldots, u^n) \neq 0$ of the system (10) exists, which is holomorphic at the origin, are

(11) $$\begin{cases} \Gamma^i_{jk} u^j u^k - 2F u^i = 0, \\ \Pi^0_{ij} u^i u^j - \dfrac{\partial F}{\partial u^i} u^i + F^2 + F = 0, \end{cases}$$

where F is the common ratio of $\Gamma^i_{jk} u^j u^k : 2u^i$. Consequently, the conditions (11) are the necessary and sufficient conditions such that the coördinates u^i be normal in Cartan's sense.

3. Question on the Coincidence of the Two Kinds of Normal Coördinates

From the analytic definitions of the two kinds of normal coördinates it is evident that they are in general not the same. The question naturally arises of investigating those projectively connected spaces for which the two definitions agree. For a given point a necessary and sufficient condition that the normal coördinates u^i of Veblen and Thomas be also normal in the sense of Cartan is that they satisfy the equation $\Pi^0_{ik} u^i u^k = 0$.

A more difficult but also more interesting problem is to characterize intrinsically the projectively connected spaces for which the two kinds of normal coördinates always coincide. For this purpose we shall follow the manner by which Cartan studies his normal coördinates. Let O be a fixed point and R_0 a reference with the origin O. Suppose u^i be a system of normal coördinates with the origin $O(u^i = 0)$. In a sufficiently small region about O in which every point can be reached by one and only one geodesic through O, we get, by displacing R_0 along all the geodesics $u^i = a^i t$, a uniquely determined reference at each point. Let

$$\omega^i = \Pi^i_k\, du^k, \qquad \omega^0_i = \Pi^0_{ik}\, du^k, \qquad \omega^j_i = \Pi^j_{ik}\, du^k$$

be the components of the projective connection with respect to this system of references. If we put $u^i = a^i t$ and vary t only, the reference undergoes a translation. That is to say, by putting $da^i = 0$ in

$$\omega^i = \Pi^i_k t\, da^k + \Pi^i_k a^k\, dt,$$

$$\omega^0_i = \Pi^0_{ik} t\, da^k + \Pi^0_{ik} a^k\, dt,$$

$$\omega^j_i = \Pi^j_{ik} t\, da^k + \Pi^j_{ik} a^k\, dt,$$

we have

$$\omega^i = a^i\, dt, \qquad \omega^0_i = \omega^j_i = 0.$$

We get thus the conditions

$$\Pi^i_k a^k = a^i, \qquad \Pi^0_{ik} a^k = 0, \qquad \Pi^j_{ik} a^k = 0.$$

Since this holds for all values of a^i, we have

(12) $$\Pi^i_k u^k = u^i, \qquad \Pi^0_{ik} u^k = 0, \qquad \Pi^j_{ik} u^k = 0.$$

The relations (12) characterize the space to be referred to a system of normal coördinates (in Cartan's sense) and to a system of references as defined above. If we define the quantities b^i_k by

(13) $$b^i_k \Pi^k_j = \delta^i_j,$$

the components of Thomas are given by

(14) $$\Gamma^j_{ik} = b^j_l \frac{\partial \Pi^l_i}{\partial u^k} + \Pi^h_i b^j_l \Pi^l_{hk} - \frac{1}{n+1}\delta^j_i \left(b^h_l \frac{\partial \Pi^l_h}{\partial u^k} + \Pi^h_{hk} \right) - \frac{1}{n+1}\delta^j_k \left(b^h_l \frac{\partial \Pi^l_h}{\partial u^i} + \Pi^h_{hi} \right).$$

The coördinates u^i are also normal in the sense of Veblen and Thomas when and only when

$$\Gamma^j_{ik} u^i u^k = 0,$$

i.e., when and only when,

$$\frac{\partial |\Pi^i_j|}{\partial u^k} u^k = 0,$$

or

$$|\Pi^i_j| = \text{constant}.$$

But $|\Pi^i_j| = 1$ at 0. Therefore the necessary and sufficient condition that the coördinates

u^i be normal in both senses is that

(15) $$[\omega^1 \dots \omega^n] = [du^1 \dots du^n].$$

In order to give the condition (15) another form let us recall how the projective connection can be found when we know the values of the tensor of curvature and torsion and of its successive covariant derivatives

$$(R^{\beta}_{\alpha kh})_O, (R^{\beta}_{\alpha kh|l})_O, (R^{\beta}_{\alpha kh|lm})_O, \dots \begin{pmatrix} \alpha, \beta = 0, 1, \dots, n \\ R^0_{0kh} = 0 \end{pmatrix}$$

at the point O. Let δ be an operation such that $\delta t = 0$, the δa^i being arbitrary. Put

$$\omega^i(\delta) = \varpi^i, \qquad \omega^j_i(\delta) = \varpi^j_i, \qquad \omega^0_i(\delta) = \varpi^0_i.$$

The system of differential equations[6]

(16) $$\begin{cases} \dfrac{\partial \varpi^i}{\partial t} = \delta a^i + a^k \varpi^i_k - \{(R^i_{0kh})_O + (R^i_{0kh|l})_O a^l t + \cdots\} a^k \varpi^h, \\ \dfrac{\partial \varpi^j_i}{\partial t} = -a^j \varpi^0_i - \delta^j_i a^k \varpi^0_k - \{(R^j_{ikh})_O + (R^j_{ikh|l})_O a^l t + \cdots\} a^k \varpi^h, \\ \dfrac{\partial \varpi^0_i}{\partial t} = -\{(R^0_{ikh})_O + (R^0_{ikh|l})_O a^l t + \cdots\} a^k \varpi^h, \end{cases}$$

where the a^i, δa^i are regarded as constants, has a solution $\varpi^i(t)$, $\varpi^j_i(t)$, $\varpi^0_i(t)$ satisfying the initial conditions

$$\varpi^i(0) = \varpi^j_i(0) = \varpi^0_i(0) = 0.$$

The components $\omega^i, \omega^j_i, \omega^0_i$ defined above are then obtained from $\varpi^i, \varpi^j_i, \varpi^0_i$ by putting $t = 1$ and by replacing the a^i by u^i. It follows that the condition (15) is equivalent to the condition

(17) $$[\varpi^1 \dots \varpi^n] = t^n [\delta a^1 \dots \delta a^n].$$

Put

$$\theta^i = a^k \varpi^i_k,$$
$$\varpi^0 = a^k \varpi^0_k.$$

The system (16) gives

(18) $$\begin{cases} \dfrac{\partial \varpi^i}{\partial t} = \delta a^i + \theta^i - \{(R^i_{0kh})_O a^k + (R^i_{0kh|l})_O a^k a^l t + \cdots\} \varpi^h, \\ \dfrac{\partial \theta^i}{\partial t} = -2a^i \varpi^0 - \{(R^i_{jkh})_O a^j a^k + (R^i_{jkh|l})_O a^j a^k a^l t + \cdots\} \varpi^h, \\ \dfrac{\partial \varpi^0}{\partial t} = -\{(R^0_{jkh})_O a^j a^k + (R^0_{jkh|l})_O a^j a^k a^l t + \cdots\} \varpi^h \end{cases}$$

[6] Cartan, Leçons, p. 223.

In order that the normal coördinates of Cartan coincide with those of Veblen and Thomas at the point O it is necessary and sufficient that the system of equations (17), (18) has a solution $\bar{\omega}^i(t)$, $\theta^i(t)$, $\bar{\omega}^0(t)$ satisfying the initial conditions

$$\bar{\omega}^i(0) = \theta^i(0) = \bar{\omega}^0(0) = 0.$$

By expanding $\bar{\omega}^i$ according to powers of t

$$\bar{\omega}^i = \delta a^i t + \frac{1}{2!}\left(\frac{\partial^2 \bar{\omega}^i}{\partial t^2}\right)_0 t^2 + \cdots$$

from the equations (18) and substituting in (17), we get, on equating the coefficients of $t^{n+1}, t^{n+2}, t^{n+3}, t^{n+4}$, the relations

$$(R^i_{0ki})_0 a^k = 0,$$

$$\{-4(R^i_{jki})_0 + (R^i_{0jp})_0(R^p_{0ki})_0\} a^j a^k = 0,$$

$$\{-2(R^i_{jki|l})_0 + (R^i_{0jp})_0(R^p_{0ki|l})_0\} a^j a^k a^l = 0,$$

$$\begin{aligned}\{4(R^i_{0jp|k})_0(R^p_{0li|m})_0 &+ 16(R^i_{jkp})_0(R^p_{0li|m})_0 - 4(R^i_{0jp|k})_0(R^p_{0lq})_0(R^q_{0mi})_0 \\ &- 8(R^i_{jkp})_0(R^p_{0q})_0(R^q_{0mi})_0 + (R^i_{0jp})_0(R^p_{0kq})_0(R^q_{0lr})_0(R^r_{0mi})_0 \\ &+ 16(R^i_{jkp})_0(R^p_{lmi})_0\} a^j a^k a^l a^m = 0.\end{aligned}$$

These relations must hold for all values of a^i and since we consider the spaces for which the two kinds of normal coördinates coincide at every point, we can drop the last subscript 0. We get thus the necessary conditions

$$(19)\quad \left\{\begin{aligned} &R^i_{0ki} = 0, \\ &\sum_{(jk)} (-4R^i_{jki} + R^i_{0jp}R^p_{0ki}) = 0, \\ &\sum_{(jklm)} (4R^i_{0jp|k}R^p_{0li|m} + 16R^i_{jkp}R^p_{0li|m} - 4R^i_{0jp|k}R^p_{0lq}R^q_{0mi} \\ &\qquad - 8R^i_{jkp}R^p_{0lp}R^q_{0mi} + R^i_{0jp}R^p_{0kq}R^q_{0lr}R^r_{0mi} + 16R^i_{jkp}R^p_{lmi}) = 0,\end{aligned}\right.$$

where the summations are extended over all the permutations of j, k and j, k, l, m respectively.

When the space is without torsion, the expansion has been carried out to t^{n+6} inclusive and the following necessary conditions are found:

$$(20)\quad \left\{\begin{aligned} &R^i_{jki} + R^i_{kji} = 0, \\ &\sum_{(pqrs)} R^i_{pqj}R^j_{rsi} = 0, \\ &\sum_{(lmpqrs)} (9R^i_{lmj|p}R^j_{qri|s} - 32R^i_{lmj}R^j_{pqk}R^k_{rsi}) = 0.\end{aligned}\right.$$

Unfortunately, there is no reason to expect that the conditions (19) or (20) be sufficient for general values of n. For $n \geq 3$ it is clear that the condition that the space be with normal connection is not sufficient.

As Professor Cartan communicated to me, he has arrived at the necessary and sufficient conditions in the case $n = 2$. In this special case, in fact, the first two equations of (19) give

$$R^1_{012} = R^2_{012} = 0, \qquad R^2_{112} = R^1_{212} = 0, \qquad R^1_{112} = R^2_{212}. \tag{21}$$

To show that the conditions (21) are also sufficient, it is only necessary to differentiate the equation (17) five times and take account of the equations (18). The fifth equation obtained is a consequence of the preceding ones. Therefore there exists a system of solutions satisfying (17), (18) and the given initial conditions. Consequently, in order that the normal coördinates of Cartan and those of Veblen and Thomas coincide at every point of a projectively connected space of two dimensions it is necessary and sufficient that the conditions (21) be satisfied. In this case, the condition that the connection be normal is sufficient. Geometrically, the conditions (21) signify that the infinitesimal displacement associated with any infinitely small cycle with origin A leaves invariant the point A and all the directions through A.

Tsing-Hua University,
Peiping, China.

Reset from
Ann. of Math. **39** (1938) 165–171

On Two Affine Connections

By SHIING-SHEN CHERN (陳省身)

Introduction

The study of geometric objects is perhaps one of the principal aims of modern differential geometry. According to Veblen(1) a geometric object in a space R_n with the coordinates $x^1, \ldots, x^n$ is defined by a set of components $A^1, \ldots, A^r$ such that these components undergo a transformation when the coordinates $x^1, \ldots, x^n$ are subjected to a general point transformation. Obviously this definition is purely analytical. In many cases it is possible to associate to the geometric object a generalized space in an intrinsic way (i.e., invariant with respect to the general group of point transformations), thus giving a geometric meaning to the geometric object in question. It is sufficient only to mention the Euclidean connection by a quadratic differential form and the projective connection by a system of paths(2).

In the present Note we propose to consider two particular geometric objects arised from the well-known problems of "textile mathematics" of Blaschke. In each case we want to show that it is possible to define an affine connection intrinsically related to the geometric object considered. At the same time the problem of equivalence is solved.

The author wishes to express his thanks to Professor Elie Cartan in Paris for the suggestion of this problem.

1. Three Webs of Surface in R_4(3). Fundamental Equations

Consider in a space of four dimensions R_4 with the coordinates $x^1, \ldots, x^4$ a three-web of surfaces (i.e., two-dimensional varieties.) By a three-web of surfaces we shall mean three families of surfaces such that through each point of a region B of R_4

(1) Cf., for example, O. Veblen and J. H. C. Whitehead, Foundations of Differential Geometry, Cambridge Tracts, pp. 46–49.

(2) E. Cartan, Sur les variétés à connexion projective, Bulletin de la Société Mathématique de France, Tome 52 (1924), pp. 205–241.

(3) A theory of invariants of three-webs of surfaces in R_4 was first established by G. Bol and later in a more general case by the author by a different method. See G. Bol: Uber Drei-Gewebe im vierdimensionalen-Raum, Mathematische Annalen, Bd. 110 (1934), pp. 431–463, and Shiing-shen Chern, Eine Invariantentheorie der Dreigewebe aus r-dimensionalen Mannigfaltigkeiten im R_{2r}, Abhandlungen aus dem Mathematischen Seminar der Hansischen Universität, Bd. 11 (1936), pp. 333–358.

there passes one and only one surface of each family. We shall further suppose that no two tangent planes to the surfaces through the same point in B belong to the same hyperplane. Analytically the three families of surfaces can be defined by three completely integrable Pfaffian systems of the form

$$\theta^1 = 0, \qquad \theta^2 = 0; \tag{1}$$

$$\theta^3 = 0, \qquad \theta^4 = 0; \tag{2}$$

$$\theta^5 = 0, \qquad \theta^6 = 0; \tag{3}$$

where θ^α ($\alpha = 1, \ldots, 6$) are Pfaffian forms in the coordinates $x^1, \ldots, x^4$. By hypothesis, the Pfaffian forms θ^1, θ^2, θ^3, θ^4 are linearly independent, so that θ^5 and θ^6 can be expressed as linear combinations of them. Since each of the pairs of Pfaffian forms θ^1, θ^2; θ^3, θ^4; θ^5, θ^6 is defined up to a linear transformation, we can make a suitable choice of the θ's so that we can write

$$\theta^5 = \theta^1 + \theta^3, \qquad \theta^6 = \theta^2 + \theta^4.$$

The third family of surfaces is then defined by the system

$$\theta^1 + \theta^3 = 0, \qquad \theta^2 + \theta^4 = 0, \tag{3'}$$

and the Pfaffian forms θ^1, θ^2, θ^3, θ^4 are defined up to the transformation

$$\begin{cases} \omega^1 = u_1^1\theta^1 + u_2^1\theta^2, & \omega^3 = u_1^1\theta^3 + u_2^1\theta^4, \\ \omega^2 = u_1^2\theta^1 + u_2^2\theta^2; & \omega^4 = u_1^2\theta^3 + u_2^2\theta^4, \end{cases} \qquad \Delta = u_1^1u_2^2 - u_2^1u_1^2 \neq 0, \tag{4}$$

where u_1^1, u_2^1, u_1^2, u_2^2 are functions of x^α. With the tensor notation of summation and agreeing that every Latin index takes the values 1 and 2, we can write (4) in the form

$$\begin{cases} \omega^i = u_j^i\theta^j. \\ \omega^{2+i} = u_j^i\theta^{2+j}. \end{cases} \tag{5}$$

We now regard the u_j^i as auxiliary variables and form the exterior derivatives of ω^i, ω^{2+i}(4). We get

$$(\omega^i)' = [du_j^i\theta^j] + u_j^i(\theta^j)',$$
$$(\omega^{2+i})' = [du_j^i\theta^{2+j}] + u_j^i(\theta^{2+j})'.$$

Noting that the $(\theta^j)'$, $(\theta^{2+j})'$ are exterior differential forms of the second order in θ^1, θ^2, θ^3, θ^4 and that the θ's can in turn be expressed in terms of the ω's, and noticing further that the Pfaffian systems

$$\omega^1 = 0, \qquad \omega^2 = 0;$$
$$\omega^3 = 0, \qquad \omega^4 = 0,$$

which are respectively equivalent to (1) and (2), are completely integrable, we see that $(\omega^i)'$, $(\omega^{2+i})'$ can be written in the form

(4) On the notion of exterior derivation, see E. Cartan, Leçons sur les invariants intégraux, Chap. VI, VII, Paris (1922).

$$(6)\quad \begin{cases} (\omega^i)' = [\vartheta^i_j\omega^j] + a^i_{jk}[\omega^j\omega^k] + b^i_{jk}[\omega^j\omega^{2+k}], \\ (\omega^{2+i})' = [\vartheta^i_j\omega^{2+j}] + c^i_{jk}[\omega^{2+j}\omega^{2+k}] + d^i_{jk}[\omega^j\omega^{2+k}], \end{cases}$$

where a^i_{jk}, b^i_{jk}, c^i_{jk}, d^i_{jk} are functions in the x's and the u's subjected to the conditions

$$(7)\quad a^i_{jk} + a^i_{kj} = 0, \qquad c^i_{jk} + c^i_{kj} = 0,$$

The symbols ϑ^i_j denote four Pfaffian forms in the x's and the u's and are not uniquely determined by the equations (6). The most general transformation on the ϑ^i_j, which leaves the form of the equations (6) unaltered, is of the form

$$(8)\quad \bar{\vartheta}^i_j = \vartheta^i_j + l^i_{jk}\omega^k + m^i_{jk}\omega^{2+k}.$$

On denoting the corresponding coefficients of the system (6) after the transformation (8) by the same letters but preceded with dashes, we have

$$(9)\quad \begin{cases} \bar{a}^i_{jk} = a^i_{jk} + \tfrac{1}{2}(l^i_{jk} - l^i_{kj}), \\ \bar{b}^i_{jk} = b^i_{jk} + m^i_{jk}, \\ \bar{c}^i_{jk} = c^i_{jk} + \tfrac{1}{2}(m^i_{jk} - m^i_{kj}), \\ \bar{d}^i_{jk} = d^i_{jk} - l^i_{kj} \end{cases}$$

It is possible to choose the l^i_{jk}, m^i_{jk} such that the equations

$$2\bar{a}^i_{jk} + \bar{d}^i_{jk} = 0, \qquad \bar{b}^i_{jk} + 2\bar{c}^i_{jk} = 0$$

hold. In fact, these equations imply the conditions

$$\begin{cases} 2a^i_{jk} + d^i_{jk} + l^i_{jk} - 2l^i_{kj} = 0, \\ d^i_{jk} + d^i_{kj} - l^i_{kj} - l^i_{jk} = 0, \\ b^i_{jk} + 2c^i_{jk} + 2m^i_{jk} - m^i_{kj} = 0, \\ b^i_{jk} + b^i_{kj} + m^i_{jk} + m^i_{kj} = 0, \end{cases}$$

which possess only one system of solutions

$$(10)\quad \begin{cases} l^i_{jk} = \tfrac{1}{3}(-2a^i_{jk} + d^i_{jk} + 2d^i_{kj}), \\ m^i_{jk} = \tfrac{1}{3}(-2c^i_{jk} - 2b^i_{jk} - b^i_{kj}). \end{cases}$$

Suppose the Pfaffian forms $\bar{\vartheta}^i_j$ so chosen and let us drop the dashes. Then we have the equations (6) with the conditions (7) and

$$(11)\quad 2a^i_{jk} + d^i_{jk} = 0, \qquad b^i_{jk} + 2c^i_{jk} = 0$$

between the coefficients. In particular, we see that all coefficients a^i_{jk}, b^i_{jk}, c^i_{jk}, d^i_{jk} are skew-symmetric in the lower indices. Since the Pfaffian system

$$\omega^1 + \omega^3 = 0, \qquad \omega^2 + \omega^4 = 0.$$

is completely integrable, it follows that

$$(12)\quad a^i_{jk} - b^i_{jk} + c^i_{jk} - d^i_{jk} = 0.$$

From equations (11) and (12) we get

(13) $$b^i_{jk} = 2a^i_{jk}, \qquad c^i_{jk} = -a^i_{jk}, \qquad d^i_{jk} = -2a^i_{jk}.$$

Consequently, the fundamental equations (6) may be written in the form

(14) $$\begin{aligned}(\omega^i)' &= [\vartheta^i_j\omega^j] + a^i_{jk}[\omega^j\omega^k] + 2a^i_{jk}[\omega^j\omega^{2+k}],\\ (\omega^{2+i})' &= [\vartheta^i_j\omega^{2+j}] - a^i_{jk}[\omega^{2+j}\omega^{2+k}] - 2a^i_{jk}[\omega^j\omega^{2+k}],\end{aligned}$$

with the conditions

(15) $$a^i_{jk} + a^i_{kj} = 0.$$

By the form of the equations (14) the Pfaffian forms ϑ^i_j are uniquely determined.

There are other possible ways of making a unique determination of the Pfaffian forms ϑ^i_j. For example, the functions l^i_{jk}, m^i_{jk} in (8) can be chosen such that

$$\bar{b}^i_{jk} = \bar{d}^i_{jk} = 0.$$

From the complete integrability of the Pfaffian system

$$\omega^1 + \omega^3 = 0, \qquad \omega^2 + \omega^4 = 0.$$

we get

$$\bar{a}^i_{jk} = -\bar{c}^i_{jk}.$$

We shall denote this choice the Pfaffian forms ϑ^i_j by φ^i_j. Therefore the fundamental equations (6) can also be put into the form

(16) $$\begin{aligned}(\omega^i)' &= [\varphi^i_j\omega^j] + c^i_{jk}[\omega^j\omega^k],(*)\\ (\omega^{2+i})' &= [\varphi^i_j\omega^{2+j}] - c^i_{jk}[\omega^{2+j}\omega^{2+k}],\end{aligned}$$

with the conditions

(17) $$c^i_{jk} + c^i_{kj} = 0,$$

by which the Pfaffian forms φ^i_j are uniquely determined.

The relations between ϑ^i_j and φ^i_j are easily found. In fact, by (8) and (10), we get

(18) $$\vartheta^i_j = \varphi^i_j - \tfrac{4}{3}c^i_{jk}\omega^k + \tfrac{2}{3}c^i_{jk}\omega^{2+k}.$$

In order to define the affine connection completely and to solve the problem of equivalence it is necessary to express the exterior derivatives of ϑ^i_j in terms of ω^i, ω^{2+i}, ϑ^i_j. As the equations (16) are simpler than (14), it is advisable first to find the expressions for $(\varphi^i_j)'$. *What we proceed to show is the existence of equations of the form*

(19) $$\begin{cases}(\vartheta^i_j)' = [\vartheta^i_k\vartheta^k_j] + f^i_{jkp}[\omega^k\omega^{2+p}] + g^i_{jkp}[\omega^k\omega^p] + h^i_{jkp}[\omega^{2+k}\omega^{2+p}],\\ g^i_{jkp} + g^i_{jpk} = 0, \qquad h^i_{jkp} + h^i_{jpk} = 0.\end{cases}$$

Let us apply the Theorem of Poincaré that the exterior derivatives of $(\omega^i)'$, $(\omega^{2+i})'$ are zero to the equations (16) and make use of these equations themselves. We get equations of the form

(*) Notice that the c^i_{jk} in (16) are not the same equations as the c^i_{jk} in (6).

$$[(\varphi_j^i)'\omega^j] - [\varphi_k^i\varphi_j^k\omega^j) = [P_{jk}^i\omega^j\omega^k]$$

$$[(\varphi_j^i)'\omega^{2+j}] - [\varphi_k^i\varphi_j^k\omega^{2+j}] = [Q_{jk}^i\omega^{2+j}\omega^{2+k}],$$

where P_{jk}^i, Q_{jk}^i are Pfaffian forms. It follows that $(\varphi_j^i)' - [\varphi_k^i\varphi_j^k]$ is of the form f_{jkp}^i $[\omega^k\omega^{2+p}]$. Hence

$$(\varphi_j^i)' = [\varphi_k^i\varphi_j^k] + f_{jkp}^i[\omega^k\omega^{2+p}]. \tag{20}$$

Applying the Theorem of Poincaré again to the first equation of (16), we get

$$dc_{jk}^i \equiv c_{jk}^p\varphi_p^i - c_{pk}^i\varphi_j^p + c_{pj}^i\varphi_k^p (\text{mod. } \omega^i, \omega^{2+i}). \tag{21}$$

By making use of (16), (18), (20), and (21) it is easily verified that $(\vartheta_j^i)' - [\vartheta_k^i\vartheta_j^k]$ is independent of ϑ_j^i. The equation (19) is thus proved.

We write together the equations (14) and (19) as follows:

$$\text{(I)}\quad \begin{cases} (\omega^i)' = [\vartheta_j^i\omega^j] + a_{jk}^i[\omega^j\omega^k] + 2a_{jk}^i[\omega^j\omega^{2+k}], \\ (\omega^{2+i})' = [\vartheta_j^i\omega^{2+j}] - a_{jk}^i[\omega^{2+j}\omega^{2+k}] - 2a_{jk}^i[\omega^j\omega^{2+k}], \\ (\vartheta_j^i)' = [\vartheta_k^i\vartheta_j^k] + f_{jkp}^i[\omega^k\omega^{2+p}] + g_{jkp}^i[\omega^k\omega^p] + h_{jkp}^i[\omega^{2+k}\omega^{2+p}], \\ a_{jk}^i + a_{kj}^i = 0, \quad g_{jkp}^i + g_{jpk}^i = 0, \quad h_{jkp}^i + h_{jpk}^i = 0. \end{cases}$$

Hence we get the

Theorem. *To a three-web of surfaces in R_4 it is possible, by the introduction of four new variables, to associate eight Pfaffian forms ω^i, ω^{2+i}, ϑ_j^i in an intrinsic way such that the fundamental equation* (I) *hold. The complete system of invariants consists of the functions a_{jk}^i, f_{jkp}^i, g_{jkp}^i, h_{jkp}^i and their covariant derivatives with respect to ω^i, ω^{2+i}, ϑ_j^i. These invariants are connected by a system of relations obtained by applying the Theorem of Poincaré to the system* (I).

When all the invariants a_{jk}^i, f_{jkp}^i, g_{jkp}^i, h_{jkp}^i are zero, the equations (I) are the equations of structure of the affine group G_8 which leaves invariant the three systems of parallel planes

$$\begin{cases} x^1 = \text{const.}, & x^2 = \text{const.}, \\ x^3 = \text{const.}, & x^4 = \text{const.}, \\ x^1 + x^3 = \text{const.}, & x^2 + x^4 = \text{const.}, \end{cases} \tag{22}$$

Hence the

Theorem. *Given a three-web of surfaces in R_4, it is possible to define in an intrinsic way in R_4 an affine connection with the fundamental group G_8. The functions a_{jk}^i measure the torsion and the f_{jkp}^i, g_{jkp}^i, h_{jkp}^i the curvature of the connection.*

It may be noted that the first two equations in (I) can be written more explicitly as follows:

$$\begin{cases} (\omega^i)' = [\vartheta_j^i\omega^j] + 2a_{12}^i\{[\omega^1\omega^2] + [\omega^1\omega^4] - [\omega^2\omega^3]\} \\ (\omega^{2+i})' = [\vartheta_j^i\omega^{2+j}] - 2a_{12}^i\{[\omega^3\omega^4] + [\omega^1\omega^4] - [\omega^2\omega^3].\} \end{cases} \tag{23}$$

2. Geometrical Interpretation

The above calculations and, in particular, the deduction of the fundamental equations (I) in the preceding section can also be treated in a geometrical manner. Consider a point P in R_4. Every line through P cuts the hyperplane at infinity R_3 in a point, every plane in a line, and every hyperplane in a plane. Hence the geometry "about" P is equivalent to the projective geometry in the hyperplane at infinity, which is a three-dimensional linear space. As every direction through P can be defined by the ratios $dx^1 : dx^2 : dx^3 : dx^4$, these ratios form a system of homogeneous projective point coordinates in R_3. It follows that $\omega^1, \omega^2\ \omega^3, \omega^4$ may serve as a system of homogeneous projective coordinates in R_3 as well. The three systems of equations

$$(24)\qquad \begin{cases} \omega^1 = 0, & \omega^2 = 0; \\ \omega^3 = 0, & \omega^4 = 0; \\ \omega^1 + \omega^3 = 0, & \omega^2 + \omega^4 = 0 \end{cases}$$

define in R_3 three lines (lines of intersection of the three tangent planes at P with R_3), lying on the quadric

$$(25)\qquad \omega^1\omega^4 - \omega^2\omega^3 = 0.$$

In fact, they are three lines of one family of generators Q_1 (half-quadric, in French *demi-quadrique*) of the quadric and can be used to define the half-quadric and the quadric.

Consider the equations (6). If $\omega^\alpha(d), \omega^\alpha(\delta)$ $(\alpha = 1, 2, 3, 4)$ are the components of two vectors through P, the quantities

$$a^i_{jk}[\omega^j\omega^k] + b^i_{jk}[\omega^j\omega^{2+k}],$$
$$c^i_{jk}[\omega^{2+j}\omega^{2+k}] + d^i_{jk}[\omega^j\omega^{2+k}]$$

define the torsion of the plane formed by the two vectors. By taking instead of the plane its line of intersection with R_3 and putting

$$(26)\qquad \begin{cases} X^1 = a^1_{jk}[\omega^j\omega^k] + b^1_{jk}[\omega^j\omega^{2+k}], \\ X^2 = a^2_{jk}[\omega^j\omega^k] + b^2_{jk}[\omega^j\omega^{2+k}], \\ X^3 = c^1_{jk}[\omega^{2+j}\omega^{2+k}] + d^1_{jk}[\omega^j\omega^{2+k}], \\ X^4 = c^2_{jk}[\omega^{2+j}\omega^{2+k}] + d^2_{jk}[\omega^j\omega^{2+k}], \end{cases}$$

we may interpret (26) as defining a line-point transformation in R_3 in which the point with the coordinates X^1, X^2, X^3, X^4 corresponds to the line joining the points $\omega^\alpha(d)$, $\omega^\alpha(\delta)$, $(\alpha = 1, 2, 3, 4)$.

The correspondence (26) is, however, not uniquely determined. A change of the ϑ^i_j according to (8) certainly changes this correspondence. In order to define the affine connection it is sufficient to impose certain geometrical properties on the correspondence (26) such that the ϑ^i_j will be uniquely determined.

On changing ϑ^i_j to $\bar{\vartheta}^i_j$ according to (8), the correspondence (26) becomes

$$(27)\quad \begin{cases} X^1 = (a^1_{jk} + l^1_{jk})[\omega^j\omega^k] + (b^1_{jk} + m^1_{jk})[\omega^j\omega^{2+k}], \\ X^2 = (a^2_{jk} + l^2_{jk})[\omega^j\omega^k] + (b^2_{jk} + m^2_{jk})[\omega^j\omega^{2+k}], \\ X^3 = (c^1_{jk} + m^1_{jk})[\omega^{2+j}\omega^{2+k}] + (d^1_{jk} - l^1_{kj})[\omega^j\omega^{2+k}], \\ X^4 = (c^2_{jk} + m^2_{jk})[\omega^{2+j}\omega^{2+k}] + (d^2_{jk} - l^2_{kj})[\omega^j\omega^{2+k}]. \end{cases}$$

According to the definition of the exterior multiplication of Grassmann, the brackets $[\omega^j\omega^k]$, $[\omega^j\omega^{2+k}]$, $[\omega^{2+j}\omega^{2+k}]$ are exactly the Plückerian line coordinates of the line joining $\omega^\alpha(d)$ and $\omega^\alpha(\delta)$, $(\alpha = 1, 2, 3, 4)$. With the usual notation we put

$$(28)\quad [\omega^\alpha\omega^\beta] = p^{\alpha\beta}, \qquad \alpha, \beta = 1, 2, 3, 4$$

Then equations (27) can be written

$$(29)\quad \begin{cases} X^1 = (2a^1_{12} + l^1_{12} - l^1_{21})p^{12} + (b^1_{11} + m^1_{11})p^{13} + (b^1_{12} + m^1_{12})p^{14} \\ \qquad + (b^1_{21} + m^1_{21})p^{23} + (b^1_{22} + m^1_{22})p^{24}, \\ X^2 = (2a^2_{12} + l^2_{12} - l^2_{21})p^{12} + (b^2_{11} + m^2_{11})p^{13} + (b^2_{12} + m^2_{12})p^{14} \\ \qquad + (b^2_{21} + m^2_{21})p^{23} + (b^2_{22} + m^2_{22})p^{24}, \\ X^3 = (2c^1_{12} + m^1_{12} - m^1_{21})p^{34} + (d^1_{11} - l^1_{11})p^{13} + (d^1_{12} - l^1_{21})p^{14} \\ \qquad + (d^1_{21} - l^1_{12})p_{23} + (d^1_{22} - l^1_{22})p^{24}, \\ X^4 = (2c^2_{12} + m^2_{12} - m^2_{21})p^{34} + (d^2_{11} - l^2_{11})p^{13} + (d^2_{12} - l^2_{21})p^{14} \\ \qquad + (d^2_{21} - l^2_{12})p_{23} + (d^2_{22} - l^2_{22})p^{24}. \end{cases}$$

To define the affine connection in question is equivalent to impose certain geometrical properties on the correspondence (29) such that the quantities l^i_{jk}, m^i_{jk} will be uniquely determined.

We require first that the points corresponding to the lines of the half-quadric Q_2 complementary to Q_1 be indeterminate. Any line of Q_2 has the equations

$$(30)\quad \omega^1 - \lambda\omega^2 = 0, \qquad \omega^3 - \lambda\omega^4 = 0,$$

λ being a parameter. Its Plückerian line coordinates are

$$p^{12} = 0, \qquad p^{13} = -\lambda^2, \qquad p^{14} = -\lambda, \qquad p^{34} = 0, \qquad p^{42} = 1, \qquad p^{23} = -\lambda.$$

Hence we get the conditions

$$(31)\quad \begin{cases} b^1_{11} + m^1_{11} = 0, & b^1_{12} + m^1_{12} + b^1_{21} + m^1_{21} = 0, & b^1_{22} + m^1_{22} = 0, \\ b^2_{11} + m^2_{11} = 0, & b^2_{12} + m^2_{12} + b^2_{21} + m^2_{21} = 0, & b^2_{22} + m^2_{22} = 0, \\ d^1_{11} - l^1_{11} = 0, & d^1_{12} - l^1_{21} + d^1_{21} - l^1_{12} = 0, & d^1_{22} - l^1_{22} = 0, \\ d^2_{11} - l^2_{11} = 0, & d^2_{12} - l^2_{21} + d^2_{21} - l^2_{12} = 0, & d^2_{22} - l^2_{22} = 0. \end{cases}$$

Since the system of linear equations (31) is compatible in l^i_{jk}, m^i_{jk}, it is possible to choose the ϑ^i_j such that the required property holds. A set of ϑ^i_j for which this property holds is characterized by the conditions

$$\begin{aligned} &b^1_{11} = b^2_{11} = b^1_{22} = b^2_{22} = 0, \quad b^1_{12} + b^1_{21} = 0, \quad b^2_{12} + b^2_{21} = 0, \\ &d^1_{11} = d^2_{11} = d^1_{22} = d^2_{22} = 0, \quad d^1_{12} + d^1_{21} = 0, \quad d^2_{12} + d^2_{21} = 0. \end{aligned} \tag{32}$$

If the given set of ϑ^i_j already possesses this property, the most general transformation on the ϑ^i_j which leaves this property unaltered is of the form

$$\begin{cases} \bar{\vartheta}^1_1 = \vartheta^1_1 + \beta_1\omega^2 + \beta_2\omega^4, \\ \bar{\vartheta}^1_2 = \vartheta^1_2 - \beta_1\omega^1 - \beta_2\omega^3, \\ \bar{\vartheta}^2_1 = \vartheta^2_1 + \beta_3\omega^2 + \beta_4\omega^4, \\ \bar{\vartheta}^2_2 = \vartheta^2_2 - \beta_3\omega^1 - \beta_4\omega^3. \end{cases} \tag{33}$$

If the Pfaffian forms ϑ^i_j are chosen such that the above property holds and if the corresponding transformation on the ϑ^i_j is the transformation (33) so that this property is left unaltered, then the correspondence (29) becomes the simpler form

$$\begin{cases} X^1 = 2(a^1_{12} + \beta_1)p^{12} + (b^1_{12} + \beta_2)(p^{14} - p^{23}), \\ X^2 = 2(a^2_{12} + \beta_3)p^{12} + (b^2_{12} + \beta_4)(p^{14} - p^{23}), \\ X^3 = 2(c^1_{12} + \beta_2)p^{34} + (d^1_{12} + \beta_1)(p^{14} - p^{23}), \\ X^4 = 2(c^2_{12} + \beta_4)p^{34} + (d^2_{12} + \beta_3)(p^{14} - p^{23}). \end{cases} \tag{34}$$

In order to determine the β_α and hence the affine connection, consider the half-quadric of lines Q_1:

$$\mu\omega^1 - \lambda\omega^3 = 0, \qquad \mu\omega^2 - \lambda\omega^4 = 0, \tag{35}$$

with the homogeneous parameters $\lambda : \mu$. The three lines in which the three tangent planes cut R_3 are given by the cubic equation

$$\lambda\mu(\lambda + \mu) = 0.$$

The Hessian of this cubic equation is

$$\lambda^2 + \lambda\mu + \mu^2 = 0.$$

which defines two lines on Q_1, invariantly connected with the above three lines. These two lines are imaginary. But they determine a linear congruence of real lines of which they are the directrices. The equations of this linear congruence are easily found to be

$$p^{12} = p^{34} = p^{23} - p^{14}. \tag{36}$$

By supposing that the points corresponding to the lines of the congruence (36) are indeterminate, we get the relations

$$\begin{aligned} &2(a^1_{12} + \beta_1) - (b^1_{12} + \beta_2) = 0, \\ &2(a^2_{12} + \beta_3) - (b^2_{12} + \beta_4) = 0, \\ &2(c^1_{12} + \beta_2) - (d^1_{12} + \beta_1) = 0, \\ &2(c^2_{12} + \beta_4) - (d^2_{12} + \beta_3) = 0, \end{aligned} \tag{37}$$

from which the β_α are uniquely determined. Hence there exists a set of Pfaffian forms ϑ^i_j possessing the property that the points corresponding to the lines of the half-quadric

(30) and the lines of the congruence (36) in the correspondence (26) are indeterminate. We see that these properties completely characterize the Pfaffian forms ϑ_j^i. Analytically this set of forms ϑ_j^i is characterized by the conditions (32) and the conditions

$$\begin{cases} 2a_{12}^1 - b_{12}^1 = 0, & 2a_{12}^2 - b_{12}^2 = 0, \\ 2c_{12}^1 - d_{12}^1 = 0, & 2c_{12}^2 - d_{12}^2 = 0. \end{cases} \tag{38}$$

With these conditions the correspondence (26) becomes

$$\begin{cases} X^1 = 2a_{12}^1\{[\omega^1\omega^2] + [\omega^1\omega^4] - [\omega^2\omega^3]\}, \\ X^2 = 2a_{12}^2\{[\omega^1\omega^2] + [\omega^1\omega^4] - [\omega^2\omega^3]\}, \\ X^3 = 2c_{12}^1\{[\omega^3\omega^4] + [\omega^1\omega^4] - [\omega^2\omega^3]\}, \\ X^4 = 2c_{12}^2\{[\omega^3\omega^4] + [\omega^1\omega^4] - [\omega^2\omega^3]\}, \end{cases} \tag{39}$$

and we get, for this set of ϑ_j^i,

$$(\omega^i)' = [\vartheta_j^i\omega^j] + 2a_{12}^i\{[\omega^1\omega^2] + [\omega^1\omega^4] - [\omega^2\omega^3]\},$$
$$(\omega^{2+i})' = [\vartheta_j^i\omega^{2+j}] + 2c_{12}^i\{[\omega^3\omega^4] + [\omega^1\omega^4] - [\omega^2\omega^3]\}.$$

We see that these equations are exactly the equations (23). That $c_{12}^i = -a_{12}^i$ follows from the fact that the system $\omega^1 + \omega^3 = 0$, $\omega^2 + \omega^4 = 0$ is completely integrable. Summing up, we get the

Theorem. *By the choice of a set of Pfaffian forms ϑ_j^i in* (6) *we can establish a line-point correspondence in the hyperplane at infinity. The ϑ_j^i in the fundamental equations* (I) *are completely characterized by the fact that the correspondence be indeterminate for all the lines of the half-quadric Q_2 and of the congruence* (36).

It is easily verified that the choice of φ_j^i as defined by the equations (16) corresponds to the fact that the line-point correspondence be indeterminate for all the lines of the half-quadric Q_2 and of the linear congruence with $\omega^1 = 0, \omega^2 = 0$ and $\omega^3 = 0, \omega^4 = 0$ as directrices.

From the geometrical meaning of the choice of the forms ϑ_j^i it is evident that the theory of invariants thus developed is symmetrical with respect to the three families of surfaces of the web.

It may also be remarked that the determination of the affine connection is in this case equivalent to a problem of classical projective geometry.

3. The Half-quadric in the Hyperplane at Infinity and its Intrinsic Affine Connection

From the preceding sections we see that given a three-web of surfaces in R_4 it is possible to define in an intrinsic way an affine connection in R_4. In the discussion of §2 use is only made of the three lines in the hyperplane at infinity, lines in which the three tangent planes through a point in space meet the hyperplane at infinity. But these three lines are not general ones. In fact, the three corresponding Pfaffian systems are in this case completely integrable. It is not difficult to see that even if the three lines in the hyperplane at infinity are general lines the definition of an intrinsic affine connection is still possible.

A more difficult problem is raised to see whether it is possible to define an intrinsic affine connection, giving in the hyperplane at infinity a half-quadric without particular reference to any of its generators. In this section we want to show that even in this case the definition of an intrinsic affine connection is possible. As we shall see, the problem may be reduced to a problem of classical projective geometry.

Suppose the lines of the given half-quadric be defined by the equations

$$\theta^1 - \lambda\theta^3 = 0, \qquad \theta^2 - \lambda\theta^4 = 0, \tag{40}$$

where $\theta^1, \theta^2, \theta^3, \theta^4$ are linearly independent Pfaffian forms in x^1, x^2, x^3, x^4 and λ is a parameter. The lines of the complementary half-quadric are then given by the equations

$$\theta^1 - \mu\theta^2 = 0, \qquad \theta^3 - \mu\theta^4 = 0, \tag{41}$$

μ being a parameter. The most general transformation which can be applied to the θ^α without changing the equations (40) is the transformation (4) of §1.

Regarding the u_j^i as auxiliary variables and forming the exterior derivatives of ω^α, we get equations of the form

$$\begin{cases} (\omega^i)' = [\vartheta_j^i\omega^j] + \tfrac{1}{2}a_{\alpha\beta}^i[\omega^\alpha\omega^\beta]^* \\ (\omega^{2+i})' = [\vartheta_j^i\omega^{2+j}] + \tfrac{1}{2}a_{\alpha\beta}^{2+i}[\omega^\alpha\omega^\beta] \end{cases} \tag{42}$$

where

$$a_{\alpha\beta}^\gamma + a_{\beta\alpha}^\gamma = 0.$$

The Pfaffian forms ϑ_j^i are Pfaffian forms in u_j^i and x^α and are not uniquely determined. For a given set of ϑ_j^i the quantities

$$X^\alpha = \tfrac{1}{2}a_{\beta\gamma}^\alpha[\omega^\beta\omega^\gamma] \tag{43}$$

define the torsion of the plane element containing the vectors $\omega^\alpha(d)$, $\omega^\alpha(\delta)$ through the point P, or, in terms of the hyperplane at infinity R_3 of R_4, the torsion of the line joining the points $\omega^\alpha(d)$ and $\omega^\alpha(\delta)$. If X^α are interpreted as the homogeneous point coordinates in R_3, equations (43) define a line-point correspondence in R_3. In order to define the intrinsic affine connection in question it is sufficient to impose a set of projective properties on the correspondence (43) so that the ϑ_j^i will then be uniquely determined.

It is possible to choose the ϑ_j^i such that the points corresponding to the lines of the half-quadric (41) are indeterminate. If ϑ_j^i has this property, the most general set of ϑ_j^i having this property is given by (33). Hence the most general correspondence (43) with this property is

$$\begin{cases} X^1 = (a_{12}^1 + 2\beta_1)[\omega^1\omega^2] + (a_{14}^1 + \beta_2)\{[\omega^1\omega^4] - [\omega^2\omega^3]\} + a_{34}^1[\omega^3\omega^4], \\ X^2 = (a_{12}^2 + 2\beta_3)[\omega^1\omega^2] + (a_{14}^2 + \beta_4)\{[\omega^1\omega^4] - [\omega^2\omega^3]\} + a_{34}^2[\omega^3\omega^4], \\ X^3 = a_{12}^3[\omega^1\omega^2] + (a_{14}^3 + \beta_1)\{[\omega^1\omega^4] - [\omega^2\omega^3]\} + (a_{34}^3 + 2\beta_2)[\omega^3\omega^4], \\ X^4 = a_{12}^4[\omega^1\omega^2] + (a_{14}^4 + \beta_3)\{[\omega^1\omega^4] - [\omega^2\omega^3]\} + (a_{34}^4 + 2\beta_4)[\omega^3\omega^4]. \end{cases} \tag{44}$$

(*) We make the convention here that every Latin index takes the values 1, 2 and every Greek index the values 1, 2, 3, 4.

In this correspondence the point corresponding to the line (40) has the coordinates

$$\text{(45)}\qquad \begin{cases} X^1 = (a^1_{12} + 2\beta_1)\lambda^2 + 2(a^1_{14} + \beta_2)\lambda + a^1_{34}, \\ X^2 = (a^2_{12} + 2\beta_3)\lambda^2 + 2(a^2_{14} + \beta_4)\lambda + a^2_{34}, \\ X^3 = a^3_{12}\lambda^2 + 2(a^3_{14} + \beta_1)\lambda + (a^3_{34} + 2\beta_2), \\ X^4 = a^4_{12}\lambda^2 + 2(a^4_{14} + \beta_3)\lambda + (a^4_{34} + 2\beta_4). \end{cases}$$

or, in writing a_α, b_α, c_α for a^α_{12}, a^α_{14}, a^α_{34} respectively,

$$\text{(46)}\qquad \begin{cases} X^1 = (a_1 + 2\beta_1)\lambda^2 + 2(b_1 + \beta_2)\lambda + c_1, \\ X^2 = (a_2 + 2\beta_3)\lambda^2 + 2(b_2 + \beta_4)\lambda + c_2, \\ X^3 = a_3\lambda^2 + 2(b_3 + \beta_1)\lambda + (c_3 + 2\beta_2), \\ X^4 = a_4\lambda^2 + 2(b_4 + \beta_3)\lambda + (c_4 + 2\beta_4). \end{cases}$$

Evidently, the conditions

$$\text{(47)}\qquad \begin{cases} (a_1 + 2\beta_1) + (b_3 + \beta_1) = 0, \\ (b_1 + \beta_2) + (c_3 + 2\beta_2) = 0, \\ (a_2 + 2\beta_3) + (b_4 + \beta_3) = 0, \\ (b_2 + \beta_4) + (c_4 + 2\beta_4) = 0, \end{cases}$$

of the "correspondence" are invariant under projective transformations.

The following geometrical interpretation of the conditions (47) may be stated: Construct the plane π_λ containing the line (40) and its corresponding point (46). As λ varies, the planes π_λ envelop a developable surface of class four. The lines (40), (41), and the characteristic of the developable surface on π_λ belong to the same pencil. In the general case, the conditions (47) are the conditions that the line joining the vertex of the pencil with the point (46) forms a cross ratio equal to 1/3 with these three lines.

From the conditions (47) the $\beta\alpha$ and the ϑ^i_j are uniquely determined. The determination of an intrinsic affine connection is thus possible.

With this choice of ϑ^i_j equations (42) take the simple form

$$\text{(48)}\qquad \begin{cases} (\omega^1)' = [\vartheta^1_1\omega^1] + [\vartheta^1_2\omega^2] + a_1[\omega^1\omega^2] - b_3\{[\omega^1\omega^4] - [\omega^2\omega^3]\} \\ \qquad + b_1[\omega^3\omega^4], \\ (\omega^2)' = [\vartheta^2_1\omega^1] + [\vartheta^2_2\omega^2] + a_2[\omega^1\omega^2] - b_4\{[\omega^1\omega^4] - [\omega^2\omega^3]\} \\ \qquad + b_2[\omega^3\omega^4], \\ (\omega^3)' = [\vartheta^1_1\omega^3] + [\vartheta^1_2\omega^4] + a_3[\omega^1\omega^2] - a_1\{[\omega^1\omega^4] - [\omega^2\omega^3]\} \\ \qquad + b_3[\omega^3\omega^4], \\ (\omega^4)' = [\vartheta^2_1\omega^3] + [\vartheta^2_2\omega^4] + a_4[\omega^1\omega^2] - a_2\{[\omega^1\omega^4] - [\omega^2\omega^3]\} \\ \qquad + b_4[\omega^3\omega^4]. \end{cases}$$

The form of the equations (48) determines uniquely the set of Pfaffian forms ϑ^i_j and hence the affine connection in question.

With the summation notation equations (48) may be written in the condensed form

$$(49)\qquad \begin{cases} (\omega^i)' = [\vartheta^i_j\omega^j] + \tfrac{1}{2}a^i_{jk}[\omega^j\omega^k] - d^i_{jk}[\omega^j\omega^{2+k}] + \tfrac{1}{2}b^i_{jk}[\omega^{2+j}\omega^{2+k}], \\ (\omega^{2+i})' = [\vartheta^i_j\omega^{2+j}] + \tfrac{1}{2}c^i_{jk}[\omega^j\omega^k] - a^i_{jk}[\omega^j\omega^{2+k}] + \tfrac{1}{2}d^i_{jk}[\omega^{2+j}\omega^{2+k}], \end{cases}$$

with the conditions

$$a^i_{jk} + a^i_{kj} = b^i_{jk} + b^i_{kj} = c^i_{jk} + c^i_{kj} = d^i_{jk} + d^i_{kj} = 0.$$

By applying the Theorem of Poincaré to (49) it may be verified that $(\vartheta^i_j)' - [\vartheta^i_k\vartheta^k_j]$ depends on ω^α only. Hence the affine connection so defined has the points of the space as its generating elements. G_8 is its fundamental group.

It may be remarked that the affine connection defined in §§1.2 is a particular case of (48). It corresponds to the case that the following relations between the components of torsion hold:

$$(50)\qquad \begin{cases} b_1 = b_2 = a_3 = a_4 = 0, \\ a_1 + b_3 = 0, \qquad a_2 + b_4 = 0. \end{cases}$$

Tsing Hua University, Kunming.
May 12, 1938.

Reset from
J. Univ. Yunnan 1 (1938) 1–18

Sur la Géométrie d'un Système d'Équations Différentielles du Second Ordre

PAR M. SHIING-SHEN CHERN,
à Yunnanfou (Chine).

1. Introduction

Dans une Note de la *Mathematische Zeitschrift* (1) M. D. D. Kosambi a étudié la géométrie d'un espace $(x^1, \dots, x^n)$ dans lequel est donné un système d'équations différentielles du second ordre

$$\frac{d^2x^i}{dt^2} + F^i\left(\frac{dx^1}{dt}, \dots, \frac{dx^n}{dt}; x^1, \dots, x^n; t\right) = 0, \qquad (i = 1, \dots, n), \tag{1}$$

où les F^i sont des fonctions arbitraires des arguments, qui sont analytiques dans un certain voisinage des valeurs initiales $\left(\frac{dx^i}{dt}\right)_0$, $(x^i)_0$, t_0, M. É. Cartan (2) a fait des observations très importantes sur ce mémoire dans une Note publiée dans le même journal. L'objet de cette Note est de placer ce problème dans la théorie générale des espaces généralisés de M. Cartan.

2. Les deux Problèmes (A) et (B)

D'après M. Cartan, étant donné le système (1), où t est un paramètre imposé (par exemple le temps), on peut formuler de deux manières différentes le problème à résoudre:

(A). Trouver les propriétés géométriques qu'on peut attacher au système (1) et qui ont un caractère intrinsèque par rapport au groupe infini des transformations

$$\begin{aligned} (x^i)' &= f^i(x^1, \dots, x^n), \qquad (i = 1, \dots, n) \\ t' &= t. \end{aligned} \tag{2}$$

(B). Trouver les propriétés géométriques qu'on peut attacher au système (1) et qui ont un caractère intrinsèque par rapport au groupe des transformations

$$\begin{aligned} (x^i)' &= f^i(x^1, \dots, x^n, t) \qquad (i = 1, \dots, n,) \\ t' &= t. \end{aligned} \tag{3}$$

(1) D. D. KOSAMBI, *Parallelism and path-spaces* (*Math. Zeitschrift*, 37, 1933, p. 608–618).
(2) É. CARTAN, *Observations sur le mémoire précédent* (*Math. Zeitschrift*, 37, 1933, p. 619–622).

On voit facilement que chacun de ces deux problèmes revient à un problème d'équivalence d'expressions de Pfaff. En effet, le système (1) étant équivalent au système de Pfaff

$$dx^i - y^i\,dt = 0, \qquad dy^i + F^i(y, x, t)\,dt = 0, \tag{4}$$

le problème (B) est un problème d'équivalence des expressions de Pfaff suivantes ([1])

$$\omega^i = \alpha^i_k(dx^k - y^k\,dt), \qquad \varpi^i = \gamma^i_k\{dy^k + F^k\,dt + \beta^k_j(dx^j - y^j\,dt)\}, \tag{5}$$

par rapport au groupe des transformations générales portant sur les variables x^i, y^i, t, où les α^i_k, β^i_k, γ^i_k sont des variables auxiliaires. Quant au problème (A), parce que la variable t n'entre pas dans les fonctions f^i des formules (2), les coefficients de dt dans ω^i et ϖ^i sont des invariants. Donc le problème (A) est un problème d'équivalence des expressions de Pfaff (5) et des invariants

$$\alpha^i_k y^k, \qquad \gamma^i_k(F^k - \beta^k_j y^j) \tag{6}$$

par rapport au groupe des transformations générales des variables x^i, y^i, t.

Dans les pages suivantes nous allons résoudre ces deux problèmes d'équivalence. On verra que dans chaque cas il est possible de définir une connexion affine dans l'espace des variables x^i, y^i, t.

3. Le Problème (B)

Dérivons les expressions de Pfaff ω^i et définissons les quantités A^i_k, B^i_k, C^i_k par les relations

$$\alpha^i_k A^k_j = \delta^i_j, \qquad \beta^i_k B^k_j = \delta^i_j, \qquad \gamma^i_k C^k_j = \delta^i_j; \tag{7}$$

nous obtenons

$$(\omega^i)' \equiv -\alpha^i_k C^k_h[\varpi^h\,dt] \qquad (\text{mod. } \omega).$$

Donc on peut choisir d'une manière intrinsèque les γ^i_k de sorte que

$$\gamma^i_k = \alpha^i_k. \tag{8}$$

La dérivée extérieure de ω^i prendra alors la forme

$$(\omega^i)' = [\theta^i_j\omega^j] + [dt\,\varpi^i], \tag{9}$$

où les θ^i_j sont des expressions de Pfaff définies à des combinaisons linéaires près des ω:

$$\theta^i_j \equiv A^h_j\,d\alpha^i_h - \alpha^i_l\beta^l_h A^h_j\,dt \qquad (\text{mod. } \omega). \tag{10}$$

En utilisant (8) et (10), on obtient la dérivée extérieure de ϖ^i

$$(\varpi^i)' \equiv [\theta^i_j\varpi^j] + \alpha^i_k\left(-2\beta^k_j + \frac{\partial F^k}{\partial y^j}\right)A^j_h[\varpi^h\,dt] \qquad (\text{mod. } \omega). \tag{11}$$

([1]) Avec la notation tensorielle toute paire d'indices répétés désigne une somme des termes où les indices prennent les valeurs de 1 jusqu'à n.

Il en résulte que le choix

$$\beta_k^i = \frac{1}{2}\frac{\partial F^i}{\partial y^k} \tag{12}$$

a un caractère intrinsèque. De plus, le coefficient de $\varpi^j \omega^k$ dans l'expression de $(\varpi^i)'$ est

$$\frac{1}{2}\alpha_l^i \frac{\partial^2 F^l}{\partial y^p \partial y^q} A_j^p A_k^q,$$

qui est symétrique en j et k. Il est donc possible, et d'une manière unique, de définir les θ_j^i de manière à satisfaire aux équations (9) et aux équations

$$(\varpi^i)' = [\theta_j^i \varpi^j] + P_j^i[dt\omega^j] + T_{jk}^i[\omega^j \omega^k], \qquad T_{jk}^i + T_{kj}^i = 0, \tag{13}$$

où les P_j^i, T_{jk}^i sont des fonctions des quantités α_k^i, x^i, y^i, t.

Pour définir complètement la connexion affine, il faut trouver les équations pour les dérivées extérieures des θ_j^i. Ces équations s'obtiennent facilement par l'application du théorème de Poincaré aux équations (9) et (13). En posant

$$\begin{cases} dP_j^i \equiv P_{j,k}^i \omega^k + P_{j|k}^i \varpi^k + P_{j|0}^i\, dt & (\text{mod. } \theta), \\ dT_{jk}^i \equiv T_{jk,l}^i \omega^l + T_{jk|l}^i \varpi^l + T_{jk|0}^i\, dt & (\text{mod. } \theta), \end{cases} \tag{14}$$

on a

$$(\theta_j^i)' = [\theta_k^i \theta_j^k] + (2T_{jk}^i - P_{k|j}^i)[dt\omega^k) - T_{kl|j}^i[\omega^k \omega^l] + R_{jkl}^i[\omega^k \varpi^l]. \tag{15}$$

Des considérations précédentes on conclut qu'en partant du système (1) il est possible de définir dans l'espace des variables x^i, y^i, t une connexion affine. Les équations (9), (13) et (15) constituent les équations de structure de cette connexion. Les quantités P_j^i, T_{jk}^i forment le tenseur de torsion et les coefficients dans (15) constituent le tenseur de courbure.

En utilisant en chaque point de l'espace un «repère naturel» de sorte qu'on ait

$$\begin{cases} \omega^i = dx^i - y^i\, dt, \\ \varpi^i = dy^i| + F^i\, dt + \dfrac{1}{2}\dfrac{\partial F^i}{\partial y^j}(dx^j - y^j\, dt), \end{cases} \tag{16}$$

il est facile de calculer les composantes de la connexion et les tenseurs de torsion et de courbure. On obtient les formules suivantes

$$\begin{cases} \theta_j^i = -\dfrac{1}{2}\dfrac{\partial^2 F^i}{\partial y^j \partial y^k}\omega^k - \dfrac{1}{2}\dfrac{\partial F^i}{\partial y^j} dt, \\ P_j^i = -\dfrac{\partial F^i}{\partial x^j} + \dfrac{1}{4}\dfrac{\partial F^i}{\partial y^k}\dfrac{\partial F^k}{\partial y^j} + \dfrac{1}{2}\dfrac{d}{dt}\dfrac{\partial F^i}{\partial y^j}, \\ T_{jk}^i = \dfrac{1}{4}\left(-\dfrac{\partial^2 F^i}{\partial y^j \partial x^k} + \dfrac{\partial 2F^i}{\partial y^k \partial x^j} + \dfrac{1}{2}\dfrac{\partial^2 F^i}{\partial y^j \partial y^l}\dfrac{\partial F^l}{\partial y^k} - \dfrac{1}{2}\dfrac{\partial^2 F^i}{\partial y^k \partial y^l}\dfrac{\partial F^l}{\partial y^j}\right), \\ R_{jkl}^i = \dfrac{1}{2}\dfrac{\partial^3 F^i}{\partial y^j \partial y^k \partial y^l}. \end{cases} \tag{17}$$

Si le système donné est le système simple

$$\frac{d^2x^i}{dt^2} = 0, \tag{18}$$

tous les invariants P^i_j, T^i_{jk}, R^i_{jkl} sont nuls. Les équations (9), (13), (15), ajoutées à l'équation

$$(dt)' = 0,$$

sont les équations de structure du groupe fini et continu suivant

$$\begin{cases} (x^{i'}) = a^i_k x^k + b^i t + c^i, \\ (y^{i'}) = a^i_k y^k + b^i, \\ t' = t. \end{cases} \tag{19}$$

Le groupe (19) est le groupe foundamental de l'espace.

4. Le Problème (A)

Comme l'on a déjà remarqué au paragraphe 2, le problème (A) est un problème d'équivalence des expressions de Pfaff (5) et des invariants (6). Une grande partie des considérations due paragraphe 3 est encore valable et l'on voit facilement que les choix (8) et (12) pour les variables auxiliaires γ^i_k, β^i_k ont un caractère intrinsèque. Le deuxième invariant dans (6) devient

$$\alpha^i_j\left(\frac{1}{2}\frac{\partial F^j}{\partial y^k}y^k - F^j\right).$$

Géométriquement l'invariance des expressions dans (6) signifie que les quantités y^i et les quantités

$$v^i = \frac{1}{2}\frac{\partial F^i}{\partial y^k}y^k - F^i \tag{20}$$

sont les composantes de deux vecteurs. D'après le théorème d'Euler, pour que le vecteur v^i soit nul, il faut et il suffit que les fonctions F^i soient nulles ou soient homogènes du second degré par rapport aux variables y^i.

Bornons-nous d'abord à ce cas spécial, mais important, où le vecteur v^i est nul. On peut poser

$$\alpha^i_k y^k = \varepsilon^i, \tag{21}$$

où les ε^i sont des valeurs numériques quelconques. Cela étant, on peut regarder les y^i comme des fonctions des α^i_k, à savoir

$$y^i = A^i_k \varepsilon^k. \tag{22}$$

Donc les formes de Pfall ϖ^i sont des combinaisons linéaires de ω^i, θ^i_j et de dt. En effet, un calcul facile montre qu'on a

$$\theta_j^i = d\alpha_p^i \, A_j^h - \frac{1}{2}\alpha_k^i \frac{\partial F^k}{\partial y^h} A_j^h \, dt - \frac{1}{2}\alpha_p^i \frac{\partial^2 F^p}{\partial y^q \, \partial y^r} A_j^q A_k^r \omega^k, \tag{23}$$

d'où

$$\varepsilon^j \theta_j^i + \varpi^i = 0. \tag{24}$$

En substituant les expressions pour ϖ^i dans (9), on obtient

$$(\omega^i)' = [\theta_j^i \omega^j] + \varepsilon^j [\theta_j^i \, dt] = [\theta_j^i(\omega^j + \varepsilon^j \, dt)].$$

En remplaçant les ω^i par $\bar{\omega}^i = \omega^i + \varepsilon^i \, dt$, on a

$$(\bar{\omega}^i)' = [\theta_j^i \bar{\omega}^j]. \tag{25}$$

Avec les formes de Pfaff nouvelles $\bar{\omega}^i$ les θ_j^i s'écrivent

$$\theta_j^i = d\alpha_h^i A_j^h - \frac{1}{2}\alpha_p^i \frac{\partial^2 F^p}{\partial y^q \, \partial y^r} A_j^q A_k^r \bar{\omega}^k. \tag{26}$$

Pour définir complètement la connexion affine il faut trouver les expressions pour les dérivées extérieures de θ_j^i. En appliquant le théorème de Poincaré au système (25) on trouve

$$(\theta_j^i)' \equiv [\theta_h^i \theta_j^h] \qquad (\text{mod. } \bar{\omega}).$$

D'autre part on vérifie que la différence $(\theta_j^i)'$—$[\theta_k^i \theta_j^k]$ ne peut contenir ni les θ ni les ϖ. Donc la dérivée extérieure de θ_j^i est de la forme

$$\begin{cases} (\theta_j^i)' = [\theta_k^i \theta_j^k] + P_{jk}^i [dt \bar{\omega}^k] + Q_{jkl}^i [\bar{\omega}^k \bar{\omega}^l] \\ \qquad (Q_{jkl}^i + Q_{jk}^i = 0). \end{cases} \tag{27}$$

Les dérivées extérieures de $\varpi^i = -\varepsilon^j \theta_j^i$ s'obtiennent des relations (27).

Il est donc possible de définir, dans ce cas particulier ($v^i = 0$) du problème (A), une connexion affine dans l'espace des variables x^i, y^i, t, dont les équations de structure sont les équations (25) et (27). Si tous les invariants P_{jk}^i, Q_{jkl}^i sont nuls, comme cela correspond au cas du système

$$\frac{d^2 x^i}{dt^2} = 0,$$

les équations (25), (27) sont les équations de structure du groupe fini et continu

$$\begin{cases} (x^i)' = a_k^i x^k + b^i, \\ (y^i)' = a_k^i y^k, \qquad t' = t \end{cases} \tag{28}$$

Le groupe (28) est donc le groupe fondamental de l'espace généralisé ainsi défini.

Il ne serait pas inutile de faire des remarques pour le cas général où le vecteur v^i n'est pas nul. Dans ce cas, on peut poser

$$\begin{cases} \alpha_k^i y^k = \varepsilon^i, \\ \alpha_k^i v^k = \eta^i, \end{cases} \tag{29}$$

où les ε^i, η^i désignent des valeurs numériques. En dérivant les relations (29) et utilisant

(23), on trouve les invariants nouveaux suivants

$$
(30)\qquad
\begin{cases}
\alpha_p^i \dfrac{\partial v^p}{\partial y^q} A_k^q, \\[2ex]
\alpha_p^i \left(\dfrac{1}{2} \dfrac{\partial^2 F^p}{\partial y^l \, \partial y^m} v^l + \dfrac{\partial v^p}{\partial x^m} - \dfrac{1}{2} \dfrac{\partial v^p}{\partial y^l} \dfrac{\partial F^l}{\partial y^m} \right) A_k^m, \\[2ex]
\alpha_k^i \left(\dfrac{1}{2} \dfrac{\partial F^k}{\partial y^j} v^j + \dfrac{\partial v^k}{\partial x^j} y^j - \dfrac{\partial v^k}{\partial y^j} F^j + \dfrac{\partial v^k}{\partial t} \right).
\end{cases}
$$

On remarque que ces expressions sont essentiellement les dérivées covariantes du vecteur v^i par rapport aux ω^i, ϖ^i, dt. En posant quelques-uns de ces invariants égaux à des valeurs numériques il est en général possible de déterminer complètement les α_k^i comme des fonctions de x^i, y^i, t. Le repère attaché à chaque point de l'espace est donc bien défini et l'on obtient une connexion affine dont le groupe fondamental est le groupe des translations

$$
(31)\qquad x^{i'} = x^i + a^i.
$$

Reset from
Bull. Sci. Math. **63** (1939) 206–212

SUR LES INVARIANTS INTÉGRAUX EN GÉOMÉTRIE

Par M. Shiing-shen Chern (陳省身)

Department of Mathematics

(Received November 25, 1939)

RÉSUME

En utilisant les composantes relatives d'un groupe de Lie on donne un critère simple pour l'existence d'une mesure invariante d'un ensemble d'éléments géometriques vis-à-vis un groupe de transformations. Ce théorème général est appliqué à deux groupes particuliers: le groupe affine unimodulaire et le group de Moebius dans le plan. On obtient ainsi des formules analogues à la formule de Crofton pour le "nombre" de droites rencontrant une courbe.

Le but de cette Note est d'exposer quelques relations entre la théorie du repère mobile de M. Elie Cartan[1] et les problèmes sur les probabilités géométriques étudiés par Crofton et Poincaré et dévelipés récemment, sous le nom de "géométrie intégrale", par MM. Blaschke et Santalo[2] avec les resultats inattendus. Ces relations étant éclaircies, beaucoup de problèmes classiques sur la géométrie intégrale peut être étendus aux espaces de Klein dont le groupe fondamental est un groupe quelconque de Lie. Comme on verra, les composantes relatives d'un groupe de Lie jouent aussi un rôle important dans la géométrie intégrale comme dans la géométrie différentielle.

§1. La mesure des éléments géométriques.

Soit donné un espace S de Klein dont le groupe fondamental est un groupe quelconque de Lie qu'on appelle G. Un ensemble d'éléments géométriques d'un corps K étant donné, le problème

1) On trouvera un exposé élémentaire de cette théorie dans "E. Cartan, La théorie des groupes finis et continus et la géométrie différentielle traitées par la méthode du repère mobile, Paris 1937". Ce livre sera cité 'comme "Cartan, groupes continus".

2) On pourra consulter "Blaschke, Vorlesungen über Integralgeometrie, Hefte I, II, Berlin 1936, 1937".

principal de la géométrie intégrale est de définir une mesure de cet ensemble dans S. Cette mesure sera donnée par une intégrale multiple étendue à l'ensemble d'éléments donné. Pour que la dé-
par rapport à G, c'est-à-dire que cet ensemble et l'ensemble obtenu
finition ait un sens, il faut et il suffit que la mesure soit invariante par rapport à G, c'est-à-dire que cet ensemble et l'ensemble obtenu après une transformation quelconque de G aient la même mesure. Supposons pour simplifier le problème que le groupe G transforme transitivement les éléments donnés. Chaque élément peut être repéré par une famille de repères de G.[3] Soit G de r paramètres et soient $\omega_1, \cdots\cdots, \omega_r$ les composantes relatives de G. Si un élément arbitraire du corps considéré dépend de m paramètres, la famille de repères attachée à lui dépend de r-m paramètres. Donc un déplacement infinitésimal dans la famille annule m combinaisons linéaires (à coefficients constants) des composantes relatives de G, qu'on peut supposer d'être $\omega_1, \cdots\cdots, \omega_m$. On dit souvent que les équations

$$\omega_1 = \omega_2 = \cdots\cdots = \omega_m = 0 \tag{1}$$

sont les équations des éléments du corps consideré de l'espace S.

Soient $u_1, \ldots, u_m$ les coordonnées des éléments dans S. On peut choisir comme les paramètres de G les $u_1, \ldots, u_m$ et $r-m$ autres paramètres $v_1, \ldots, v_{r-m}$. La mesure d'un ensemble $\mathcal{E}$ de K est donnée par l'intégrale multiple

$$\int_{\mathcal{E}} f(u_1, \cdots, u_m, v_1, \cdots, v_{r-m}) [\omega_1 \cdots \omega_m], \tag{2}$$

étendue à $\mathcal{E}$. Pour que l'intégrale (2) puisse servir comme la mesure de $\mathcal{E}$, il faut qu'elle soit invariante par rapport à G pour un ensemble $\mathcal{E}$ arbitraire, c'est-à-dire que la forme extérieure suivante

$$f(u_1, \cdots, u_m, v_1, \cdots, v_{r-m}) [\omega_1 \cdots \omega_m \tag{3}$$

soit invariante par rapport à G. Le problème principal de la géométrie intégrale revient donc à la recherche des invariants intégraux de la forme (3). Or chacune des formes de Pfaff $\omega_1, \cdots \cdots, \omega_m$ est invariante par rapport à G et il existe toujours une transformation (et une seule) de G transformant un repère donné correspondant aux paramètres $(u_1^{(0)}, \cdots\cdots, u_m^{(0)}, v_1^{(0)}, \cdots\cdots, v_{r-m}^{(0)})$ à

3) Cartan, groupes continus, Chap. VI.

un repère quelconque de paramètres $(u_1, \ldots, u_m, v_1, \ldots, v_{m-r})$. On conclut donc qu'un invariant intégral de la forme (3) est nécessairement de la forme

(4) $$c\,[\omega_1 \cdots \omega_m]$$

où c' est une constante.

Mais un tel invariant intégral ne donne pas toujours lieu à la mesure d'un ensemble du corps K. Pour qu'il soit ainsi, il faut encors qu'il possède "la propriété de l'invariance du choix (Wahlinvarianz). Cela veut dire, que la forme différentielle extérieure (4) doit être reductible à la forme

(5) $$F(u_1, \cdots, u_m)\,[du_1 \cdots du_m],$$

où les paramètres $v_1, \ldots, v_{r-m}$, qui peuvent intervenir dans F, disparaissent. Cette propriété peut être formulée, d'après M. Blaschke, dans la forme suivante: Choisissons pour chaque élément du corps un repère de la famille qui lui est attachée. La forme (4) devient donc la forme (5). La propriété de l'invariance du choix exprime que cette forme resultée (5) est indépendante du choix des repères qui sont attachés aux éléments du corps.

Nous avons donc démontré le théorème suivant:

La condition nécessaire et suffisants pour que tout ensemble d'éléments géométriques d'un corps K possède une mesure invariante par rapport au groupe G est que la forme (4) ait la propriété de l'invariance du choix. La mesure est ainsi égale à 1 intégrale

(6) $$c \int [\omega_1 \cdots \omega_m]$$

étendue à l'ensemble consideré.

Si le corps K est constitué des repères de G, l'invariance du choix est trivials. Un ensemble de repères de G possède donc toujours une mesure donnée par l'intégrale (6) avec $m=r$. Cette mesure généralise la "mesure cinématique" de Poincaré.

Nous allons appliquer ces remarques générales à des groupes particuliers.

§2. La géométrie intégrale dans le plan affine.

Prenons un plan affine dont le groupe fondamental est le groupe de transformations affines unimodulaires.[4] Ce groupe G_5 est défini par les équations

4) Cartan, groupes continus, p. 156.

$$(7) \qquad \begin{cases} x' = ax + by + c, \\ y' = a'x + b'y + c', \end{cases} \qquad \begin{vmatrix} a & b \\ a' & b' \end{vmatrix} = 1$$

Un repère R de G_5 se composera de deux vecteurs quelconques $\vec{I}_1$ et $\vec{I}_2$ qui ont même origine A et qui vérifient la relation

$$(8) \qquad \vec{I}_1 \Lambda \vec{I}_2 = 1.$$

Les composantes relatives d'un repère A $\vec{I}_1 \vec{I}_2$ sont les cinq formes ω que définissent les formules

$$(9) \qquad \begin{cases} d\vec{A} = \omega_1 \vec{I}_1 + \omega_2 \vec{I}_2, \\ d\vec{I}_1 = \omega_{11}\vec{I}_1 + \omega_{12}\vec{I}_2, \\ d\vec{I}_2 = \omega_{21}\vec{I}_1 - \omega_{11}\vec{I}_2. \end{cases}$$

D'après notre théorème général, l'invariant intégral pour le corps de repères de G_5 est donné par l'expression

$$(10) \qquad c\,[\omega_1\,\omega_2\,\omega_{11}\,\omega_{12}\,\omega_{21}].$$

Il est évident que le corps de points A du plan possède l'invariant intégral

$$(11) \qquad c\,[\omega_1\,\omega_2]$$

Pour chercher s'il existe d'invariant intégral pour les droites du plan, attachons à une droite la famille de repères A $\vec{I}_1 \vec{I}_2$ de sorte que A soit situé sur la droite et que le vecteur $\vec{I}_1$ soit le long de la droite. Les équations des droites du plan sont donc

$$\omega_2 = \omega_{12} = 0.$$

La relation entre deux repères $A\,\vec{I}_1\,\vec{I}_2$ et $B\vec{J}_1\,\vec{J}_2$ attachés à la même droite est donnée par les équations

$$(12) \qquad \begin{cases} A = B + \lambda \vec{J}_1, \\ \vec{I}_1 = \mu \vec{J}_1, \\ \vec{I}_2 = \dfrac{1}{\mu}(\vec{J}_2 + \nu \vec{J}_1). \end{cases}$$

Désignons par $\overline{\omega}_1$, $\overline{\omega}_2$, $\overline{\omega}_{11}$, $\overline{\omega}_{12}$, $\overline{\omega}_{21}$ les composantes relatives de la famille des repères $B\,\vec{J}_1\,\vec{J}_2$; on obtient

$$\omega_2 = \mu(\overline{\omega}_2 + \lambda\,\overline{\omega}_{12}),$$
$$\omega_{12} = \mu^2\overline{\omega}_{12},$$

d'où

$$[\omega_2\,\omega_{12}] = \mu^3\,[\overline{\omega}_2\,\overline{\omega}_{12}].$$

Donc l'invariant intégral (11) n'a pas la propriété de l'invariance du choix. Par conséquent, un ensemble de droites du plan ne possède pas une mesure invariante par rapport à G_5.

Nous allons montrer qu'il existe pour le corps de paraboles un invariant intégral (défini jusqu'à un facteur constant près) par rapport à G_5. Attachons-nous en effet à chaque parabole la famille de repères de sorte que la parabole ait l'équation

$$y = \tfrac{1}{2}x^2. \tag{13}$$

La relation entre deux repères $A\,\vec{I}_1\,\vec{I}_2$ et $B\,\vec{J}_1\,\vec{J}_2$ de la famille est donnée par les équations

$$\left\{\begin{aligned} A &= B + \lambda\vec{J}_1 + \tfrac{1}{2}\lambda^2\vec{J}_2,\\ \vec{I}_1 &= \vec{J}_1 + \lambda\vec{J}_2,\\ \vec{I}_2 &= \vec{J}_2.\end{aligned}\right. \tag{14}$$

Désignons-nous, comme dans le cas précédent, par $\overline{\omega}_1$, $\overline{\omega}_2$, $\overline{\omega}_{11}$, $\overline{\omega}_{12}$, $\overline{\omega}_{21}$ les composantes relatives de la famille des repères $B\,\vec{J}_1\,\vec{J}_2$; on obtient

$$\left\{\begin{aligned} \omega_1 &= d\lambda + \overline{\omega}_1 + \lambda\overline{\omega}_{11} + \tfrac{1}{2}\lambda^2\overline{\omega}_{21},\\ \omega_2 &= \overline{\omega}_2 + \lambda\overline{\omega}_{12} - \lambda\overline{\omega}_1 - \frac{3}{2}\lambda^2\overline{\omega}_{11} - \frac{1}{2}\lambda^3\overline{\omega}_{21},\\ \omega_{11} &= \overline{\omega}_{11} + \lambda\overline{\omega}_{21},\\ \omega_{12} &= d\lambda + \overline{\omega}_{12} - 2\lambda\overline{\omega}_{11} - \lambda^2\overline{\omega}_{21},\\ \omega_{21} &= \overline{\omega}_{21}.\end{aligned}\right. \tag{15}$$

Il en résulte que les équations des paraboles sont

$$\omega_{21} = 0, \quad \omega_{11} = 0, \quad \omega_2 = 0, \quad \omega_{12} = \omega_1, \tag{16}$$

et qu'*il existe in invariant intégral pour les paraboles, qui est donné par l'expression*

$$c\,[\omega_2\, \omega_{11}\, \omega_{21} (\omega_1 - \omega_{12})]. \tag{17}$$

Il sera utile de transformer l'expression (17) en une autre forme. Considérons une courbe du plan et la famille des repères de Frenet $B\, \vec{J}_1\, \vec{J}_2$ de cette courbe. Un déplacement infinitésimal de ces repères est donné par les formules de Frenet:

$$\left\{\begin{aligned} d\vec{B} &= ds\, \vec{J}_1, \\ d\vec{J}_1 &= ds\, \vec{J}_2, \\ d\vec{J}_2 &= k\, ds\, \vec{J}_1, \end{aligned}\right. \tag{18}$$

où s est l'arc affine de la courbe. Pour déterminer une parabole qui rencontre la courbe au point B il suffit de se donner le repère relativement auquel la parabole a l'équation (13). A cause de l'invariance du choix de l'expression (17) on peut prendre le repère A $\vec{I}_1\, \vec{I}_2$ dont l'origine A se confond avec B. Le repère $A\, \vec{I}_1\, \vec{I}_2$ étant un repère d'ordre O de la courbe, on a

$$\left\{\begin{aligned} \vec{I}_1 &= \lambda \vec{J}_1 + \mu\, \vec{J}_2, \\ \vec{I}_2 &= \varrho \vec{J}_1 + \sigma\, \vec{J}_2, \end{aligned}\right. \qquad \lambda\sigma - \mu\varrho = 1. \tag{19}$$

Le déplacement infinitésimal des repères $A\, \vec{I}_1\, \vec{I}_2$ est donné par les équations

$$\left\{\begin{aligned} d\vec{A} &= ds(\sigma\, \vec{I}_1 - \mu\, \vec{I}_2), \\ d\vec{I}_1 &= (\sigma d\lambda - \varrho d\mu + \overline{\sigma\mu k - \varrho\lambda\, ds})\, \vec{I}_1 \\ &\qquad + (-\mu d\lambda + \lambda d\mu + \overline{-\mu^2 k + \lambda^2\, ds})\, \vec{I}_2, \\ d\vec{I}_2 &= (\sigma d\varrho - \varrho d\sigma + \overline{\sigma^2 k - \varrho^2\, ds})\, \vec{I}_1 \\ &\qquad + (-\mu d\varrho + \lambda d\sigma + \overline{-\mu\sigma k + \lambda\varrho\, ds})\, \vec{I}_2. \end{aligned}\right. \tag{20}$$

5) Cartan, groupes continus, p. 160.

Donc l'expression (17) peut être écrite sous la forme

$$-\,c\,[ds\,d\lambda\,d\mu\,d\sigma].$$

Nous supprimerons dans les discussions suivantes la constante c. La mesure d'un ensemble de paraboles étant toujours supposée d'être positive, on peut écrire les formules suivantes pour l'invariant intégral des paraboles:

$$(21)\qquad \dot{P} = |\,[\omega_2\,\omega_{11}\,\omega_{21}(\omega_1 - \omega_{12})]\,| = |\,[ds\,d\lambda\,d\mu\,d\sigma]\,|.$$

On peut établir une correspondance biunivoque entre les paraboles passant par A et les points de l'espace (λ, μ, σ). Choisissons un domaine D du volume V (au sens euclidien) dans l'espace (λ, μ, σ). On déduit de (21) la formule suivante:

$$(22)\qquad \int P = Vs$$

où s est la longeur affine de la courbe et l'intégrale au gauche est étendue à toutes les paraboles qui rencontrent la courbe et dont les images dans l'espace (λ, μ, σ) sont situées dans D. Il est bien-entendu que chaque parabole sera comptée un nombre de fois égal au nombre de ses points d'intersection avec la courbe relativement auxquels son image est située dans D.

Il faut remarquer que, les repères de Frenet $B\,\vec{J}_1\,\vec{J}_2$ étant intrinséquement liés à la courbe, la condition que l'image de la parabole soit située dans D a une signification géométrique invariante.

§3. La géométrie intégrale dans le plan de Moebius.

Appliquons encore la méthode gènerale à la géométrie de Moebius du plan. C'est la géométrie dont le groupe fondamental G est un groupe de transformations ponctuelles conservant les cercles. Un repère R de G se composera de deux cercles orientés orthogonaux entre eux A_1, A_2, et des deux points A_0, A_3 communs à ces cercles dont on imposera la condition que le produit scalaire $A_0\,A_3$ soit égal à 1.[6] On aura donc

$$(23)\qquad A_1^2 = A_2^2 = A_0\,A_3 = 1,$$

6) E. Cartan, Les espaces â connexion conforme, Annales Soc. Pol. de Math. 1923, pp. 171-174.

tous les autres produits scalaires étant nuls. Un déplacement infinitésimal des repères $A_0 A_1 A_2 A_3$ est donné par les equations

$$(24) \quad \begin{cases} dA_0 = \omega_0^0 A_0 + \omega_0^1 A_1 + \omega_0^2 A_2, \\ dA_1 = \omega_1^0 A_0 \qquad\qquad + \omega_1^2 A_2 - \omega_0^1 A_3, \\ dA_2 = \omega_2^0 A_0 - \omega_1^2 A_1 \qquad\qquad - \omega_0^2 A_3, \\ dA_3 = \qquad\quad - \omega_1^0 A_1 - \omega_2^0 A_2 - \omega_0^0 A_3, \end{cases}$$

où les ω_0^0, ω_0^1, ω_0^2, ω_1^0, ω_2^0, ω_1^2 sont les composantes relatives indépendantes du groupe G, qui dépend de six paramètres. L'invariant intégral du corps de repères est

$$(25) \quad c\,[\omega_0^0\, \omega_0^1\, \omega_0^2\, \omega_1^0\, \omega_2^0\, \omega_1^2].$$

Il est facile de voir qu'il n'existe pas d'invariant intégral pour les points. En effet, les équations des points A_0 du plan étant

$$(26) \quad \omega_0^1 = \omega_0^2 = 0,$$

la transformation

$$(27) \quad \overline{A_0} = \lambda A_0, \quad \overline{A_1} = A_1, \quad \overline{A_2} = A_2, \quad \overline{A_3} = \frac{1}{\lambda} A_3,$$

donne

$$(28) \quad \begin{cases} d\overline{A_0} \cdot \overline{A_1} = \lambda\,(dA_0 \cdot A_1) = \lambda\, \omega_0^1, \\ d\overline{A_0} \cdot \overline{A_2} = \lambda\,(dA_0 \cdot A_2) = \lambda\, \omega_0^2, \end{cases}$$

d'où

$$[(d\overline{A_0} \cdot \overline{A_1})(d\overline{A_0} \cdot \overline{A_2})] = \lambda^2\,[\omega_0^1\, \omega_0^2],$$

ce qui montre que la forme $[\,\omega_0^1 \;\; \omega_0^2\,]$ ne possède pas la propriété de l'invariance du choix.

Considérons maintenant les cercles orientés du plan. Les équations des cercles A_1 sont

(29) $$\omega_1^0 = \omega_1^2 = \omega_0^1 = 0.$$

Un calcul facile montre que le repère le plus général $\overline{A}_0\,\overline{A}_1\,\overline{A}_2\,\overline{A}_3$ avec $\overline{A}_1 = A_1$ est donné par les équations

(30) $$\begin{cases} \overline{A}_0 = \varrho\left(\frac{1}{2}A_0 + \lambda A_2 - \lambda^2 A_3\right), \\ \overline{A}_1 = A_1, \\ \overline{A}_2 = \frac{1}{2}\varrho\sigma(\lambda-\mu)\left(A_0 + \overline{\lambda+\mu}\,A_2 - 2\lambda\mu A_3\right), \\ \overline{A}_3 = \sigma\left(\frac{1}{2}A_0 + \mu A_2 - \mu^2 A_3\right), \\ (\lambda-\mu)^2 = -\dfrac{2}{\varrho\sigma}. \end{cases}$$

En désignant par $\overline{\omega}_0^0, \ldots\ldots, \overline{\omega}_1^2$ les composantes relatives des repères $\overline{A}_0\,\overline{A}_1\,\overline{A}_2\,\overline{A}_3$, on obtient

$$\overline{\omega}_0^1 = -\overline{A}_0\,d\overline{A}_1 = -\varrho\left(-\tfrac{1}{2}\omega_0^1 + \lambda\,\omega_1^2 - \lambda^2\omega_1^0\right),$$

$$\overline{\omega}_1^2 = \overline{A}_2\,d\overline{A}_1 = \tfrac{1}{2}\varrho\sigma(\lambda-\mu)\left(-\omega_0^1 + \overline{\lambda+\mu}\,\omega_1^2 - 2\lambda\,\mu\omega_1^0\right),$$

$$\overline{\omega}_1^0 = \overline{A}_3\,d\overline{A}_1 = \sigma\left(-\tfrac{1}{2}\omega_0^1 + \mu\,\omega_1^2 - \mu^2\,\omega_1^0\right),$$

d'où

$$[\overline{\omega}_0^1\,\overline{\omega}_1^2\,\overline{\omega}_1^0] = \frac{1}{4}\varrho^2\sigma^2(\lambda-\mu)^4\,[\omega_0^1\,\omega_1^2\,\omega_1^0] = [\omega_0^1\,\omega_1^2\,\omega_1^0].$$

Cela montre que les cercles du plan ont un invariant intégral par rapport à G, qui est donné par l'expression

(31) $$[\omega_0^1\,\omega_1^2\,\omega\ ].$$

Nous allons écrire l'expression (31) dans une autre forme. Considérons une courbe du plan. Si la courbe n'est pas un cercle, on peut attacher à chaque point de la courbe un repère de Frenet $B_0\,B_1\,B_2\,B_3$, de sorte que le déplacement infinitésimal de ces repères soit donné par les équations

(32) $$\begin{cases} dB_0 = ds\,B_1, \\ dB_1 = k\,ds\,B_0 - ds\,B_3, \\ dB_2 = ds\ B_0, \\ dB_3 = -k\,ds\,B_1 - ds\,B_2. \end{cases}$$

où s est un paramètre intrinsèque (l'arc généralisé) de la courbe. Pour déterminer un cercle A_1 rencontrant la courbe il suffit de se donner le repère $A_0A_1A_2A_3$, A_0 étant le point d'intersection. Le repère $A_0A_1A_2A_3$ étant évidemment le repère le plus général d'ordre 0, la relation entre $A_0A_1A_2A_3$ et $B_0B_1B_2B_3$ est donnée par les équations

$$(33)\quad \begin{cases} A_0 = \lambda B_0, \\ A_1 = \varrho B_0 + \cos\varphi\, B_1 - \sin\varphi\, B_2, \\ A_2 = \sigma B_0 + \sin\varphi\, B_1 + \cos\varphi\, B_2, \\ A_3 = -\dfrac{1}{\lambda}\Big\{\tfrac{1}{2}(\varrho^2+\sigma^2)B_0 + (\varrho\cos\varphi + \sigma\sin\varphi)B_1 + (-\varrho\sin\varphi \\ \qquad + \sigma\cos\varphi)B_2 - B_3\Big\}. \end{cases}$$

On peut prendre comme les coordonnées du cercle A_1 le paramètre intrinsèque s et les deux quantités ϱ, φ. On déduit de (33) et (32) que

$$dA_1 = (d\varrho + \overline{k\cos\varphi - \sin\varphi}\, ds)B_0 + (-\sin\varphi\, d\varphi + \varrho\, ds)B_1 \\ - \cos\varphi\, d\varphi\, B_2 - \cos\varphi\, ds\, B_3,$$

d'où

$$\begin{cases} \omega_0^1 = -A_0\, dA_1 = \lambda\cos\varphi\, ds, \\ \omega_1^2 = A_2\, dA_1 = -d\varphi + {*}ds, \\ \omega_1^0 = A_3\, dA_1 = \dfrac{1}{\lambda} d\varrho + {*}d\varphi + {*}ds. \end{cases}$$

Par conséquent, on a comme l'invariant intégral des cercles:

$$(34)\qquad K = [\omega_0^1\, \omega_1^2\, \omega_1^0] = \cos\varphi\, d\varphi\, ds\, d\varrho.$$

En supposant que la valeur de ϱ soit située entre les limites a et $a+b$, on déduit de (34) la formule suivante:

$$(35)\qquad \int \dot{K} = 4\, bs,$$

où s est le paramètre intrinsèque de la courbe et l'intégrale au gauche est étendue à tous les cercles qui rencontrent la courbe et pour lesquels les valeurs de ϱ sont situées entre a et $a+b$. Il est

bien-entendu que chaque cercle sera compté un nombre de fois égal au nombre de ses points d'intersection avec la courbe relativement auxquels la valeur de ϱ est entre a et $a+b$. Il faut remarquer que cette condition pour ϱ a une signification géométrique invariante.

Reprinted from
Science Reports Nat. Tsing Hua Univ.
4 (1940) 85–95

Académie des Sciences

Séance Du Lundi 3 Juin 1940

Présidence de M. Georges Perrier

Correspondence

THÉORIE DES ENSEMBLES.—*Sur une généralisation d'une formule de Crofton.* Note de **M. Shiing-shen Chern**, présentée par M. Élie Cartan.

Dans cette Note nous allons donner une généralisation de la formule de Crofton ([1])

$$\iint n\, dp\, d\varphi = c\sigma, \tag{1}$$

où le premier membre est la measure de l'ensemble des droites du plan rencontrant une courbe Γ et le second membre est l'arc σ de Γ multiplié par une constante c.

Pour généraliser cette formule, considérons une courbe Γ dans un espace à n dimensions E dont le groupe fondamental G est un groupe quelconque de Lie à r paramètres. Soient $\omega_1, \ldots, \omega_r$ les composantes relatives de G et soient

$$\omega_1 = 0, \qquad \ldots, \qquad \omega_n = 0 \tag{2}$$

les équations des points dans E. Nous nous plaçons dans le cas général, où l'on peut introduire sur Γ un paramètre intrinsèque σ (l'invariant de M. Pick), de manière que

$$\omega_i = c_i\, d\sigma \qquad (i = 1, \ldots, n), \tag{3}$$

où les c_i sont des constantes.

Cela étant, considérons un second corps N d'éléments défini par les équations

$$\varpi_1 = 0, \qquad \ldots, \qquad \varpi_m = 0, \tag{4}$$

où les $\varpi_k (k = 1, \ldots, m)$ sont des combinaisons linéaires à coefficients constants des ω_i et où le système (4) est complètement intégrable. Supposons que le groupe G transforme transitivement les points de E et les éléments du corps N. Par un procédé donné par M. É. Cartan, on peut repérer un point de E ou un élément de N par une famille de repères de G. On dit qu'un point de E et un élément de N sont *incidents* si les deux familles de repères qui leur sont attachées ont un repère en commun, alors elles ont en commun une famille de repères dépendant de $r - s$ paramètres, où s désigne le nombre d'équations linéairement indépendantes dans le système simultané (2), (4). Il est facile

---

([1]) Voir R. Deltheil, *Probabilitiés géométriques*, Paris, 1926, p. 60.

de voir que cette définition d'incidence est d'accord avec les définitions habituelles d'incidence dans la géométrie euclidienne, la géométrie projective etc. On voit aussi facilement que les éléments de N incidents avec un point de E dépendent de $s - n$ paramètres.

Supposons encore que le corps N possède une mesure invariante par rapport à G. Cette mesure sera donnée par l'intégrale multiple

$$\int \varpi_1 \varpi_2 \ldots \varpi_m, \tag{5}$$

étendue à l'ensemble des éléments de N considérés. Si $s - n = m - 1$, les éléments de N incidents aux points de Γ dépendent de m paramètres. Si la mesure des éléments de N autour d'un point est finie, comme dans le cas où le groupe d'isotropie est clos, on a la formule généralisée de Crofton

$$\int \varpi_1 \varpi_2 \ldots \varpi_m = c\sigma \qquad (c = \text{const.}), \tag{6}$$

où σ est l'invariant de M. Pick de Γ et où l'intégrale à gauche est étendue à tous les éléments de N incidents aux points de Γ, chaque élément étant compté un nombre de fois égal à son nombre de points d'incidence avec Γ. Si la mesure des éléments de N autour d'un point n'est pas finie, il faut ajouter une condition auxiliaire pour les éléments de N entrant en l'intégrale (6) pour que la formule (6) soit encore valable.

La démonstration de (6) s'appuie sur une transformation de (5) en une forme où figure $d\sigma$. La même méthode permet d'étendre la formule (6) dans les cas où $s - < m - 1$.

Reset from
C. R. Acad. Sci. Paris **210** (1940) 757–758

Sulla Formula Principale Cinematica Dello Spazio ad *n* dimensioni

Nota di Shiing-shen Chern e Chih-ta Yien (Kunming, China)

Sunto.–*L'Autore dà un'estensione della formula principale cinematica di* Blaschke.

In una Nota recente ([1]) W. Blaschke ha dato la formula seguente che egli ha chiamato «la formula principale cinematica»:

$$\int K(R_0 \cdot R_1)\dot{R}_1 = 8\pi^2\{V_0 K_1 + A_0 M_1 + M_0 A_1 + K_0 V_1\}. \tag{1}$$

In questa formula R_0 indica un campo fisso, R_1 un campo mobile e $K(R_0 \cdot R_1)$ la curvatura totale di Gauss. L'integrale a sinistra è esteso a tutte le posizioni di R_1 nel senso di Poincaré.

A destra V_i, A_i, M_i, K_i $(i = 0, 1)$ indicano rispettivamente il volume, l'area, l'integrale della curvatura media, la curvatura totale di R_i $(i = 0, 1)$. Quasi tutte le formule importanti della geometria integrale sono conseguenze di questa formula generale.

Scopo di questa Nota è di dare la formula corrispondente nello spazio euclideo ad n dimensioni. Siano R_0, R_1 due domini nello spazio, R_0 fisso ed R_1 mobile. Sia $\dot{R}_1$ la «densità cinematica» di Poincaré di R_1 e $K(R_0 \cdot R_1)$ la curvatura totale dell'intersezione di R_0 e R_1. Essendo la frontiera di R_i $(i = 0, 1)$ una ipersuperficie di S_i si hanno, in ogni punto di S_i, $n - 1$ curvature principali $\frac{1}{r_1^{(i)}}, \ldots, \frac{1}{r_{n-1}^{(i)}}$.

Sia

$$M_k^{(i)} = \frac{1}{\binom{n-1}{k}} \int \left(\sum \frac{1}{r_{\nu_1}^{(i)}} \cdots \frac{1}{r_{\nu_k}^{(i)}}\right) \Omega^{(i)}, \qquad \begin{matrix} i = 0, 1 \\ k = 0, \ldots, n-2 \end{matrix} \tag{2}$$

l'integrale della k-ma funzione elementare simmetrica delle curvature principali di S_i, ove Ω_i è l'elemento di superficie di S_i. La formula principale cinematica dello spazio ad n dimensioni è la seguente:

$$\int K(R_0 \cdot R_1)\dot{R}_1 = J_n\left\{K_0 V_1 + K_1 \bar{V}_0 + \frac{1}{n}\sum_{k=0}^{n-2}\binom{n}{k+1} M_k^0 M_{n-2-k}^{(1)}\right\}, \tag{3}$$

([1]) «Rendiconti Acc. dei Lincei», 1936, pp. 546–547. V. anche il volume: W. Blaschke, *Vorlesungen ueber Integralgeometrie*, «Zweites Heft», p. 99.

dove J_n è la misura cinematica attorno a un punto:

$$J_n = O_{n-1} O_{n-2} \dots O_1,$$

O_{p-1} essendo l'area della sfera di raggio 1 nello spazio euclideo a p dimensioni, che è data dalle formule

$$O_{2p-1} = \frac{2\pi^p}{(p-1)!}, \qquad O_{2p} = 2^{p+1} \frac{\pi^p}{1 \cdot 3 \dots (2p-1)}. \tag{4}$$

Noi indicheremo la dimostrazione della formula (3), che si appoggia and un'idea di W. Blaschke ([2]). Sia V_n una varietà a m dimensioni nello spazio euclideo a n dimensioni E_n $(m < n)$.

Consideriamo un punto P su $\overline{V}_m$ e un n-edro rettangolare $P\vec{e}_1 \dots \vec{e}_n$ in modo che i primi m vettori $\vec{e}_1, \dots, \vec{e}_m$ siano i vettori tangenti di V_m in P. Se si prendóno come elementi gli h-edri $P\vec{e}_1 \dots \vec{e}_h$ $(h \leq m)$ su V_m, si definisce la sua densità come la forma differenziale esteriore seguente:

$$\dot{L}_{h,m} = \prod_{i=1,\dots,m} (dP\vec{e}_i) \prod_{\substack{i=1,\dots,h \\ j=i+1,\dots,m}} (d\vec{e}_i \dots \vec{e}_j) \tag{5}$$

che è di grado $\frac{1}{2}(h+1)(2m-h)$.

Consideriamo ora due ipersuperficie S_0, S_1 in E_n. Sia S_0 fissa ed S_1 mobile. Per ogni posizione di S_1 le due ipersuperficie hanno una varietà d'intersezione V_{n-2} ad $n-2$ dimensioni. Prendiamo un $(n-2)$-edro su V_{n-2}. Questo $(n-2)$-edro, considerato come su V_{n-2}, S_0, S_1, ha rispettivamente le densità $\dot{L}_V$, $\dot{L}_0$, $\dot{L}_1$ definite dalla (5). Si verifica che vale la formula seguente

$$\dot{L}_V \dot{S}_1 = \sin^{n-1} \varphi \dot{L}_0 \dot{L}_1 \dot{\varphi}, \tag{6}$$

dove $\dot{S}_1$ è la densità cinematica di S_1 e φ è l'angolo fra le normali di S_0 e S_1.

Per valutare l'integrale nella (3) notiamo che la curvatura totale $K(R_0 \cdot R_1)$ si compone di tre parti: 1) della curvatura totale K_{10} dell'ipersuperficie $S_0 R_1$ di S_0 in R_1; 2) della curvatura totale K_{01} dell'ipersuperficie $S_1 R_0$ di S_1 in R_0; 3) di una quantità B dedotta dall'intersezione V_{n-2} di S_0 e S_1.

Con un ragionamento ben noto si possono stabilire le seguenti formule

$$\int K_{10}(R_0 \cdot R_1) \dot{R}_1 = J_n C_0 V_1, \tag{7}$$

$$\int K_{01}(R_0 \cdot R_1) \dot{R}_1 = J_n C_1 V_0. \tag{8}$$

Non resta che da valutare l'integrale

$$\int B \dot{R}_1. \tag{9}$$

([2]) W. Blaschke, *Integralgeometrie*, 17, *Uber Kinematik*, «Deltion», 17, 2, Atene 1936, pp. 3–14.

A questo scopo designiamo con $P\vec{\xi}_1 \ldots \vec{\xi}_{n-2}$ l'$(n-2)$-edro su $V_{n-2} = S_0$, S_1 e con $\vec{\xi}_n$, $\vec{\eta}_n$ le normali di S_0, S_1 in P. Se $\vec{v}$, $\vec{rv}$ sono i due vettori unità che dividono l'angolo di $\vec{\xi}_n$, $\vec{\eta}_n$ in due parti uguali, si ottiene

$$(10) \qquad B = \int \prod_{i=1,\ldots,n-2} (\cos \sigma d\vec{v} \cdot \vec{\xi}_i + \sin \sigma d\vec{rv} \cdot \vec{\xi}_i)\, d\sigma,$$

dove l'integrale è esteso a tutti i punti di V_{n-2} e all'intervallo $-\dfrac{\varphi}{2} \leq \sigma \leq \dfrac{\varphi}{2}$. Designiamo con $\vec{e}_1^{(i)}, \ldots, \vec{e}_{n-1}^{(i)}$ i vettori unità nelle direzioni principali di S_i in P $(i = 0, 1)$. Si ha

$$(11) \qquad \vec{\xi}_j = \sum_{k=1}^{n-1} C_{jk}^{(i)} \vec{e}_k^{(i)} \qquad \begin{matrix} i = 0, 1 \\ j = 1, \ldots, n-2, \end{matrix}$$

da cui si deduce

$$(12) \qquad B = \frac{1}{\sin^{n-2}\varphi} \int_{-\varphi/2}^{\varphi/2} d\sigma \int D\dot{\Omega},$$

dove $\dot{\Omega}$ è l'elemento di superficie di V_{n-2} e D il determinante di ordine $n-2$:

$$(13) \quad (-1)^{n-2} D = \left| \sum_{k=1}^{n-2} \frac{C_{pk}^{(0)} C_{qk}^{(0)}}{V_k^{(0)}} \sin\left(\frac{\varphi}{2} - \sigma\right) + \sum_{k=1}^{n-2} \frac{C_{ph}^{(1)} C_{qk}^{(1)}}{V_k^{(1)}} \sin\left(\frac{\varphi}{2} + \sigma\right) \right|.$$

E utilizzando un teorema sui determinanti si ha

$$(14) \quad (-1)^{n-2} D = H_0 \sin^{n-2}\left(\frac{\varphi}{2} - \sigma\right) + H_1 \sin^{n-3}\left(\frac{\varphi}{2} - \sigma\right) \sin\left(\frac{\varphi}{2} + \sigma\right) + \ldots$$
$$+ H_{n-2} \sin^{n-2}\left(\frac{\varphi}{2} + \sigma\right),$$

dove

$$H_p = \sum \begin{vmatrix} C_{1\nu_1}^{(0)} & \ldots & C_{1\nu_q}^{(0)} & C_{1\mu_1}^{(1)} & \ldots & C_{1\mu_p}^{(1)} \\ \ldots & \ldots & \ldots & \ldots & \ldots & \ldots \\ C_{n-2,\nu_1}^{(0)} & \ldots & C_{n-2,\nu_q}^{(0)} & C_{n-2,\mu_1}^{(1)} & \ldots & C_{n-2,\mu_p}^{(1)} \end{vmatrix}^2 \frac{1}{r_{\nu_1}^{(0)} \ldots r_{\nu_q}^{(0)} r_{\mu_1}^{(1)} \ldots r_{\mu_p}^{(1)}}$$

$$(p + q = n - 2),$$

le somme essendo estese a tutte le combinazioni indipendenti $\mu_1, \ldots, \mu_p, \nu_1, \ldots, \nu_q$ di $1, \ldots, n-1$.

Dalle (12), (6), (14) si deduce

$$\int B\dot{R}_1 = \frac{1}{J_{n-2}} \int \frac{1}{\sin^{n-2}\varphi} D\, d\sigma\, \dot{L}_v \dot{R}_1 = \frac{1}{J_{n-2}} \int \sin\varphi\, D\, d\sigma\, \dot{L}_0 \dot{L}_1 \dot{\varphi}$$
$$= \frac{1}{J_{n-2}} \left\{ b_{n-2} \int H_0 \dot{L}_0 \dot{L}_1 + \cdots + b_0 \int H_{n-2} \dot{L}_0 \dot{L}_1 \right\}$$
$$= J_n (a_{n-2} M_{n-2}^{(0)} M_0^{(1)} + \cdots + a_0 M_0^{(0)} M_{n-2}^{(1)}),$$

dove $b_0, \ldots, b_{n-1}, a_0, \ldots, a_{n-2}$ sono costanti. Le costanti $a_0, \ldots, a_{n-2}$ possono determinarsi considerando il caso particolare in cui R_0 ed R_1 sono due sfere di raggi 1 ed h. Così si dimostra la formula enunciata (3).

Per $n = 3$ si deduce dalla (3) la formula (1) di BLASCHKE. Il caso $n = 4$ dà

(16)
$$\int K(R_0 \cdot R_1)\dot{R}_1 = 16\pi^4\left(K_0\bar{V}_1 + K_1\bar{V}_0 + M_0^{(0)}M_2^{(1)} + M_2^{(0)}M_0^{(1)} + \frac{3}{2}M_1^{(0)}M_1^{(1)}\right).$$

Reset from
Boll. Un. Mat. Ital. **2** (1940) 434–437

Sur les Invariants de Contact en Géométrie Projective Différentielle (*)

SHIING-SHEN CHERN

SVMMARIVM.—Auctor demonstrat quomodo Cartanica methodus, quae vocatur «du repère mobile», quaestionibus invariantia contactus varietatum tangentium respicientibus applicari possit.

Introduction

On connait la fécondité de la méthode du repère mobile ([1]) de M. E. CARTAN pour étudier les questions diverses dans la géométrie projective différentielle. L'étude faite par M. CARTAN lui-même de la déformation projective des surfaces ([2]) fournit un exemple célèbre. Le but de cette Note est de montrer comment cette méthode s'applique à des questions concernant les invariants de contact des variétés tangentes.

§1. Les Invariants de Contact des Courbes Planes

Commençons par considérer le cas le plus simple, c'est la géométrie différentielle projective des courbes planes ([3]). Le repère projectif dans le plan se compose de trois points *analytiques* linéairement indépendants A, A_1, A_2, définis à un facteur commun près. Une famille de repères étant donnée, elle satisfait aux *équations du déplacement infinitesimal* de la forme

[1]
$$\begin{aligned} dA &= \omega_0^0 A + \omega_0^1 A_1 + \omega_0^2 A_2, \\ dA_1 &= \omega_1^0 A + \omega_1^1 A_1 + \omega_1^2 A_2, \\ dA_2 &= \omega_2^0 A + \omega_2^1 A_1 + \omega_2^2 A_2, \end{aligned}$$

où les formes de PFAFF $\omega_\alpha^\beta(\alpha, \beta = 0, 1, 2)$ s'appellent les *composantes relatives*. Soit C

(*) Nota presentata dall'Accademico Pontificio Ugo Amaldi il 27 giugno 1941.

([1]) On trouve les généralités de cette méthode et quelques-unes de ses applications dans le livre: E. CARTAN, *La théorie des groupes finis et continus et la géométrie différentielle traitée par la méthode du repère mobile.* Paris, 1937.

([2]) Voir CARTAN, *Sur la déformation projective des surfaces*, «Annales scientifiques de l'Ecole Normale Supérieure, série 3, t. 37 (1920), pp. 259–356.

([3]) Voir CARTAN, *Leçons sur la théorie des espaces à connexion projective.* Paris, 1936, spécialement pag. 91-111. Ce livre sera cité dans le suivant comme: CARTAN, *Connexion projective.*

une courbe décrite par le point *géométrique* A. On attache à chaque point A de C un repère AA_1A_2 tel que AA_1 soit la tangente de C au point A. La famille de repères AA_1A_2 ainsi obtenus est dite du premier ordre et est caractérisée analytiquement par la condition

[2] $$\omega_0^2 = 0.$$

Le repère du premier ordre le plus général $\bar{A}\bar{A}_1\bar{A}_2$ relatif au point A est alors donné par les équations

[3] $$\begin{aligned} \bar{A} &= A, \\ \bar{A}_1 &= \rho(A_1 + \lambda A), \\ \bar{A}_2 &= \sigma(A_2 + \mu A_1 + \nu A), \end{aligned}$$

où les quantités $\rho, \sigma, \lambda, \mu, \nu$, qu'on appelle les paramètres sécondaires, sont arbitraires. En désignant par $\bar{\omega}_\alpha^\beta(\alpha, \beta = 0, 1, 2)$ les composantes relatives de la famille de repères $\bar{A}\bar{A}_1\bar{A}_2$, on trouve

$$\bar{\omega}_0^1 = \frac{\omega_0^1}{\rho}, \qquad \bar{\omega}_1^2 = \frac{\rho\omega_1^2}{\sigma},$$

d'où

[4] $$\frac{\bar{\omega}_1^2}{\bar{\omega}_0^1} = \frac{\rho^2}{\sigma}\cdot\frac{\omega_1^2}{\omega_0^1}.$$

Cela étant, on dit que deux courbes C et C^* ayant lè point commun A sont tangentes au point A, si elles ont un repère du premier ordre commun en A. Alors les deux familles de repères du premier ordre de C *et* C^* relatifs au point A se confondent. Considérons deux familles de repères du premier ordre, attachés aux differents points de C^*, telles que l'une contienne le repère AA_1A_2, et l'autre $\bar{A}\bar{A}_1\bar{A}_2$.

Soient θ_α^β et $\bar{\theta}_\alpha^\beta(\alpha, \beta = 0, 1, 2)$ les composantes relatives de chacune de ces deux familles. Au point A considéré, on a, correspondant à la formule [4],

$$\frac{\bar{\theta}_1^2}{\bar{\theta}_0^1} = \frac{\rho^2}{\sigma}\cdot\frac{\theta_1^2}{\theta_0^1}.$$

Il en résulte que

$$\frac{\bar{\theta}_1^2}{\bar{\theta}_0^1}\bigg/\frac{\bar{\omega}_1^2}{\bar{\omega}_0^1} = \frac{\theta_1^2}{\theta_0^1}\bigg/\frac{\omega_1^2}{\omega_0^1},$$

de sorte que la valeur de

[5] $$I = \frac{\theta_1^2}{\theta_0^1}\bigg/\frac{\omega_1^2}{\omega_0^1}$$

au point A soit indépendante du choix du repère du premier ordre attaché à A. Il en suit que I est un invariant projectif des deux courbes C et C^*.

Montrons que l'invariant I ainsi défini se confond avec l'invariant de contact bien connu de SMITH-MEHMKE [1]. A cet effet definissons les coordonnées non homogènes x, y d'un point M relatives au repère AA_1A_2 par l'équation

$$M = A + xA_1 + yA_2. \tag{6}$$

La droite AA_1 étant la tangente en, A, l'équation de la courbe C au voisinage de A s'écrit

$$y = \frac{1}{2}mx^2 + \cdots, \tag{7}$$

où les termes non écrits sont au moins du troisième ordre par rapport à x. De même, l'équation de C^* par rapport AA_1A_2 est de la forme

$$y = \frac{1}{2}m^*x^2 + \cdots. \tag{8}$$

En utilisant les conditions suivantes pour la fixitè du point M [2]:

$$\begin{aligned} &dx + \omega_0^1 + x(\omega_1^1 - \omega_0^0) + y\omega_2^1 - x^2\omega_1^0 - xy\omega_2^0 = 0,\\ &dy + \omega_0^2 + x\omega_1^2 + y(\omega_2^2 - \omega_0^0) - xy\omega_1^0 - y^2\omega_2^0 = 0, \end{aligned} \tag{9}$$

on trouve immédiatement que

$$\frac{\omega_1^2}{\omega_0^1} = m,$$

et, de la même manière, que

$$\frac{\theta_1^2}{\theta_0^1} = m^*.$$

Il en résulte que

$$I = \frac{m^*}{m}, \tag{10}$$

qui démontre notre énoncé.

Avec la définition précédente de l'invariant I on peut démontrer très simplement les théorèmes de C. SEGRE et de B. SEGRE. Prenons sur la tangente commune AA_1 un point $P = A + \varepsilon A_1$ infiniment voisin à A. Un point sur la droit PA_2 étant de la forme $P + \tau A_2$, les points M et N où PA_2 recontre les courbes C et C^* sont donnés par les valeurs suivantes de τ:

$$\tau = \frac{1}{2}m\varepsilon^2 + \cdots,$$

$$\tau^* = \frac{1}{2}m^*\varepsilon^2 + \cdots.$$

(1) FUBINI-ČECH, *Introduction à la géométrie projetive différentielle des surfaces.* Paris, 1931, pag. 17–20.
(2) CARTAN, *Connexion projective*, pag. 88.

Il en suit que le rapport anharmonique (MN, A_2P) tend vers I lorsque ε tend vers zéro. En remarquant de [3] que le point A_2 est arbitraire, on obtient le théorème de C. SEGRE: *Prenons un point arbitraire* A_2 *dans le plan. Soient* P, M, N *les points d'intersection d'une droite passant par* A_2 *avec la tangente commune* AA_1 *et les deux courbes. La limite du rapport anharmonique* (MN, A_2P) *lorsque la droite tend vers* A_2A *est indépendante du choix de* A_2 *et est égale à* I.

On démontre le thérème de B. SEGERE d'une manière analogue. L'équation de la conique tangente aux droites AA_1, A_1A_2 respectivement aux points A et A_2 est de la forme

$$y = lx^2,$$

où l est un paramètre. Il en suit que parmi ces coniques celle qui a un contact du second ordre avec C en A a l'équation

$$y = \frac{1}{2}mx^2,$$

et celle qui a avec C^* un contact du second ordre en A a l'equation

$$y = \frac{1}{2}m^*x^2.$$

Ces deux coniques ont un contact double, leur invariant étant manifestement égal à I. En remarquant que le point A_2 et la droite A_2A_1 passant par A_2 son arbitraires, on a le théorème de B. SEGRE: *L'invariant des deux coniques qui passent par le point* A_2 *et y ont une droite tangente* A_2A_1 *et qui sont osculatrices en* A *aux courbes* C *et* C^* *ne dépend ni du point* A_2, *ni de la droite* A_2A_1 *et il est égal à* I.

Des considérations analogues conduisent à un invariant de contact nouveau. Supposons que le point commun A ne soit pas un point d'inflexion des deux courbes C et C^* et que ces courbes aient un contact du quatrième ordre en A. Cela signifie que les familles de repères du quatrième ordre de C et C^* au point A se confondent. Pour les composantes relatives d'une famille de repères du quatrième ordre on a les relations

[11] $\quad \omega_0^2 = 0, \qquad \omega_1^2 - \omega_0^1 = 0, \qquad \omega_2^2 - 2\omega_1^1 + \omega_0^0 = 0, \qquad \omega_2^1 - \omega_1^0 = 0.$

Si AA_1A_2 est un repère du quatrième ordre attaché au point A, le repère le plus général $\bar{A}\bar{A}_1\bar{A}$ du quatrième ordre en A est donné par les équations

[12]
$$\begin{aligned} \bar{A} &= A, \\ \bar{A}_1 &= \rho(A_1 + \lambda A), \\ \bar{A}_2 &= \rho^2\left(A_2 + \lambda A_1 + \frac{1}{2}\lambda^2 A\right). \end{aligned}$$

En désignant par $\bar{\omega}_\alpha^\beta(\alpha, \beta = 0, 1, 2)$ les composantes relatives de la famille $\bar{A}\bar{A}_1\bar{A}_2$, on trouve

$$\bar{\omega}_0^1 = \frac{\omega_0^1}{\rho}, \qquad \bar{\omega}_2^0 = \rho^2\omega_2^0,$$

de sort que

[13] $$\frac{\bar{\omega}_2^0}{\bar{\omega}_0^1} = \rho^3 \frac{\omega_2^0}{\omega_0^1}.$$

Soient $\theta_\alpha^\beta(\alpha, \beta = 0, 1, 2)$ les composantes relatives d'une famille de repères du quadrième ordre de C^*, qui contient le repère AA_1A_2. L'équation [13] et l'équation analogue pour l'effet sur θ_2^0/θ_0^1 d'un changement de repère [12] montrent que la valeur de

[14] $$J = \frac{\theta_2^0}{\theta_0^1}\bigg/\frac{\omega_2^0}{\omega_0^1}$$

au point A est un invariant projectif. Il est un invariant de contact pour deux courbes ayant un contact du quatrième ordre.

On peut donner une interprétation géométrique à l'invariant J. Soit AA_1A_2 un repère commun du quatrième ordre de C et C^* en A. Relative au repère AA_1A_2 l'équation de C s'écrit

[15] $$y = \frac{1}{2}x^2 + px^5 + \cdots,$$

et celle de C^* s'écrit

[16] $$y = \frac{1}{2}x^2 + p^*x^5 + \cdots.$$

En utilisant les conditions de fixité du point [9], on trouve

$$\frac{\omega_2^0}{\omega_0^1} = 2_0p, \quad \frac{\theta_2^0}{\theta_0^1} = 2_0p^*,$$

de sorte que

[17] $$J = \frac{p^*}{p}.$$

Les deux courbes C et C^* ont en A la même conique osculatrice, qui a l'équation

[18] $$y = \frac{1}{2}x^2.$$

Cela étant, on constate facilement la signification suivante pour J: *Prenons sur la conique osculatrice comune* [18] *un point quelconque* Q *et joignons-le à un point* P *de* [18] *infiniment voisin de* A. *La droite* QP *rencontre* C *et* C^* *aux points* M, N, *dont le rapport anharmonique avec* Q, P *tend vers* J *lorsque la droite* QP *tend vers* QA. La démonstration de ce théorème est immédiate, si l'on identifie Q avec le point A_2 du repère.

§2. L'extension à L'espace Projectif à *n* Dimension

Nous allons étendre les considérations précédentes aux courbes tangentes dans l'espace projectif à n dimensions.

Un repère projectif dans l'espace à n dimensions se compose de $n+1$ points analytiques linéairement indépendants $A, A_1, \ldots, A_n$, définis à un facteur commun près. Les équations du déplacement infinitésimal d'une famille de repères $AA_1 \ldots A_n$ son de la forme

[19] $$dA_\alpha = \sum_{\beta=0}^{n} \omega_\beta^\alpha A_\beta \qquad (\alpha = 0, 1, \ldots, n)$$

où ω_β^α sont les composantes relatives et où l'on écrit A_0 pour A. Les conditions pour la fixité du point

[20] $$M = A + x^1 A_1 + \ldots + x^n A_n$$

dans l'espace s'écrivent alors

[21] $$dx^i + \omega_0^i + \sum_{k=1}^{n} x^k \omega_k^i = x^i\left(\omega_0^0 + \sum_{k=1}^{n} x^k \omega_k^0\right) \qquad i = 1, \ldots, n.$$

Cela étant, considérons dans l'espace une courbe C décrite par le point A. Soit k un entier $\leq n-1$. Attachons à chaque point A de C un repère $AA_1 \ldots A_n$ tel que AA_1 soit la tangente de C en A, AA_1A_2 la variété plane osculatrice à deux dimensions, et, plus généralement, $AA_1 \ldots A_l$ la variété plane osculatrice à l dimensions, $(1 = 1, 2, \ldots, k)$. On appelle un tel repère un *repère d'ordre* k au point A. Pour une famille de repères d'ordre k de la courbe C on a

[22] $$\omega_l^m = 0 \qquad (l = 0, 1, \ldots, k-1; m = l+2, l+2, \ldots, n).$$

Si $AA_1 \ldots A_n$ est un repère d'ordre k au point A, le repère le plus général $\bar{A}\bar{A}_1 \ldots \bar{A}_n$ d'ordre k en A est donné par les équations

[23] $$\begin{aligned} \bar{A} &= A_1, \\ \bar{A}_1 &= a_1^0 A + a_1^1 A_1, \\ &\ldots \\ \bar{A}_k &= a_k^0 A + a_k^1 A_1 + \cdots + a_k^k A_k, \\ \bar{A}_{k+1} &= a_{k+1}^0 A + a_{k+1}^1 A_1 + \cdots + a_{k+1}^n A_n, \\ &\ldots \\ \bar{A}_n &= a_n^0 A + a_n^1 A_1 + \cdots + a_n^n A_n, \end{aligned}$$

où les ceefficients a sont arbitraires. En effectuant le changement du repère [23], on trouve, $\bar{\omega}_\alpha^\beta(\alpha, \beta = 0, 1, \ldots, n)$ étant les composantes relatives de la famille de repères $\bar{A}\bar{A}_1 \ldots \bar{A}_n$,

$$\bar{\omega}_0 = \frac{1}{a_1^1}\omega_0^1, \qquad \bar{\omega}_{k-1}^k = \frac{a_{k-1}^{k-1}}{a_k^k}\omega_{k-1}^k,$$

de sorte que

[24] $$\frac{\bar{\omega}_{k-1}^k}{\bar{\omega}_0^1} = \frac{a_1^1 a_{k-1}^{k-1}}{a_k^h}\frac{\omega_{k-1}^k}{\omega_0^1}$$

Soient C et C^* deux courbes qui ont en A un contact d'ordre $k(1 \leq k \leq n-1)$ et dont les variétés planes osculatrices à l dimensions $(l = 1, 2, \ldots, k)$ en A sont bien déterminées. Par dèfinition, cela signifie que les familles de repères d'ordre k de C et C^* en A se confondent. En désignant par $\theta_\alpha^\beta(\alpha, \beta = 0, 1, \ldots, n)$ les composantes relatives d'une famille de repères d'ordre k de C^*, qui contient $AA_1 \ldots A_n$ et par $\bar{\theta}_\alpha^\beta$ celles d'une famille contenant $\bar{A}\bar{A}_1 \ldots \bar{A}_n$, on à, au point A considéré

$$\frac{\bar{\theta}_{k-1}^k}{\bar{\theta}_0^1} = \frac{a_1^1 a_{k-1}^{k-1}}{a_k^k} \frac{\theta_{k-1}^k}{\theta_0^1}.$$

Il en résulte que la valeur de

[25] $$I_k = \frac{\theta_{k-1}^k}{\theta_0^1} \Big/ \frac{\omega_{k-1}^k}{\omega_0^1}$$

est indépendante du choix du repère $AA_1 \ldots A_n$ et qu'elle est un invariant projectif des courbes C et C^* ayant un contact d'ordre k en A. Nous l'appellerons *l'invariant de contact d'ordre* k de C et C^*. On voit donc que deux courbes ayant en A un contact d'ordre k ont $k-1$ invariants de contact, à savoir $I_2, I_3, \ldots, I_k$.

Dans le cas particulier où $k = n-1$, on peut donner un invariant de plus. Dans ce cas, en effet, la transformation [23] transforme la composante relative ω_{n-1}^n d'après la formule

$$\bar{\omega}_{n-1}^n = \frac{a_{n-1}^{n-1}}{a_n^n} \omega_{n-1}^n.$$

La transformation sur θ_{n-1}^n étant la même, on en déduit que

[26] $$I_n = \frac{\theta_{n-1}^n}{\theta_0^1} \Big/ \frac{\omega_{n+1}^n}{\omega_0^1}$$

est un invariant de contact, que nous appellerons *l'invariant généralisé de* SMITH-MEHMKE. Pour $n = 2$, cet invariant I_n est précisément l'invariant de SMITH-MEHMKE.

Nous allons donner une interprétation géométrique de l'invariant I_k, qui est analogue au thèorème de C. SEGRE. Le repère $AA_1 \ldots A_n$ étant choisi comme indiqué plus haut, les équations de C au voisinage de A peuvent être écrites dans la forme

[27] $$\begin{aligned} x^1 &= t, \\ x^2 &= b_2^2(t)^2 + b_3^2(t)^3 + \cdots, \\ &\cdots \\ x^{k-1} &= b_{k-1}^{k-1}(t)^{k-1} + \cdots, \\ x^k &= b_k^k(t)^k + \cdots, \\ x^{k+1} &= b_{k+1}^{k+1}(t)^{k+1} + \cdots, \\ &\cdots \\ x^n &= b_{k+1}^n(t)^{k+1} + \cdots, \end{aligned}$$

où t est le paramètre sur C. Si l'on différencie l'équation pour x^k et on égale les coefficientes de $(t)^{k-1}$ dans les deux membres, on obtient, en tenant compte des conditions de fixité du point [21],

[28] $$\frac{\omega_{k-1}^k}{\omega_0^1} = k\frac{b_k^k}{b_{k-1}^{k-1}}$$

Ecrivons ensuite les équations de C^* par rapport au repère $AA_1 \ldots A_x$ dans la forme

[29] $$\begin{aligned} x^1 &= \tau, \\ x^2 &= c_2^2(\tau)^2 + (c_3^2(\tau)^3 + \cdots, \\ &\cdots \\ x^{k-1} &= c_{k-1}^{k-1}(\tau)^{k-1} + \cdots, \\ x^k &= c_k^k(\tau)^k + \cdots, \\ x^{k+1} &= c_{k+1}^{k+1}(\tau)^{k+1} + \cdots, \\ &\cdots \\ x^n &= c_{k+1}^n(\tau)^{k+1} + \cdots, \end{aligned}$$

τ étant le paramètre. On trouve, analoguement à [28],

[30] $$\frac{\theta_{k-1}^k}{\theta_0^1} = k\frac{c_k^k}{c_{k-1}^{k-1}}.$$

Il en résulte que

[31] $$\mathrm{I}_k = \frac{c_k^k}{c_{k-1}^{k-1}}\bigg/\frac{b_k^k}{c_{k-1}^{k-1}} = \frac{c_k^k}{b_k^k}\bigg/\frac{c_{k-1}^{k-1}}{b_{k-1}^{k-1}}.$$

Dans le cas $k = n - 1$, on trouve, dans la même manière, l'expression suivante pour l'invariant I_n:

[32] $$\mathrm{I}_n = \frac{c_n^n}{c_{n-1}^{n-1}}\bigg/\frac{b_n^n}{b_{n-1}^{n-1}} = \frac{c_n^n}{c_n^n}\bigg/\frac{c_{n-1}^{n-1}}{b_{n-1}^{n-1}}.$$

Des expression [31] et [32] pour I_k, I_n il est facile d'eux donner une interprétation géométrique simple. En effet, les points M de C et les points N de C^* peuvent être mis en correspondance biunivoque per la condition que la droite MN joignant les points correspondants rencontre la variété plane $A_2 \ldots A_n$. Cette correspondance est définie par l'équation $\tau = t$. Désignons par P_l le point où MN recontre la variété plane $AA_1 \ldots A_{l-1}A_{l+1} \ldots A_n$, $(l = 0, 1, \ldots, n)$. Il est facile de vérifier que l'invariant I_k est égal à la limite du rapport anharmonique $(MN, P_{k-1}P_k)$ lorsque M, N tendent vers A. Dans le cas $k = n - 1$, l'invariant I_n est égal à la limite de $(MN, P_{n-1}P_n)$. D'une manière précise, on a le théorème:

Soient C et C deux courbes qui ont en* A *un contact d'ordre* $k(1 \leq k \leq 1)$ *et soit* $AA_1 \ldots A_n$ *un repère d'ordre k commun. Mettons les points* M *de C et les points* N *de C* en correspondance par la condition que la droite* MN *rencontre la variété plane* $A_2 \ldots A_n$. *Si* P_l *est le point où* MN *recontre la variété plane* $AA_1 \ldots A_{l-1}A_{l+1} \ldots A_n$, $(l = 0, 1 \ldots n)$, *la limite du rapport anharmonique* $(MN, P_{k-1}P_k)$ *lorsque* M *tend* *vers* A *est indépendante du choix du repère* $AA_1 \dots A_n$ *et est égale à* I_k. *Dans le cas* $k = n-1$, *la limite de* $(MN, P_{n+1}P_n)$ *est égale à* I_n *pour tout choix de* $AA_1 \dots A_n$.

Des invariants I_k et I_n on dèduit que les quantités

$$J_l = I_2 I_3 \dots I_l = \frac{c_l^l}{b_l^l}, \qquad (l = 2, \dots, k, \text{ si } k \leq n-2),$$

[33]

$$J_l = I_2 I_3 \dots I_l = \frac{c_l^l}{b_l^l}, \qquad (l = 2, \dots, n, \text{ si } k = n-1)$$

sont aussi des invariants de contact. Dans le cas $k = n-1$ ces invariants one été donnés par M.B. SEGRE ([1]), tandis que l'invariance de J_l pour $k \leq n-2$ a été signalée par M. BUCHIN SU ([2]). En employant les notations précédentes, on peut énoncer le théorème suivant: *L'invariant* J_k *est égal à la limite du rapport anharmonique* (MN, P_0P_k) *lorsque* M *tend vers* A, *tandis que, dans le cas* $k = n-1$, *l'invariant* J_k *est égal à la limite de* (MN, P_0P_n).

Il serait intéressant d'indiquer une généralisation possible de l'invariant I_n pour deux courbes ayant un contact d'ordre $n-1$. Pour une famille de repères $AA_1 \dots A_n$ d'ordre $k \leq n-1$ de C les composantes relatives $\omega_k^{k+1}, \dots, \omega_k^n$ sont aussi des multiples de ω_0^1. Posons

$$\xi_k^\varrho = \frac{\omega_k^\varrho}{\omega_0^1}, \qquad \rho = k+1, \dots, n. \tag{34}$$

Après le changement de repères [23] les quantités $\bar{\xi}_k^\varrho$ correspondantes à la famille de repères $\bar{A}\bar{A}_1 \dots \bar{A}_n$ sont liées aux ξ_k^ϱ par les relations

$$a_1^1 a_k^k \xi_k^\varrho = \sum_{\sigma=k+1}^{n} a_\sigma^\varrho \bar{\xi}_k^\sigma \qquad (\rho = k+1, \dots, n). \tag{35}$$

On voit que la transformation sur ξ_k^ϱ est la transformation homographique la plus générale possible. Nous nous contentons ici de faire la remarque qu'il serait possible de définir, au moyen des ξ_k^ϱ, des invariants de contact de $n-k+1$ courbes ayant un contact d'ordre k les unes avec les autres au même point A.

Ajoutons encore une remarque concernant l'invariance de I_k par rapport aux projections. Projetons, en effet, les courbes C et C^* de la variété plane $A_{l+1} \dots A_n$ $(1 \geq k)$ à $n-l-1$ dimensions dans $AA_1 \dots A_l$. Il est facile de vérifier que les projections de C et C^* ont en A un contact d'ordre k si $l \geq k+1$ et qu'elles ont au moins un contact d'ordre $k-1$ si $1 = k$. De plus, dans le cas $l \geq k+1$, l'invariant d'ordre k de C et C^* est égal à celui de leurs projections, tandis que, pour $1 = k$, il est égal à l'invariant généralisé de SMITH-MEHMKE de leurs projections. En remarquant que la variété plane $A_{l+1} \dots A_n$ est arbitraire, on a le théorème: *L'invariant de contact d'ordre k de deux courbes reste inaltéré, si l'on les projette d'une variété plane arbitraire à* $n-l-1$

([1]) B. SEGRE, *Sugli elementi curvilinei che hanno comuni le origini ed i relativi spazi osculatori*, «Rendiconti Accad. dei Lincei», (VI) 22 (1925_2), pag. 392–399.

([2]) B. SU, *Some arithmetical invariants of a curve in projective space of n dimensions*, à paraître dans le «Journal de Science Mathématique de l'Université de Tucuman».

dimensions $l \geq k + 1$, *qui ne recontre pas la variété plane osculatrice à k dimensions, commune* $AA_1 \ldots A_k$, *dans une variété plane à l dimensions, qui contient* $AA_1 \ldots A_k$. *Cet invariant est aussi égal à l'invariant généralisé de* Smith-Mehmke *des projections des deux courbes dans* $AA_1 \ldots A_k$ *d'une variété plane à* $n - k - 1$ *dimensions, qui ne recontre pas* $AA_1 \ldots A_k$.

§3. Quelques Invariants de Contact des Surfaces dans L'espace Ordinaire

Indiquons comment nos considérations peuvent être appliquées aux surfaces tangentes dans l'espace projectif ordinaire.

A cet effet, considérons deux surfaces tangentes au point commun A. Prenons le repère $AA_1A_2A_3$ tel que le plan AA_1A_2 soit le plan tangent commun en A. En définissant les coordonnées non homogènes x, y, z d'un poins M par rapport au repère $AA_1A_2A_3$ par la relation

[36] $$M = A + xA_1 + yA_2 + zA_3,$$

on peut écrire les équations des deux surfaces respectivement dans les formes

[37] $$z = \frac{1}{2}(ax^2 + 2bxy + cy^2) + \cdots,$$

[38] $$z = \frac{1}{2}(a_1x^2 + 2b_1xy + c_1y^2) + \cdots.$$

Il serait facile de vérifier, avec les méthodes employées plus haut, le fait bien connu que les quantités

[39] $$\frac{a_1c_1 - b_1^2}{ac - b^2}, \quad \frac{ac_1 - 2bb_1 + ca_1}{ac - b^2}$$

sont des invariants de contact.

Cependant, les méthodes précédentes conduisent, pour deux surfaces ayant un contact du deuxième ordre au point A, à deux invariants de contact qui me semblent nouveaux. On peut supposer les équations des deux surfaces dans les formes

[40] $$z = xy + \frac{1}{3}(px^3 + 3qx^2y + 3rxy^2 + sy^3) + \cdots,$$

[41] $$z = xy + \frac{1}{3}(p_1x^3 + 3q_1x^2y + 3r_1xy^2 + s_1y^3) + \cdots,$$

où les termes non écrits sont au moins du quatrième ordre par rapport à x, y Si $AA_1A_2A_3$ est le repère par rapport auquel les surfaces ont les équations des formes [40], [41], le repère le plus général $\bar{A}\bar{A}_1\bar{A}_2\bar{A}_3$ ayant la même propriété est donné par les équations

[42] $$\bar{A} = A, \quad \bar{A}_1 = \rho A + \alpha A_1, \quad \bar{A}_2 = \sigma A + \beta A_2,$$
$$\bar{A}_3 = \tau A + \lambda A_1 + \mu A_2 + \alpha\beta A_3,$$

ou

$$[43]\qquad \bar{A} = A, \quad \bar{A}_1 = \rho A + \alpha A_2, \quad \bar{A}_2 = \sigma A + \beta A_1, \quad \bar{A}_3 = \tau A + \mu A_1 + \lambda A_2 + \alpha\beta A_3,$$

où le changement [43] peut être regardé comme le produit de [42] et du changement

$$[44]\qquad \bar{A} = A, \quad \bar{A}_1 = A_2, \quad \bar{A}_2 = A_1, \quad \bar{A}_3 = A_3.$$

Si les équations des surfaces [40], [41] par rapport au repère $\bar{A}\bar{A}_1\bar{A}_2\bar{A}_3$ défini par [42] sont

$$[45]\qquad z = xy + \frac{1}{3}(\bar{p}x^3 + 3\bar{p}x^2y + 3\bar{r}xy^2 + \bar{s}y^3) + \cdots,$$

$$[46]\qquad z = xy + \frac{1}{3}(\bar{p}_1x^3 + 3\bar{q}_1x^2y + 3\bar{r}_1xy^2 + \bar{s}_1y^3) + \cdots,$$

on obtient

$$[47]\qquad \bar{p} = \frac{\alpha^2}{\beta}p, \quad \bar{s} = \frac{\beta^2}{\alpha}s, \qquad \bar{p}_1 = \frac{\alpha^2}{\beta}p_1, \quad \bar{s}_1 = \frac{\beta^2}{\alpha}s_1,$$

d'où

$$[48]\qquad \frac{\bar{p}_1}{\bar{p}} = \frac{p_1}{p}, \quad \frac{\bar{s}_1}{\bar{s}} = \frac{s_1}{s}.$$

D'autre part, le changement du repère [44] transforme les quantités p_1/p et s_1/s d'après les formules

$$[49]\qquad \frac{\bar{p}_1}{\bar{p}} = \frac{s_1}{s}, \quad \frac{\bar{s}_1}{\bar{s}} = \frac{p_1}{p}.$$

Il en résulte que les quantités

$$[50]\qquad H = \frac{1}{2}\left(\frac{p_1}{p} + \frac{s_1}{s}\right), \qquad K = \frac{p_1}{p}\cdot\frac{s_1}{s}$$

sont indépendantes du choix du repère $AA_1A_2A_3$. Elles sont deux invariants de contact des surfaces [40], [41] au point A.

Pour donner une interprétation géométrique à H, K, considérons une quadrique Σ ayant un contact du deuxième ordre avec les deux surfaces au point A et prenons sur elle un point quelconque Q. Par un un choix convenable du repère $AA_1A_2A_3$ on peut supposer la quadrique Σ d'avoir l'équation

$$[51]\qquad z = xy$$

et le point Q de confondre avec A_3. Cela posé, prenons une droite quelconque passant par A_3 et infiniment voisine de A_3A. Soient P, M, N see points d'intersection avec la quadrique Σ et les deux surfaces respectivement. On a donc

$$P = A + \delta A_1 + \varepsilon A_2 + \delta\varepsilon A_3$$

$$M = P + \left\{\frac{1}{3}(p\delta^3 + 3q\delta^2\varepsilon + 3r\delta\varepsilon^2 + s\varepsilon^3) + \cdots\right\} A_3,$$

$$N = P + \left\{\frac{1}{3}(p_1\delta^3 + 3q_1\delta^2\varepsilon + 3r_1\delta\varepsilon^2 + s_1\varepsilon^3) + \cdots,\right\} A_3,$$

où δ, ε sont des infiniment petits. On en déduit que

[52] $$(\mathrm{MN}, A_3 p) = \frac{p_1\delta^3 + 3q_1\delta^2\varepsilon + 3r_1\delta\varepsilon^2 + s_1\varepsilon^3 + \cdots}{p\delta^3 + 3q\delta^2\varepsilon + 3r\delta\varepsilon^2 + s\varepsilon^3 + \cdots}.$$

Si la droite A_3P tend vers A_3A le long des deux directions asymptotiques, les limites du rapport anharmonique (MN, A_3P) sont p_1/p et s_1/s. On arrive donc à l'interprétation gèométrique suivante de p_1/p et s_1/s (c'est-à-dire de H et K): *Prenons une quadrique* Σ *ayant un contact du deuxième ordre avec les deux surfaces en* A *et un point quelconque* Q *sur* Σ. *Une droite quelconque passant par* Q *rencontre* Σ *et les deux surfaces aux points* P, M, N. *Lorsque cette droite tend vers* QA *le long des directions asymptotiques, le rapport anharmonique* (MN, QP) *tend vers* p_1/p *et* s_1/s *pour tout choix de la quadrique* Σ *et du point* Q *sar* Σ.

Reset from
Acta Pontif. Acad. Sci. **5** (1941) 123–140

ANNALS OF MATHEMATICS
Vol. 43, No. 3, July, 1942

THE GEOMETRY OF ISOTROPIC SURFACES

BY SHIING-SHEN CHERN

(Received May 23, 1941; revised October 10, 1941)

Introduction

The geometry in a space in which there is given a family of varieties has been the object of extensive researches. The geometry of paths[1] corresponds to the case of a $(2n-2)$-parameter family of curves in an n-dimensional space, defined by a system of differential equations of the second order of a certain type. It may be regarded as a natural generalization of projective geometry, since Cartan has proved that of all the projective connections having the same geodesics there exists one, and only one, which is characterized by intrinsic properties and which he called normal.[2] Recently, M. Hachtroudi[3] proved that in a three-dimensional space with a three-parameter family of surfaces there can be defined, in an intrinsic way, one and only one projective connection with the contact elements as the elements of the space. It is the aim of the present paper to study the geometry of a space in which there is given a two-parameter family of surfaces. Our main results may be summarized in the following two theorems:

THEOREM 1. *Suppose there be given in a three-dimensional space a two-parameter family of surfaces $\{\Sigma\}$ such that the tangent planes at a point of the surfaces of the family through the point do not pass through a fixed direction. Then there can be defined in the space one and only one four-dimensional Weyl geometry, which has the contact elements of the surfaces as its "points" and possesses certain intrinsic properties.*

THEOREM 2. *There exists a point transformation carrying one family of surfaces $\{\Sigma\}$ into another $\{\bar{\Sigma}\}$ when and only when the two four-dimensional Weyl geometries are equivalent.*

The paper is divided into four sections. In §1 we give the definition of a Weyl space of elements[4] and of some of its fundamental notions. The determination of the Weyl space from the family of surfaces is then given in §2. In §3 we show that the so-defined Weyl geometry naturally leads to a solution of the problem of equivalence, i.e., the problem of deciding when two such families are equivalent under point transformations. The important particular

---

[1] Cf., for instance, O. Veblen and T. Y. Thomas, *The geometry of paths*, Trans. Amer. Math. Soc., vol. 25 (1923), pp. 551–608.

[2] E. Cartan, *Sur les variétés à connexion projective*, Bull. de la Soc. Math. de France, t. 52 (1924), pp. 205–241.

[3] M. Hachtroudi, *Les espaces d'éléments à connexion projective normale*, Actualités Scientifiques et Industrielles, no. 565, Paris, 1937.

[4] Cf. E. Cartan, *La méthode du repère mobile, la théorie des groupes continus, et les espaces généralisés*, Actualités Scientifiques et Industrielles, Paris, 1935.

case by which the Weyl space reduces to a three-dimensional point space is discussed in §4.

The extension of these results to a space of n-dimensions will be treated in a subsequent paper.

1. Weyl Space of Elements

Consider a three-dimensional space S with the coordinates x, y, z. To each point M of the space we attach three vectors I_1, I_2, I_3, such that their scalar products satisfy the conditions

$$(1)\qquad \begin{cases} I_1 I_3 = -I_2^2 \neq 0, \\ I_1^2 = I_3^2 = I_1 I_2 = I_2 I_3 = 0. \end{cases}$$

The totality of the point M and the vectors I_1, I_2, I_3 through M satisfying (1) will be called a *frame*. We say that the space is a Weyl space or that a Weyl connection is defined in the space, if a law of infinitesimal displacement of the following form is given:

$$\begin{cases} dM = \omega_1 I_1 + \omega_2 I_2 + \omega_3 I_3, \\ dI_1 = \omega_{11} I_1 + \omega_{12} I_2 + \omega_{13} I_3, \\ dI_2 = \omega_{21} I_1 + \omega_{22} I_2 + \omega_{23} I_3, \\ dI_3 = \omega_{31} I_1 + \omega_{32} I_2 + \omega_{33} I_3, \end{cases}$$

where the ω's are Pfaffian forms in x, y, z. On differentiating (1) and making use of the above equations, we find

$$\omega_{13} = 0, \qquad \omega_{31} = 0, \qquad \omega_{11} + \omega_{33} - 2\omega_{22} = 0,$$
$$\omega_{12} - \omega_{23} = 0, \qquad \omega_{21} - \omega_{32} = 0.$$

Hence it turns out to be more convenient to put

$$\tau = \omega_{12}, \qquad \pi_1 = \omega_{11}, \qquad \pi_2 = \omega_{22}, \qquad \pi_3 = \omega_{21},$$

and to write the equations of infinitestimal displacement in the form:

$$(2)\qquad \begin{cases} dM = \omega_1 I_1 + \omega_2 I_2 + \omega_3 I_3, \\ dI_1 = \pi_1 I_1 + \tau I_2, \\ dI_2 = \pi_3 I_1 + \pi_2 I_2 + \tau I_3, \\ dI_3 = \qquad\quad \pi_3 I_2 + (-\pi_1 + 2\pi_2) I_3. \end{cases}$$

To serve our later purpose we shall generalize the notion of a Weyl space to that of a *Weyl space of elements*, by which we shall mean the following: We attach to each point of the space a one-parameter family of contact elements with the point as origin. A contact element is then defined by four coordinates x, y, z, t, where x, y, z are the coordinates of the point and t fixes the plane. By attaching a frame to each contact element and supposing the ω_1, ω_2, ω_3, τ, π_1, π_2, π_3 in (2) to be Pfaffian forms in x, y, z, t, we shall get a Weyl space of elements. To define a Weyl space of elements it is thus necessary to give the one-parameter family of contact elements through each point and the seven Pfaffian forms in (2).

It will be advantageous to interpret the four-parameter family of contact elements to be the "points" of a four-dimensional auxiliary space S'. Given a curve in S' defined by the equations

$$x = x(\lambda), \qquad y = y(\lambda), \qquad z = z(\lambda), \qquad t = t(\lambda), \tag{3}$$

we can express the Pfaffian forms in (2) as Pfaffian forms in λ. Then equations (2) become a system of ordinary differential equations. By a well-known theorem in the theory of differential equations, we see that when the initial frame $M^{(0)}I_1^{(0)}I_2^{(0)}I_3^{(0)}$ corresponding to an initial value $\lambda = \lambda_0$ is given, the family of frames $MI_1I_2I_3$ satisfying (2) is uniquely determined (with respect to $M^{(0)}I_1^{(0)}I_2^{(0)}I_3^{(0)}$). In this case, we say that *the family of frames is developed, along the curve* (3), *in the Euclidean space defined by the frame* $M^{(0)}I_1^{(0)}I_2^{(0)}I_3^{(0)}$. If we join a point P_0 of S' to a point P of S' by two different curves and develop, in the Euclidean space of the frame attached to P_0, the two families of frames along these two curves, the frames obtained corresponding to the frame at P will in general be different.

Instead of studying the development of the frame at P on the frame at P_0 along two different curves, we shall consider an infinitesimal parallelogram with P_0 as a vertex and with two directions d and δ through P_0 as its two sides. Such a parallelogram is called a *cycle*. If we develop the family of frames first along the side d and then along δ or vice versa, the frames obtained will be different, so that there exists an infinitesimal transformation carrying one frame to the other. This infinitesimal transformation is called the *infinitesimal transformation associated to the cycle*. By a classical method,[5] it is easy to show that the infinitesimal transformation associated to a cycle formed by the directions d and δ is given by equations of the form

$$\tag{4} \begin{cases} \nabla M = d\delta M - \delta d M = \Omega_1 I_1 + \Omega_2 I_2 + \Omega_3 I_3, \\ \nabla I_1 = d\delta I_1 - \delta d I_1 = \Pi_1 I_1 + \mathrm{T} I_2, \\ \nabla I_2 = d\delta I_2 - \delta d I_2 = \Pi_3 I_1 + \Pi_2 I_2 + \mathrm{T} I_3, \\ \nabla I_3 = d\delta I_3 - \delta d I_3 = \qquad \Pi_3 I_2 + (-\Pi_1 + 2\Pi_2) I_3, \end{cases}$$

[5] Cf., for example, E. Cartan, *Leçons sur la géométrie des espaces de Riemann*, Paris, 1928, Chap. VII.

where, with the notations of Cartan's exterior calculus,[6] we have

$$(5)\quad \begin{cases} \Omega_1 = \omega_1' + [\pi_1\omega_1] + [\pi_3\omega_2], \\ \Omega_2 = \omega_2' + [\tau\omega_1] + [\pi_2\omega_2] + [\pi_3\omega_3], \\ \Omega_3 = \omega_3' + [\tau\omega_2] - [\pi_1\omega_3] + 2[\pi_2\omega_3], \\ \mathrm{T} = \tau' - [\pi_1\tau] + [\pi_2\tau], \\ \Pi_1 = \pi_1' + [\pi_3\tau], \\ \Pi_2 = \pi_2', \\ \Pi_3 = \pi_3' + [\pi_1\pi_3] - [\pi_2\pi_3], \end{cases}$$

so that T, Ω_i, Π_i ($i = 1, 2, 3$) are exterior quadratic differential forms in x, y, z, t.

Suppose the frame attached to each contact element be so chosen that M coincides with the point of the element and I_1, I_2 belong to the plane of the element. Then the contact element will be fixed, if and only if

$$\omega_1 = \omega_2 = \omega_3 = \tau = 0.$$

It follows that ω_1, ω_2, ω_3, τ are linear combinations of dx, dy, dz, dt and are themselves linearly independent. We can therefore put

$$(6)\quad \begin{cases} \Omega_i = P_{i1}[\omega_2\,\omega_3] + P_{i2}[\omega_3\,\omega_1] + P_{i3}[\omega_1\,\omega_2] + \sum_{k=1}^{3} Q_{ik}[\omega_k\,\tau], \qquad i = 1, 2, 3, \\ \Pi_i = R_{i1}[\omega_2\,\omega_3] + R_{i2}[\omega_3\,\omega_1] + R_{i3}[\omega_1\,\omega_2] + \sum_{k=1}^{3} S_{ik}[\omega_k\,\tau], \qquad i = 1, 2, 3, \\ \mathrm{T} = P_1[\omega_2\,\omega_3] + P_2[\omega_3\,\omega_1] + P_3[\omega_1\,\omega_2] + \sum_{k=1}^{3} Q_k[\omega_k\,\tau]. \end{cases}$$

The 42 coefficients in (6) are the components of the "tensor of curvature and torsion" of the space.

The Weyl space of elements as defined above may be studied from two different points of view. We may either regard the space as a genuine four-dimensional space (x, y, z, t) and study its properties invariant under the group (or, more precisely, the pseudo-group) of transformations

$$(7)\qquad \bar{x} = \bar{x}(x, y, z, t),\quad \bar{y} = \bar{y}(x, y, z, t),\quad \bar{z} = \bar{z}(x, y, z, t),\quad \bar{t} = \bar{t}(x, y, z, t),$$

or we may study the properties invariant under the group of transformations

$$(8)\qquad \bar{x} = \bar{x}(x, y, z),\qquad \bar{y} = \bar{y}(x, y, z),\qquad \bar{z} = \bar{z}(x, y, z),\qquad \bar{t} = \bar{t}(x, y, z, t).$$

[6] An introduction to the notions of exterior multiplication and exterior differentiation is given in: E. Cartan, *Leçons sur les invariants intégraux*, Paris, 1922.

For our purpose we shall restrict ourselves to the latter point of view and shall call a property *intrinsic*, if it is invariant under transformations of the form (8). In particular, it follows that the concept of a point in S is an intrinsic property.

2. The Two-Parameter Family of Surfaces and Its Intrinsic Weyl Geometry

Suppose there be given in the space S a family $\{\Sigma\}$ of surfaces depending on two parameters. We shall study the properties of $\{\Sigma\}$, which are not affected by an arbitrary transformation of coordinates

$$\bar{x} = \bar{x}(x, y, z), \qquad \bar{y} = \bar{y}(x, y, z), \qquad \bar{z} = \bar{z}(x, y, z), \tag{9}$$

To make this study it will be convenient to introduce the four-dimensional auxiliary space S' formed by the contact elements of the surfaces of the family $\{\Sigma\}$. By the use of a parameter t the contact elements of $\{\Sigma\}$ having the same origin (x, y, z) can be defined by an equation of the form

$$\theta \equiv A(x, y, z, t)dx + B(x, y, z, t)dy + C(x, y, z, t)dz = 0, \tag{10}$$

where A, B, C are not all zero. The parameter t serves to distinguish the contact elements through the point and is arbtrary to the degree that it may be submitted to a transformation of the form

$$\bar{t} = \bar{t}(x, y, z, t). \tag{11}$$

In the auxiliary space S' of our contact elements with the coordinates x, y, z, t, each surface Σ of the family is represented by a two-dimensional variety. All these two-dimensional variaties fill up the space S' in the sense that through every point (in a certain neighborhood) of S' there passes one, and only one, such variety. The family $\{\Sigma\}$ of surfaces can therefore be defined by a completely integrable Pfaffian system of the form

$$\theta = 0, \qquad \theta_1 \equiv A_1dx + B_1dy + C_1dz + Edt = 0, \qquad E \neq 0, \tag{12}$$

in the four variables x, y, z, t. The complete integrability of the system is expressed by the fact that the exterior derivatives θ', θ_1' are congruent to zero, mod. θ, θ_1.

By this representation of the family $\{\Sigma\}$ of surfaces in S as a family of surfaces in S' the properties of the original family unaffected by transformations of the form (9) are represented by properties of the corresponding family in S', which remain unaltered under the transformations (9), (11), and conversely. But these are exactly the intrinsic properties in S' in the sense defined in §1.

Suppose, without loss of generality, that $C \neq 0$, so that we may simplify the system (12) to the form

$$\begin{cases} \theta \equiv dz - pdx - qdy = 0, \\ \theta_1 \equiv dt - rdx - sdy = 0, \end{cases} \tag{13}$$

where p, q, r, s are functions of x, y, z, t. We introduce the notation

$$(14)\qquad \begin{cases} \dfrac{d}{dx} = \dfrac{\partial}{\partial x} + p\dfrac{\partial}{\partial z} + r\dfrac{\partial}{\partial t}, \\[2ex] \dfrac{d}{dy} = \dfrac{\partial}{\partial y} + q\dfrac{\partial}{\partial z} + s\dfrac{\partial}{\partial t}. \end{cases}$$

With this notation we have, for any function F in the variables x, y, z, t, the formula

$$(15)\qquad dF = \frac{dF}{dx}dx + \frac{dF}{dy}dy + \frac{\partial F}{\partial z}\theta + \frac{\partial F}{\partial t}\theta_1,$$

and the integrability conditions of the system (13) can be written in the following simple form:

$$(16)\qquad \frac{dp}{dy} - \frac{dq}{dx} = 0, \qquad \frac{dr}{dy} - \frac{ds}{dx} = 0.$$

We now proceed to prove that when a two-parameter family of surfaces (13) is given in S such that

$$(17)\qquad P \equiv p_t q_{tt} - q_t p_{tt} \neq 0,$$

there can be defined, in an intrinsic way, a Weyl geometry of elements in S. As it is easy to see, *the vanishing of P signifies that the contact elements through a point of S pass through a fixed direction.*

To each contact element formed by the point (x, y, z) and the plane $\theta = 0$ we attach a frame $MI_1I_2I_3$ *adapted to it* by supposing the following conditions to be satisfied:

a) M coincides with the point (x, y, z).

b) I_1 is along the line of intersection of the plane and a neighboring plane through M, i.e., the line defined by the equations

$$(18)\qquad \theta = 0, \qquad p_t dx + q_t dy = 0.$$

We shall call this line the *characteristic* of the element.

c) I_2 lies on the plane $\theta = 0$.

If $MI_1I_2I_3$ is such a frame, the most general frame $M^*I_1^*I_2^*I_3^*$ satisfying the same conditions is given by

$$(19)\qquad \begin{cases} M^* = M, \qquad I_1 = uI_1^*, \qquad I_2 = wI_2^* + \dfrac{uv}{w}I_1^*, \\[2ex] I_3 = \dfrac{w^2}{u}I_3^* + vI_2^* + \dfrac{1}{2}\dfrac{uv^2}{w^2}I_1^*, \end{cases}$$

where u, v, w are parameters, with $uw \neq 0$.

We now suppose our Weyl geometry to have the following two intrinsic properties:

1) Points are developed into points, i.e., the contact elements having the same origin are developed into the contact elements having the same origin.

2) The contact elements belonging to a fixed surface of the family are developed into contact elements having the same plane.

Let us express the conditions for these two properties analytically and see how far the Weyl geometry is thus defined. When the point M is fixed in space, we have

$$dx = dy = dz = 0.$$

Hence condition 1 is expressed by the relation

$$dM = 0 \pmod{dx, dy, dz},$$

i.e., ω_1, ω_2, ω_3 are linear combinations of dx, dy, θ. Similarly, condition 2 is expressed by the fact that ω_3, τ are linear combinations of θ and θ_1. It follows in particular that ω_3 is a multiple of θ. On the other hand, the condition b for the choice of the frame means that ω_2 is a linear combination of θ and $p_t\,dx + q_t\,dy$.

With a definite system of coordinates x, y, z, t we may make use of the parameters u, v, w in (19) to simplify the expressions for ω_1, ω_2, ω_3. It is easy to verify that we may choose the frame attached to each contact element such that we have

$$\begin{cases} \omega_2 = -(p_t\,dx + q_t\,dy) + \eta\theta, \\ \omega_3 = \theta, \end{cases} \tag{20}$$

where η remains undetermined. The most general change of frame leaving the conditions a, b, c and the form of the equations (20) unaltered is given by

$$I_1 = I_1^*, \qquad I_2 = I_2^* + vI_1^*, \qquad I_3 = I_3^* + vI_2^* + \tfrac{1}{2}v^2I_1^*, \tag{21}$$

v being a parameter.

In order to impose further conditions on the Weyl space, we are led to consider two particular kinds of cycles. The first kind is one whose directions d and δ displace the point of the contact element in the same direction and is analytically characterized by the conditions

$$\frac{\omega_1(d)}{\omega_1(\delta)} = \frac{\omega_2(d)}{\omega_2(\delta)} = \frac{\omega_3(d)}{\omega_3(\delta)},$$

or by

$$[\omega_1\omega_2] = [\omega_2\omega_3] = [\omega_3\omega_1] = 0. \tag{22}$$

The second may be described as one displacing the surface of the contact element in the same direction and will be defined by

$$\frac{\omega_3(d)}{\omega_3(\delta)} = \frac{\tau(d)}{\tau(\delta)},$$

or by

$$[\omega_3\tau] = 0. \tag{23}$$

The property of a cycle to be either one of the two kinds is intrinsic.

We now suppose our Weyl space to possess the following three further properties:

3) The infinitesimal transformation associated to every cycle displaces the point M in the direction of I_1.

4) The infinitesimal transformation associated to every cycle whose directions d and δ displace the surface of the contact element in the same direction leaves the point M invariant.

5) The infinitesimal transformation associated to every cycle whose directions d and δ displace the point of the contact element in the same direction leaves the plane MI_1I_2 invariant.

We shall prove that the conditions 1 to 5 completely define the Weyl space.

To prove this, it is sufficient to determine the seven Pfaffian forms in (2). In the meantime we can still normalize the frame (with respect to the coordinates x, y, z, t) by making use of the transformation (21). Before doing this, we remark that the property 3 is expressed analytically by the fact that ∇M is equal to a multiple of I_1, i.e., by the equations

$$(24)\qquad \begin{cases} \omega_2' + [\tau\omega_1] + [\pi_2\,\omega_2] + [\pi_3\,\omega_3] = 0, \\ \omega_3' + [\tau\omega_2] - [\pi_1\,\omega_3] + 2[\pi_2\,\omega_3] = 0. \end{cases}$$

When these relations are satisfied, the property 4 is characterized by the condition that $\Omega_1 = 0$ when (23) holds, i.e., by

$$(25)\qquad \omega_1' + [\pi_1\omega_1] + [\pi_3\omega_2] + a[\tau\omega_3] = 0,$$

where a remains undetermined. Similarly, the property 5 is given by an equation of the form

$$(26)\qquad \tau' - [\pi_1\tau] + [\pi_2\tau] - b[\omega_1\omega_3] - c[\omega_2\omega_3] = 0.$$

From the second equation of (24) we find, by making use of (20),

$$[(\tau + \theta_1)(p_t dx + q_t dy)] \equiv 0, \qquad \text{mod } \theta.$$

Since τ is a linear combination of θ and θ_1, it follows that τ is of the form

$$(27)\qquad \tau = -\theta_1 + \zeta\theta,$$

where ζ is undetermined. We calculate next the first equation of (24), mod ω_2, ω_3. We find then

$$[(\omega_1 + p_{tt}dx + q_{tt}dy)\theta_1] \equiv 0, \qquad \text{mod } \omega_2, \omega_3.$$

But ω_1 is a linear combination of dx, dy, θ only, so that it must be of the form

$$\omega_1 = -(p_{tt}dx + q_{tt}dy) + \sigma(p_t dx + q_t dy) + \xi\theta.$$

By applying the change of frame (21), we find that ω_1 is then changed into $\omega_1 + v\omega_2 + \frac{1}{2}v^2\theta$. We may therefore assume the frame so chosen that

$$(28)\qquad \omega_1 = -(p_{tt}dx + q_{tt}dy) + \xi\theta.$$

When ω_1, ω_2, ω_3 are of the forms given by (20), (28), the frame attached to each contact element is uniquely determined.

It now remains to determine the functions ξ, η, ζ and the Pfaffian forms π_1, π_2, π_3 from the conditions (24), (25), (26) and to show that they are thus completely determined. For this purpose we have to calculate the exterior derivatives of ω_1, ω_2, ω_3, τ. We remark that the calculation of every such exterior derivative is equivalent to the calculation of the exterior derivative of a Pfaffian form of the form

$$\omega = l\theta - m\,dx - n\,dy, \tag{29}$$

where l, m, n are functions of x, y, z, t. We find, by making use of (20), (27), (28),

$$(30)\quad \left\{\begin{aligned} P\omega' = &-\left(\frac{dm}{dy}-\frac{dn}{dx}\right)[\omega_1\omega_2] + \left\{\left(\frac{dm}{dy}-\frac{dn}{dx}\right)\eta + (m_t q_t - n_t p_t)\zeta\right.\\ &\left.+\left(\frac{dl}{dx}q_t - \frac{dl}{dy}p_t\right) + (m_z q_t - n_z p_t) + l(p_r q_t - q_z p_t)\right\}[\omega_1\omega_3]\\ &+\left\{-\left(\frac{dm}{dy}-\frac{dn}{dx}\right)\xi + (-m_t q_{tt} + n_t p_{tt} - lP)\zeta + \left(-\frac{dl}{dx}q_{tt} + \frac{dl}{dy}p_{tt}\right)\right.\\ &\left.+(-m_z q_{tt} + n_z p_{tt}) + l(-p_z q_{tt} + q_z p_{tt})\right\}[\omega_2\omega_3]\\ &+(n_t p_t - m_t q_t)[\omega_1\tau] + (lP + m_t q_{tt} - n_t p_{tt})[\omega_2\tau]\\ &+\{(m_t q_t - n_t p_t)\xi + (-m_t q_{tt} + n_t p_{tt} - lP)\,\eta + l_t P\}[\omega_3\tau]. \end{aligned}\right.$$

On substituting into (24), (25), (26), for ω_1', ω_2', ω_3', τ their expressions in terms of the exterior products of ω_1, ω_2, ω_3, τ, we shall get relations of the form

$$(31)\quad \left\{\begin{aligned} -[\pi_1\omega_1] - [\pi_3\omega_2] &= \lambda_1[\omega_1\omega_2] + \lambda_2[\omega_1\omega_3] + \lambda_3[\omega_2\omega_3] + \lambda_4[\omega_1\tau] + \lambda_5[\omega_2\tau],\\ -[\pi_2\omega_2] - [\pi_3\omega_3] &= \mu_1[\omega_1\omega_2] + \mu_2[\omega_1\omega_3] + \mu_3[\omega_2\omega_3] + \mu_4[\omega_2\tau] + \mu_5[\omega_3\tau],\\ [\pi_1\omega_3] - 2[\pi_2\omega_3] &= \nu_1[\omega_1\omega_3] + \nu_2[\omega_2\omega_3] + \nu_3[\omega_3\tau],\\ [\pi_1\tau] - [\pi_2\tau] &= \rho_1[\omega_1\tau] + \rho_2[\omega_2\tau] + \rho_3[\omega_3\tau], \end{aligned}\right.$$

where we have, in particular,

$$(32)\quad \left\{\begin{aligned} &\lambda_1 = \frac{1}{P}\left(\frac{dq_{tt}}{dx} - \frac{dP_{tt}}{dy}\right), \qquad \lambda_4 = \frac{1}{P}(p_t q_{ttt} - q_t p_{ttt}),\\ &\qquad\qquad\qquad\qquad \lambda_5 = \xi - \frac{1}{P}(p_{tt} q_{ttt} - q_{tt} p_{ttt}),\\ &\mu_1 = \frac{1}{P}\left(\frac{dq_t}{dx} - \frac{dp_t}{dy}\right), \qquad \mu_4 = \eta, \qquad \mu_5 = -\xi - \eta^2 + \eta_t,\\ &\mu_2 = \frac{1}{P}\left(\frac{d\eta}{dx}q_t - \frac{d\eta}{dy}p_t\right) + \frac{\eta}{P}\left(\frac{dp_t}{dy} - \frac{dq_t}{dx} + p_z q_t - q_z p_t\right) - \zeta\\ &\qquad\qquad + \frac{1}{P}(-p_t q_{tz} + q_t p_{tz}),\\ &\nu_1 = \frac{1}{P}(p_z q_t - q_z p_t), \qquad \nu_2 = \frac{1}{P}(-p_z q_{tt} + q_z p_{tt}) - \zeta, \qquad \nu_3 = -\eta,\\ &\rho_1 = \frac{1}{P}(r_t q_t - s_t p_t), \qquad \rho_2 = \zeta + \frac{1}{P}(-r_t q_{tt} + s_t p_{tt}). \end{aligned}\right.$$

In order that the system of equations (31) have a set of solutions for π_1, π_2, π_3 it is necessary and sufficient that the following relations hold:

$$(33)\qquad \begin{cases} \lambda_1 - \mu_2 + \nu_2 - 2\rho_2 = 0, & \lambda_4 + \nu_3 - 2\mu_4 = 0, \\ \lambda_5 - \mu_5 = 0, & \mu_1 - \nu_1 + \rho_1 = 0. \end{cases}$$

The set of solutions is then given by

$$(34)\qquad \begin{cases} \pi_1 = (-\nu_1 + 2\rho_1)\omega_1 + (-\nu_2 + 2\rho_2)\omega_2 + \lambda_2\omega_3 + (-\nu_3 + 2\mu_4)\tau, \\ \pi_2 = (-\nu_1 + \rho_1)\omega_1 + (-\nu_2 + \rho_2)\omega_2 + (\lambda_2 - \rho_3)\omega_3 + \mu_4\tau, \\ \pi_3 = -\mu_2\omega_1 + (\lambda_2 - \mu_3 - \rho_3)\omega_2 + \lambda_3\omega_3 + \mu_5\tau, \end{cases}$$

and is uniquely determined. Of the relations (33) the last one is identically satisfied, while the other three equations determine ξ, η, ζ. The Weyl connection is thus completely determined.

On summing up our results, we get the theorem:

Suppose that there is given in the space (x, y, z) a two-parameter family of surfaces such that the tangent planes to the surfaces through a point do not pass through a fixed direction. Then there can be defined in the space, in an intrinsic way, one and only one Weyl connection of elements having the following properties:

1. *Points are developed into points.*
2. *Surfaces of the family are developed into planes.*
3. *The infinitesimal transformation associated to every cycle displaces the point of the element in the direction of the characteristic of the element.*
4. *The infinitesimal transformation associated to every cycle by which the directions d and δ displace the surface of the element in the same "direction" leaves the element invariant.*
5. *The infinitesimal transformation associated to every cycle by which the directions d and δ displace the origin of the element in the same direction leaves the plane of the element invariant.*

3. The Problem of Equivalence

In the study of the geometry of a two-parameter family of surfaces in space under the group of transformations (9) the fundamental problem is the following: Given a two-parameter family $\{\Sigma\}$ of surfaces in S and another such family $\{\bar{\Sigma}\}$ in a space $\bar{S}$ with the coordinates $\bar{x}, \bar{y}, \bar{z}$. When are they *equivalent*, i.e., when can the one be carried into the other by a transformation (9)? We shall show in this section that the solution of this problem follows as a consequence of the theorem of the last section, e.g., the possibility of defining an intrinsic Weyl connection in the space.

Before discussing this so-called "problem of equivalence," we shall give some further properties of the Weyl geometry defined in the last section. Our choice of the frames in §2 adapted to each contact element made use of the coordinates of the space. If we suppose only that the point M of the frame $MI_1I_2I_3$ coincides with the point of the contact element and that the plane MI_1I_2 with the plane of the element, the frame attached to each contact element will depend on three parameters and the relation between two frames attached to the same element will be given by (19). By taking the parameters u, v, w in (19) as independent variables, we get a family of frames depending on the seven parameters x, y, z, t, u, v, w. Then the seven Pfaffian forms ω_1, ω_2, ω_3, τ, π_1, π_2, π_3 are linearly independent and it is easy to verify that they satisfy the system of equations (5), (6), which are called *the equations of structure of the Weyl space*. For this general family of frames the conditions

$$\Omega_1 = a[\omega_3\tau], \qquad \Omega_2 = \Omega_3 = 0, \qquad \mathrm{T} = b[\omega_1\omega_3] + c[\omega_2\omega_3] \tag{35}$$

are still satisfied, because they have an intrinsic geometric meaning.

On account of these forms of the expressions for Ω_1, Ω_2, Ω_3, T the expressions for Π_1, Π_2, Π_3 can be simplified. In fact, by applying to the first four equations of (5) the theorem that the exterior derivative of the exterior derivative of a Pfaffian form is zero (the so-called "Poincaré's Theorem") and noticing that the resulting equations are identically satisfied if the expressions in (6) are zero, we get

$$\begin{cases} \Omega_1' = [\Pi_1\omega_1] + [\Pi_3\omega_2] - [\Omega_1\pi_1] - [\Omega_2\pi_3], \\ \Omega_2' = [\mathrm{T}\omega_1] + [\Pi_2\omega_2] + [\Pi_3\omega_3] - [\Omega_1\tau] - [\Omega_2\pi_2] - [\Omega_3\pi_3], \\ \Omega_3' = [\mathrm{T}\omega_2] - [\Pi_1\omega_3] + 2[\Pi_2\omega_3] - [\Omega_2\tau] + [\Omega_3\pi_1] - 2[\Omega_3\pi_2], \\ \mathrm{T}' = [\mathrm{T}\pi_1] - [\mathrm{T}\pi_2] - [\Pi_1\tau] + [\Pi_2\tau]. \end{cases} \tag{36}$$

These are some of the "Bianchi Identities" of the Weyl space. When the conditions (35) are satisfied, these equations can be simplified to the form

$$\begin{cases} [\Pi_1\omega_1] + [\Pi_3\omega_2] - [(da + 3a\pi_1 - 3a\pi_2)\omega_3\tau] = 0, \\ [\Pi_2\omega_2] + [\Pi_3\omega_3] + c[\omega_1\omega_2\omega_3] = 0, \\ [\Pi_1\omega_3] - 2[\Pi_2\omega_3] + b[\omega_1\omega_2\omega_3] = 0, \\ [\Pi_1\tau] - [\Pi_2\tau] - b[\omega_1\omega_2\tau] + [(b\omega_1 + c\omega_2)'\omega_3] + [(b\omega_1 + c\omega_2)\pi_2\omega_3] = 0. \end{cases} \tag{37}$$

From the third equation we see that every term of $\Pi_1 - 2\Pi_2 + b[\omega_1\omega_2]$ must contain ω_3. On the other hand, we get, by multiplying the fourth equation by ω_3,

$$[(\Pi_1 - \Pi_2 - b\omega_1\omega_2)\tau\omega_3] = 0.$$

Similarly, the second and the third equations give respectively

$$[\Pi_2\omega_2\omega_3] = 0, \qquad [(\Pi_1 - 2\Pi_2)\omega_2\omega_3] = 0,$$

from which it follows that

$$[(\Pi_1 - \Pi_2 - b\omega_1\omega_2)\omega_2\omega_3] = 0.$$

Again, by multiplying the right-hand sides of the first equation by ω_3, the second by ω_2, the third by $-\omega_1$, and adding, we get

$$[(\Pi_1 - \Pi_2 - b\omega_1\omega_2)\omega_1\omega_3] = 0.$$

It follows from the three equations that every term of $\Pi_1 - \Pi_2 - b[\omega_1\omega_2]$ must contain ω_3. We can therefore put

$$(38) \qquad \begin{cases} \Pi_1 = 3b[\omega_1\omega_2] + e[\omega_1\omega_3] + f[\omega_2\omega_3] + g[\omega_3\tau], \\ \Pi_2 = 2b[\omega_1\omega_2] + h[\omega_1\omega_3] + i[\omega_2\omega_3] + j[\omega_3\tau]. \end{cases}$$

From (37) it then follows that Π_3 is of the form

$$(39) \qquad \Pi_3 = (h - c)[\omega_1\omega_2] + f[\omega_1\omega_3] + j[\omega_2\tau] + k[\omega_2\omega_3] + l[\omega_3\tau].$$

Consequently, for our Weyl space, the seven Pfaffian forms ω_1, ω_2, ω_3, τ, π_1, π_2, π_3 satisfy the equations of structure (5), with Ω_1, Ω_2, Ω_3, T, Π_1, Π_2, Π_3 given by (35), (38), (39). In these equations there are introduced 11 quantities $a, b, c, e, f, g, h, i, j, k, l$, which are functions of x, y, z, t, u, v, w and which constitute the tensor of curvature and torsion of the space.

When two such Weyl spaces of elements are given, with the coordinates (x, y, z, t) and $(\bar{x}, \bar{y}, \bar{z}, \bar{t})$, we say that they are equivalent, if to every family of frames of the one (with one frame attached to each contact element), there is a corresponding family of frames of the other, such that a transformation of the form (7) exists, which satisfies the relations

$$(40) \qquad \bar{\omega}_i = \omega_i, \qquad \bar{\tau} = \tau, \qquad \bar{\pi}_i = \pi_i, \qquad i = 1, 2, 3.$$

If we attach to each contact element (x, y, z, t) the most general family of frames depending on three parameters u, v, w and to $(\bar{x}, \bar{y}, \bar{z}, \bar{t})$ the family of frames with the parameters $\bar{u}, \bar{v}, \bar{w}$, it is evident that the two Weyl spaces are equivalent, when and only when there exists a transformation of the form

$$(41) \qquad \begin{cases} \bar{x} = \bar{x}(x, y, z, t, u, v, w), \cdots, & \bar{t} = \bar{t}(x, y, z, t, u, v, w), \\ \bar{u} = \bar{u}(x, y, z, t, u, v, w), \cdots, & \bar{w} = \bar{w}(x, y, z, t, u, v, w), \end{cases}$$

satisfying the relations (40).

With these preparations we now establish the theorem:

There exists a point transformation of the form (9) *carrying one family of surfaces* $\{\Sigma\}$ *into another* $\{\bar{\Sigma}\}$ *when and only when the two corresponding Weyl spaces of elements are equivalent.*

The proof of this theorem is immediate. In fact, when the two families of surfaces are equivalent, the corresponding Weyl spaces are equivalent, since our definition of the Weyl space is intrinsic. Conversely, when the Weyl spaces are equivalent, the transformation (41) will realize the equality of the sets of Pfaffian forms in (40). From the first three equations in (40), we see that in the transformation (41) the coordinates $\bar{x}, \bar{y}, \bar{z}$ are functions of x, y, z alone, so that the transformation in question is a point transformation. From

$$\bar{\omega}_3 = \omega_3, \qquad \bar{\tau} = \tau$$

we conclude that the transformation carries $\{\Sigma\}$ to $\{\bar{\Sigma}\}$. Hence the theorem is proved.

Analytically this result may be formulated as follows: *To solve the problem of equivalence for a two-parameter family of surfaces (whose tangent planes at a point do not pass through a fixed direction) in a three-dimensional space* (x, y, z), *we introduce four auxiliary variables* t, u, v, w *and seven linearly independent Pfaffian forms* ω_i, τ, π_i $(i = 1, 2, 3)$ *in the coordinates and in these variables. All these Pfaffian forms are invariant in the sense that the two families of surfaces* $\{\Sigma\}$ *and* $\{\bar{\Sigma}\}$ *are equivalent, when and only when there exists a transformation of the form* (41) *satisfying* (40).

Since the process of exterior differentiation is invariant under the point transformations (41), it follows that the 11 quantities $a, \cdots, l$ in (35), (38), (39) are invariants. We shall call them the *fundamental invariants* of the family $\{\Sigma\}$. From a given invariant F the equation

$$dF = F_1\omega_1 + F_2\omega_2 + F_3\omega_3 + F_0\tau + F_1'\pi_1 + F_2'\pi_2 + F_3'\pi_3 \tag{42}$$

defines the new invariants $F_1, F_2, F_3, F_0, F_1', F_2', F_3'$, called the *covariant derivatives* of F. According to a theorem of E. Cartan,[7] *the complete set of invariants of the family of surfaces consists of the fundamental invariants and their successive covariant derivatives.* This is to be understood as follows: By eliminating the independent variables x, y, z, t, u, v, w, we may set up relations of the form

$$\Phi(F_1, \cdots, F_m) = 0 \tag{43}$$

between the invariants $F_1, \cdots, F_m$ of our complete set. *Then the two families of surfaces* $\{\Sigma\}$ *and* $\{\bar{\Sigma}\}$ *are equivalent if and only if the relations* (43) *and*

$$\Phi(\bar{F}_1, \cdots, \bar{F}_m) = 0$$

hold at the same time for $F_1, \cdots, F_m$ *and for the corresponding invariants* $\bar{F}_1, \cdots, \bar{F}_m$ *of* $\{\bar{\Sigma}\}$. Moreover, it is also well-known that the equivalence of $\{\Sigma\}$ and $\{\bar{\Sigma}\}$ can be decided by only a finite number of such relations (43).

4. The Weyl Space of Points

An important particular case of our geometry is the case by which the infinitesimal transformation associated to every cycle depends only on the displacements of the origins of the contact elements. The space of elements is

[7] E. Cartan, *Les sousgroupes des groupes continus de transformations*, Annales de l'Ecole Normale Supérieure, série 3, t. 25 (1908), pp. 57–194.

then identical with the three-dimensional point space (x, y, z) with a Weyl connection. The conditions for this are, from (35), (38), (39),

$$a = g = j = l = 0.$$

These conditions are, however, not independent. In fact, when $a = 0$, the first equation of (37) will give

$$g = l = 0.$$

It follows that *the conditions*

$$a = j = 0 \tag{44}$$

are the necessary and sufficient conditions for the Weyl space to be a Weyl space of points.

We may give a geometrical interpretation to the condition $a = 0$. In fact, *it signifies that the tangent planes to the surfaces of the family through a fixed point in space envelop a cone of the second order.* To prove this, let δ be an operation under which the point remains fixed, so that

$$\omega_1(\delta) = \omega_2(\delta) = \omega_3(\delta) = 0.$$

Since t is now the only variable, we may determine the auxiliary variables u, v, w as functions of t such that the conditions

$$\pi_1(\delta) = \pi_2(\delta) = \pi_3(\delta) = 0$$

are also satisfied. We put $\tau(\delta) = e$, and we can assume, by changing the parameter t when necessary, that $\delta e = 0$.

If δ is an operation so chosen and d is an operation by which all the variables vary, we get, from (5) and (35),

$$\begin{cases} \delta\omega_1 = -ae\omega_3, \\ \delta\omega_2 = -e\omega_1, \\ \delta\omega_3 = -e\omega_2. \end{cases} \tag{45}$$

From (45) we get

$$\begin{cases} \delta^2\omega_3 = e^2\omega_1, \\ \delta^3\omega_3 = -ae^3\omega_3, \\ \delta^4\omega_3 = ae^4\omega_2 + (\cdots)\omega_3, \\ \delta^5\omega_3 = -ae^5\omega_1 + (\cdots)\omega_2 + (\cdots)\omega_3. \end{cases} \tag{46}$$

If we define the "plane coordinates" l_1, l_2, l_3 of a plane through the point by the relation

$$l_1\omega_1 + l_2\omega_2 + l_3\omega_3 = 0, \tag{47}$$

we shall get, when the coordinates l_1, l_2, l_3 of the tangent plane of the family are expanded in powers of e,

$$(48)\qquad \begin{cases} l_1 = \frac{1}{2}e^2 - \frac{1}{120}ae^5 + \cdots, \\ l_2 = -e + \frac{1}{24}ae^4 + \cdots, \\ l_3 = 1 - \frac{1}{6}ae^3 + \cdots. \end{cases}$$

From (48) it follows that

$$(49)\qquad l_2^2 - 2l_1 l_3 = \frac{1}{10}ae^5 + \cdots.$$

Hence the cone

$$(50)\qquad l_2^2 - 2l_1l_3 = 0$$

is the osculating quadric cone of the cone enveloped by the tangent planes of the family $\{\Sigma\}$ and the condition $a = 0$ is a necessary and sufficient condition that these tangent planes themselves envelop a quadric cone.

Added October 10, 1941. The author has succeeded in extending the results of this paper to a space of $n(\geqq 3)$ dimensions. The following theorem has been established: *Given in a space of n dimensions a family of hypersurfaces depending on $n - 1$ parameters such that the tangent hyperplanes at a point to the hypersurfaces of the family through the point envelop a hypercone of $n - 1$ dimensions. Then it is possible to define in the space, in an intrinsic way, a Weyl geometry.* The elements of this Weyl space are in general more complicated for $n \geqq 4$ than for $n = 3$. The problem of equivalence is solved as a consequence of this result.

NATIONAL TSING HUA UNIVERSITY
KUNMING, CHINA

On a Weyl Geometry Defined from an $(n-1)$ Parameter Family of Hypersurfaces in a Space of n Dimensions

By SHIING-SHEN CHERN (陳省身)
Department of Mathematics, National Tsing Hua University
(Received September 15, 1941)

(*from Science Record Vol. 1. Nos. 1–2, pp. 7–10. August 1942*)

A generalized space in the sense of Cartan is an n-dimensional number space with a space of Klein attached to each point and with an infinitesimal transformation of a given Lie group G, which allows us to map the Klein spaces attached to two neighboring points on each other. According as the group G is the projective group, the group of similarity transformations, or any other Lie group, we have a generalized projective geometry, a Weyl geometry, or any other generalized geometry. From this point of view the fundamental problem of modern differential geometry is to define from a given geometric object (in the sense of Veblen) an intrinsic generalized space. It is in this sense that the geometry of paths of the Princeton school is a generalized projective geometry. An important generalization of the geometry of paths was recently made by M. Hachtroudi[1], who showed that the geometry in a space of n-dimensions with an n-parameter family of hypersurfaces is also a generalized projective geometry. We shall prove in this note that the geometry in a space of $n(\geq 3)$ dimensions with an $(n-1)$ parameter family of hypersurfaces is a Weyl geometry.

Let $x^1, \ldots, x^{n-1}, z$ be the coordinates of the space. By the use of $n-2$ auxiliary variables $t_1, \ldots, t_{n-2}$ a family of hypersurfaces in the space depending on $n-1$ parameters can be defined by a completely integrable Pfaffian system of the form*

$$\text{(1)} \qquad \begin{cases} \theta \equiv dz - p_1\, dx^1 - \cdots - p_{n-1}\, dx^{n-1} = 0, \\ \theta_\alpha \equiv dt_\alpha - r_{\alpha 1}\, dx^1 - \cdots - r_{\alpha, n-1}\, dx^{n-1} = 0, \end{cases}$$

where pi, $r_{\alpha i}$ are functions of x^i, z, t_α. For any function F in x^i, z, t_α we define

$$\text{(2)} \qquad \frac{dF}{dx^i} = \frac{\partial F}{\partial x^i} + \frac{\partial F}{\partial z} p_i + \frac{\partial F}{\partial t_\alpha} r_{\alpha i}.$$

The integrability conditions of the system can then be written

$$\text{(3)} \qquad \frac{dp_i}{dx^j} = \frac{dp_j}{dx^i}, \qquad \frac{dr_{\alpha i}}{dx^j} = \frac{dr_{\alpha j}}{dx^i}.$$

* We agree, throughout this paper, that Latin indices run from 1 to $n-1$ and Greek indices from 1 to $n-2$.

To define a Weyl geometry which is invariant under point transformations in the space is to introduce

$$\left\{\frac{n(n+1)}{2}+1\right\}-(2n-2)=\frac{n^2-3n+6}{2}$$

further auxiliary variables and a set of $\frac{n(n+1)}{2}+1$ Pfaffian forms, uniquely determined, such that they fulfill the equations of structure of a Weyl space. Of these Pfaffian forms $2n-2$ are of the form

$$(4)\qquad \begin{cases} \omega^i = u^i_j\,dx^j + v^i\theta. \\ \omega = u\theta. \\ \pi_\alpha = \omega_\alpha\theta + \xi^\beta_\alpha\theta_\beta \end{cases}$$

where u^i_j, u, v^i, ω_α ξ^α_β are arbitrary and some of them are to be determined by the properties of the Weyl space in question. The further discussion of the problem depends on imposing on the Pfaffian forms in (4) some conditions of intrinsic nature and this is achieved by forming the exterior derivatives of these Pfaffian forms.

First of all, by introducing a Pfaffian form π, determined up to an additive term in ω, we may assume that the following equation holds:

$$(5)\qquad \omega = [\pi\omega] + [\omega^\alpha\pi_\alpha],$$

which implies the conditions

$$(6)\qquad \xi^\alpha_\beta u^\beta_i = \frac{\partial p_i}{\partial t_\alpha},$$

and the number of the auxiliary variables in (4) is thus reduced.

Next, let v^i_j and η^α_β be defined by the relations

$$(7)\qquad \begin{cases} u^i_j v^j_k = v^i_j u^j_k = \delta^i_k, \\ \xi^\alpha_\beta\eta^\beta_\gamma = \eta^\alpha_\beta\xi^\beta_\gamma = \delta^\alpha_\gamma, \end{cases}$$

where δ^i_k, δ^α_γ are the Kronecker symbols. Then, from (6), we have

$$(8)\qquad \frac{\partial p_i}{\partial t_\alpha} v^i_{n-1} = 0.$$

The coefficient of $[\pi_\beta\omega^{n-1}]$ in $(\omega^\alpha)'$ is then found to be

$$u\eta^\alpha_\rho\eta^\beta_\sigma\frac{\partial^2 p_i}{\partial t_\rho\partial t_\sigma}v^i_{n-1}.$$

In order to make this coefficient equal to $\delta^{\alpha\beta}$ it is necessary to assume that

$$(9)\qquad \left|\frac{\partial^2 p_i}{\partial t_\rho\partial t_\sigma}v^i_{n-1}\right| \neq 0,$$

which condition we suppose to be satisfied. By proceeding in this manner, we can determine v^i and ω_α. The remaining auxiliary variables in (4) are then $(n^2-3n+6)/2$

in number and we can determine one and only one set of $(n^2 - 3n + 6)/2$ Pfaffian forms, which, together with the Pfaffian forms in (4), satisfy the equations of structure of a Weyl space. This is the way that we define the Weyl geometry in the space. It is evident that the definition is intrinsic.

To this problem there is associated a problem of projective differential geometry. In fact, the tangent hyperplanes at a point P of the hypersurfaces of the family through the point depend on $n-2$ parameters. Since the group of point transformations in space induces a projective group on the directions through P, all projective properties of the family of tangent hyperplanes through P are intrinsic properties of the given family of hypersurfaces. We may cut the tangent hyperplanes by a hyperplane not through P, obtaining thus in a projective space of $n-1$ dimensions a family of hyperplanes depending on $n-2$ parameters, whose projective geometry forms therefore a part of our problem. With this geometrical interpretation it may be shown that the condition (9) means that this $(n-2)$-parameter family of hyperplanes envelops a hypersurface (i.e., a point variety of $n-2$ dimensions).

The general problem which we have in view is the geometry of an m-parameter family of hypersurfaces in an n-dimensional space. As mentioned above, the case $m = n$ was solved by M. Hachtroudi and the case $m = n - 1$ in this note. It is evident that the case $m = n + k$ can be reduced to the case $m = n - k$, so that the next problem in question is one for which $m = n - 2$. When $n = 4$, $m = 2$, i.e., with a two-parameter family of hypersurfaces in four-dimensional space, we are able to show that, when a certain relative invariant vanishes, it is possible to define in the space a generalized geometry having a certain 8-parameter group for fundamental group. We hope to consider this case in a later paper.

The above results for $n = 3$ will be published in a paper in the Annals of Mathematics.

REFERENCES

1. M. Hachtroudi, *Les espaces d'éléments à connexion projective normale, Actualités Scientifiques et Industrielles, no. 565*, Paris 1937.

Reset from
Acad. Scinica Sci. Record **1** (1942) 7–10

Reprinted from the Proceedings of the NATIONAL ACADEMY OF SCIENCES,
Vol. 29, No. 1, pp. 33–37. January, 1943

ON THE EUCLIDEAN CONNECTIONS IN A FINSLER SPACE

BY SHIING-SHEN CHERN

TSING HUA UNIVERSITY AND ACADEMIA SINICA

Communicated November 2, 1942

The generalization of the parallelism of Levi-Civita in a Riemann space to a Finsler space has been regarded as one of the most important problems of Finslerian Geometry. For its solution different suggestions were made by J. L. Synge,[1] J. H. Taylor,[2] L. Berwald[3] and E. Cartan.[4] In this note we shall study the problem by employing a different method—the method of equivalence. We shall prove that in a general Finsler space an infinite number of Euclidean connections can be defined, in which the connections defined by other authors are included as particular cases.

Let x^i[5] be the coordinates of an n-dimensional Finsler space, whose fundamental integral is

$$s = \int_{t_0}^{t_1} F\left(x^i, \frac{dx^i}{dt}\right) dt, \tag{1}$$

where F is positively homogeneous of the first order in the last n arguments:

$$F(x^i, \lambda y^i) = \lambda F(x^i, y^i),\ y^i = \frac{dx^i}{dt}, \lambda > 0. \tag{2}$$

It is well known that the Pfaffian form

$$\omega = \frac{\partial F}{\partial y^i} dx^i \tag{3}$$

is invariant (under a general point transformation). We adjoin to the coordinates x^i of the space $n(n-1)$ auxiliary variables v_k^α, subjected to the conditions

$$v_k^\alpha y^k = 0, \tag{4}$$

and put

$$v_k^n = \frac{\partial F}{\partial y^k}. \tag{5}$$

Then in the space of all the variables x^i, y^i, v_k^α we have n linearly independent invariant Pfaffian forms, namely,

$$\omega^i = v_k^i dx^k, \tag{6}$$

where $\omega^n = \omega$. It is from the Pfaffian forms ω^i that we shall develop our invariant theory of Finsler spaces, from which the Euclidean connections in the space are derived as consequences.

We introduce the elements u^i_k of the inverse matrix of (v^i_k), so that we have

$$u^i_j v^j_k = v^i_j u^j_k = \delta^i_k. \tag{7}$$

The elements u^k_n do not depend on the auxiliary variables, since a comparison of equations (4), (5), (7) gives

$$u^k_n = \frac{1}{F} y^k. \tag{8}$$

If we form the exterior derivative of ω^n, we see that we can write

$$(\omega^n)' = [\omega^\alpha \omega_\alpha{}^n], \tag{9}$$

where

$$\omega_{\alpha}{}^{n} = -u^k_\alpha \frac{\partial^2 F}{\partial y^j \partial y^k} dy^j + \frac{1}{F} u^j_\alpha \left(\frac{\partial F}{\partial x^j} - \frac{\partial^2 F}{\partial x^k \partial y^j} y^k \right) \omega^n +$$

$$u^j_\alpha u^k_\beta \frac{\partial^2 F}{\partial x^j \partial y^k} \omega^\beta + \lambda_{\alpha\beta}\omega^\beta, \qquad \lambda_{\alpha\beta} = \lambda_{\beta\alpha}, \tag{10}$$

the $n(n-1)/2$ quantities $\lambda_{\alpha\beta}$ being arbitrary.

We shall suppose that the fundamental integral (1) leads to a "regular problem" of the calculus of variations. As is well known, this amounts to assuming that the matrix

$$\left(\frac{\partial^2 F}{\partial y^i \partial y^j} \right)$$

is of rank $n-1$, or, in our notation, that the Pfaffian forms ω^i, ω^n_α. are linearly independent.

If we form the exterior derivative of ω^α, we see that the following invariant conditions can be imposed:

$$(\omega^\alpha)' \equiv \delta^{\alpha\beta}[\omega_\beta{}^n \omega^n], \text{ mod. } \omega^\gamma, \tag{11}$$

where $\delta^{\alpha\beta}$ is Kronecker's symbol. The conditions (11) are equivalent to

$$u^j_\alpha u^k_\beta \left(F \frac{\partial^2 F}{\partial y^j \partial y^k} \right) = \delta_{\alpha\beta}, \tag{12}$$

under which the variables u^i_α are not all independent. By carrying out the calculation of $(\omega^\alpha)'$, we get

$$(\omega^\alpha)' - \delta^{\alpha\beta}[\omega_\beta{}^n \omega^n] = [\omega^\beta \omega_\beta{}^\alpha], \tag{13}$$

where

$$\omega_\beta{}^\alpha = v^\alpha_k du^k_\beta - \delta^{\alpha\gamma} \left(u^j_\gamma u^k_\beta \frac{\partial^2 F}{\partial x^j \partial y^k} + \lambda_{\beta\gamma} \right) \omega^n + \mu^\alpha_{\beta\gamma} \omega^\gamma, \tag{14}$$

the quantities $\mu^{\alpha}_{\beta\gamma}$ being symmetric in β, γ and being introduced to get the most general expression for $\omega^{\beta\alpha}$.

Since the variables u^i_α are connected by the relations (12), the Pfaffian forms $\omega_{\beta\cdot}^{\alpha}$ are not linearly independent. In fact, we find that they satisfy the relations

$$\omega_{\alpha\beta} + \omega_{\beta\alpha} \equiv 0, \text{ mod. } \omega^i, \ \omega_{\alpha\cdot}^{n},$$

where the quantities $\delta_{\alpha\beta}$ are used to raise and lower indices, thus

$$\omega_{\alpha\beta} = \delta_{\beta\gamma}\omega_{\alpha\cdot}^{\gamma}, \ \omega_{\alpha}^{\ \beta} = \delta_{\alpha}^{\ \gamma}\omega_{\gamma\cdot}^{\beta}, \text{ etc.} \tag{15}$$

Our fundamental result is that we can, by a proper choice of $\lambda_{\alpha\beta}$, $\mu^{\alpha}_{\beta\gamma}$, arrive at the conditions

$$\omega_{\alpha\beta} + \omega_{\beta\alpha} = H_{\alpha\beta\cdot}^{\ \ \gamma}\omega_{\gamma\cdot}^{n}, \tag{16}$$

and that under these conditions the quantities $\lambda_{\alpha\beta}$, $\mu^{\alpha}_{\beta\gamma}$, and hence the Pfaffian forms $\omega_{\alpha\cdot}^{n}$, $\omega_{\beta\cdot}^{\alpha}$ are completely determined. These Pfaffian forms are therefore invariant Pfaffian forms. The quantities $H_{\alpha\beta\gamma}$ constitute the first set of invariants of the Finsler space. Their vanishing signifies that the Finsler space is a Riemann space.

The expressions for the exterior derivatives of $\omega_{\beta\cdot}^{\alpha}$, $\omega_{\alpha\cdot}^{n}$ will lead to further invariants of the space. To find them we write the equations (9), (13), (16) in the condensed form

$$\left.\begin{aligned} (\omega^i)' &= [\omega^j\omega_{j\cdot}^{i}], \\ \omega_{ij} + \omega_{ji} &= H_{ijk}\omega^{kn}, \end{aligned}\right\} \tag{17}$$

with the understanding that H_{ijk} is zero when any one of its indices is n. Putting

$$\Omega_{k\cdot}^{i} = (\omega_{k\cdot}^{i})' - [\omega_{k\cdot}^{j}\omega_{j\cdot}^{i}], \tag{18}$$

we find, by simply applying the "theorem of Poincaré" that the exterior derivative of the exterior derivative of a Pfaffian form is zero, that $\Omega_{k\cdot}^{i}$ are of the form

$$\Omega_{k\cdot}^{i} = R_{k\cdot jl}^{\ i}[\omega^l\omega^j] + P_{k\cdot j\cdot}^{\ i\ \alpha}[\omega_{\alpha\cdot}^{n}\omega^j], \tag{19}$$

where

$$R_{k\cdot jl}^{\ i} + R_{k\cdot lj}^{\ i} = 0 \tag{20}$$

For a function F in our variables its differential dF is a linear combination of ω^i, $\omega_{j\cdot}^{i}$, the coefficients of the linear combination being the "covariant derivatives." The invariants H_{ijk}, $R_{k\cdot jl}^{\ i}$, $P_{k\cdot j\cdot}^{\ i\ \alpha}$ and their covariant derivatives form a complete system of invariants in the sense that they are sufficient to determine a Finsler space up to a point transformation.

Now we shall enter into the geometrical interpretation of our results. For this purpose we put

$$\pi_{ij} = \omega_{ij} + \gamma_{ij\alpha}\omega^{\alpha n}, \tag{21}$$

and impose the conditions

$$\gamma_{ij\alpha} + \gamma_{ji\alpha} = -H_{ij\alpha}, \tag{22}$$

in order to have

$$\pi_{ij} + \pi_{ji} = 0. \tag{23}$$

We also put

$$\pi^i = \omega^i. \tag{24}$$

The Pfaffian forms π^i, $\pi_{j.}^{\ i}$, of which the latter have not yet been completely determined, will then be employed to define the Euclidean connection in the space.

To each set of variables x^i, y^i, v_k^α we attach a Euclidean space of n dimensions with the frame of reference $M\vec{e}_1 \ldots \vec{e}_n$, where M is a point and $\vec{e}_1, \ldots, \vec{e}_n$ are n mutually perpendicular unit vectors through M. The equations

$$\left.\begin{aligned} d\vec{M} &= \pi^i \vec{e}_i, \\ d\vec{e}_i &= \pi_{i.}^{\ j}\vec{e}_j \end{aligned}\right\} \tag{25}$$

then determine the infinitesimal displacement between two neighboring Euclidean spaces or a Euclidean connection. The property of the Euclidean connection depends on the expressions for the following exterior quadratic differential forms

$$\left.\begin{aligned} \Pi^i &= (\pi^i)' - [\pi^j \pi_{j.}^{\ i}] \\ \Pi_{j.}^{\ i} &= (\pi_{j.}^{\ i})' - [\pi_{j.}^{\ k}\pi_{k.}^{\ i}]. \end{aligned}\right\} \tag{26}$$

And we find

$$\left.\begin{aligned} \Pi^i &= \Omega^i - \gamma_{j..}^{\ i\alpha}[\omega^j \omega_{\alpha.}^{\ n}], \\ \Pi_{j.}^{\ i} &= \Omega_{j.}^{\ i} + \gamma_{j..}^{\ i\alpha}\Omega_{\alpha.}^{\ n} - \gamma_{j..}^{\ \alpha\beta}\gamma_{\alpha..}^{\ i\rho}[\omega_{\beta.}^{\ n}\omega_{\rho.}^{\ n}] + [(d\gamma_{j..}^{\ i\alpha} + \\ &\qquad \gamma_{j..}^{\ i\beta}\omega_{\beta.}^{\ \alpha} - \gamma_{\beta..}^{\ i\alpha}\omega_{j.}^{\ \beta} + \gamma_{j..}^{\ \beta\alpha}\omega_{\beta.}^{\ i})\omega_{\alpha.}^{\ n}]. \end{aligned}\right\} \tag{27}$$

Due to a reason which we shall give later, it is important to impose the condition that Π^i, $\Pi_{j.}^{\ i}$ be exterior quadratic differential forms in ω^i, $\omega_{\alpha.}^{\ n}$. This gives

$$d\gamma_{j..}^{\ i\alpha} + \gamma_{j..}^{\ i\beta}\omega_{\beta.}^{\ \alpha} - \gamma_{\beta..}^{\ i\alpha}\omega_{j.}^{\ \beta} + \gamma_{j..}^{\ \beta\alpha}\omega_{\beta.}^{\ i} \equiv 0, \text{ mod. } \omega^k, \omega_{\alpha.}^{\ n}. \tag{28}$$

Summing up, the conditions on $\gamma_{ij\alpha}$ are (22), (28). We shall satisfy (22) by supposing

$$\left.\begin{aligned} \gamma_{in\alpha} &= \gamma_{ni\alpha} = 0, \\ \gamma_{\rho\sigma\alpha} + \gamma_{\sigma\rho\alpha} &= -H_{\rho\sigma\alpha}. \end{aligned}\right\} \tag{29}$$

Then (28) is reduced to the following more symmetrical form:

$$d\gamma_{\rho\sigma\alpha} + \gamma_{\rho\sigma\beta}\omega^{\beta}_{.\alpha} + \gamma_{\beta\sigma\alpha}\omega^{\beta}_{.\rho} + \gamma_{\rho\beta\alpha}\omega^{\beta}_{.\sigma} \equiv 0, \text{ mod. } \omega^{i}, \omega_{\alpha}^{n}{}_{..} \qquad (30)$$

It is important to note that $\gamma_{\rho\sigma\alpha}$ have naturally to be invariants. An example of such a set of invariants is furnished by $H_{\rho\sigma\alpha}$. We can easily verify that the condition (30) for $\gamma_{\rho\sigma\alpha}$ is equivalent to saying that $\gamma_{\rho\sigma\alpha}$ is of the form

$$\gamma_{\rho\sigma\alpha} = G_{ijk}u^{i}_{\rho}u^{j}_{\sigma}u^{k}_{\alpha}, \qquad (31)$$

where G_{ijk} are functions of x^m, y^m only.

To each set of invariants $\gamma_{\rho\sigma\alpha}$ satisfying the conditions (29), (30) we have the set of Pfaffian forms π^i, $\pi_{i.}^{j}$. With these Pfaffian forms the equations (25) define a Euclidean connection, which gives the infinitesimal displacement between the Euclidean spaces attached to two neighboring sets of variables x^i, y^i, v^{α}_{k}. Owing to the property that Π^i, $\Pi_{j.}^{i}$ are exterior quadratic differential forms in ω^i, $\omega_{\alpha}^{n}{}_{.}$, it follows that Π^i, $\Pi_{j.}^{i}$ are zero, when x^i, y^i are given as functions of a parameter t:

$$x^i = x^i(t), \; y^i = y^i(t). \qquad (32)$$

Hence along a one-parameter family of contact elements (x^i, y^i) the system of equations (25) is completely integrable and the Euclidean spaces attached can be developed in one and the same Euclidean space. This is essentially the generalization of the well-known theorem of Fermi to a Finsler space. From this fact we see that we can regard the Finsler space as formed by the $(2n - 1)$-parameter family of its contact elements and define a Euclidean connection in the space, with a "tangent Euclidean space" attached to each contact element. There are as many Euclidean connections in the space as there are invariants satisfying our conditions. If we take

$$\gamma_{\rho\sigma\alpha} = -{}^1/_2 H_{\rho\sigma\alpha}, \qquad (33)$$

we get the Euclidean connection defined by Cartan.

In conclusion, we remark that each of our Euclidean connections has the property that the equivalence of the Euclidean connection is a necessary and sufficient condition for the equivalence of the Finsler spaces (under point transformations).

[1] Synge, J. L., *Trans. Amer. Math. Soc.*, **27**, 61–67 (1925).
[2] Taylor, J. H., *Ibid.*, **27**, 246–264 (1925).
[3] Berwald, L., *Atti Congr. Mat., Bologna*, **4**, 263–270 (1928).
[4] Cartan, E., *Les espaces de Finsler*, Paris, 1934.
[5] It is agreed, throughout this paper, that a Latin index runs from 1 to n and a Greek index from 1 to $n - 1$.

Reprinted from the Proceedings of the NATIONAL ACADEMY OF SCIENCES,
Vol. 30, No. 4, pp. 95–97. April, 1944

LAPLACE TRANSFORMS OF A CLASS OF HIGHER DIMENSIONAL VARIETIES IN A PROJECTIVE SPACE OF n DIMENSIONS

BY SHIING-SHEN CHERN

INSTITUTE FOR ADVANCED STUDY

Communicated February 15, 1944

The theory of conjugate nets and their Laplace transforms has been the object of extensive study. Its extension to r-dimensional varieties in a projective space of n dimensions, $r < n$, has so far not been very successful. The object of this note is to indicate a possible way that the theory can be extended.

Consider a projective space of n dimensions. To develop the differential geometry in the space we employ Cartan's method of moving frames, taking as projective frame the set of $n + 1$ linearly independent vectors $A, A_1, \ldots, A_n$ (each with $n + 1$ components), determined up to a common factor. The equations of infinitesimal displacement are of the form

$$\left.\begin{aligned} dA &= \omega_{00}A + \omega_1 A_1 + \ldots + \omega_n A_n \\ dA_1 &= \omega_{10}A + \omega_{11}A_1 + \ldots + \omega_{1n}A_n \\ &\vdots \\ dA_n &= \omega_{n0}A + \omega_{n1}A_1 + \ldots + \omega_{nn}A_n \end{aligned}\right\} \qquad (1)$$

where the ω's are Pfaffian forms in the parameters on which depend the frames. By remarking that the exterior derivatives of the left-hand sides of (1) are zero, we get the so-called equations of structure of the space:

$$\left.\begin{aligned} \omega'_{00} &= [\omega_1\omega_{10}] + \ldots + [\omega_n\omega_{n0}] \\ \omega'_i &= [\omega_{00}\omega_i] + [\omega_1\omega_{1i}] + \ldots + [\omega_n\omega_{ni}] \\ \omega'_{ij} &= [\omega_{i0}\omega_j] + [\omega_{i1}\omega_{1j}] + \ldots + [\omega_{in}\omega_{nj}],\ i, j = 1, \ldots, n \\ \omega'_{i0} &= [\omega_{i0}\omega_{00}] + [\omega_{i1}\omega_{10}] + \ldots + [\omega_{in}\omega_{n0}] \end{aligned}\right\} \quad (2)$$

Let V_r be a variety of r dimensions in the space. We make the convention that a linear combination of the vectors of a frame represents a point whose homogeneous coördinates are its components. To each point of V_r we attach a frame such that A coincides with the point, $A_1, \ldots, A_r$ lie in the tangent r-plane, and $A_{r+1}, \ldots, A_n$ $n - r$ other points of the space. For such a family of frames over V_r we have

$$\begin{cases} \omega_{r+1} = \ldots = \omega_n = 0, \\ \omega_{i\alpha} = \sum_{j=1}^{r} g_{ij\alpha}\omega_j,\ i = 1, \ldots, r,\ \alpha = r + 1, \ldots, n \end{cases} \quad (3)$$

where $g_{ij\alpha}$ are functions of parameters on the variety and are symmetric with respect to i, j. Construct the quadratic forms

$$\Phi_\alpha = \omega_1\omega_{1\alpha} + \ldots + \omega_r\omega_{r\alpha},\ \alpha = r + 1, \ldots, n \quad (4)$$

and suppose q of them be linearly independent. This number q is an arithmetic invariant. We can choose the frame attached to the point A such that $\Phi_{r+1}, \ldots, \Phi_{r+q}$ are linearly independent and $\Phi_{r+q+1}, \ldots, \Phi_n$ are identically zero. We shall call the net of quadric cones of vertex A defined by the equation

$$\lambda_{r+1}\Phi_{r+1} + \ldots + \lambda_{r+q}\Phi_{r+q} = 0 \quad (5)$$

the asymptotic net at A. It generalizes the asymptotic tangents of a surface.

E. Cartan[1] studied the varieties of r dimensions whose asymptotic net is reducible to the form

$$\lambda_1\omega_1^2 + \ldots + \lambda_r\omega_r^2 = 0 \quad (6)$$

Cartan proved that such varieties depend on $r(r - 1)$ arbitrary functions in two variables. It is not difficult to prove that for $r = 2$ these varieties are in general the surfaces sustaining conjugate nets.

From the last remark we can deduce a notable geometrical property of these varieties. Suppose the frames attached to the points of V_r are so chosen that the asymptotic net has the equation (6). Then the lines AA_i, $i = 1, \ldots, r$, are invariantly related to the variety. They define on the variety r families of curves such that through every point passes one and only one curve of each family. If the tangent r-plane is displaced along a curve of the ith family, its intersection with a neighboring tangent

r-plane is the tangent $(r-1)$-plane $AA_1 \ldots A_{i-1}A_{i+1} \ldots A_r$. This is clearly a generalization of the well-known property of conjugate nets.

It turns out that this property is characteristic to the varieties under consideration. In fact, suppose that a variety of r dimensions possesses r families of curves such that when the tangent r-plane is displaced along a curve of one family, its intersection with a neighboring tangent r-plane is the tangent $(r-1)$-plane tangent to the other $r-1$ curves. Then, under a certain generality assumption, the variety has the property that its asymptotic net is reducible to the equation (6). The generality assumption has to be made, because there are some degenerate cases, which are not interesting. A simple and reasonable assumption of this kind is that the osculating planes of the variety are of $2r$ dimensions.

The above results can be described as a geometrical interpretation of the varieties studied by Cartan. To define for such varieties a kind of transformation generalizing the transformation of Laplace, we proceed as follows:

Let V_r be such a variety whose r families of curves have at the point A the tangents AA_i, $i = 1, \ldots, r$. Then the following theorem can be proved: On each tangent AA_i there exist $r-1$ points A_{ij} $(j \neq i)$ such that when the s-plane $AA_{i_1} \ldots A_{i_s}$ is displaced along AA_j, $j \neq i_1, \ldots, i_s$, its intersection with a neighboring s-plane is the $(s-1)$-plane $A_{i_1j} \ldots A_{i_sj}$.

Associated to each point A we have therefore $r(r-1)$ points A_{ij}. When A describes our variety, each point A_{ij} describes a variety of dimension $\leq r$. It can be proved that if the locus of A_{ij} is of dimension r, it also has the property described above. It is natural to call these varieties the Laplace transforms of V_r, there being altogether $r(r-1)$ Laplace transforms. The Laplace transforms of a Laplace transform of V_r will consist of V_r, some of the Laplace transforms of V_r, and some new varieties. There is a great number of interesting problems which generalize the classical theory of periodic Laplace sequences. The problems seem to be particularly fascinating, because it is probable to expect some results which are not direct generalizations of the theory of surfaces.

The proofs of the above results offer no difficulty and are therefore omitted. For long time there was the opinion that the projective differential geometry of r-dimensional varieties in a space of n dimensions has serious analytical difficulties. We wish to point out here that Cartan's method of moving frames offers a workable scheme which, at least theoretically, can push to an end the problems which are of geometrical interest.

[1] Cartan, E., "Sur les variétés de courbure constante d'un espace euclidien ou non-euclidien," *Bull. Soc. math. France*, **47**, 125–160 (1919); **48**, 132–208 (1920).

Reprinted from the Proceedings of the NATIONAL ACADEMY OF SCIENCES,
Vol. 30, No. 9, pp. 269–273. September, 1944

INTEGRAL FORMULAS FOR THE CHARACTERISTIC CLASSES OF SPHERE BUNDLES

BY SHIING-SHEN CHERN

INSTITUTE FOR ADVANCED STUDY, PRINCETON, NEW JERSEY AND TSING HUA UNIVERSITY, KUNMING, CHINA

Communicated July 13, 1944

The notion of sphere bundles[1] of H. Whitney has proved to be very fruitful in the applications of topology. So far in this theory an important rôle has been played by a class of topological invariants in the base space, called the characteristic cohomology classes and which can be regarded as generalizations of the Euler-Poincaré characteristic. By assuming the base space to be a differentiable manifold with a Riemannian metric, on which the bundles of vectors and p-vectors are defined in a natural way, we shall give in this note various relations which exist between these topological invariants and certain differential invariants of the Riemannian manifold. In fact, the relations are such that the former are expressed as integrals of the latter. In the formulae the generalized Gauss-Bonnet formula due to Allendoerfer and Weil[2] is included as a particular case.

1. At the basis of our considerations is a closed orientable n-dimensional Riemannian manifold of class C^m, $m \geqq 4$, which we denote by R^n. In R^n

we consider the frames $P\,\mathfrak{e}_1 \ldots \mathfrak{e}_n$, each of which consists of a point P and n mutually perpendicular unit vectors $\mathfrak{e}_1, \ldots, \mathfrak{e}_n$ through P, arranged in an order coherent with the orientation of R^n. Using the notations of E. Cartan,[3] the parallelism of Levi-Civita is defined by equations of the form

$$dP = \omega_i \mathfrak{e}_i, \qquad d\mathfrak{e}_i = \omega_{ij}\mathfrak{e}_j, \qquad \omega_{ij} + \omega_{ji} = 0, \tag{1}$$

where the ω's are Pfaffian forms. In terms of these Pfaffian forms the equations of structure of the space are

$$d\omega_i = \omega_j\omega_{ji}, \qquad d\omega_{ij} = -\omega_{ik}\omega_{jk} + \Omega_{ij}, \tag{2}$$

and the identities of Bianchi are

$$\omega_j\Omega_{ji} = 0, \qquad d\Omega_{ij} = -\omega_{ik}\Omega_{jk} + \omega_{jk}\Omega_{ik}. \tag{3}$$

It is to be mentioned that Ω_{ij} are exterior quadratic differential forms which give the curvature properties of R^n.

Consider now the tangent sphere bundle of R^n in the sense of Whitney and, in particular, the problem of defining over R^n a field of p-vectors, $1 \leqq p \leqq n$, where by a p-vector is meant an ordered set of p vectors (and not the figure determined by the Grassmannian coördinates). As is well known, the characteristic classes arise from the points of R^n at which the p vectors are linearly dependent. Topologically there is no loss of generality to assume the p vectors, when linearly independent, to be mutually perpendicular unit vectors.

We are thus led to consider in R^n the figures formed by a point P and an ordered set of p mutually perpendicular unit vectors through P. With P fixed, all these figures form, according to a natural topology, a finite manifold $V_{n,p}$ of dimension $p(2n - p - 1)/2$. The Betti group $B^{n-p}(V_{n,p})$ of dimension $n - p$ (with integer coefficients) of $V_{n,p}$ is an infinite cyclic group or a cyclic group of order two, according as one of the following conditions is satisfied or not: $p = 1$ or $n - p$ is even. We consider the sphere bundle $\mathfrak{S}(p)$ with R^n as the base space and the $V_{n,p}$ as fibres and proceed to study the relations between the characteristic classes arising therefrom and certain differential invariants of R^n.

2. With our problem formulated in this way, it is convenient to adopt the language of the method of moving frames. To a figure of the above description we attach a frame $P\mathfrak{e}_1 \ldots \mathfrak{e}_n$ such that P is the origin and $\mathfrak{e}_{n-p+1}, \ldots, \mathfrak{e}_n$ the p vectors of the figure in question. With the intervention of the family of frames thus attached, Pfaffian forms ω_i, ω_{ij} and differential invariants can be constructed from (1, 2). From them it is possible to construct exterior differential forms of higher degree. Whether they belong to $\mathfrak{S}(p)$ or not can be decided by the following simple criterion: *A differential form* Π *belongs to* $\mathfrak{S}(p)$ *if and only if its exterior derivative* $d\Pi$ *is a form in* ω_i, ω_{iA} $A = n - p + 1, \ldots n$, *only*. A differential form belonging to $\mathfrak{S}(p)$ is called intrinsic.

3. For our considerations we shall suppose either $p = 1$ or $n - p$ be even. The essential step consists of establishing a formula for differential forms which we shall reproduce here for the case that $n - p$ is even. We consider the following intrinsic differential forms

$$(2k;\ h_1, \ldots, h_p) = \epsilon_{\alpha_1 \ldots \alpha_{n-p}} \Omega_{\alpha_1\alpha_2} \ldots \Omega_{\alpha_{2k-1}\alpha_{2k}}\, \omega_{\alpha_{2k+1}\, n-p+1} \cdots$$

$$\omega_{\alpha_{2k+h_1}\, n-p+1}\, \omega_{\alpha_{2k+h_1+1}\, n-p+2} \ldots \omega_{\alpha_{2k+h_1+h_2}\, n-p+2} \cdots$$

$$\omega_{\alpha_{2k+h_1+\ldots+h_{p-1}+1}\, n} \cdots \omega_{\alpha_{2k+h_1+\ldots+h_p}\, n}, \quad (4)$$

where each index α runs from 1 to $n - p = 2k + h_1 + \ldots + h_p$ and where $\epsilon_{\alpha_1 \ldots \alpha_{n-p}}$ denotes the Kronecker symbol which is $+1$ or -1 according as $\epsilon_{\alpha_1, \ldots, \alpha_{n-p}}$ form an even or odd permutation of $1, \ldots, n - p$, and is otherwise zero. We put also

$$\Phi_k = \sum \frac{2.4 \ldots (n - p - 2k)}{(2.4 \ldots 2q_1) \ldots (2.4 \ldots 2q_p)} (2k;\ 2q_1, \ldots, 2q_p) \quad (5)$$

where the summation is extended over all partitions of $\frac{1}{2}(n - p) - k$ as a sum of $q_1, \ldots, q_p$. Further let

$$a_k = (-1)^k \frac{(n - p - 2k + 2) \ldots (n - p)}{2.4 \ldots 2k}, \quad k \geqq 1, \quad (6)$$

$$a_0 = 1$$

Then we have the formula

$$d\left\{\sum_{k=0}^{1/2(n-p)} a_k \Phi_k\right\} = 0. \quad (7)$$

The proof of this formula can be carried out in a purely algebraic way.

4. Suppose we consider the p-vectors having the same origin, i.e., those belonging to the same $V_{n,p}$. Under this condition we have evidently

$$\omega_1 = \ldots = \omega_n = 0,$$

and formula (7) becomes

$$d\Phi_0 = 0$$

This shows that Φ_0 is a closed intrinsic form in $V_{n,p}$. Noticing that $B^{n-p}(V_{n,p})$ is an infinite cyclic group, it follows from the well-known results of de Rham[4] that the value of the integral

$$\int_Z \Phi_0$$

where Z is an $(n - p)$-dimensional cycle of $V_{n,p}$, will give the homology class to which Z belongs.

The theory of p-fields in R^n leads, when $n - p$ is even, to an integral characteristic class W^{n-p+1}. To derive from it a numerical topological invariant we consider the scalar product $W^{n-p+1}M^{n-p+1}$, where M^{n-p+1} is a submanifold of dimension $n - p + 1$ in R^n, closed or with boundary. When M^{n-p+1} is closed, we derive from (7) and the above remark that

$$W^{n-p+1}.M^{n-p+1} = 0,$$

which corresponds to the known fact that W^{n-p+1} is of finite order.

If, however, M^{n-p+1} has a boundary B, we shall get in this way a formula which links the integral of a differential invariant over B with a topological invariant of M^{n-p+1}. For definiteness we suppose $p = 2$ (so that n is even). B is of dimension $n - 2$ and we suppose for simplicity that B possesses a tangent $(n - 2)$-plane at each of its points. Both M^{n-1} and B are supposed to be orientable, so that M^{n-1} is two-sided in R^n and B two-sided in M^{n-1}. At each point of B we take as $\mathfrak{e}_n$ the outside (or inside) normal of M^{n-1} and as $\mathfrak{e}_{n-1}$ the outside (or inside) normal of B in the tangent hyperplane of M^{n-1}. We have thus defined over B a field $\mathfrak{F}$ of two vectors and the form

$$\Pi = \sum_{k=0}^{1/2(n-2)} a_k\Phi_k,$$

being intrinsic in $\mathfrak{S}(2)$, is defined on B. Then we have

$$\int_B \Pi = c.M^{n-1}.LL^{n-1}$$

where c is a numerical constant. This formula is the analog of the Gauss-Bonnet formula for a polyhedron. The corresponding formulation for a general p is without difficulty.

5. As we have shown in another occasion,[5] these discussions lead for $p = 1$ to a proof of the Gauss-Bonnet formula by which it is not necessary to imbed the cells of R^n in a Euclidean space of sufficiently high dimension. There the formula corresponding to (7) is of the form

$$d\Pi = \Omega,$$

where $\Omega = 0$ when n is odd. The differential form Π, of degree $n - 1$, is defined in the space of vectors and has some interest in itself. It is defined, whenever an orientable submanifold M^r, of dimension r, is given in R^n. For, supposing that M^r has a tangent r-plane at each of its points, we attach to a point of M^r all the unit normal vectors at that point if $r \neq n - 1$ and the outside (or inside) unit normal vector at that point if $r = n - 1$. In this way the set of unit vectors attached to the points of M^r forms, according to the natural topology, a manifold of dimension $n - 1$. This process can also be carried out for M^r which ceases to have tangent r-

planes at certain points. Then we attach to each of these points the unit normal vectors belonging to the corresponding dual angle. By such a convention the differential form II is defined over M^r and generalizes the notion of the element of geodesic curvature for $n = 1$. We can prove that, if r is even,

$$\int_{M^r} \text{II} = c\chi(M^r),$$

where $\chi(M^r)$ stands for the Euler-Poincaré characteristic of M^r and c is a constant. The proof of this formula is somewhat complicated.

6. It is evident that the same considerations can be carried out for the normal sphere bundles of a submanifold imbedded in a differentiable manifold. To avoid duplication we shall not give the results here. We only mention that the corresponding differential invariants are those which give the so-called Gaussian torsion of the submanifold.

7. Finally, we wish to make the following remark. The process employed here bears a close resemblance to that used by Gysin[6] in his work on the homology theory of fibre spaces. Our differential forms take the place of cochains and we have taken the group of real numbers as the coefficient group of the homology theory. While we thus lose such topological invariants like torsion, we are compensated by the fact that the differential forms are intrinsic in the sense of differential geometry and have thus arrived at results which connect differential invariants with topological invariants.

[1] Whitney, H., "On the Topology of Differentiable Manifolds," *Lectures in Topology* Michigan, 1941, pp. 101–141. Also: Stiefel, E., "Richtungsfelder und Fernparallelismus in n-dimensionalen Mannigfaltigkeiten," *Comm. Math. Helv.*, **8**, 305–353 (1936).

[2] Allendoerfer, C. B., and Weil, André, "The Gauss-Bonnet Theorem for Riemannian Polyhedra," *Trans. Amer. Math. Soc.*, **53**, 101–129 (1943).

[3] Cartan, E., *Leçons sur la géometrie des espaces de Riemann*, Paris, 1928.

[4] de Rahm, G., "Sur l'analysis situs des variétés a n dimensions," *J. Math. Pures Appl.*, **10** (9), 115–200 (1931).

[5] Chern, S., "A Simple Intrinsic Proof of the Gauss-Bonnet Formula for Closed Riemannian Manifolds," to appear in *Annals of Math.*

[6] Gysin, W., "Zur Homologietheorie der Abbildungen und Faserungen von Mannigfaltigkeiten," *Comm. Math. Helv.*, **14**, 61–122 (1941–1942).

SOME NEW CHARACTERIZATIONS OF THE EUCLIDEAN SPHERE

By Shiing-shen Chern

1. Introduction. Among convex surfaces in three-dimensional Euclidean space the sphere can be characterized in various ways. In this note we shall give some further characterizations of the sphere based on the consideration of differential-geometric properties of convex surfaces. The characterizations are of the nature that a certain local property holds throughout the surface and the theorems in question are not valid locally. Some theorems on non-convex closed surfaces will also be given.

In a Euclidean space of three dimensions we consider a closed surface S, differentiable of class C^m, $m \geq 3$, and with the property that the tangent plane to S is well defined at every point. At a point of S let r_1 and r_2 denote the principal curvatures. S is called convex if the Gaussian curvature $K = r_1r_2$ is everywhere positive. It is called a W-surface if dr_1 and dr_2 are linearly dependent, that is, if functions λ_1, λ_2 exist, not both zero, such that

$$\lambda_1 dr_1 + \lambda_2 dr_2 = 0. \tag{1}$$

This means either that both r_1 and r_2 are constant or that r_1 and r_2 are connected by a functional relation

$$F(r_1, r_2) = 0. \tag{2}$$

We shall call our W-surface *special*, if the functions λ_i in (1) can be chosen to be positive: $\lambda_i > 0$, $i = 1, 2$. With exception of the case $r_i =$ constant, $i = 1, 2$, a special W-surface is one for which one principal curvature is a strictly monotone decreasing function of the other. Examples of such special W-surfaces are given by: (a) $r_1r_2 =$ constant > 0; (b) $r_1 + r_2 =$ constant, etc. With these definitions our first theorem can be stated as follows:

THEOREM 1. *A convex special W-surface is a sphere.*

The proof of Theorem 1 will be given in the next sections. It is perhaps interesting to give some of its consequences. By introducing the Gaussian curvature K and the mean curvature H of the surface according to the formulas

$$K = r_1r_2, \tag{3}$$

$$2H = r_1 + r_2, \tag{4}$$

we have immediately the corollaries (see [1; 195–199]):

COROLLARY 1. *A closed surface of constant Gaussian curvature is a sphere.*

Received January 18, 1945.

COROLLARY 2. *A convex surface of constant mean curvature is a sphere.*

Both corollaries are well known, the first one known to be the indeformability of the sphere ("Unverbiegbarkeit der Kugel"). Notice that the convexity condition is not imposed in Corollary 1, because the constant Gaussian curvature of a closed surface is necessarily positive.

It is easy to give other consequences of Theorem 1, among which we state the following, which were first given by S. Nakajima [5]:

COROLLARY 3. *A convex surface is a sphere, if $K = cH^2$, c being a constant not equal to zero.*

COROLLARY 4. *A convex surface is a sphere if*

$$aK + bH + c = 0,$$

a, b, c being constants not all zero.

Corollary 3 is an immediate consequence. Corollary 4 follows from the observation that the condition in question holds only when $b^2 - 4ac \geq 0$. In fact, the surface has at least an umbilic which is a singular point of the net of lines of curvature and at which the normal curvature r is a root of the equation

$$ar^2 + br + c = 0.$$

This equation must therefore have real roots, which gives $b^2 - 4ac \geq 0$. The case $b^2 - 4ac > 0$ is covered by our Theorem 1 and the case $b^2 - 4ac = 0$ will be proved by a remark after the proof of Theorem 1.

The conditions imposed in Theorem 1 and its corollaries are of two types, first a local condition and then the condition that the local property holds throughout the surface. With the local conditions alone the theorems are evidently not valid. We shall now give another theorem of the same nature, which is related to a problem studied by O. Bonnet (see [3]). The problem of Bonnet can be stated as follows: Can a surface (that is, a neighborhood of a surface) be isometrically mapped onto another with conservation of the principal curvatures such that the mapping is neither a motion nor a reflection? Bonnet's problem is entirely local and it is well known that a general surface does not possess this local property. We shall call a surface with this property a surface of Bonnet, which includes, for instance, the surfaces of constant mean curvature. With these explanations our second theorem can be stated as follows:

THEOREM 2. *A convex surface of Bonnet is a sphere.*

The motivation to the results presented here is an attempt to get more information on local differential-geometric properties which can hold throughout a closed surface and on how far the latter is determined. The local problem is solved, at least theoretically, by the theory of partial differential equations due to Riquier, Delassus, and Cartan (for a modern treatment, cf. [6]). It seems in most cases not clear when a locally compatible property can hold in the large and it seems that there lies a field of geometrical researches which is not yet explored.

2. Formulas of the theory of surfaces. For the proofs of our theorems we need the fundamental formulas of the theory of surfaces. It is perhaps most convenient for our purpose to write them in the notation of E. Cartan (for details see [2; 218–227]).

The basic geometrical idea to derive these formulas is to attach right-handed rectangular trihedrals to points of the surface. This will now be done for the whole surface S, which is assumed to be differentiable, closed, and orientable. We consider a point P of S, which is not an umbilic. At P there are two principal directions d_i , $i = 1, 2$, with the corresponding principal curvatures r_i , $i = 1, 2$. These directions are then fixed by the convention $r_1 > r_2$. To P right-handed rectangular trihedrals $Pe_1e_2e_3$ are attached, satisfying the conditions: (1) e_3 is the unit vector along the outward normal; (2) e_1 , e_2 are unit vectors along d_1 and d_2 , respectively. There are two such trihedrals at each non-umbilical point, differing from each other by the signs of e_1 and e_2 . In a sufficiently small neighborhood of P we have two disconnected families of rectangular trihedrals, exactly two through each point. Between these trihedrals we have the vectorial differential equations

$$\text{(5)} \qquad \begin{cases} dP = \omega_1 e_1 + \omega_2 e_2 , \\ de_1 = \omega_{12} e_2 + \omega_{13} e_3 , \\ de_2 = -\omega_{12} e_1 + \omega_{23} e_3 , \\ de_3 = -\omega_{13} e_1 - \omega_{23} e_2 , \end{cases}$$

where ω_1 , ω_2 , ω_{12} , ω_{13} , ω_{23} are Pfaffian forms on S, the first two being linearly independent. It is also to be remarked that condition (2) is expressed by the equations

$$\text{(6)} \qquad \omega_{13} = r_1\omega_1 , \qquad \omega_{23} = r_2\omega_2 ,$$

where r_1 and r_2 are precisely the principal curvatures.

The same equations (5) and (6) hold in the neighborhood of an umbilic, at which $r_1 = r_2$. At an umbilic P we have thus attached all the right-handed rectangular trihedrals $Pe_1e_2e_3$ having the property that e_3 is the outward unit normal vector.

The Pfaffian forms ω are connected by the so-called equations of structure, which express the condition that the exterior derivatives of the right members of (5) are zero. Exterior differentiation of (6), with due regard of the equations of structure, gives

$$
(7) \quad \begin{cases} [dr_1\omega_1] = (r_2 - r_1)[\omega_{12}\omega_2], \\ [dr_2\omega_2] = (r_1 - r_2)[\omega_1\omega_{12}]. \end{cases}
$$

In fact, equations (7) are essentially the well-known equations of Codazzi.

Let us again restrict ourselves to a neighborhood of P, which contains no umbilic. Equations (7) show that ω_{12} is a linear combination of ω_1, ω_2, which we write

$$
(8) \quad \omega_{12} = h\omega_1 + k\omega_2 .
$$

We also define, for any function f defined on the surface, its covariant derivatives by means of the equation

$$
(9) \quad df = f_1\omega_1 + f_2\omega_2 .
$$

Then we have, from the equations of structure and (7), the equations

$$
(10) \quad \begin{cases} \omega_1' = h[\omega_1\omega_2], \qquad \omega_2' = k[\omega_1\omega_2], \\ [dr_1\omega_1] = h(r_2 - r_1)[\omega_1\omega_2], \\ [dr_2\omega_2] = k(r_1 - r_2)[\omega_1\omega_2], \\ h_2 - k_1 = r_1r_2 + h^2 + k^2. \end{cases}
$$

These equations play a fundamental rôle in the theory of surfaces.

To give the conditions for a surface of Bonnet we put

$$
(11) \quad L = r_1 - r_2
$$

and

$$
(12) \quad dH = \frac{L}{2}(u\omega_1 + v\omega_2).
$$

By making use of (10), we get

$$
(13) \quad dL = L\{(u - 2k)\omega_1 - (v - 2h)\omega_2\}.
$$

The local conditions for a surface to be a surface of Bonnet are then, according to E. Cartan [3; 67],

$$
(14) \quad \begin{cases} h_1 + k_2 = 0, \\ u_1 - v_2 = u^2 + v^2 - hv - ku, \\ u_2 = v(u - k), \\ v_1 = u(h - v). \end{cases}
$$

3. Proof of the theorems.

Proof of Theorem 1. Let 0 be a point on S at which r_1 attains its maximum. Suppose 0 be an umbilic and denote the normal curvature at 0 by r_0. If r_1 and r_2 are the principal curvatures at any other point P of S, we have

$$r_0 \geq r_1 \geq r_2 .$$

Since r_2 is either a monotone decreasing function of r_1 or a constant, we have also

$$r_0 \leq r_2 .$$

Combining the two sets of inequalities, we get

$$r_0 = r_1 = r_2 .$$

It follows that every point on S is an umbilic and that S is a sphere.

We consider next the case that 0 is not an umbilic and restrict ourselves to a neighborhood of 0 free from umbilics. From (1) we have

$$dr_1 = \lambda dr_2 \qquad (\lambda < 0).$$

The third and fourth equations of (10) then give

$$[dr_1\omega_1] = h(r_2 - r_1)[\omega_1\omega_2],$$
$$[dr_1\omega_2] = \lambda[dr_2\omega_2] = \lambda k(r_1 - r_2)[\omega_1\omega_2].$$

It follows that

$$dr_1 = (r_1 - r_2)(\lambda k\omega_1 + h\omega_2).$$

Since r_1 attains a maximum at 0 and 0 is not an umbilic, we have, at 0,

$$h = k = 0. \tag{15}$$

Again, the second differential d^2r_1 at 0 is given by

$$d^2r_1 \,|_0 = (r_1 - r_2)_0(\lambda dk\omega_1 + dh\omega_2)_0 .$$

That r_1 attains a maximum at 0 gives therefore the conditions, at 0,

$$\lambda k_1 \leq 0, \qquad h_2 \leq 0,$$

or, since $\lambda < 0$,

$$k_1 \geq 0, \qquad h_2 \leq 0. \tag{16}$$

Conditions (15) and (16) contradict the last equation of (10), for $r_1r_2 > 0$. Hence the assumption that r_1 attains a maximum at a non-umbilical point is impossible. Our theorem is therefore proved.

Proof of Corollary 4. According to our remark on Corollary 4 it only remains to prove the statement for the case that K and H are connected by the equation

$$aK + 2(ac)^{\frac{1}{2}}H + c = 0$$

or

$$(a^{\frac{1}{2}}r_1 + c^{\frac{1}{2}})(a^{\frac{1}{2}}r_2 + c^{\frac{1}{2}}) = 0.$$

We maintain that r_1 must be a constant. For, if not, r_1 must attain a maximum, which is not equal to $-(c/a)^{\frac{1}{2}}$, since otherwise there would exist a point at which $r_1 < -(c/a)^{\frac{1}{2}} = r_2$. On the other hand, if r_1 attains a maximum $> -(c/a)^{\frac{1}{2}}$, the value of r_2 in a neighborhood of that point is $-(c/a)^{\frac{1}{2}}$. An argument similar to the above will show that this is impossible.

If the constant value of $r_1 \neq -(c/a)^{\frac{1}{2}}$, we have $r_2 = -(c/a)^{\frac{1}{2}}$. But this is impossible, since they are equal at an umbilic. It follows that $r_1 = -(c/a)^{\frac{1}{2}}$. By considering the minimum of r_2, we easily show that $r_2 = -(c/a)^{\frac{1}{2}}$.

Proof of Theorem 2. The proof of Theorem 2 is based on the same idea used above. We consider the function $L = r_1 - r_2 \geq 0$ and the point 0 at which it attains its maximum. The theorem is already proved if this point is an umbilic. Suppose that 0 is not an umbilic. From (13) we have

$$\frac{dL}{L} = (u - 2k)\omega_1 - (v - 2h)\omega_2 .$$

Hence we have, at 0,

$$u - 2k = 0, \qquad v - 2h = 0. \tag{17}$$

A consideration of the second differential d^2L at 0 gives

$$u_1 - 2k_1 \leq 0, \qquad -(v_2 - 2h_2) \leq 0, \tag{18}$$

and hence

$$u_1 - v_2 - 2(k_1 - h_2) \leq 0. \tag{19}$$

But from the last equation of (10) and the second equation of (14) we have

$$u_1 - v_2 - 2(k_1 - h_2) = u^2 + v^2 - hv - ku + 2(r_1r_2 + h^2 + k^2) \leq 0.$$

Combining with (17), we get, at 0,

$$u^2 + v^2 + 2K \leq 0,$$

which is a contradiction. Hence our theorem is proved.

It is interesting to note here that Theorem 1 is no longer valid, if the W-surface is not assumed to be special. A simple example to illustrate this is given by an ellipsoid of revolution. Take an ellipse with axes of lengths $2a$, $2b$, $a > b$, and rotate it about its shorter axis. It is well known that of the two principal radii of curvature at a point of the ellipsoid (see [4; 111]) one is equal to the radius of curvature of the meridian curve and the other is equal to the length of the normal of the meridian curve cut out by the axis of rotation. If we write the parametric equations of the ellipse as

$$x = a \cos t, \qquad y = b \sin t \qquad (0 \leq t \leq \pi) \tag{20}$$

an elementary calculation will give

$$(21)\qquad \begin{cases} r_1 = \dfrac{a}{b}\,(b^2\cos^2 t + a^2\sin^2 t)^{-\frac{1}{2}}\,, \\ \\ r_2 = ab(b^2\cos^2 t + a^2\sin^2 t)^{-3/2} \end{cases}$$

from which we see that $r_1 \geq r_2$. From (21) we get

$$(22)\qquad r_2 = \frac{b^4}{a^2}\,r_1^3\,,$$

so that r_2 is a monotone increasing function of r_1.

On the other hand, I have not been able to ascertain whether Theorem 1 remains true when the surface is only supposed to be closed. The following question remains therefore unanswered: Is a two-sided closed differentiable surface a sphere if it is a special W-surface?

Deleted because of error.

THEOREM 4. *A closed analytic surface of revolution of genus zero is a sphere, if it is a special W-surface.*

Proof. Since the surface is of genus zero, the meridian curve must meet the axis of rotation and is orthogonal to it. At the neighborhood of such a point of intersection (which is usually called a pole of the surface of revolution), we shall show that our assumption is only possible when $r_1 = r_2$.

In fact, let us take the meridian curve in the xy-plane, with the y-axis for axis of rotation. At the neighborhood of the y-axis write the equation of the curve as

$$(27)\qquad y = \varphi(x),$$

where $\varphi(x)$ is analytic and satisfies the condition

$$\varphi'(0) = 0.$$

The expressions for the principal curvatures of the surface are

$$(28)\qquad r_1 = \frac{d^2\varphi}{dx^2}\left\{1 + \left(\frac{d\varphi}{dx}\right)^2\right\}^{-3/2}, \qquad r_2 = \frac{1}{x}\frac{d\varphi}{dx}\left\{1 + \left(\frac{d\varphi}{dx}\right)^2\right\}^{-\frac{1}{2}},$$

from which we immediately get

$$(29)\qquad r_1 = \frac{d}{dx}\,(r_2 x).$$

If r_2' does not vanish identically in a neighborhood of the pole, our condition can be written as

$$\frac{dr_1}{dr_2} = \frac{d^2}{dx^2}(r_2 x) \Big/ \frac{dr_2}{dx} < 0,$$

or

$$x\,\frac{d^2 r_2}{dx^2} \Big/ \frac{dr_2}{dx} + 2 < 0.$$

Suppose we write

$$\frac{dr_2}{dx} = c_0 + c_m x^m + \cdots \qquad (m \geq 1,\ c_m \neq 0),$$

where the dots denote terms of higher order than m. Then we have

$$x\,\frac{d^2 r_2}{dx^2} \Big/ \frac{dr_2}{dx} + 2 = \frac{m c_m x^m + \cdots}{c_0 + c_m x^m + \cdots} + 2,$$

which cannot be negative, when x is sufficiently small. It follows that $r_2' \equiv 0$ in a neighborhood of the pole or that $r_2 =$ constant. But then equation (29) shows that r_1 is equal to the same constant. Since our surface is analytic, it is a sphere.

5. **An intrinsic upper bound for the mean curvature of an embedded convex surface** (Added March 10, 1945).

Professor H. Weyl called my attention to the relations of the formulas established here with some of his early results given in the paper [7].

Following Weyl let us denote by $l(K)$ the second differential parameter of the Gaussian curvature K. Then the results in question can be stated as follows:

THEOREM 5. (*See* [7].) *The mean curvature H of an imbedded convex surface has an upper bound the maximum of $K - l(K)/4K$.*

THEOREM 6. (*See* [7].) *At a non-umbilical point of a convex surface it never occurs that the mean curvature has a relative maximum and the Gaussian curvature has at the same time a relative minimum.*

To prove these theorems we form the exterior derivative of the equation (24), which gives

$$\text{(30)}\qquad \begin{aligned}&[\{-dr_{12} + (2r_{21} - r_{11})\omega_{12}\}\omega_1] + [\{-dr_{21} + (r_{22} - 2r_{12})\omega_{12}\}\omega_2]\\ &\qquad -K(r_1 - r_2)[\omega_1\omega_2] = 0.\end{aligned}$$

It follows that we can put

$$(31)\qquad \begin{aligned} dr_{12} + (r_{11} - 2r_{21})\omega_{12} &= r_{121}\omega_1 + r_{122}\omega_2 , \\ dr_{21} + (2r_{12} - r_{22})\omega_{12} &= r_{211}\omega_1 + r_{212}\omega_2 , \end{aligned}$$

and that between the new coefficients in the right sides of these equations we have

$$(32)\qquad r_{122} - r_{211} = K(r_1 - r_2).$$

This equation is essentially the equation (34) in Weyl's paper quoted above.

Let us put

$$(33)\qquad dK = K_1\omega_1 + K_2\omega_2 ,$$

whose exterior differentiation gives

$$[(dK_1 - K_2\omega_{12})\omega_1] + [(dK_2 + K_1\omega_{12})\omega_2] = 0.$$

This allows us to put

$$(34)\qquad \begin{aligned} dK_1 - K_2\omega_{12} &= K_{11}\omega_1 + K_{12}\omega_2 , \\ dK_2 + K_1\omega_{12} &= K_{21}\omega_1 + K_{22}\omega_2 , \end{aligned}$$

where

$$(35)\qquad K_{12} = K_{21}.$$

The expression

$$(36)\qquad l(K) = K_{11} + K_{22}$$

is then the second differential parameter of K.

To utilize the above formulas we take the expression

$$K = r_1 r_2 .$$

Successive differentiations of this equation give

$$(37)\qquad \begin{cases} K_1 = r_1 r_{21} + r_2 r_{11} , \\ K_2 = r_1 r_{22} + r_2 r_{12} , \\ K_{11} = r_1 r_{211} + 2r_{11}r_{21} + r_2 r_{111} , \\ K_{22} = r_1 r_{222} + 2r_{12}r_{22} + r_2 r_{122} , \end{cases}$$

and therefore

$$(38)\qquad l(K) = r_1(r_{211} + r_{222}) + r_2(r_{111} + r_{122}) + 2r_{11}r_{21} + 2r_{12}r_{22} .$$

Consider now a point at which the mean curvature H assumes a relative maximum. At this point we have

$$r_{11} + r_{21} = 0, \qquad r_{12} + r_{22} = 0,$$

$$r_{111} + r_{211} = p \leq 0, \qquad r_{122} + r_{222} = q \leq 0.$$

Therefore

$$l(K) = (r_2 - r_1)(r_{122} - r_{211}) + r_2 p + r_1 q - 2r_{11}^2 - 2r_{12}^2,$$

or, by using (32),

$$l(K) = -K(r_1 - r_2)^2 + r_2 p + r_1 q - 2r_{11}^2 - 2r_{12}^2 \leq 0. \tag{39}$$

Consequently, we have

$$H^2 = K + \tfrac{1}{4}(r_1 - r_2)^2 = K - \frac{l(K)}{4K} + \frac{1}{4K}(r_2 p + r_1 q - 2r_{11}^2 - 2r_{12}^2).$$

Since the expression in the parentheses is ≤ 0, it follows that

$$H^2 \leq K - \frac{l(K)}{4K}. \tag{40}$$

This inequality holds at a point at which H has a relative maximum. If therefore M denotes an upper bound of the function $K - l(K)/4K$ over the convex surface, we shall have $H^2 \leq M$. This proves Theorem 5. It is to be remarked that from the rigidity of the imbedding of the convex surface in a three-dimensional Euclidean space follows the existence of an upper bound for H. But this derivation gives an explicit upper bound and is also simpler.

If the point at which H attains a relative maximum is a non-umbilical point, we have, at that point,

$$l(K) = K_{11} + K_{22} < 0.$$

Hence K cannot have a relative minimum at that point, for otherwise $K_{11} \geq 0$, $K_{22} \geq 0$. This proves Theorem 6.

References

1. W. Blaschke, *Vorlesungen über Differentialgeometrie* I, Dritte Auflage, Berlin.
2. E. Cartan, *La théorie des groupes finis et continus et la géométrie différentielle traitées par la méthode du repère mobile*, Paris, 1937.
3. E. Cartan, *Sur les couples de surfaces applicables avec conservation des courbures principales*, Bulletin des Sciences Mathématiques, Serie II, Tome 66(1942), pp. 55–72, 74–85.
4. V. Kommerell and K. Kommerell, *Theorie der Raumkurven und krummen Flächen*, I, Berlin, 1931.
5. S. Nakajima, *Über charakteristische Eigenschaften der Kugel*, Tohoku Mathematical Journal, vol. 26(1926), pp. 361–364.

6. J. M. THOMAS, *Differential Systems*, New York, 1937.
7. H. WEYL, *Über die Bestimmung einer geschlossenen konvexen Fläche durch ihr Linienelement*, Vierteljahrschrift der Naturforschenden Gesellschaft in Zürich, Jahrgang 61(1916), pp. 40–72.

INSTITUTE FOR ADVANCED STUDY, PRINCETON, AND TSING HUA UNIVERSITY, CHINA.

Reprinted from
Duke Math. J. **12** (1945) 279–290

SOME NEW VIEWPOINTS IN DIFFERENTIAL GEOMETRY IN THE LARGE[1]

SHIING-SHEN CHERN

1. Introduction. Differential geometry in the large is concerned with relations which exist between the local properties of a geometric being[2] given in a manifold and the properties of the manifold as a whole. The manifold is differentiable in the sense that it is covered by a set of coordinate neighborhoods, each having the same number of coordinates, and that two different systems of coordinates in a common region are related by a differentiable transformation of class not less than 1. The latter assumption allows the use of differentiation in studying the local geometry and thus leads to a number of geometric properties the study of which was initiated by Euler, Gauss, and Monge.

To give an example of a problem of this nature we consider a closed surface S, differentiably imbedded (of class not less than 2) in a Euclidean space of three dimensions. Let K be the Gaussian curvature and dA the surface element of S. Then the classical Gauss-Bonnet formula asserts that

$$(1)\qquad \frac{1}{2\pi}\int\int_S KdA = 2(1-p),$$

where p is the genus of S. This formula expresses the genus p of S, a topological invariant, in terms of a differential invariant. In other words, p is completely determined by the local properties of S.

As another example consider a closed curve C imbedded in the Euclidean plane. If C is rectifiable with length l and bounds a region of area A, then

$$(2)\qquad l^2 - 4\pi A \geqq 0.$$

The equality sign holds only when C is a circle. This so-called isoperimetric inequality has recently been derived in an unexpected

An address delivered before the Summer Meeting of the Society in New Brunswick on September 15, 1945, by invitation of the Program Committee; received by the editors June 22, 1945.

[1] Dedicated to Professor Elie Cartan.

[2] The term *geometric object* was first suggested by Professor Oswald Veblen, who now prefers to use *geometric being*, a translation of the French term "être géométrique" due to Cartan.

way in integral geometry, which deals with the integrals of differential invariants.

In problems of this nature there are two aspects, a local one and a global one. During several decades of intensive work the local aspect has been extensively studied and developed, culminating in the doctrine of tensor analysis. While tensor analysis gives an adequate tool for handling most local problems, it is natural that the study of global problems will necessitate the introduction of new concepts and the modifications of the classical treatment which has so far been followed. It is the main aim of the present article to emphasize the importance, for the global study of manifolds with a geometric being, of drawing into consideration new topological spaces associated with the manifold. In fact, this idea is important for both local and global aspects of differential geometry. For local problems it is certainly very familiar to Elie Cartan, who introduced the notion of tangent space ("espace tangent") for his general theory of geometric beings (affine connections, projective connections, conformal connections, and so on). The tangent space in the sense of Cartan is not always the space of tangent vectors and therefore constitutes one of the sources of difficulty for the understanding of his work. On the other hand, recent works on fibre bundles in topology (Stiefel, Whitney, Feldbau, Ehresmann, Pontrjagin, Steenrod, and so on) seem to lay the foundation for a global theory of the ideas of Cartan. It is a conviction of the author that a mingling of these two streams of thought will give the ideas and tools better adaptable to the study of differential geometry in the large than hitherto achieved. The present article will be devoted to a discussion of different aspects of the problem arising from this viewpoint.

Before going into details, we shall give a brief summary of the main points of our discussion. Our problem will be the study, on a differentiable manifold, of a geometric being given in terms of each local coordinate system by components which obey a definite transformation law under change of the local coordinates. We emphasize that for each such problem it is in general possible to define in a certain sense a natural fibre bundle associated with the manifold. The geometric being then defines in a unique way a set of linear differential forms in the fibre bundle, which gives all the local properties of the geometric being. For Riemannian manifolds the natural fibre bundle is the space of all frames on the manifold and the corresponding linear differential forms give what is essentially known as the parallelism of Levi-Civita. The nature of the fibre bundle to be associated is best decided by the solution of the so-called problem of equivalence, which is the form problem in the case of Riemannian geometry. With the set of linear differential forms in the fibre bundle, differential forms of higher degree can be obtained by operations of Grassmann analysis. Such differential forms define cochains and also cocycles, when they are exact. A study of the interrelations between the cochains and cocycles in the given manifold and the associated fibre bundles, which arise from the projection of the latter into the former, will give a deeper understanding of the global theory of the geometric being. An attempt is made in the following to illustrate in the simplest cases the results that can be derived from this idea. Although this will be our main concern in the present paper, it does not imply that this is the only service that the theory of fibre bundles could render to differential geometry. In fact, there are indications that other possible applications could be fruitfully made. The whole field seems therefore to deserve a more thorough exploitation.

2. Grassmann algebra. At the fore of the Cartan scheme is the so-called Grassmann algebra. In a formal algebraic way this can be defined as follows: Let K be a field of characteristic zero and let $V(n, K)$ be a vector space of n dimensions over K. The Grassmann algebra H over $V(n, K)$ is a hypercomplex system over K satisfying the following conditions:

(1) H contains a unit element 1 and all elements of $V(n, K)$ and is generated by these elements (by operations of the hypercomplex system).

(2) If x, y belong to $V(n, K)$, their multiplication satisfies the alternating rule:

$$xy = -yx. \tag{3}$$

(3) The elements of H satisfy no other relations than those derived from (1) and (2).

Let $e_1, \cdots, e_n$ be a set of basis elements of $V(n, K)$. From the above conditions the following conclusion can be drawn: Every element of H can be expressed in one and only one way as a linear combination, with coefficients in K, of the elements 1, $e_{i_1} \cdots e_{i_m}$, $i_1 < \cdots < i_m$, $i_1, \cdots, i_m = 1, \cdots, n$. In symbols we write, if x belongs to H,

$$x = \sum_{m=0}^{n} x^{(m)}, \tag{4}$$

where

$$x^{(m)} = \sum_{i_1, \cdots, i_m = 1; i_1 < \cdots < i_m}^{n} a_{i_1 \cdots i_m} e_{i_1} \cdots e_{i_m}. \tag{5}$$

The element x is said to be of degree m, if

$$x^{(m)} \neq 0, \qquad x^{(m+1)} = \cdots = x^{(n)} = 0.$$

It is called an exterior form, or simply a form, if $x^{(i)}=0$, $i\neq m$. The degree of an element is independent of the choice of the set of basis elements, and the same is true of the property that an element be a form.

Let W be a vector subspace of dimension m of $V(n, K)$. We take a set of basis vectors $f_1, \cdots, f_m$ in W and form the product $w=f_1 \cdots f_m$. It is easy to prove that w is determined by W up to a nonzero factor of K. The form w is called the associated form of W. If w_1 and w_2 are the associated forms of two vector subspaces W_1 and W_2, of degrees m and p respectively, then W_1 and W_2 have a nonzero intersection when and only when $w_1w_2=0$. Moreover, if $w_1w_2\neq 0$, it is the associated form of the ambient vector subspace of dimension $m+p$ of W_1, W_2. Such are simple instances of ways in which the Grassmann algebra can be applied to the study of the geometry in a vector space.

3. **The differential calculus of Elie Cartan.** The Grassmann algebra was applied with success to the Pfaffian problem by Frobenius and Darboux. The so-called bilinear covariant is an exterior differential form of degree two. But it was Elie Cartan who started the systematic use of a calculus closely related to the Grassmann algebra and pertaining to exterior differential forms of higher degree.

Let M^n be a differentiable manifold of dimension n and class not less than 2. At a point P of M^n the contravariant vectors and the covariant vectors constitute two vector spaces (over the real field) which are dual to each other in the sense that one is the space of the linear forms in the other. The notion of a linear differential form at P is identical with that of a covariant vector. For let $x^1, \cdots, x^n$ be local coordinates at P. With respect to the coordinates x^i a covariant vector will have the components X_i. The differential form

$$\omega = \sum_{i=1}^{n} X_i dx^i \tag{6}$$

is then intrinsic in the sense that it is independent of the choice of the local coordinate system. Conversely, an intrinsic form ω has a set of components in each local coordinate system, which constitute a covariant vector.

Over the vector space of linear differential forms at a point P of M^n we construct the Grassmann algebra. A form of this Grassmann algebra is called an exterior differential form or simply a differential form. In a local coordinate system x^i an exterior differential form can be written

$$(7) \qquad \omega = \sum_{i_1, \cdots, i_m = 1}^{n} a_{i_1 \cdots i_m}(x^1, \cdots, x^n) dx^{i_1} \cdots dx^{i_m},$$

where we can assume the coefficients $a_{i_1 \cdots i_m}$ to be skew-symmetric in their indices. In the terminology of tensor analysis the components $a_{i_1 \cdots i_m}$ are the components of an alternating covariant tensor of order m.

In a neighborhood of P in which the local coordinates x^i are valid let $a_{i_1 \cdots i_m}$ be of class not less than 2. We define a differential form of degree $m+1$, called the exterior derivative of ω, by means of the relation

$$(8) \qquad d\omega = \sum_{i_1, \cdots, i_m = 1}^{n} da_{i_1 \cdots i_m} dx^{i_1} \cdots dx^{i_m},$$

where $da_{i_1 \cdots i_m}$ is an ordinary differential. This process of exterior differentiation has the following properties, which can be easily verified:

(1) If ω, θ are two differential forms of degrees m and p respectively, then

$$(9) \qquad d(\omega\theta) = d\omega \cdot \theta + (-1)^m \omega d\theta.$$

(2) For any ω:

$$(10) \qquad d(d\omega) = 0.$$

(3) The process of exterior differentiation is invariant under coordinate substitutions, by which we mean the following: Let the local coordinates x^i in a neighborhood of P represent a set of points S in an open set D in a Euclidean space of n dimensions. Let an open set E in a Euclidean space of r dimensions with the coordinates $y^1, \cdots, y^r$ be mapped into D by the mapping

$$(11) \qquad x^i = f^i(y^1, \cdots, y^r), \qquad i = 1, \cdots, n,$$

such that S is the image of a set $T \subset E$, the functions f^i being supposed to be of class not less than 2. By the definition

$$F(x^1, \cdots, x^n) \to F(x^1(y), \cdots, x^n(y)), \qquad dx^i \to \sum_{k=1}^{r} \frac{\partial x^i}{\partial y^k} dy^k;$$

this mapping induces in a natural way (that is, with the multiplication of the Grassmann algebra preserved) a mapping of a differential form ω into a form of the space $(y^1, \cdots, y^r)$, which we denote by $\tilde{\omega}$. Then we have

$$d\tilde{\omega} = \widetilde{d\omega}. \tag{12}$$

In particular, when $r = n$, this signifies that the process of exterior differentiation is intrinsic, that is, independent of the local coordinate system under which it is performed.

The exterior differential forms are the forms under the integral sign in the theory of multiple integrals. In fact, let ω be a form of degree m defined on the manifold M^n. Let K^m be a chain of dimension m (in the sense of combinatorial topology) on M^n, which is a sum, with integral, rational, or real coefficients, of images of simplexes in a Euclidean space by mappings which are differentiable of class not less than 2. Then it is possible to define the integral of ω over K^m. If ∂K^{m+1} denotes the boundary of a chain K^{m+1} of dimension $m+1$, the theorem of Stokes can be written

$$\int_{\partial K^{m+1}} \omega = \int_{K^{m+1}} d\omega. \tag{13}$$

To see the connection between this formulation of the Stokes' Theorem and the usual one, notice that

$$d(Pdx + Qdy) = \left(\frac{\partial Q}{\partial x} - \frac{\partial P}{\partial y}\right) dxdy, \tag{14a}$$

$$\begin{aligned} &d(Pdx + Qdy + Rdz) \\ &\quad = \left(\frac{\partial R}{\partial y} - \frac{\partial Q}{\partial z}\right) dydz + \left(\frac{\partial P}{\partial z} - \frac{\partial R}{\partial x}\right) dzdx \\ &\quad\quad + \left(\frac{\partial Q}{\partial x} - \frac{\partial P}{\partial y}\right) dxdy. \end{aligned} \tag{14b}$$

4. **Fibre bundles.** A simple example of a fibre bundle is the manifold formed by all the nonzero vectors tangent to a sphere of three-dimensional Euclidean space. It is a topological manifold, but a very special one. However, it turns out that manifolds with similar properties play an important rôle in the application of topology to differential geometry. In the example of vectors tangent to a sphere the following three facts deserve attention:

(1) The spaces formed by the vectors with the same origin are homeomorphic to each other, and therefore to a fixed space which we call F_0.

(2) The tangent vectors drawn at all points of a neighborhood U

are homeomorphic to a topological product of U and F_0. In particular, if $P \in U$, there is a homeomorphism T which carries the tangent vectors with the origin P to F_0, T depending on P and U.

(3) In the notation of (2) let us write $T(P, U)$ for T. If P belongs to a second neighborhood V, the mapping $T(P, V)T^{-1}(P, U)$ is a homeomorphism of F_0 onto itself. It is possible to show that T can be chosen so that $T(P, V)T^{-1}(P, U) \in G$, where G is either the rotation group or the affine group in F_0.

The last statement can be demonstrated as follows: Let the neighborhoods be coordinate neighborhoods, such that U is the set of all points whose local coordinates u^1, u^2 satisfy the inequalities $|u^i| < \epsilon$, $i = 1, 2$. With respect to the local coordinate system u^1, u^2 a vector $\mathfrak{x}$ has the components X^1, X^2. The vectors with origins in U are thus decomposed into a topological product in an obvious way. If a point belongs to a neighborhood V with the local coordinates v^1, v^2 and the vector $\mathfrak{x}$ has the components Y^1, Y^2 in the coordinate system v^1, v^2, then it is well known that

$$Y^1 = \frac{\partial v^1}{\partial u^1} X^1 + \frac{\partial v^1}{\partial u^2} X^2, \qquad Y^2 = \frac{\partial v^2}{\partial u^1} X^1 + \frac{\partial v^2}{\partial u^2} X^2,$$

which is an affine transformation.

Guided by this particular example, we shall give the definition of a general fibre bundle. For the terms and notation which will be used consistently later the following table of reference is given:

General case	Special example	Notation
Fibre bundle	Manifold of tangent vectors of a sphere	$\mathfrak{F}$
Point of fibre bundle	Tangent vector	P
Base space	Sphere	M
Fibre	Vectors with same origin	F
Projection	Mapping of a tangent vector into its origin	$\pi(P) \in M$
Transformation group in fibre	Affine or rotation group in space of vectors with same origin	G

We define a fibre bundle $\mathfrak{F}$ as a topological space which has the following properties:

(1) There exists a (continuous) mapping π of F onto another topological space M: $\pi(F) = M$. The space M is called the base space and the mapping π is called the projection. The complete inverse image

$\pi^{-1}(p)$ of a point $p \in M$ is called the fibre at p. It will follow from assumptions made later that all the fibres are homeomorphic to each other and hence to a definite topological space F_0.

(2) There exists a family of neighborhoods which cover M. If U is a neighborhood of the family, the inverse image $\pi^{-1}(U)$ is a topological product, which means that there exists a homeomorphism ψ_U, depending on U, such that

$$\psi_U\{\pi^{-1}(U)\} = U \times F_0,$$

and that for every $p \in U$

$$\psi_U\{\pi^{-1}(p)\} = p \times F_0.$$

(3) Let U, V be two such neighborhoods of M, and let $p \in U \cap V$. The mapping $\psi_V\{\psi_U^{-1}(p \times F_0)\}$ is a homeomorphism of $p \times F_0$, and hence of F_0, into itself. This homeomorphism belongs to a group G given in advance in F_0.

For the cases in which we are interested both M and $\mathfrak{F}$ will be supposed to be differentiable manifolds of class not less than 1, and the group G will be a Lie group.

In order to show the scope of the notion of a fibre bundle, the following further examples are given—all pertaining to a differentiable manifold M of dimension n and class not less than 1:

(1) The tangent vectors of M constitute a fibre bundle with M as the base space.

(2) Let $\mathfrak{e}_1, \cdots, \mathfrak{e}_p$, $1 \leq p \leq n$, be tangent vectors of M with the common origin P. The elements $(P; \mathfrak{e}_1, \cdots, \mathfrak{e}_p)$ constitute a fibre bundle over M. Another fibre bundle is constituted by the elements satisfying the condition that $\mathfrak{e}_1, \cdots, \mathfrak{e}_p$ are linearly independent. These fibre bundles have been studied by Stiefel and Whitney [26, 33].[3]

(3) We call a scalar density of weight k a geometrical being, which has a component in each local coordinate system and whose components f and f^* in the local coordinate systems x^i and x^{*i} are connected by the relation

$$f^* = f \cdot J^k, \tag{15}$$

where

$$J = \partial(x^1, \cdots, x^n)/\partial(x^{*1}, \cdots, x^{*n}). \tag{15a}$$

The scalar densities or the nonzero scalar densities in M constitute a fibre bundle over M.

---

[3] Numbers in brackets refer to the Bibliography at the end of the paper.

(4) A general procedure of defining an important class of fibre bundles is as follows: Let M be imbedded in a Euclidean space E^{n+N} of dimension $n+N$. At a fixed point O of E^{n+N} let $H(n, N)$ be the Grassmann manifold formed by all the oriented linear subspaces of dimension n through O. A mapping $T(M)\subset H(n, N)$ defines a fibre bundle over M such that the fibre at each point P of M is the n-dimensional linear subspace through P parallel to $T(P)$.

One of the important results in the theory of fibre bundles is the introduction of the so-called characteristic cocycle. To give its definition we assume M to be a polyhedron, and let K be a simplicial decomposition of M. Let K^r be the r-dimensional skeleton of K, that is, the subcomplex of K consisting of all simplexes of dimension not greater than r. We suppose the fibre bundle to be orientable in the sense explained by Steenrod [24]. Let F_0 be connected and have an abelian fundamental group. Under these assumptions the following theorems can be established:

THEOREM 4.1. *Let $H^i(F_0)$ be the ith homology group (with integral efficients) of F_0. If*

$$H^1(F_0) = \cdots = H^{r-1}(F_0) = 0,$$

it is possible to define a continuous mapping ϕ of K^r into $\mathfrak{F}$ such that $\pi\phi(p)=p$, $p\in K^r$.

THEOREM 4.2. *With the same notation as in Theorem 4.1, if*

$$H^1(F_0) = \cdots = H^{r-1}(F_0) = 0, \qquad H^r(F_0) \neq 0,$$

there exists a cohomology class (with $H^r(F_0)$ as the coefficient group) γ^{r+1} of dimension $r+1$, whose vanishing is a necessary and sufficient condition that a continuous mapping ϕ of K^{r+1} into $\mathfrak{F}$ can be defined such that $\pi\phi(p)=p$, $p\in K^{r+1}$. This cohomology class is a topological invariant of $\mathfrak{F}$.

The cohomology class γ^{r+1} is called the characteristic cohomology class and any one of its cocycles a characteristic cocycle. If M is an orientable manifold, the dual of γ^{r+1} is a homology class of dimension $n-r-1$ and is called the characteristic homology class. In particular, if M is a differentiable manifold and $\mathfrak{F}$ the fibre bundle of nonzero tangent vectors of M, the characteristic cocycle γ is of dimension n and its value for the fundamental cycle of M is the Euler-Poincaré characteristic of M. In this way we see that the theory of characteristic class generalizes the classical theory of vector fields on a differentiable manifold.

5. Riemannian geometry. In the preceding sections are sketched the preliminaries for the theory of a geometric being. As an illustration we shall apply the tools so established to Riemannian geometry, the "simplest" and most important among the geometric beings. It will be seen that the results are essentially those given by Cartan and their relations with the now classical treatment by tensor analysis will also be indicated; but we have put ourselves in a more general viewpoint, which will lead to further developments.

Let M be a differentiable manifold of dimension n and class not less than 4. In M suppose a Riemannian metric be given, which, in terms of a local coordinate system x^i, is defined by a positive definite quadratic differential form[4]

$$ds^2 = \sum_{i,j} g_{ij}(x^k)(dx^i dx^j), \qquad g_{ij} = g_{ji}. \tag{16}$$

The manifold M is then called a Riemannian manifold.

Consider a point P of M and the contravariant tangent vectors of M having the origin P. With the differential form (16) (or the tensor g_{ij}) the scalar product of two vectors $\mathfrak{e}$, $\mathfrak{f}$ can be defined, which we denote by $\mathfrak{e}\cdot\mathfrak{f}$. A vector $\mathfrak{e}$ is called a unit vector if $\mathfrak{e}^2=1$. An ordered set of n vectors $\mathfrak{e}_1, \cdots, \mathfrak{e}_n$ is said to constitute a frame (ennuple), if

$$\mathfrak{e}_i\mathfrak{e}_j = \delta_{ij} = \begin{cases} 1, & i = j, \\ 0, & i \neq j, \end{cases} \qquad i, j = 1, \cdots, n. \tag{17}$$

A differentiable curve through P is defined by the equations

$$x^i = x^i(s), \tag{18}$$

where s is the arc length of the curve and the functions are of class not less than 1. We denote by dP/ds the unit tangent vector of the curve at P. It is easy to show, by referring to a local coordinate system, that there exists a vector (which we shall denote by dP) whose components with respect to a frame are linear differential forms and which is equal to the product of ds and the unit tangent vector along a curve. This vector is intrinsic, that is, independent of the choice of coordinates. Referred to a frame $\mathfrak{e}_1, \cdots, \mathfrak{e}_n$, we can write

$$dP = \omega_1\mathfrak{e}_1 + \cdots + \omega_n\mathfrak{e}_n, \tag{19}$$

where $\omega_1, \cdots, \omega_n$ are linear differential forms. Then we have, along a curve, $(dP/ds)^2 = 1$ or

[4] The differential form in question is an ordinary and not an exterior differential form. As a means of distinction, we insert a parenthesis about the differentials when the former is the case.

(20) $$ds^2 = (\omega_1)^2 + \cdots + (\omega_n)^2.$$

The forms ω_i, $i=1, \cdots, n$, are therefore the forms the sum of whose squares is equal to the given (ordinary) quadratic differential form.

As the origin P runs over M, the frames $e_1, \cdots, e_n$ constitute a fibre bundle with M as the base space. Each fibre is topologically the space of all frames having the same origin. The fibre bundle is of dimension $n(n+1)/2$ and each fibre is of dimension $n(n-1)/2$. The group which operates in each fibre is the group of orthogonal transformations.

There is yet no intrinsic meaning attached to the vectors de_i. We want to see whether it is possible to define intrinsically certain differential forms ω_{ij} such that

(21) $$de_i = \sum_j \omega_{ij} e_j.$$

From (17) it follows that ω_{ij} has to be skew-symmetric in its indices:

(22) $$\omega_{ij} + \omega_{ji} = 0.$$

The answer to our question is given by the following theorem:

THEOREM 5.1. *There exists one, and only one, set of linear differential forms ω_{ij} in the fibre bundle such that the equation* (21) *and the equation*

(23) $$d(dP) = 0$$

hold.

In fact, the condition (23) gives, when expanded,

(24) $$d\omega_i - \sum_j \omega_j \omega_{ji} = 0.$$

To calculate $d\omega_i$ we make use of a local coordinate system x^i and take a definite decomposition of ds^2 as a sum of squares:

(25) $$ds^2 = (\theta_1)^2 + \cdots + (\theta_n)^2,$$

where

(25a) $$\theta_i = \sum_i a_{ij}(x) dx^j.$$

Since (20) and (25) give two decompositions of the same ordinary quadratic differential form as sums of squares, we have

(26) $$\omega_i = \sum u_{ij}\theta_j,$$

where u_{ij} are the elements of an orthogonal matrix:

(26a) $$U = (u_{ij}), \qquad UU' = I = \text{identity matrix}.$$

Forming the exterior derivative of the equation (26), we get

$$d\omega_i = \sum du_{ij}\theta_j + \sum u_{ij} d\theta_j,$$

which shows that $d\omega_i$ is of the form

(27) $$d\omega_i = \sum \omega_j \phi_{ji},$$

where ϕ_{ji} are linear differential forms in the variables x^i, u_{ij} and can be explicitly calculated. Substitution into (24) gives

$$\sum \omega_j(\omega_{ji} - \phi_{ji}) = 0,$$

which gives in turn

(28) $$\omega_{ji} - \phi_{ji} = \sum \lambda_{jik}\omega_k, \qquad \lambda_{jik} = \lambda_{kij}.$$

In this equation the forms ϕ_{ji} are known and the quantities λ_{jik} will be determined in order that ω_{ji} be determined. Since ω_{ij} is skew-symmetric in its indices, we have

$$-(\phi_{ij} + \phi_{ji}) = \sum (\lambda_{ijk} + \lambda_{jik})\omega_k.$$

This equation shows that $\phi_{ij}+\phi_{ji}$ is of the form

(29) $$\phi_{ij} + \phi_{ji} = -\sum A_{ijk}\omega_k, \qquad A_{ijk} = A_{jik},$$

and that then

(30) $$\lambda_{ijk} + \lambda_{jik} = A_{ijk}.$$

The quantities λ_{ijk}, being symmetric in the first and third indices, are then uniquely determined and are given by

(31) $$\lambda_{ijk} = (-A_{ikj} + A_{jki} + A_{ijk})/2.$$

This proves our theorem.

The equation (21) may be interpreted as giving a means of transporting a vector at an infinitesimally near point of P to a vector at P. It is essentially the notion known as the parallelism of Levi-Civita.

By forming the exterior derivative of (24) and making use of these equations themselves, we get

$$\sum_j \omega_j\left(\sum_k \omega_{ik}\omega_{kj} - d\omega_{ij}\right) = 0.$$

It follows easily from this that we can write

(32) $$d\omega_{ij} = \sum_k \omega_{ik}\omega_{kj} + \Omega_{ij},$$

where Ω_{ij} are of the form

$$\Omega_{ij} = \sum R_{ij,kl}\omega_k\omega_l, \qquad R_{ij,kl} + R_{ij,lk} = 0, \tag{33}$$
$$\Omega_{ij} + \Omega_{ji} = 0.$$

We shall call the forms Ω_{ij} the curvature forms.

In a slightly different version our results may be summarized in the theorem:

THEOREM 5.2. *Let M be a Riemannian manifold of dimension n and let $\mathfrak{F}$ be the fibre bundle of all frames of M. There is a unique way to define in $\mathfrak{F}$ (of dimension $n(n+1)/2$) a set of $n(n+1)/2$ linearly independent linear differential forms ω_i, ω_{ij}, which satisfy the equations* (19), (21), (24).

Let us indicate briefly the relations of these considerations to the ordinary treatment by means of tensor analysis. In tensor analysis emphasis is laid particularly on the local coordinate system. The procedure is as follows: With respect to the local coordinates x^i let $\mathfrak{f}_k$ be the velocity vector of the coordinate curve $x^i = \text{const.}$, $i \neq k$, with x^k as the time. Then we have

$$\mathfrak{f}_i \cdot \mathfrak{f}_k = g_{ik}, \tag{34}$$

and

$$dP = dx^1 \cdot \mathfrak{f}_1 + \cdots + dx^n \cdot \mathfrak{f}_n. \tag{35}$$

The components X^i of a vector $\mathfrak{x}$ with respect to the vectors $\mathfrak{f}_k$ are defined by the equation

$$\mathfrak{x} = \sum X^i \mathfrak{f}_i. \tag{36}$$

By an argument analogous to the proof of Theorem 5.1 it can be shown that there exists a set of linear differential forms such that

$$d\mathfrak{f}_i = \sum \theta_i^j \mathfrak{f}_j, \tag{37}$$

and

$$d(dP) = 0.$$

The vector $\mathfrak{x} + d\mathfrak{x}$ is called parallel to $\mathfrak{x}$ if $d\mathfrak{x} = 0$, which can be written, by using equations (36) and (37),

$$dX^i + \sum \theta_j^i X^j = 0. \tag{38}$$

These equations are easily recognized to be the well known equations which define the parallelism of Levi-Civita.

The important fact in Riemannian geometry is the necessity of taking into consideration not only the Riemannian manifold itself but the fibre bundle over it. In our approach we consider directly the fibre bundle. The usual tensor approach avoids it by laying emphasis on the local coordinate system. Quantities of geometrical interest are those which obey a simple transformation law under changes of the local coordinate systems. For Riemannian geometry the two methods are essentially the same, so far as local problems are concerned. It is, however, to be remarked that even for local problems their generalizations to other geometric beings will lead to different formulations. We shall give in the next section a systematic way of treating the local theory of a geometric being.

6. Method of equivalence. So far the introduction given in the above section of the fibre bundle over a Riemannian manifold M seems to be accidental. The underlying reason will be most clear, if we start with the following problem:

Let two Riemannian metrics be given by the positive definite quadratic differential forms

$$ds^2 = \sum_{i,j} g_{ij}(x)(dx^i dx^j), \qquad g_{ij} = g_{ji}, \tag{39}$$

$$ds^{*2} = \sum_{i,j} g_{ij}^*(x^*)(dx^{*i} dx^{*j}), \qquad g_{ij}^* = g_{ji}^*, \tag{40}$$

in two coordinate neighborhoods U, U^* respectively. We determine the conditions that a neighborhood $V \subset U$ can be mapped by a differentiable mapping (of sufficiently large class) with nonvanishing functional determinant into U^* such that

$$ds^{*2} = ds^2.$$

This so-called form problem was solved by Christoffel and Lipschitz. It is clear that if the problem has a solution, the local theories of the two Riemannian metrics are essentially the same.

To give a different formulation of our problem we determine two sets of linear differential forms $\theta_i(x, dx)$ and $\theta_i^*(x^*, dx^*)$ such that

$$ds^2 = (\theta_1)^2 + \cdots + (\theta_n)^2, \qquad ds^{*2} = (\theta_1^*)^2 + \cdots + (\theta_n^*)^2.$$

Our problem will have a solution if and only if the differentiable mapping in question and the functions $u_{ij}(x)$ can be determined in a neighborhood $V \subset U$ such that

$$\theta_i^*(x^*, dx^*) = \sum_j u_{ij}(x)\theta_j(x, dx), \tag{41}$$

where the matrix

$$U = (u_{ij}) \tag{42}$$

is an orthogonal matrix.

In this way our problem, originally formulated in terms of quadratic differential forms, is reduced to one in linear differential forms. In this formulation the problem is a particular case of a more general problem, which was first stated and solved by Elie Cartan. Cartan's problem is as follows:

Let $\theta_i(x, dx)$ and $\theta_i^*(x^*, dx^*)$ be two sets of n linearly independent linear differential forms in the coordinate neighborhoods U, U^* of the coordinates $x^i, x^{*i}, i=1, \cdots, n$. Let Γ be a linear group in the vector space of n dimensions. We wish to determine the conditions that there exists a differentiable mapping with a nonvanishing functional determinant of a neighborhood $V \subset U$ into U^* such that

$$\theta_i^* = \sum_j u_{ij}(x)\theta_j, \tag{43}$$

where the linear transformation belongs to the given linear group Γ.

In our case Γ is the orthogonal group.

The discussion of this local problem requires Cartan's theory of Pfaffian systems in involution and is not simple. For details we refer to [3].

We only remark that if Γ consists of the identity only, the solution of this problem is relatively simple. The general procedure is to reduce the problem to this particular case by the introduction of new variables. We shall illustrate this method by considering our form problem. Let u_{ij} and u_{ij}^* be two sets of new variables which are elements of the orthogonal matrices

$$U = (u_{ij}), \qquad U^* = (u_{ij}^*),$$

and let us put

$$\omega_i = \sum u_{ij}\theta_j, \qquad \omega_i^* = \sum u_{ij}^*\theta_j^*. \tag{44}$$

Our form problem will have a solution if and only if there exists a differentiable mapping of a neighborhood $V \subset U$ into U^* and ar orthogonal matrix X, whose elements are functions in V, such that under this mapping and the mapping between (u_{ij}) and (u_{ij}^*) defined by

$$U^*X_2 = U, \tag{45}$$

we have

$$\omega_i^* = \omega_i. \tag{46}$$

From the last equation we derive

$$d\omega_i^* = d\omega_i. \tag{47}$$

By an argument exactly identical to one used in the last section it can be shown that there exists one and only one set of linear differential forms ω_{ij} in x^i, u_{ij}, which are skew-symmetric in their indices and which satisfy the equations

$$d\omega_i = \sum \omega_j \omega_{ji}.$$

Similarly, we find a set ω_{ij}^* $(= -\omega_{ji}^*)$ of linear differential forms such that

$$d\omega_i^* = \sum \omega_j^* \omega_{ji}^*.$$

When (46) and (47) hold, it is easy to derive

$$\omega_{ij}^* = \omega_{ij}. \tag{48}$$

Let V be a sufficiently small neighborhood of U and let $O(n)$ be the group manifold of the orthogonal group. We consider the topological product $V \times O(n)$. Similarly, we have the topological product $V^* \times O^*(n)$. In order that our form problem possess a solution it is necessary and sufficient that there exist a differentiable mapping between $V \times O(n)$ and $V^* \times O^*(n)$ such that under this mapping the equations (46) and (48) hold. It is easy to see that the forms ω_i, ω_{ij}, whose number $n(n+1)/2$ is equal to the dimension of $V \times O(n)$, are linearly independent. Thus our problem is reduced to Cartan's general problem with Γ consisting of the identity only.

It can therefore be seen that the discussions of the last section are nothing else than a geometrical treatment of the above analytic considerations. We summarize these results as follows:

Given a Riemannian manifold M of dimension n with a positive definite quadratic differential form ds^2. To M is associated the fibre bundle, with M as its base space, of all sets of linear differential forms (that is, covariant vectors) the sum of whose squares is equal to ds^2. The fibre bundle is of dimension $n(n+1)/2$ and there exists in it a set of $n(n+1)/2$ linearly independent linear differential forms ω_i, ω_{ij} which are invariant. There exists locally a differentiable mapping which carries one ds^2 into another when and only when such a mapping exists in the corresponding fibre bundles under which the forms ω_i, ω_{ij} are respectively equal.

We may say that the deep reason that this very fibre bundle (namely, the one whose fibres are topologically the manifolds $O(n)$) is attached to a Riemannian manifold comes as a result of the solution of the form problem. From the case of the Riemannian manifold it is clear that, for the theory of other kinds of geometric beings, the solution of a corresponding local "equivalence problem" is a natural preliminary step for the establishment of a proper theory in the large. In fact, from the solution of the equivalence problem the nature of the fibre bundle to be attached to the manifold can be determined.

There is a great variety of geometric beings which are of interest in differential geometry. In order to illustrate our method we shall take as a further example the so-called geometry of paths. Let M be a differentiable manifold of dimension n. In a neighborhood with the local coordinates x^i let there be given a system of differential equations of the second order:

$$\frac{d^2x^i}{dt^2} + \sum \Gamma^i_{jk}(x) \frac{dx^j}{dt} \frac{dx^k}{dt} = F(x, t) \frac{dx^i}{dt}, \tag{49}$$

where the functions Γ^i_{jk}, F are supposed to be differentiable of class not less than 3. The integral curves of the differential system (49) are called the paths. They have the property that through every point of M and tangent to every vector through that point there passes one and only one path. The independent variable t is a parameter on the path. It is irrelevant in the sense that a change of parameter $t=\phi(\tau)$, where $\phi(\tau)$ is differentiable, gives the same path. In short, our manifold M carries a system of paths which is defined in each local coordinate system by a differential system of the form (49) and our problem is to study the geometrical properties in M arising from the paths. An important particular case of this problem is the case that M is the projective space of n dimensions and the paths are the straight lines in M, which, in a suitable local coordinate system, are defined by the system (49), with $\Gamma^i_{jk}=0$.

Following the spirit of our solution of the form problem we start by considering the following problem of equivalence: In another coordinate neighborhood with the local coordinates x^{*i} let a system of paths be given by a differential system analogous to (49). Determine the conditions under which a differentiable mapping between the coordinates x^i and x^{*i}, which carries one system of paths into another, exists.

To apply the method of equivalence we shall write the system (49) in a system of total differential equations. For simplicity let us suppose that, for the integral curves of (49) under consideration, x^n is not a constant. We can then take x^n to be the parameter along the curves, and the system (49) can be written

$$\frac{dx^\alpha}{dx^n}=y^\alpha,\qquad \frac{dy^\alpha}{dx^n}+\sum\Gamma^\alpha_{jk}y^jy^k=Fy^\alpha,\qquad \sum\Gamma^n_{jk}y^jy^k=F,$$

where the index α runs from 1 to $n-1$ and $y^n=1$. This system can be written as a system of total differential equations as follows:

$$(50)\qquad \begin{aligned}\theta^\alpha &= dx^\alpha - y^\alpha dx^n = 0,\\ \pi^\alpha &= dy^\alpha + \{\sum\Gamma^\alpha_{jk}y^jy^k - y^\alpha\sum\Gamma^n_{jk}y^jy^k\}dx^n = 0.\end{aligned}$$

We introduce also the form dx^n, so that θ^α, dx^n constitute a set of linearly independent linear differential forms in the coordinates x^i. On the other manifold with the local coordinates x^{*i} we shall introduce in a similar way the new variables $y^{*\alpha}$ and the forms $\theta^{*\alpha}$, dx^{*n}, $\pi^{*\alpha}$. Let us write

$$(51)\qquad \Theta=\begin{pmatrix}dx^n\\ \theta^\alpha\\ \pi^\alpha\end{pmatrix},\qquad \Theta^*=\begin{pmatrix}dx^{*n}\\ \theta^{*\alpha}\\ \pi^{*\alpha}\end{pmatrix},$$

and let us introduce the matrix

$$(52)\qquad U=\begin{pmatrix}A(1,1) & B(1,n-1) & 0\\ 0 & C(n-1,n-1) & 0\\ 0 & D(n-1,n-1) & E(n-1,n-1)\end{pmatrix},$$

whose nonzero elements are independent variables, and where the numbers in the parentheses denote the numbers of rows and columns of the sub-matrices in question. It is easy to show that our problem of equivalence possesses a solution when and only when there is a differentiable mapping between two neighborhoods (x^i, y^α) and $(x^{*i}, y^{*\alpha})$ and the elements of U can be determined as functions of x^i, y^α, such that under this mapping the matrix equation

$$(53)\qquad \Theta^*=U\Theta$$

holds. Now in the linear vector space of $2n-1$ dimensions all the matrices U form a group. Hence in this formulation our problem of equivalence is a particular case of the problem of equivalence of Cartan stated above.

The problem can be solved by applying Cartan's general procedure. We shall not give the details here. The solution of a problem which contains this problem as a particular case was given in [12].

Our final result can be stated in the theorem:

THEOREM 6.1. *In a differentiable manifold M of dimension n and* *class not less than 3 let a system of paths be given, defined locally by the differential equations* (49). *It is possible to introduce in a neighborhood of M with the local coordinates* x^i, $n(n+1)$ *other variables* u_A, $A=1, \cdots, n(n+1)$, *and to define* $n(n+2)$ *linearly independent linear differential forms* ω_B, $B=1, \cdots, n(n+2)$, *in the variables* x^i, u_A, *having the following property: A necessary and sufficient condition that a local differentiable mappings exists, which carries one system of paths into another with the local coordinates* x^{*i} *and the new variables* u_A^*, *and so on, is that there exists a differentiable mapping between two neighborhoods of the spaces* (x^i, u_A) *and* (x^{*i}, u_A^*) *under which the forms* ω_B *and* ω_B^* *are mapped into each other.*

With respect to a local coordinate system x^i the variables x^i, u_A, $A=1, \cdots, n(n+1)$, determine a topological product. We are thus led to consider the fibre bundle over M, which is of dimension $n(n+2)$ and is locally given by the coordinates x^i, u_A. In the fibre bundle there are defined exactly $n(n+2)$ intrinsic linearly independent linear differential forms. The significance of this fibre bundle to the differential geometry of paths is evident. It is important to point out that in the geometry of paths it is, in view of our Theorem 6.1, this fibre bundle, and not the fibre bundle of tangent vectors of M, which plays the fundamental rôle.

From the above discussions of Riemannian geometry and the geometry of paths it can be seen that the method of equivalence offers the best weapon for a frontal attack on the problem of a geometric being. Only from the outcome of the solution of the problem of equivalence can it be decided the most important kind of fibre bundle to be defined over the manifold. It is of course possible to define, in a more or less intuitive way, other kinds of fibre bundles over the manifold, but they will play a minor rôle. We wish also to remark that the solution of this local problem of equivalence is not always simple. So far as the writer is aware, the problem of equivalence for a skew-symmetric covariant tensor of order two, for instance, has not been explicitly solved (although the solution is theoretically always possible).

7. Relations between the fibre bundle and its base manifold. We have attempted to show in the above, in the cases of Riemannian geometry and the geometry of paths, how the local theory of a geometric being leads naturally to the fibre bundle to be associated to the manifold on which the geometric being is defined. The fact remains true of any geometric being, but its proof requires the fundamental theorem on partial differential equations to the effect that every system of partial differential equations can be "prolonged" to a system in involution, and we shall not enter into its discussion. It is to be remarked that for problems of this sort, as for any problem concerned with both pure and applied mathematics, the general result and the particular cases offer entirely different problems, one being an existence theorem and the other being an explicit solution. In our examples we solve the problems explicitly, thereby insuring the existence of the solution.

After defining over the given manifold M a fibre bundle $\mathfrak{F}$, the next step is to study their relationship. Before proceeding farther, let us remark that the fibre bundle $\mathfrak{F}$ will give rise to new fibre bundles over M by identification. Consider, for instance, the case that M is a Riemannian manifold. Let $P\mathfrak{e}_1 \cdots \mathfrak{e}_n$ be a frame in M. For any integer p, $1 \leq p \leq n$, we shall define a class of frames to be all frames $P\mathfrak{e}_1 \cdots \mathfrak{e}_n$ such that the point P and the vectors $\mathfrak{e}_1, \cdots, \mathfrak{e}_p$ are identical. These classes of frames constitute, with a natural topology, a fibre bundle over M, which we denote by $\mathfrak{F}^{(p)}$. $\mathfrak{F}^{(p)}$ is the fibre bundle of all ordered sets of p mutually perpendicular unit vectors of M. In particular, $\mathfrak{F}^{(1)}$ is the fibre bundle of unit vectors of M and $\mathfrak{F}^{(n)}$ is the fibre bundle $\mathfrak{F}$ itself. For distinction we shall call $\mathfrak{F}$ the principal fibre bundle of M and all others obtained by identification the associated fibre bundles. We shall consistently assume that the identification is made only in each fibre, so that the projection of the associated bundle onto the base manifold is the one induced by that of the principal bundle. It is seen that this process is very general.

Let $\mathfrak{F}$ be the principal fibre bundle over M and $\mathfrak{G}$ an associated fibre bundle over M. There is a projection $\pi(\mathfrak{F}) \subset M$ and a projection $\pi_1(\mathfrak{G}) \subset M$, and also a projection $\pi_2(\mathfrak{F}) \subset \mathfrak{G}$, defined by assigning to a point of $\mathfrak{F}$ the class to which it belongs during the identification. From the remark made at the end of the last paragraph it is clear that $\pi = \pi_1\pi_2$. We also remember that a set of intrinsic linearly independent linear differential forms is defined in $\mathfrak{F}$ by the geometric being, whose number is equal to the dimension of $\mathfrak{F}$. Our main aim will be to see how these differential forms behave under the projections π, π_1, π_2.

In order to understand this relationship more clearly we shall make a digression to recall some elementary facts in combinatorial topology. Let K be a finite complex of dimension n, whose simplexes we denote by σ_i^r, $0 \leq r \leq n$, $1 \leq i \leq \alpha_r$. Let R be a commutative ring. A chain of dimension r or an r-chain is a sum

$$\overset{r}{C} = \sum_{i=1}^{\alpha_r} \lambda_i \overset{r}{\sigma}_i, \qquad \lambda_i \in R. \tag{54}$$

To the chains C^r we introduce a boundary operator ∂, which is linear:

(55) $$\partial(C_1^r + C_2^r) = \partial C_1^r + \partial C_2^r, \qquad \partial(\lambda C^r) = \lambda\partial C^r, \qquad \lambda \in R,$$

and is defined for the simplexes σ_i^r by the so-called incidence relations:

(56) $$\partial\sigma_i^r = \sum_{j=1}^{\alpha_{r-1}} \eta_{ij}^{(r)} \sigma_j^{r-1}.$$

The relations (55) and (56) completely determine the boundary operator ∂ and the chain ∂C^r of dimension $r-1$ is called the boundary of C^r. It is easy to verify that

(57) $$\partial\partial C^r = 0.$$

A chain whose boundary is zero is called a cycle. Equation (57) shows that a chain which is the boundary of another chain is a cycle, called a bounding cycle. The cycles form an abelian group and the bounding cycles a subgroup of it. Their quotient group is called the r-dimensional homology group of K (with the additive group of R as its coefficient group).

A cochain γ^r of dimension r is a linear function of the r-chains. We shall denote the function by the notation $\gamma^r \cdot C^r$, which has therefore the properties:

(58) $$\gamma^r \cdot (C_1^r + C_2^r) = \gamma^r \cdot C_1^r + \gamma^r \cdot C_2^r, \quad \gamma^r \cdot (\lambda C^r) = \lambda \gamma^r \cdot C^r, \qquad \lambda \in R.$$

From a cochain γ^r of dimension r we define a cochain $\delta\gamma^r$ of dimension $r+1$, called its coboundary, by means of the relation

(59) $$\delta\gamma^r \cdot C^{r+1} = \gamma^r \cdot (\partial C^{r+1}).$$

It then follows from (57) and (59) that

(60) $$\delta\delta\gamma^r = 0.$$

A cochain whose coboundary is zero is called a cocycle. All the cocycles of dimension r form an abelian group, which contains as subgroup the group of all cocycles which are coboundaries. Their quotient group is called the r-dimensional cohomology group of K.

Now, let M be a differentiable manifold of dimension n and class not less than 2. Let ω be a differential form of degree r in M, whose coefficients are of class not less than 2. As we have remarked in §3, it is possible to define in a rigorous manner the integral of ω over an r-simplex and also the integral of ω over an r-chain. Moreover, the functional $\int_{C^r}\omega$ is linear in the r-chains C^r. Hence it defines an r-cochain with the ring of real numbers as the coefficient ring. When there is no confusion, we shall speak of ω as an r-cochain.

From (13) the Theorem of Stokes can be written in the form

(61) $$\int_{C^{r+1}} d\omega = \int_{\partial C^{r+1}} \omega.$$

Comparison with (59) shows that if ω defines an r-cochain, then $d\omega$ defines its coboundary. It follows that the r-cochain ω is a cocycle if and only if $d\omega = 0$ (in which case ω is called exact), and that ω is a coboundary if there exists a form θ whose exterior derivative $d\theta$ is equal to ω (in which case ω is called derived).

It was de Rham [22] who proved the important theorem that to every r-cochain γ (with real or rational coefficient group) there exists a differential form ω of degree r which defines γ according to the above process. We see from §4 that from the fibre bundles over M there are the characteristic cocycles in M which are invariants of the fibre bundles. A theorem in the differential geometry in the large will be obtained, whenever it is possible to define a characteristic cocycle by means of a differential form constructed locally from the geometric being.

Suppose now that a simplicial mapping f exists, which maps a complex K into a complex K^*. The mapping f induces a mapping ϕ of the chains of K into the chains of K^*. From ϕ a mapping ψ of the cochains of K^* into the cochains of K can be defined. In fact, if γ^* is a cochain of dimension r of K^*, we define $\psi\gamma^* \cdot \sigma^r = \gamma^* \cdot \phi\sigma^r$ for every simplex σ^r of K. This so-called inverse mapping ψ maps cocycles of K^* into cocycles of K and cocycles of the same cohomology class into cocycles of the same cohomology class. Therefore it induces a mapping of each cohomology group of K^* into the cohomology group of the same dimension of K and it can be proved that it is a homomorphism. It can also be proved that the homomorphism remains unchanged, if f is replaced by a mapping homotopic to f. The result is that we can define from a continuous mapping of a polyhedron P into a polyhedron P^* a homomorphism of a cohomology group of P^* into one of the same dimension of P.

In the case of a fibre bundle $\mathfrak{F}$ over M in which we are interested, there is a continuous mapping of $\mathfrak{F}$ into M, namely the projection. We assume that both $\mathfrak{F}$ and M are differentiable manifolds. According to the above, the projection induces a mapping of a cohomology group of M into one of the same dimension of $\mathfrak{F}$. In particular, an exact differential form of M will be mapped into an exact differential form of $\mathfrak{F}$.

Consider the case that M is a differentiable manifold in which a geometric being is given. Let $\mathfrak{F}$ be the principal fibre bundle and $\mathfrak{G}$ an associated fibre bundle over M. It is known that there is a projection $\pi_2(\mathfrak{F}) \subset \mathfrak{G}$. There are also linear differential forms defined in $\mathfrak{F}$

by the geometric being in M. From the linear differential forms it is possible to construct, by operations of the Grassmann algebra and the Cartan calculus, differential forms of higher degree. For any such differential form it is important to decide whether it is the image of a differential form of $\mathfrak{G}$ under the inverse mapping induced by π_2. There are several criterions for this problem. We shall give a criterion for the case that M is a Riemannian manifold, but the process can be readily generalized to other geometric beings.

Let $\mathfrak{G}=\mathfrak{F}^{(p)}$ and let the forms ω_i, ω_{ij}, Ω_{ij} be defined in the principal fibre bundle $\mathfrak{F}$. Let Π be constructed from ω_i, ω_{ij}, Ω_{ij} by operations of the Grassmann algebra. We wish to decide whether Π is the image of a form in $\mathfrak{G}$.

For this purpose let us use in this paragraph the following ranges of indices:

$$1 \leqq \alpha, \beta \leqq p, \qquad p+1 \leqq r, s \leqq n, \qquad 1 \leqq i, k \leqq n.$$

Let δ be an operation (not coboundary) under which the figure $Pe_1 \cdots e_p$ remains unchanged, so that

$$\omega_i(\delta) = 0, \qquad \omega_{\alpha i}(\delta) = 0. \tag{62}$$

We also put

$$\omega_{rs}(\delta) = e_{rs}. \tag{62a}$$

From (24) and (32) we have

$$\begin{aligned} &\delta\omega_\alpha = 0, \quad \delta\omega_r = \sum e_{rs}\omega_s, \quad \delta\omega_{\alpha\beta} = 0, \quad \delta\omega_{\alpha r} = -\sum e_{sr}\omega_{\alpha s}, \\ &\delta\Omega_{\alpha\beta} = 0, \quad \delta\Omega_{\alpha r} = \sum e_{rs}\Omega_{\alpha s}, \quad \delta\Omega_{rs} = \sum (e_{rt}\Omega_{ts} - \Omega_{rt}e_{ts}). \end{aligned} \tag{63}$$

Then we have the following theorem:

THEOREM 7.1. *A form Π is the image of a differential form in $\mathfrak{F}^{(p)}$ if and only if $\delta\Pi=0$.*

This theorem is also true for $p=0$, if we make the convention $\mathfrak{F}^{(0)}=M$.

If a form in $\mathfrak{F}$ is the image of a form in $\mathfrak{F}^{(p)}$, we shall say simply that it is a form of $\mathfrak{F}^{(p)}$.

It can easily be verified, by means of our criterion, that the following forms are forms of M:

$$\begin{aligned} &\Delta_m = \sum \Omega_{i_1i_2}\Omega_{i_2i_3} \cdots \Omega_{i_{2m}i_1}, \quad 1 \leqq m \leqq n/4, \\ &\Delta_0 = \sum \epsilon_{i_1\cdots i_n}\Omega_{i_1i_2} \cdots \Omega_{i_{n-1}i_n}, \quad \text{if } n \text{ is even}, \end{aligned} \tag{64}$$

where $\epsilon_{i_1\cdots i_n}$ is the Kronecker index, which is $+1$ or -1, according as $i_1, \cdots, i_n$ forms an even or odd permutation of $1, \cdots, n$, and is otherwise zero.

Important is the case of an exact form of M which is not derived in M, but has an image in a fibre bundle which is derived. The form Δ_0 in (64) has, for instance, an image in $\mathfrak{F}^{(1)}$, which is derived. It is on this fact that the generalized Gauss-Bonnet formula was proved in a simple way [10]. A further example to illustrate this idea will be given for Riemannian manifolds of four dimensions. (Compare [14].)

8. **Theorem of Pontrjagin.** From a topological viewpoint Pontrjagin has in two recent notes [20, 21] studied a problem related to ours. We shall summarize here his results in a slightly more general version which will allow its generalization to complex analytic Hermitian manifolds.

Let E^{n+N} be a real oriented Euclidean space of dimension $n+N$. In E^{n+N} we consider the Grassmann manifold $H(n, N)$, formed by all the oriented n-dimensional linear spaces through the origin. To each submanifold M' of $H(n, N)$ there is defined in a natural way a sphere bundle with M' as the base space, that is, a fibre bundle whose fibres consist of spheres cut on the unit hypersphere about the origin of E^{n+N} by the n-dimensional linear subspaces of M'. Given any manifold M, and a mapping $f(M) \subset H(n, N)$, a sphere bundle can be defined over M as base space by taking as the fibre attached to a point P of M the unit hypersphere in $f(P)$. With a proper definition of equivalence of sphere bundles, Whitney and Steenrod have proved the following theorem [25, 33]:

THEOREM 8.1. *To a given bundle of spheres of dimension $n-1$ over a compact manifold M there exists a mapping $f(M) \subset H(n, N)$, which defines a sphere bundle over M equivalent to the given one, provided that* dim $M \leq N$. *Two sphere bundles over M defined by the mappings $f_i(M) \subset H(n, N)$, $i=1, 2$, are equivalent, when and only when the mappings f_1 and f_2 are homotopic.*

From the mapping $f(M) \subset H(n, N)$ the cocycles of dimension not greater than dim M of $H(n, N)$ are mapped by the inverse mapping into cocycles of M. The latter part of Theorem 8.1 asserts that the image cocycles of M are cohomologous, if the sphere bundles are equivalent. The image of a cohomology class of $H(n, N)$ under the inverse mapping is therefore independent of the choice of the mapping f, provided that the sphere bundle induced remains equivalent. Such a cohomology class in M is called a generalized characteristic class.

If M is a differentiable manifold and $\mathfrak{F}$ the tangent sphere bundle over M, we can find the mapping f in question by imbedding M in a Euclidean space E^{n+N} and defining $f(P)$ to be the n-plane of $H(n, N)$ parallel to the tangent n-plane of M at P. M has then a Riemannian metric induced by the Euclidean metric of E^{n+N}.

In E^{n+N} consider the frames $Pe_1 \cdots e_{n+N}$. We use the ranges of indices

$$1 \leqq i, k \leqq n, \qquad n+1 \leqq r, s \leqq n+N, \qquad 1 \leqq A, B, C \leqq n+N,$$

and put

$$\omega_{AB} = de_A \cdot e_B.$$

Then we have

$$d\omega_{AB} = \sum_C \omega_{AC}\omega_{CB}.$$

To each n-plane of $H(n, N)$ we attach the frames $Oe_1 \cdots e_n e_{n+1} \cdots e_{n+N}$ such that $e_1, \cdots, e_n$ belong to the n-plane. Then the nN linear differential forms $\omega_{\alpha r}$ constitute a set of linearly independent forms of $H(n, N)$. To represent an integral cocycle of $H(n, N)$ in the sense of §7 we define

$$\Omega_{ik} = -\sum_r \omega_{ir}\omega_{kr}. \tag{65}$$

It is then a consequence of the first main theorem in the theory of vector invariants for orthogonal groups [31] that every integral cocycle of dimension not greater than n of $H(n, N)$ can be represented by a differential form which is a linear combination, with constant coefficients, of the products of the forms Δ_m, Δ_0 defined by (64), where the Ω_{ik} are here given by (65).

The forms Ω_{ik} defined in (65) are forms of $H(n, N)$. But their images in M induced by the mapping f are the curvature forms of M derived from the Riemannian metric of M induced by E^{n+N}. In fact, let us attach to each point P of M all frames $Pe_1 \cdots e_{n+N}$ such that $e_1, \cdots, e_n$ are tangent vectors. On M we have then

$$\omega_r = 0,$$

and

$$\begin{aligned} d\omega_i &= \sum_k \omega_k\omega_{ki}, \\ d\omega_{ik} &= \sum_j \omega_{ij}\omega_{jk} - \sum_r \omega_{ir}\omega_{kr} = \sum_j \omega_{ij}\omega_{jk} + \Omega_{ik}. \end{aligned} \tag{66}$$

From (66) it is seen that the images of the Ω_{ik} in (65) are the curvature forms of the Riemannian metric induced on M.

In this way we can prove the following theorem of Pontrjagin:

THEOREM 8.2. *Let M be a compact orientable differentiable manifold imbedded in a Euclidean space and let Ω_{ik} be the curvature forms derived from the induced Riemannian metric. An integral characterisic cocycle of M can be represented by a differential form which is a linear combination, with constant coefficients, of the products of the forms Δ_m, $0 \leq m \leq n/4$.*

It is highly probable that the relations asserted in Theorem 8.2 remain true, if the Ω_{ik} are the curvature forms of a Riemannian metric given intrinsically on M. But so far this has only been established for the case of the form Δ_0 and for a few other cases. The result will follow if it can be proved that a Riemannian manifold can be isometrically imbedded in a Euclidean space. Unfortunately this imbedding question is not settled, even for $n=2$.

9. **Hermitian geometry.** Our results are more satisfactory in the case of Hermitian geometry. By an Hermitian manifold M we shall mean a complex analytic manifold in which there is given an Hermitian metric:

$$ds^2 = \sum_{i,k=1}^{n} g_{ik}(z, \bar{z})(dz_i d\bar{z}_k), \qquad \bar{g}_{ik} = g_{ki}, \tag{67}$$

where z_i are the local complex coordinates and where the bar denotes the operation of taking the conjugate complex.

As in the case of Riemannian manifolds, the principal fibre bundle for an Hermitian manifold is one whose fibres consist of a set of complex vectors $e_1, \cdots, e_n$, such that

$$e_i \cdot \bar{e}_j = \delta_{ij}, \tag{68}$$

where the scalar product is understood in the sense of Hermitian geometry. Like the Riemannian case, the fundamental formulas of Hermitian geometry can be established to be

$$ds^2 = \sum (\omega_i \bar{\omega}_i), \quad d\omega_i = \sum \omega_j \omega_{ji}, \quad d\omega_{ik} = \sum \omega_{ij}\omega_{jk} + \Omega_{ik}, \tag{69}$$

where

$$\omega_{ij} + \bar{\omega}_{ji} = 0, \quad \Omega_{ij} + \bar{\Omega}_{ji} = 0, \quad \Omega_{ij} = \sum_{k,l} R_{ij,kl}\omega_k \bar{\omega}_l. \tag{70}$$

From Ω_{ij} we construct the differential forms

$$\Delta_m = \sum_{i_1,\cdots,i_m} \Omega_{i_1 i_2}\Omega_{i_2 i_3}\cdots\Omega_{i_m i_1}, \qquad 1 \leq m \leq n, \tag{71}$$

and also the differential forms

$$\Psi_m = \sum \delta(i_1, \cdots, i_m; j_1, \cdots, j_m)\Omega_{i_1 j_1}\cdots\Omega_{i_m j_m}, \qquad 1 \leq m \leq n, \tag{72}$$

where $\delta(i_1, \cdots, i_m; j_1, \cdots, j_m)$ is the Kronecker symbol which is equal to $+1$ or -1 according as $j_1, \cdots, j_m$ constitutes, an even or odd permutation of $i_1, \cdots, i_m$, and is otherwise zero, and where the summation is extended to all $i_1, \cdots, i_m$ from 1 to n. It is easy to see that every Ψ_m can be expressed as a polynomial of $\Delta_1, \cdots, \Delta_m$ with constant coefficients, and conversely.

Over the compact Hermitian manifold M we can construct the fibre bundle $\mathfrak{F}^{(p)}$ such that the fibre over a point P of M consists of ordered sets of vectors $\mathfrak{e}_1, \cdots, \mathfrak{e}_p$, satisfying the relations

$$\mathfrak{e}_i - \bar{\mathfrak{e}}_j = \delta_{ij}, \qquad 1 \leq i, j \leq p. \tag{73}$$

Then we can prove the following theorem:

THEOREM 9.1. *The characteristic cocycle of $\mathfrak{F}^{(p)}$ in M in the sense of §4 is of dimension $2n-2p+2$ and can be defined, up to a constant factor, by the form Ψ_{n-p+1}.*

These characteristic cocycles, altogether n of them, corresponding to the values $p=1, \cdots, n$, we shall call the basic characteristic cocycles. On the other hand, the imbedding process of §8 can be carried over to the present case, resulting in a correspondence between the equivalent fibre bundles over M (in the sense of Steenrod) and the homotopy classes of mappings of M into $H(n, N, C)$, where $H(n, N, C)$ is the Grassmann manifold of all n-dimensional complex linear spaces through the origin in a complex Euclidean space of $n+N$ (complex) dimensions. This allows us to define a generalized characteristic cohomology class of M as the inverse image of a cohomology class of dimension not greater than $2n$ of $H(n, N, C)$. But in the complex case the first main theorem on vector invariants for the unitary group is of simpler form than the corresponding theorem for the orthogonal group, and we can prove the following theorem:

THEOREM 9.2. *Every cohomology class of $H(n, N, C)$ of dimension not greater than $2n$ can be obtained by operations of the cohomology ring from n basic classes, namely, the classes whose inverse images are the basic characteristic classes of M.*

It is in this sense that we can say that for complex analytic Hermitian manifolds the characteristic classes arising from the consideration of the principal fibre bundle are completely determined by the Hermitian metric. We also remark that complex analytic Hermitian manifolds play an important rôle in algebraic geometry and in the theory of analytic functions of several complex variables.

The proofs of the results announced in this section will be published elsewhere.

10. The kinematic principal formula in integral geometry. In the above sections we have discussed the interrelations between local differential-geometric properties and topology. There are, however, other geometric properties in the large which are equally of interest and which are closely related to the ideas set forth above. Moreover, they also illustrate the important rôle played by the fibre bundle in the theory of a geometric being.

Let M be a Riemannian manifold of n dimensions, and let $\mathfrak{F}$ be its principal fibre bundle. We have defined in the space $\mathfrak{F}$ of dimension $n(n+1)/2$ the forms ω_i, ω_{ik}. The differential form of degree $n(n+1)/2$:

$$\mathfrak{P} = \prod \omega_i \cdot \prod_{i<k} \omega_{ik} \tag{74}$$

can serve as a kind of volume element in $\mathfrak{F}$. If M is the Euclidean space of n dimensions, $\mathfrak{P}$ is called the kinematic density, which was first introduced by Poincaré for the case $n=2$.

To illustrate how the kinematic density can be utilized in geometrical problems suppose M to be the Euclidean space of n dimensions. Then $\mathfrak{F}$ is the space of frames in M and is topologically a product. In M we consider two closed hypersurfaces S_1, S_2 of class not less than 2. For each S_i let $r_1^{(i)}, \cdots, r_{n-1}^{(i)}$ be the principal curvatures and let

$$\frac{1}{C_{n-1,k}} \int_{S_i} \sum r_1^{(i)} \cdots r_k^{(i)} dO_i = H_k^{(i)}, \quad i = 1, 2;\ k = 0, \cdots, n-1, \tag{75}$$

where the integrand is the kth elementary symmetric function of the principal curvatures and dO_i is the surface element of S_i. Denote by $V^{(i)}$ the volume bounded by S_i in M. Let S_1 be fixed in the space and let S_2 take all possible positions. For each position of S_2 denote by $\chi(S_1S_2)$ the Euler-Poincaré characteristic of the intersection of S_1, S_2. Then we have the following so-called kinematic principal formula:

$$\begin{aligned} \int_{S_2} \chi(S_1 \cdot S_2)\mathfrak{P} = J_n \Big\{ & H_{n-1}^{(1)} V^{(2)} + V^{(1)} H_{n-1}^{(2)} \\ & + \frac{1}{n} \sum_{k=0}^{n-2} C_{n,k+1} H_k^{(1)} H_{n-2-k}^{(2)} \Big\}, \end{aligned} \tag{76}$$

where J_n is the kinematic measure about a point in M and is given by

$$J_n = O_{n-1}O_{n-2} \cdots O_1, \tag{76a}$$

O_{p-1} being the area of the unit sphere in a Euclidean space of p dimensions.

The formula (76) was proved by Blaschke for $n=2, 3$ [2], and was established for general n by Yien and the present writer [13]. It has numerous interesting consequences. In particular, when $n=2$ and when both S_1 and S_2 are convex curves, it becomes

$$\int_{S_1 \cdot S_2 \neq 0} \mathfrak{P} = 2\pi(A_1 + A_2) + L_1L_2, \tag{77}$$

where A_i is the area bounded by S_i and L_i is the length of S_i, $i=1, 2$. Formula (77) was first established by Santalò and can be used to derive a sharpening of the isoperimetric inequality (2) [23]. It is thus through all the round-about discussions on differential geometry and topology that we arrived at the relation between the theory of fibre bundles and the isoperimetric inequality given at the beginning of this article.

Bibliography

1. Allendoerfer, C. B., and Weil, A., *The Gauss-Bonnet theorem for Riemannian polyhedra*, Trans. Amer. Math. Soc. vol. 53 (1943) pp. 101–129.

2. Blaschke, W., *Vorlesungen über Integralgeometrie*, Part 1, 2d ed., Leipzig, 1936; Part 2, Leipzig, 1937.

3. Cartan, E., *Les sousgroupes des groupes continus de transformations*, Ann. École Norm. (3) vol. 25 (1908) pp. 57–194.

4. ———, *La théorie des groupes finis et continus et la géométrie différentielle traitée par la méthode du repère mobile*, Paris, 1937.

5. ———, *Leçons sur les invariants intégraux*, Paris, 1922.

6. ———, *Sur les variétés à connexion affine et la théorie de la relativité généralisée*, Ann. École Norm. (3) vol. 40 (1923) pp. 325–412; vol. 41 (1924) pp. 1–25; vol. 42 (1925) pp. 17–88.

7. ———, *Le principe de dualité et certaines intégrales multiples de l'espace tangentiel et de l'espace réglé*, Bull. Soc. Math. France vol. 24 (1896) pp. 140–177.

8. ———, *Sur les variétés à connexion projective*, Bull. Soc. Math. France vol. 52 (1924) pp. 205–241.

9. ———, *Sur les invariants intégraux de certains espaces homogènes clos et les propriétés topologiques de ces espaces*, Annales de la Société Polonaise de Mathématique vol. 8 (1929) pp. 181–225.

10. Chern, S., *A simple intrinsic proof of the Gauss-Bonnet formula for closed Riemannian manifolds*, Ann. of Math. vol. 45 (1944) pp. 747–752.

11. ———, *Integral formulas for the characteristic classes of sphere bundles*, Proc. Nat. Acad. Sci. U.S.A. vol. 30 (1944) pp. 269–273.

12. ———, *A generalization of the projective geometry of linear spaces*, Proc. Nat. Acad. Sci. U.S.A. vol. 29 (1943) pp. 38–43.

13. Chern, S., and Yien, C. T., *Sulla formula principale cinematica dello spazio ad n dimensioni*, Bollettino della Unione Matematica Italiana (2) vol. 2 (1940) pp. 432–437.

14. Chern, S., *On Riemannian manifolds of four dimensions*, Bull. Amer. Math. Soc. vol. 51 (1945) pp. 964–971.

15. Ehresmann, C., *Sur la topologie de certains espaces homogènes*, Ann. of Math. vol. 35 (1934) pp. 396–443.

16. ———, Various notes on fibre spaces in C. R. Acad. Sci. Paris vol. 213 (1941) pp. 762–764; vol. 214 (1942) pp. 144–147; vol. 216 (1943) pp. 628–630.

17. Ehresmann, C., and Feldbau, J., *Sur les propriétés d'homotopie des espaces fibrés*, C. R. Acad. Sci. Paris vol. 212 (1941) pp. 945–948.

18. Feldbau, J. See Ehresmann, C.

19. Paris Seminar, *Notes on Elie Cartan's scientific work*, Paris, 1936.

20. Pontrjagin, L., *Characteristic cycles on manifolds*, C. R. (Doklady) Acad. Sci. USSR N.S. vol. 35 (1942) pp. 34–37.

21. ———, *On some topologic invariants of Riemannian manifolds*, C. R. (Doklady) Acad. Sci. USSR N.S. vol. 43 (1944) pp. 91–94.

22. de Rham, G., *Sur l'analysis situs des variétés à n dimensions*, J. Math. Pures Appl. vol. 10 (1931) pp. 115–200.

23. Santalò, L. A., *Sobre la medida cinematica en el plano*, Abh. Math. Sem. Hamburgischen Univ. vol. 11 (1935) pp. 222–236.

24. Steenrod, N., *Topological methods for the construction of tensor functions*, Ann. of Math. vol. 43 (1942) pp. 116–131.

25. ———, *The classification of sphere bundles*, Ann. of Math. vol. 45 (1944) pp 294–311.

26. Stiefel, E., *Richtungsfelder und Fernparallelismus in n-dimensionalen Mannigfaltigkeiten*, Comment. Math. Helv. vol. 8 (1936) pp. 305–343.

27. Thomas, T. Y., and Veblen, O., *The geometry of paths*, Trans. Amer. Math. Soc. vol. 25 (1923) pp. 551–608.

28. Veblen, O. See Thomas, T. Y.

29. Veblen, O., and Whitehead, J. H. C., *Foundations of differential geometry*, Cambridge, 1932.

30. Weil, A. See Allendoerfer, C. B.

31. Weyl, H., *The classical groups*, Princeton, 1939.

32. Whitehead, J. H. C. See Veblen, O.

33. Whitney, H., *Topological properties of differentiable manifolds*, Bull. Amer. Math. Soc. vol. 43 (1937) pp. 785–805.

34. ———, *On the topology of differentiable manifolds*, *Lectures in topology*, Michigan, 1941, pp. 101–141.

35. Yien, C. T. See Chern, S.

INSTITUTE FOR ADVANCED STUDY AND
TSING HUA UNIVERSITY

Reprinted from
Bull. Amer. Math. Soc. **52** (1946) 1–30

DIFFERENTIAL GEOMETRY IN SYMPLECTIC SPACE I

By Shiing-shen Chern (陳省身) and Hsien-chung Wang (王憲鍾).

Department of Mathematics

(*Received October 5, 1940*)

ABSTRACT

The subgroup of the group of projective transformations in a $(2n+1)$*-dimensional space leaving an absolute null system invariant is called the symplectic group by Weyl. Our object is to study the point varieties in the space with the symplectic group as fundamental group. In this space the flat subspaces are not always equivalent, nor are the elements of contact. The varieties are divided into as many classes as there are classes of elements of contact. For each class we give the degree of generality of the varieties in question by applying the theory of Pfaffian systems in involution. The theory of curves and hypersurfaces is also developed.*

INTRODUCTION

The object of this paper is to study the point varieties in a $(2n+1)$-dimensional space $(n \geq 1)$ in which the fundamental group is the subgroup of the projective group leaving an absolute null system invariant. This group, called by Weyl the symplectic group, depends on $(n+1)(2n+3)$ parameters and is a simple group in the sense of Lie. We shall call the space with the symplectic group as fundamental group the symplectic space. In the case $n=1$ the element of arc and the essential differential invariants of curves were determined by Noth[1] by applying the method of Lie. So far as the authors are aware, no further study of the differential geometry in symplectic space has yet been made, with perhaps the exception of some special results which may be deduced, from theorems in projective differential geometry.

In order to study the differential geometry in symplectic space for itself, we define in §1 the moving frame in the space and derive the equations of structure of Cartan, which form a basis of all further considerations. The study of point varieties turns out to be more complicated than expected, as the p-flats and the p-dimensional elements of contact (i. e., a point and a p-flat through the point) possess an arithmetic invariant and the p-dimensional varieties are to be classified according to the nature of its tangent p-flats. We give in §2 a classification of the p-flats and the elements of contact. The corresponding classification of the p-dimensional varieties is carried out in §3 and for each class the problem on the existence and the degree of generality of the varieties in question is studied as an application of the theory of Pfaffian systems in involution. For instance, we prove that there exists no variety of dimension $p>n$ such that its tangent p-flat at every point is contained in the null hyperplane of that point. Further, the n-dimensional varieties having the last property exist and depend on one function in n variables.

In §4 we derive the Frenet formulas for the curves which do not belong to a flat space of lower dimension and whose osculating elements of contact are "general". We then give in §5 some preliminary considerations for the theory of hypersurfaces. The geometrical interpretations of these results and the problem on the deformation of varieties with respect to the symplectic group are reserved for a next paper.

§1. MOVING FRAME. EQUATIONS OF STRUCTURE.

Let S be a $(2n+1)$-dimensional projective space and let $(x_0, x_1, \ldots, x_n, y_0, y_1, \ldots, y_n)$ be its homogeneous projective coordinates. By a null system[2] in the space we shall mean a non-singular correlation such that the null hyperplane of every point (i. e., the hyperplane corresponding to the point) contains the point. Let $(\xi_0, \xi_1, \ldots, \xi_n, \eta_0, \eta_1, \ldots, \eta_n)$ be a second set of coordinates. Then, by choosing properly the system of reference in S, we may assume the null system to be given by the equation[3]

$$\sum_{i=0}^{n} (x_i \eta_i - y_i \xi_i) = 0. \tag{1}$$

All projective transformations leaving this equation invariant form a group, called the symplectic group and the space with the symplectic group as fundamental group will be called the symplectic space. It can easily be shown that the symplectic group depends on $(n+1)(2n+3)$ parameters.

As basis of all following discussions we shall define the frame of the symplectic group. The projective frame in S commonly used consists of $2n+2$ analytic points $A_0, A_1, \ldots, A_n, B_0, B_1, \ldots, B_n$, determined up to a common factor. It will thus be natural to define the symplectic frame in S to be a projective frame, referred to which the absolute null system has the equation (1). Let AB denote the value of the expression in the left-hand side of (1) for the points $A(x, y)$, $B(\xi, \eta)$, so that we have

$$AB = -BA, \quad AA = 0. \tag{2}$$

Then *the projective frame* $A_0, A_1, \ldots, A_n, B_0, B_1, \ldots, B_n$ *satisfying the conditions*

$$\left.\begin{aligned} &A_i\, A_j = 0, \quad B_i\, B_j = 0, \\ &A_i\, B_j = 0, \quad i \neq j, \\ &A_0\, B_0 = A_1\, B_1 = \ldots = A_n\, B_n \neq 0, \end{aligned}\right\} \tag{3}$$

can serve as a symplectic frame.

To prove this, let us first find the equation of the absolute null system referred to the frame (A_i, B_i). Suppose the coordinates of the points M, N be defined by

$$\left.\begin{aligned} M &= \sum_i x_i\, A_i + \sum_j y_j\, B_j, \\ N &= \sum_i \xi_i\, A_i + \sum_j \eta_j\, B_j. \end{aligned}\right\} \tag{4}$$

The equation of the absolute null system is then

$$MN = 0, \tag{5}$$

which reduces to (1) on account of the conditions (3). If, therefore, (A_i^*, B_i^*) is a projective frame satisfying to set of conditions corresponding to (3), the projective transformation which carries (A_i, B_i) to (A_i^*, B_i^*) will leave the equation (1) invariant and hence belongs to the symplectic group. It is also unique. Thus our statement is proved.

It is important to remark here that the assumption on the linear independence of the points A_i, B_i is not necessary. For, by considering the determinant formed by the "scalar products" in (3), we have

$$| A_0 \dots A_n \, B_0 \dots B_n |^2 = (A_0 \, B_0)^{2n+2} \neq 0,$$

where the left-hand member denotes the square of the determinant formed by the analytic points A_i, B_i. Therefore any $2n+2$ analytic points (determined up to a common factor) satisfying the conditions (3) can serve as a symplectic frame.

Two points M and N are called *conjugate* if they satisfy the condition (5). With this definition the null hyperplane of a point M consists of all points in space conjugate to M.

The equations of infinitesimal displacement of the frame (A_i, B_i) can be written in the form

$$\left.\begin{aligned} d\,A_i &= \sum_k \omega_{ik} A_k + \sum_k \theta_{ik} B_k, \\ d\,B_i &= \sum_k \varphi_{ik} A_k + \sum_k \pi_{ik} B_k, \end{aligned}\right\} \tag{6}$$

where ω_{ik}, θ_{ik}, φ_{ik}, π_{ik}, called relative components of the frame, are Pfaffian forms in the parameters of the frame. These relative components are not independent. In fact, by forming the differentials of the equations (3) and taking account of these equations themselves, we find

$$\theta_{ij} = \theta_{ij}\,, \quad \varphi_{ji} = \varphi_{ji}\,, \quad \pi_{ij} + \omega_{ji} = 0\,, \quad i \neq j, \quad \pi_{00} + \omega_{00} = \dots = \pi_{nn} + \omega_{nn}. \tag{7}$$

With these relations there remain $(n+1)(2n+3)$ indpendent relative components[4]. There exists no other relation between the relative components, since the number of parameters of the symplectic group is $(n+1)(2n+3)$.

The relative components satisfy, however, a set of quadratic relations. These relations can be found by forming the exterior derivatives of the equations (6). They are

$$
\left.\begin{aligned}
\omega'_{ik} &= \sum_j [\omega_{ij}\,\omega_{jk}] + \sum_j [\theta_{ij}\,\varphi_{jk}],\\
\theta'_{ik} &= \sum_j [\omega_{ij}\,\theta_{jk}] + \sum_j [\theta_{ij}\,\pi_{jk}],\\
\varphi'_{ik} &= \sum_j [\varphi_{ij}\,\omega_{jk}] + \sum_j [\pi_{ij}\,\varphi_{jk}],\\
\pi'_{ik} &= \sum_j [\varphi_{ij}\,\theta_{jk}] + \sum_j [\pi_{ij}\,\pi_{jk}],
\end{aligned}\right\} \tag{8}
$$

which are called the *equations of structure* of the space.

§2. CLASSIFICATION OF p-FLATS AND OF ELEMENTS OF CONTACT

Consider a p-flat L_p, determined by $p+1$ linearly independent points $M_0, M_1, \cdots, M_p$. Since the equations

$$M_0M=0, \quad M_1M=0, \cdots, M_pM=0 \tag{9}$$

are linearly independent, they define a $(2n-p)$-flat L_{2n-p}, which we call the *conjugate flat* of L_p. L_{2n-p} consists of all the points conjugate to all points of L_p. The relation between L_p and L_{2n-p} is evidently reciprocal.

The p-flats in symplectic space are not always equivalent. To see this, it is only necessary to find the intersection of L_p and L_{2n-p}. A point on L_p is given by

$$\lambda_0M_0+\lambda_1M_1+\cdots+\lambda_pM_p. \tag{10}$$

The conditions that it also lies on L_{2n-p} are

$$
\left.\begin{aligned}
&(M_0M_0)\lambda_0+(M_0M_1)\lambda_1+\ldots.+(M_0M_p)\lambda_p=0,\\
&\ldots.\\
&(M_pM_0)\lambda_0+(M_pM_1)\lambda_1+\ldots.+(M_pM_p)\lambda_p=0.
\end{aligned}\right\} \tag{11}
$$

The matrix of this system of linear equations in $\lambda_0, \ldots., \lambda_p$:

$$
\begin{pmatrix}
(M_0M_0) & (M_0M_1) \ldots. & (M_0M_p)\\
\ldots. & & \\
(M_pM_0) & (M_pM_1) \ldots. & (M_pM_p)
\end{pmatrix} \tag{12}
$$

is a skew-symmetric square matrix, so that its rank is an even integer $2s \leqq p+1$. The system (11) has therefore $p-2s+1$ linearly independent solutions. The points corresponding to these solutions determine a $(p-2s)$–flat L_{p-2s}, which is evidently the intersection of L_p and L_{2n-p}.

Consider first the case that p is odd and that the matrix (12) has the maximum rank $p+1$. In this case, the conjugate flats L_p and L_{2n-p} have no common point and there is no point of L_p conjugate to all points of L_p. Let $N_0, N_1, \ldots, N_{2n-p}$ be the points determining L_{2n-p}, so that the equations of L_p are

$$N_0 M=0, \ N_1 M=0, \ \ldots, \ N_{2n-p} M=0.$$

If A_1, B_1 are two non-conjugate points of L_p, then the points $N_0, N_1, \ldots, N_{2n-p}$ A_1, B_1 are linearly independent, as the property that B_1 is a linear combination of A_1, $N_0, \ldots, N_{2n-p}$ would imply $A_1 B_1 = 0$. The flat defined by the equations

$$N_0 M=0, \ N_1 M=0, \ \ldots, \ N_{2n-p} M=0, \quad A_1 M=0, \ B_1 M=0$$

is of dimension $p-2$ and belongs to L_p. It also has the property that none of its points is conjugate to all its points. Proceeding on this way, we see that if a flat of $p=2s-1$ dimensions has the property that its intersection with the conjugate flat is empty, it is possible to choose in it $2s=p+1$ points $A_1, \ldots, A_s$, $B_1, \ldots, B_s$ such that

$$A_l A_m = B_l B_m = 0, \quad l, m=1, \ldots, s,$$

$$A_l B_m = 0, \qquad l \neq m; \ l, m=1, \ldots, s,$$

$$A_1 B_1 = \ldots = A_s B_s \neq 0.$$

We shall now come back to the general p-flat L_p and try to attach to it a properly chosen frame. We choose first in L_{p-2s} the points $A_0, A_1, \ldots, A_{p-2s}$ such that

$$A_l A_m = 0, \quad l, m=0, 1, \ldots, p-2s.$$

The p-flat L_p can then be regarded as being determined by L_{p-2s} and a $(2s-1)$-flat L_{2s-1} having no point in common with L_{p-2s}. The flat L_{2s-1} contains evidently no point conjugate to all its points. We may therefore choose in L_{2s-1} $2s$ points $A_{p-2s+1}, \ldots, A_{p-s}$, $B_{p-2s+1}, \ldots, B_{p-s}$ such that

$$A_l A_m = B_l B_m = 0, \quad l, m = p-2s+1, \ldots, p-s,$$

$$A_l A_m = 0, \qquad l \neq m;\ l, m = p-2s+1, \ldots, p-s,$$

$$A_{p-2s+1} B_{p-2s+1} = \ldots = A_{p-s} B_{p-s} \neq 0.$$

All these $2s$ points are obviously conjugate to $A_0, A_1, \ldots, A_{p-2s}$. Similarly, by considering L_{2n-p}, we can choose in it $2(n-p+s)$ points $A_{p-s+1}, \ldots, A_n$, $B_{p-s+1}, \ldots, B_n$, such that the conditions (3) are satisfied.

It remains to give the points $B_0, B_1, \ldots, B_{p-2s}$. For this purpose consider the $(2n-2p+4s-1)$-flat determined by $A_{p-2s+1}, \ldots, A_n, B_{p-2s+1}, \ldots, B_n$. This flat has no point in common with its conjugate flat L' of dimension $2p-4s+1$. By modifying properly the points $A_0, \ldots, A_{p-2s}$, it follows from our previous result that we can choose the points $A_0, \ldots, A_{p-2s}, B_0, \ldots, B_{p-2s}$ such that the conditions (3) imposed on them are satisfied. The points $A_0, \ldots, A_n, B_0, \ldots, B_n$ then form a symplectic frame.

Since the flat L_{p-2s} belongs also to L_{2n-p}, there exist between p, n, s the inequalities

$$p-n \leq s \leq \tfrac{1}{2}(p+1). \tag{13}$$

By summarizing the above discussions, we get the theorem:

A p-flat L_p possesses only an arithmetic invariant s equal to half the rank of the skew-symmetric matrix (12) and satisfying the inequalities (13). If $p \leq n$, the p-flats are divided into $\frac{1}{2}(p+3)$ or $\frac{1}{2}(p+2)$ classes according as p is odd or even. If $p > n$, the p-flats are divided into $\frac{1}{2}(2n-p+3)$ or $\frac{1}{2}(2n-p+2)$ classes according as p is odd or even.

We shall call s the rank of L_p.

The figure formed by a point and a p-flat L_p through the point is called a p-dimensional element of contact or simply a p-element. From the above discussions we have the theorem:

A p-element is characterized by the rank of its p-flat L_p and the fact that its point is conjugate to all points of L_p or not. The p-element will be called one of the first kind or of the second kind according as the former or the latter case occurs.

To a p-element of the first kind of rank s we attach the frame (A_i, B_i) such that A_0 is the point and that $A_0, \ldots, A_{p-s}$ $B_1, \ldots, B_s$ belong to the p-flat. To one of the second kind of rank s the frame (A_i, B_i) is so attached that A_0 is the point of the p-element and that $A_0, \ldots, A_{p-s}, B_0, \ldots, B_{s-1}$ belong to the p-flat.

§3. THE EXISTENCE AND DEGREE OF GENERALITY OF VARIETIES ENVELOPED BY ELEMENTS OF CONTACT OF A GIVEN TYPE.

Let V_p be a p-dimensional variety in space. The figure formed by a point A_0 of V_p and the tangent p-flat at A_0 is a p-element. Suppose that these p-elements are of the first kind of rank s. We can then attach to every point of V_p a frame (A_i, B_i) such that A_0 coincides with the point and that $A_0, \ldots, A_{p-s}$, $B_1, \ldots, B_s$ belong to the p-flat. For the family of frames thus attached to V_p we have

$$\left.\begin{aligned} &\omega_{p-s+1}=\ldots=\omega_n=0,\\ &\theta_{00}=\theta_{s+1}=\ldots=\theta_n=0. \end{aligned}\right\} \tag{14}$$

where we write $\omega_\alpha, \theta_\alpha$ for $\omega_{0\alpha}, \theta_{0\alpha}$. Forming the exterior derivatives of these equations, we get

$$\left.\begin{aligned} &[\omega_1\,\omega_{1\lambda}]+\ldots+[\omega_{p-s}\,\omega_{p-s\lambda}]+[\theta_1\,\varphi_{1\lambda}]+\ldots+[\theta_s\,\varphi_{s\lambda}]=0,\ \lambda=p-s+1,\ldots,n.\\ &[\omega_1\,\theta_1]+\ldots+[\omega_{p-s}\,\theta_{p-s}]=0,\\ &[\omega_1\,\theta_{1\nu}]+\ldots+[\omega_{p-s}\,\theta_{p-s\nu}]-[\theta_1\,\omega_{\nu 1}]-\ldots-[\theta_s\,\omega_{\nu s}]=0,\ \nu=s+1,\ldots,n. \end{aligned}\right\} \tag{15}$$

The second equation of (15) holds only when

$$s=o \quad \text{or} \quad s=p.$$

If $s=p$, the inequality $2s\leqq p+1$ gives $p=1$, so that the variety under consideration is a curve. By excluding this case, we see that *if a p-dimensional variety V_p, $p\geqq 2$, is enveloped by a family of p-elements of the first kind, then $p\leqq n$ and the tangent p-flats are of rank zero.*

Consider now such a variety V_p $(2\leqq p\leqq n)$, enveloped by a family of p-elements of the first kind of rank zero. In this case, the equations (14) and (15) become

$$\left.\begin{aligned} &\omega_{p+1}=\ldots\ldots=\omega_n=0,\\ &\theta_{00}=\theta_1=\ldots\ldots=\theta_n=0,\\ &[\omega_1\,\omega_{1\lambda}]+\ldots\ldots+[\omega_p\,\omega_{p\lambda}]=0,\ \lambda=p+1,\ \ldots\ldots,\ n,\\ &[\omega_1\,\theta_{1\alpha}]+\ldots\ldots+[\omega_p\,\theta_{p\alpha}]=0. \end{aligned}\right\}\qquad(16)$$

To this system we shall apply the theory of Pfaffian systems in involution[6]. With the notation of Kähler we observe that each of the equations

$$[\omega_1\,\omega_{1\lambda}]+\ldots\ldots+[\omega_p\,\omega_{p\lambda}]=0,$$

$$\lambda=p+1,\ \ldots\ldots,\ n,$$

$$[\omega_1\,\theta_{1\lambda}]+\ldots\ldots+[\omega_p\,\theta_{p\lambda}]=0,$$

is involutive, with

$$\sigma_1=\sigma_2=\ldots\ldots=\sigma_p=1.$$

It is also well known that the system

$$[\omega_1\,\theta_{1\nu}]+\ldots\ldots+[\omega_p\,\theta_{p\nu}]=0,\quad \nu=1,\ \ldots\ldots,\ p$$

is involutive with

$$\sigma_1=p,\ \sigma_2=p-1,\ \ldots\ldots,\ \sigma_p=1.$$

It follows that *the Pfaffian system (16) is in involution* and that the integers σ_i $(i=1,\ \ldots\ldots,\ p)$ are given by

$$\sigma_1=2n-p,\ \sigma_2=2n-p-1,\ \ldots\ldots,\ \sigma_p=2n-2p+1.$$

Therefore *the varieties V_p $(2\leqq p\leqq n)$ enveloped by a family of p-elements of the first kind of rank zero depend on $2n-2p+1$ functions in p variables. If $p=n$, they depend on one function in n variables.*

Suppose now that V_p $(p\geqq 2)$ is enveloped by a family of p-elements of the second kind of rank s. From the discussion of the flats in our space we must have $p\leqq n+s$. To every point of V_p we may attach a frame $(A_i,\ B_i)$ such that A_0 coincides with the point and that $A_0,\ \ldots\ldots,\ A_{p-s},\ B_0,\ \ldots\ldots,\ B_{s-1}$ belong to the tangent p-flat at A_0. For the family of frames so attached we have

$$\left.\begin{aligned}&\omega_{p-s+1}=\dots=\omega_n=0,\\&\theta_s=\dots=\theta_n=0.\end{aligned}\right\}\tag{17}$$

Forming the exterior derivatives of these equations, we get

$$\left.\begin{aligned}&[\omega_1\,\omega_{1\lambda}]+\dots+[\omega_{p-s}\,\omega_{p-s\lambda}]+[\theta_{00}\,\varphi_{0\lambda}]+[\theta_1\,\varphi_{1\lambda}]+\dots+[\theta_{s-1}\,\varphi_{s-1\lambda}]=0,\\&\qquad\qquad\qquad\qquad\lambda=p-s+1,\dots,n,\\&[\omega_1\,\theta_{1\nu}]+\dots+[\omega_{p-s}\,\theta_{p-s\nu}]-[\theta_{00}\,\omega_{\nu0}]-[\theta_1\,\omega_{\nu1}]-\dots-[\theta_{s-1}\,\omega_{\nu s-1}]=0,\\&\qquad\qquad\qquad\qquad\nu=s,\dots,n.\end{aligned}\right\}\tag{18}$$

In the system formed by the equations (17), (18), the principal components are

$$\omega_1,\ \dots,\ \omega_{p-s},\ \theta_{00},\ \theta_1,\ \dots,\ \theta_{s-1},$$

the remaining Pfaffian forms being the secondary components. To see whether this system is in involution, we notice that each of the equations (18_1) and of the last $n-p+s$ equations (18_2) for which $\nu=p-s+1,\dots,n$ contains secondary components which do not occur in the other equations. Each of these $2n-2p+2s$ equations is thus involutive, with

$$\sigma_1=\sigma_2=\dots=\sigma_p=1.$$

As for the remaining $p-2s+1$ equations

$$[\omega_1\,\theta_{1\nu}]+\dots+[\omega_{p-s}\,\theta_{p-s\nu}]-[\theta_{00}\,\omega_{\nu0}]-[\theta_1\,\omega_{\nu1}]-\dots-[\theta_{s-1}\,\omega_{\nu s-1}]=0,$$
$$\nu=s,\dots,p-s,$$

which exist only when $2s<p+1$, we see easily that they also form an involutive system, with

$$\sigma_1=\dots=\sigma_{2s}=p-2s+1,\ \sigma_{2s+1}=p-2s,\ \dots,\ \sigma_p=1.$$

Hence the whole system (18) is involutive, with

$$\sigma_p=2n-2p+2s+1,\quad\text{for}\quad 2s\leqq p,$$
$$\sigma_p=2n-p+1,\quad\text{for}\quad 2s=p+1.$$

We get thus the theorem:

The varieties V_p $(2\leqq p\leqq n+s)$ enveloped by a family of p-elements of the second kind of rank s depend on $2n-2p+2s+1$ functions or $2n-p+1$ functions in p variables according as $2s\leqq p$ or $2s=p+1$.

§4. FRENET FORMULAS FOR A GENERAL CURVE

By a "general" curve in symplectic space we shall mean a curve which does not belong to a flat space of lower dimension and whose osculating elements of contact are of the second kind and of maximum rank. We shall establish in this section the so-called "Frenet Formulas" for a general curve, from which the differential invariants of the curve are readily obtained. Our method of obtaining these formulas is to attach a frame in an invariant way to every point of the curve.

Consider a fixed point of the curve. We attach to it the family of frames (A_i, B_i) such that A_0 coincides with the point and that $A_0 B_0$ is the tangent at A_0. Let δ denote the variation within the family of frames attached to a fixed point and d the variation within the family of all frames attached to different points of the curve. Let also

$$\left.\begin{aligned}&\omega_{ij}=\omega_{ij}(d), \quad \theta_{ij}=\theta_{ij}(d), \quad \varphi_{ij}=\varphi_{ij}(d), \quad \pi_{ij}=\pi_{ij}(d),\\ &e_{ij}=\omega_{ij}(\delta), \quad f_{ij}=\theta_{ij}(\delta), \quad g_{ij}=\varphi_{ij}(\delta), \quad h_{ij}=\pi_{ij}(\delta).\end{aligned}\right\} \tag{19}$$

Then we have

$$\omega_\alpha=\theta_\alpha=0. \tag{20}$$

By forming the exterior derivatives of these equations and making use of the equations of structure, we get

$$[\theta\ \varphi_{0\alpha}]=0, \quad [\theta\ \omega_{\alpha 0}]=0,$$

where we put $\theta=\theta_{00}$. These equations show that $\varphi_{0\alpha}$ and $\omega_{\alpha 0}$ are multiples of θ or that

$$e_{\alpha 0}=g_{0\alpha}=0.$$

Now the equations of structure give

$$\omega'_{\alpha_0}=[\omega_{\alpha_0}\ \omega_{00}]+\sum_\beta[\omega_{\alpha\beta}\ \omega_{\beta 0}]+\sum_\beta[\theta_{\alpha\beta}\ \varphi_{0\beta}],$$

$$\varphi'_{0\alpha}=\sum_\beta[\varphi_{\alpha\beta}\ \omega_{\beta 0}]-\sum_\beta[\omega_{\beta\alpha}\ \varphi_{0\beta}]+[\pi_{00}\ \varphi_{0\alpha}],$$

or

$$\delta\,\omega_{\alpha 0}=-e_{00}\ \omega_{\alpha 0}+\sum_\beta e_{\alpha\beta}\ \omega_{\beta 0}+\sum_\beta f_{\alpha\beta}\ \varphi_{0\beta},$$

$$\delta\,\varphi_{0\alpha}=\sum_\beta g_{\alpha\beta}\omega_{\beta 0}+h_{00}\ \varphi_{0\alpha}-\sum_\beta e_{\beta\alpha}\,\varphi_{0\beta}.$$

From these equations we see that the conditions

$$\omega_{\alpha 0}=\varphi_{0\alpha}=0$$

are independent of the choice of the frames attached to the curve. Geometrically these conditions signify that the curve is a straight line. When this case is excluded, we can choose the frames such that we have

$$\varphi_{01}=0,\quad \varphi_{0\lambda_2}=0,\quad \omega_{\alpha 0}=0,\quad \lambda_2=2,\ \ldots,\ n. \tag{21}$$

We shall call the family of frames for which the conditions (20), (21) hold the family of frames of the second order.

From the equations for $\delta\omega_{\alpha 0}$, $\delta\varphi_{0\alpha}$ and the conditions (21) we find

$$e_{1\lambda_2}=f_{1\alpha}=e_{00}+e_{11}-2h_{00}=0.$$

The equations of structure then give

$$(\omega_{00}+\omega_{11}-2\omega_{00})'=3[\theta\ \varphi_{00}]+\sum_{\lambda_2}[\omega_{1\lambda_2}\ \omega_{\lambda_2 1}]+\sum_\alpha[\theta_{1\alpha}\ \varphi_{1\alpha}],$$

$$\theta'_{1\alpha}=\sum_{\lambda_2}[\omega_{1\lambda_2}\ \theta_{\lambda_2\alpha}]-[\theta_{1\alpha}\ \omega_{11}]+\sum_\beta[\theta_{1\beta}\ \pi_{\beta\alpha}],$$

$$\omega'_{1\lambda_2}=-[\omega_{1\lambda_2}\ \omega_{11}]+\sum_{\mu_2}[\omega_{1\mu_2}\ \omega_{\mu_2\lambda_2}]+\sum_\alpha[\theta_{1\alpha}\ \varphi_{\alpha\lambda_2}],\ \mu_2=2,\ \ldots,\ n,$$

or

$$\delta(\omega_{00}+\omega_{11}-2\pi_{00})=-3g_{00}\,\theta-\sum_{\lambda_2} e_{\lambda_2 1}\,\omega_{1\lambda_2}-\sum_{\alpha} g_{1\alpha}\,\theta_{1\alpha},$$

$$\delta\,\theta_{1\alpha}=-\sum_{\lambda_2} f_{\alpha\lambda_2}\,\omega_{1\lambda_2}+e_{11}\,\theta_{1\alpha}-\sum_{\beta} h_{\beta\alpha}\,\theta_{1\beta},$$

$$\delta\,\omega_{1\lambda_2}=e_{11}\,\omega_{1\lambda_2}-\sum_{\mu_2} e_{\mu_2\lambda_2}\,\omega_{1\mu_2}-\sum_{\alpha} g_{\alpha\lambda_2}\,\theta_{1\alpha}.$$

In particular, the second equation gives, for $\alpha=1$,

$$\delta\theta_{11}=(l_{11}-h_{11})\theta_{11},$$

which shows that the condition $\theta_{11}=0$ is invariant. In fact, it signifies geometrically that the curve belongs to a 2-flat or the osculating 3-flat of the curve is of rank one. Hence for a general curve we have $\theta_{11}\neq 0$ and we can make

$$\theta_{11}=\theta.$$

Moreover, it is still possible to choose the frames such that we have

$$\omega_{00}+\omega_{11}-2\,\pi_{00}=0,\ \theta_{11}=\theta,\ \ \theta_{1\lambda_2}=0,\ \ \omega_{1\lambda_2}=0. \tag{22}$$

The frames attached to the curve for which the conditions (20), (21), (22) are satisfied are called frames of the third order.

When the conditions (20), (21), (22) are satisfied, we have

$$3\,g_{00}+g_{11}=0,\ \ e_{11}-e_{00}=0,\ \ e_{\lambda_2 1}=0,\ \ g_{1\lambda_2}=0.$$

From the equations of structure we have

$$\theta'=\tfrac{1}{2}[\theta(\omega_{11}-\omega_{00})],$$

$$(3\,\varphi_{00}+\varphi_{11})'=-\tfrac{3}{2}[(\varphi_{00}-\varphi_{11})(\omega_{11}-\omega_{00})],$$

$$(\omega_{11}-\omega_{00})'=-4\,[\theta\,\varphi_{00}],$$

$$\omega'_{\lambda_2 1}=-\ [\omega_{11}\omega_{\lambda_2 1}\ ]+\sum_{\mu_2}[\omega_{\lambda_2\mu_2}\omega_{\mu_2 1}\ ]+\sum_{\mu_2}[\theta_{\lambda_2\mu_2}\ \varphi_{1\mu_2}\ ],$$

$$\varphi'_{1\lambda_2}\ -\sum_{\mu_2}[\varphi_{\mu_2\lambda_2}\omega_{\mu_2 1}]+[\pi_{11}\varphi_{1\lambda_2}]-\sum_{\mu_2}[\omega_{\mu_2\lambda_2}\ \varphi_{1\mu_2}\ ],$$

which give

$$\delta\theta = 0,$$

$$\delta(3\varphi_{00}+\varphi_{11}) = -6\, g_{00}\,(\omega_{11}-\omega_{00}),$$

$$\delta(\omega_{11}-\omega_{00}) = 4\, g_{00}\,\theta,$$

$$\delta\omega_{\lambda_2 1} = -e_{11}\omega_{\lambda_2 1} + \sum_{\mu_2} e_{\lambda_2\mu_2}\,\omega_{\mu_2 1} + \sum_{\mu_2} f_{\lambda_2\mu_2}\,\varphi_{1\mu_2},$$

$$\delta\varphi_{1\lambda_2} = \sum_{\mu_2} g_{\lambda_2\mu_2}\omega_{\mu_2 1} + h_{11}\,\varphi_{1\lambda_2} - \sum_{\mu_2} e_{\mu_2\lambda_2}\,\varphi_{1\mu_2}.$$

From the first equation we see that θ is independent of the choice of the frames of the third order. It is the *generalized arc* in symplectic space. On the other hand, the equations

$$\omega_{\lambda_2 1} = \varphi_{1\lambda_2} = 0$$

signify that the osculating 3-flat $A_0B_0A_1B_1$ is fixed, so that the curve belongs to a 3-flat. Since this case is to be excluded, we can make

$$\left.\begin{array}{l} \varphi_{12}=0, \quad \varphi_{1\lambda_3} = 0, \quad \omega_{\lambda_3 1} = 0, \quad \lambda_3 = 3, \ldots, n, \\ \omega_{11}=\omega_{00}. \end{array}\right\} \tag{23}$$

From (7), (22), (23), we get

$$\omega_{00} = \omega_{11} = \pi_{00} = \pi_{11}. \tag{24}$$

Frames for which the conditions (20) - (24) are verified are called frames of the fourth order.

When the conditions (20)–(24) are satisfied, we have

$$g_{00}=0, \quad e_{22}-e_{00}=0, \quad f_{2\lambda_3} = 0, \quad e_{2\lambda_3} = 0.$$

It follows that $3\varphi_{00}+\varphi_{11}/\theta$ is an *invariant of the fourth order*, which is the invariant of lowest order of the curve. As above, we find, from the equations o structure,

$$\varphi'_{00} = 0,$$

$$(\omega_{22}-\omega_{00})' = -\sum_{\lambda_3}[\omega_{\lambda_3 2}\omega_{2\lambda_3}] - \sum_{\lambda_3}[\varphi_{2\lambda_3}\theta_{2\lambda_3}].$$

$$\omega'_{2\lambda_3} = [\omega_{22}\omega_{2\lambda_3}] - \sum_{\mu_3} [\omega_{\mu_3\lambda_3}\omega_{2\mu_3}] - \sum_{\lambda_2} [\varphi_{\lambda_2\lambda_3}\theta_{2\lambda_2}],$$

$$\theta'_{2\lambda_2} = [\omega_{22}\theta_{2\lambda_2}] - \sum_{\lambda_3} [\theta_{\lambda_2\lambda_3}\omega_{2\lambda_3}] - \sum_{\mu_2} [\pi_{\mu_2\lambda_2}\theta_{2\mu_2}],$$

and

$$\delta\varphi_{00} = 0,$$

$$\delta(\omega_{22} - \omega_{00}) = -\sum_{\lambda_3} e_{\lambda_3 2}\,\omega_{2\lambda_3} - \sum g_{2\lambda_2}\theta_{2\lambda_2},$$

$$\delta\omega_{2\lambda_3} = e_{22}\omega_{2\lambda_3} - \sum_{\mu_3} e_{\mu_3\lambda_3}\,\omega_{2\mu_3} - \sum_{\lambda_2} g_{\lambda_2\lambda_3}\,\theta_{2\lambda_2},$$

$$\delta\,\theta_{2\lambda_2} = -\sum_{\lambda_3} f_{\lambda_2\lambda_3}\,\omega_{2\lambda_3} + e_{22}\,\theta_{2\lambda_2} - \sum_{\mu_2} h_{\mu_2\lambda_2}\,\theta_{2\mu_2}.$$

In particular, the last equation gives, for $\lambda_2 = 2$,

$$\delta\,\theta_{22} = 0,$$

which shows that θ_{22} is an invariant. The vanishing of θ_{22} characterizes the curves belonging to a 4-flat or having osculating 5-flats of rank 2. Thus for a general curve we have $\theta_{22} \neq 0$ and we can make

$$\omega_{22} = \omega_{00}, \quad \omega_{2\lambda_3} = 0, \quad \theta_{2\lambda_3} = 0. \tag{25}$$

When the conditions (20)—(25) are satisfied we have

$$e_{\lambda_3 2} = 0, \quad g_{2\lambda_2} = 0,$$

From the equations of structure we find $\delta\,\varphi_{22} = 0$, so that φ_{22} is an invariant. If $n = 2$, we see easily that we have thus attached only one frame (or, more exactly, a finite number of frames) to each point of the curve. The formulas giving the infinitesimal displacement of the one-parameter family of frames so attached are the formulas of Frenet required.

Suppose that we are in the general ease $n \geqq 3$. To derive the formulas of Frenet we proceed by induction. Thus we assume that the following conditions hold:

$$\left.\begin{aligned}
&\omega_{00}=\omega_{11}=\dots=\omega_{ll}=\pi_{00}=\dots=\pi_{ll},\\
&\omega_{\alpha}=\omega_{\alpha 0}=\omega_{1\lambda_2}=\omega_{\lambda_2 1}=\dots=\omega_{l\lambda_{l+1}}=0,\\
&\theta_{11}=\theta,\ \theta_{\alpha}=\theta_{1\lambda_2}=\theta_{2\lambda_3}=\dots=\theta_{l\lambda_{l+1}}=0,\\
&\varphi_{01}=\varphi_{12}=\dots=\varphi_{l-1\,l}=\theta,\\
&\varphi_{0\lambda_2}=\varphi_{1\lambda_3}=\dots=\varphi_{l-1\,\lambda_{l+1}}=0,\ \lambda_{l+1}=l+1,\ \dots,\ n,\\
&\theta_{22},\ \dots,\ \theta_{ll},\ \varphi_{00},\ \dots,\ \varphi_{l-1\,l-1}=\text{multiples of } \theta \text{ by invariants},\\
&\theta_{22}\neq 0,\ \dots,\ \theta_{ll}\neq 0,
\end{aligned}\right\} \qquad (26)$$

where $l \geqq 2$. By forming the exterior derivatives of these equations and making use of the equations of structure, we get

$$\omega'_{ll}=[\theta_{ll}\ \varphi_{ll}]=0,$$

$$\omega'_{l\lambda_{l+1}}=[\theta_{ll}\ \varphi_{l\lambda_{l+1}}]=0,$$

$$\theta'_{l\lambda_{l+1}}=-[\theta_{ll}\ \omega_{\lambda_{l+1}l}]=0,$$

These equations show that $\omega_{\lambda_{l+1}l}$, $\varphi_{l\lambda_l}$ are multiples of θ, or that

$$e_{\lambda_{l+1}l}=0, \quad g_{l\lambda_l}=0.$$

Now we have

$$\omega'_{\lambda_{l+1}l}=-[\omega_{00}\,\omega_{\lambda_{l+1}l}]+[\omega_{\lambda_{l+1}l+1}\,\omega_{l+1\,l}]+\dots+[\omega_{\lambda_{l+1}n}\ \omega_{nl}]+[\theta_{l+1\lambda_{l+1}}\ \varphi_{l\,l+1}]$$
$$+\dots+[\theta_{n\lambda_{l+1}}\ \varphi_{ln}],$$

$$\varphi'_{l\lambda_l}=[\varphi_{l+1\lambda_l}\ \omega_{l+1\,l}]+\dots+[\varphi_{n\lambda_l}\ \omega_{nl}]+[\omega_{00}\ \varphi_{l\lambda_l}]-[\omega_{l\lambda_l}\ \varphi_{ll}]$$
$$-\dots-[\omega_{n\lambda_l}\ \varphi_{ln}],$$

and

$$\delta\omega_{\lambda_{l+1}l} = -e_{00}\,\omega_{\lambda_{l+1}l} + e_{\lambda_{l+1}l+1}\,\omega_{l+1l} + \dots + e_{\lambda_{l+1}n}\,\omega_{nl} + f_{l+1\lambda_{l+1}}\,\varphi_{ll+1}$$
$$+ \dots + f_{n\lambda_{l+1}}\,\varphi_{ln},$$

$$\delta\varphi_{l\lambda_l} = g_{l+1\lambda_l}\,\omega_{l+1l} + \dots + g_{n\lambda_l}\,\omega_{nl} + e_{00}\,\varphi_{l\lambda_l} - e_{l\lambda_l}\,\varphi_{ll} - \dots - e_{n\lambda_l}\,\varphi_{ln}.$$

In particular, the last equation gives, for $\lambda_l = l$,

$$\delta\varphi_{ll} = 0,$$

which shows that φ_{ll} is a invariant. On the other hand, the remaining equations show tnat the conditions

$$\omega_{\lambda_{l+1}l} = 0, \quad \varphi_{l\lambda_{l+1}} = 0$$

are invariant. They signify that the osculating $(2l+1)-$flat $A_0\, B_0\, A_1\, B_1 \dots A_l\, B_l$ is fixed, so that the curve belongs to a $(2l+1)-$flat. As this case is to be excluded when $l<n$, we can choose the frames to make

$$\omega_{\lambda_{l+1}l} = 0, \quad \varphi_{ll+1} = 0, \quad \varphi_{l\lambda_{l+2}} = 0. \tag{27}$$

When the conditions (27) are satisfied, we have

$$e_{l+1,\,l+1} - e_{00} = 0, \quad e_{l+1,\,\lambda_{l+2}} = 0, \quad f_{l+2\,\lambda_{l+1}} = 0.$$

The equations of structure now give

$$(\omega_{l+1\,l+1} - \omega_{00})' = -[\omega_{l+2\,l+1}\,\omega_{l+1\,l+2}] - \dots - [\omega_{nl+1}\,\omega_{l+1n}] - [\varphi_{l+1\,l+1}\,\theta_{l+1\,l+1}]$$
$$- \dots - [\varphi_{l+1n}\,\theta_{l+1n}],$$

$$\omega'_{l+1\,\lambda_{l+2}} = [\omega_{l+1\,l+1}\,\omega_{l+1\,\lambda_{l+2}}] - [\omega_{l+2\,\lambda_{l+2}}\,\omega_{l+1\,l+2}]$$
$$- \dots - [\omega_{n\lambda_{l+2}}\,\omega_{l+1n}] - [\varphi_{l+1\,\lambda_{l+2}}\,\theta_{l+1\,l+1}] - \dots - [\varphi_{n\lambda_{l+2}}\,\theta_{l+1n}],$$

$$\theta'_{l+1\,\lambda_{l+1}} = [\omega_{l+1\,l+1}\,\theta_{l+1\,\lambda_{l+1}}] - [\theta_{l+2\,\lambda_{l+1}}\,\omega_{l+1\,l+2}]$$
$$- \dots - [\theta_{n\lambda_{l+1}}\,\omega_{l+1n}] - [\pi_{l+1\,\lambda_{l+1}}\,\theta_{l+1\,l+1}] - \dots - [\pi_{n\lambda_{l+1}}\,\theta_{l+1n}].$$

and

$$\delta(\omega_{l+1\,l+1}-\omega_{00})=-e_{l+2\,l+1}\,\omega_{l+1\,l+2}-\dots-e_{n\,l+1}\,\omega_{l+1\,n}-g_{l+1\,l+1}\,\theta_{l+1\,l+1}$$

$$-\dots-g_{l+1\,n}\,\theta_{l+1\,n},$$

$$\delta\omega_{l+1\,\lambda_{l+2}}=e_{l+1\,l+1}\,\omega_{l+1\,\lambda_{l+2}}-e_{l+2\,\lambda_{l+2}}\,\omega_{l+1\,l+2}-\dots-e_{n\,\lambda_{l+1}}\,\omega_{l+1\,n}$$

$$-g_{l+1\,\lambda_{l+2}}\,\theta_{l+1\,l+1}-\dots-g_{n\,\lambda_{l+2}}\,\theta_{l+1\,n},$$

$$\delta\theta_{l+1\,\lambda_{l+1}}=e_{l+1\,l+1}\,\theta_{l+1\,\lambda_{l+1}}-f_{l+2\,\lambda_{l+1}}\,\omega_{l+1\,l+2}-\dots-f_{n\,l\,l+1}\,\omega_{l+1\,n}$$

$$-h_{l+1\,\lambda_{l+1}}\,\theta_{l+1\,l+1}-\dots-h_{n\,\lambda_{l+1}}\,\theta_{l+1\,n}.$$

In particular, the last equation gives, for $\lambda_{l+1}=l+1$

$$\delta\,\theta_{l+1\,l+1}=0,$$

which shows that $\theta_{l+1\,l+1}$ is an invariant. It cannot be zero, for otherwise either the curve would belong to a $(2l+2)$—flat or the osculating $(2l+3)$—flats would be of rank $l+1$. Since $\theta_{l+1\,l+1}\neq 0$, we can arrive at the conditions

$$\omega_{l+1\,l+1}-\omega_{00}=0,\quad \omega_{l+1\,\lambda_{l+2}}=0,\quad \theta_{l+1\,\lambda_{l+2}}=0. \tag{28}$$

It is easily seen that the set of conditions (26), (27), (28) is identical with the set of conditions (26) when l is replaced by $l+1$. Hence the induction is complete.

By continuing this process, we shall arrive at a family of frames attached to the curve such that the conditions (26) with $l=n$ hold. We then find that φ_{nn} is an invariant and that when these conditions are satisfied, the frame attached to each point of the curve is unique (or finite in number), Since the analytic points A_i, B_i of the frame are determined up to a common factor, we may suitably determine this factor of that

$$\omega_{00}=0.$$

The family of frames thus attached to the curve is called the family of *Frenet frames*. The equations of infinitesimal displacement in the family of Frenet frames are

$$\left.\begin{aligned} &dA_0=\theta B_0,\ dA_1=\theta B_1,\ dA_{\lambda_2} = \theta_{\lambda_2\ \lambda_2}\ B_{\lambda_2} \\ &dB_0=\varphi_{00}A_0+\theta A_1,\ dB_\alpha=\theta A_{\alpha-1}+\varphi_{\alpha\alpha}\ A_\alpha+\theta\ A_{\alpha+1}\ (A_{n+1}\equiv 0). \end{aligned}\right\} \tag{29}$$

By putting

$$\left.\begin{aligned} &ds=\theta,\ \ K_\lambda\ ds=\theta_{\lambda\lambda},\ \ \lambda=2,\ \ldots,\ n, \\ &K_i\ ds=\varphi_{ii}, \end{aligned}\right\} \tag{30}$$

we may write the above equations in the form

$$\left.\begin{aligned} &\frac{dA_0}{ds}=B_0,\ \frac{dA_1}{ds}=B_1,\ \frac{dA_\lambda}{ds}=H_\lambda B_\lambda,\ \ \lambda=2,\ \ldots,\ n, \\ &\frac{dB_0}{ds}=K_0A_0+A_1,\ \frac{dB_\alpha}{ds}A_{\alpha-1}+K_\alpha A_\alpha+A_{\alpha+1}(A_{n+1}\equiv 0). \end{aligned}\right\} \tag{31}$$

Formulas (29) or (31) are the *formulas of Frenet* sought. The functions H_λ, K_i are differential invariants of the curve.

When $n=1$, the above induction process is not necessary. In this case, the Frenet formulas for a general curve are found to be

$$\left.\begin{aligned} &\frac{dA_0}{ds}=B_0,\ \frac{dA_1}{ds}=B_1, \\ &\frac{dB_0}{ds}=K_0A_0+A_1,\ \ \frac{dB_1}{ds}=A_0+K_1A_1, \end{aligned}\right\} \tag{32}$$

which are also included in the formulas (31).

§5. HYPERSURFACES

In this section we shall develop the fundamental equations for hpyersurfaces (i. e., varieties of $2n$ dimensions) in space. Let Σ be a hypersurface. To every point of Σ we attach the family of frames $(A_i,\ B_i)$ such that A_0 coincides with the point and that the $2n$-flat $A_0A_1\ldots A_nB_0B_1\ldots B_{n-1}$ is the tangent hyperplane at A_0. We have then

$$\theta_n=0. \tag{33}$$

The tangent hyperplane at A_0 has A_n as its null point. As A_0 describes the hypersurface Σ, the point A_n describes in general a hypersurface $\overline{\Sigma}$, whose tangent hyperplane at A_n is $A_0A_1....A_nB_1....B_n$. The relation between Σ and $\overline{\Sigma}$ is evidently reciprocal and we say that the two hypersurfaces are ***conjugate.***

By forming the exterior derivative of the equation (33), we get

$$\sum_{\alpha} [\omega_\alpha \, \theta_{\alpha n}] - \sum_{\sigma} [\theta_{0\sigma} \, \omega_{n\sigma}] = 0. \tag{34}$$

By Cartan's lemma we see that $\theta_{\alpha n}$, $\omega_{n\sigma}$ are linear combinations of ω_α, $\theta_{0\sigma}$, the latter being the principal components on the hypersurface. We may therefore put

$$\left.\begin{aligned} \theta_{\alpha n} &= \sum_{\beta} E_{\alpha\beta} \, \omega_\beta + \sum_{\varrho} F_{\alpha\varrho} \, \theta_{0\varrho}, \\ -\omega_{n\sigma} &= \sum_{\beta} F_{\beta\sigma} \, \omega_\beta + \sum_{\varrho} G_{\sigma\varrho} \, \theta_{0\varrho} \end{aligned}\right\} \tag{35}$$

where

$$E_{\alpha\beta} = E_{\beta\alpha}, \quad G_{\sigma\varrho} = G_{\varrho\sigma}. \tag{36}$$

On the other hand, the Pfaffian forms $\theta_{\alpha n}$, $\omega_{n\sigma}$ are the principal components on $\overline{\Sigma}$, which we also assume to be a hypersurface. Therefore they are linearly independent and the symmetric matrix

$$\begin{pmatrix} E_{\alpha\beta} & F_{\alpha\varrho} \\ F_{\beta\sigma} & G_{\sigma\varrho} \end{pmatrix} \tag{37}$$

is non-singular. Let its inveres matrix be

$$\begin{pmatrix} H_{\alpha\beta} & I_{\varrho\alpha} \\ I_{\sigma\beta} & J_{\varrho\sigma} \end{pmatrix} \tag{38}$$

so that we have

$$H_{\alpha\beta}=H_{\beta\alpha},\quad J_{\rho\sigma}=J_{\sigma\rho},\tag{39}$$

and

$$\left.\begin{aligned}
&\sum_{\beta} E_{\alpha\beta} H_{\beta\gamma}+\sum_{\rho} F_{\alpha\rho} I_{\rho\gamma}=\delta_{\alpha\gamma},\\
&\sum_{\beta} E_{\alpha\beta} I_{\rho\beta}+\sum_{\sigma} F_{\alpha\sigma} J_{\rho\sigma}=0,\\
&\sum_{\beta} F_{\beta\rho} H_{\beta\alpha}+\sum_{\sigma} G_{\rho\sigma} I_{\sigma\alpha}=0,\\
&\sum_{\beta} F_{\beta\rho} I_{\sigma\beta}+\sum_{\tau} G_{\rho\tau} J_{\sigma\tau}=\delta_{\rho\sigma},
\end{aligned}\right\}\tag{40}$$

where $\delta_{\alpha\gamma}$ and $\delta_{\rho\sigma}$ are the *Kronecker* indices.

If we want to determine the curves on the hypersurface whose osculating 2-flat at A_0 is contained in the tangent hyperplane $A_0A_1\dots A_nB_0\dots B_{n-1}$, we are led to a quadratic form in the principal components, whose vanishing gives the "asymptotic directions" on the hypersurface. Now we have

$$dA_0=\sum_i \omega_i A_i+\sum_\sigma \theta_{0\sigma} B_\sigma,$$

$$d^2A_0=\dots+(\sum_\alpha \omega_\alpha\,\theta_{\alpha n}-\sum_\sigma \theta_{0\sigma}\,\omega_{n\sigma})\,B_n,$$

where the terms represented by dots contain A_i, B_σ only. From the last equation we see that the asymptotic directions or Σ are defined by the equation

$$\sum_\alpha \omega_\alpha\,\theta_{\alpha n}-\sum_\sigma \theta_{0\sigma}\,\omega_{n\sigma}=0.\tag{41}$$

Similarly, we find that the asymptotic directions on Σ are defined by the same equation.

It is important to remark that E_{nn} and J_{00} are relative invariants. In fact, we see from (41) that their vanishing signifies that the line A_0A_n is an asymptotic direction on Σ and $\overline{\Sigma}$ repectively. We assume that such cases do not occur, so that

$$E_{nn} \neq 0, \quad J_{00} \neq 0.$$

With these assumptions we change the frame attaehed to A_0 by putting

$$\left.\begin{aligned} A_0^* &= a_0 A_0, \quad A_n^* = a_n A_n, \\ B_0^* &= \sum_i a_{0i} A_i + \sum_\sigma b_{0\sigma} B_\sigma, \\ B_n^* &= \sum_i a_{ni} A_i + \sum_\alpha b_{n\alpha} B_\alpha, \end{aligned}\right\} \tag{42}$$

where

$$\left.\begin{aligned} & a_0 b_{00} = a_n b_{nn}, \\ & \sum_\alpha a_{0\alpha} b_{n\alpha} \sum_\sigma a_{n\sigma} b_{0\sigma}, \\ & a_0 a_n b_{00} b_{nn} \neq 0. \end{aligned}\right\} \tag{43}$$

Put again

$$c = A_0 B_0, \quad c^* = A_0^* B_0^*. \tag{44}$$

Then the effect of the change of frame (42) on the relative components ω_n, ω_{n0} θ_{00}, θ_{nn} is given by

$$\left.\begin{aligned} c^* \omega_n^* &= dA_0^* B_n^* = c a_0 \left(\sum_\alpha b_{n\alpha} \omega_\alpha - \sum_\sigma a_{n\sigma} \theta_{0\sigma}\right), \\ c^* \omega_{n0}^* &= dA_n^* B_0^* = c a_n \left(\sum_\sigma b_{0\sigma} \omega_{n\sigma} - \sum_\alpha a_{0\alpha} \theta_{n\alpha}\right), \\ c^* \theta_{00}^* &= A_0^* dA_0^* = c a_0^2 \theta_{00}, \\ c^* \theta_{nn}^* &= A_n^* dA_n^* = c a_n^2 \theta_{nn}. \end{aligned}\right\} \tag{45}$$

It follows that by choosing

$$\left.\begin{aligned} a_{0\alpha} &= -\frac{a_0^2}{a_n} I_{0\alpha}, & b_{00} &= -\frac{a_0^2}{a_n} J_{00}, \\ a_{n\sigma} &= -\frac{a_n^2}{a_0} F_{n\sigma} & b_{n\alpha} &= \frac{a_n^2}{a_0} E_{n\alpha}, \end{aligned}\right\} \tag{46}$$

with

$$a_0^4 J_{00} = -a_n^4 E_{nn}, \tag{47}$$

we can arrive at the conditions

$$\theta^*_{nn} = \omega^*_n, \quad \theta^*_{00} = \omega^*_{n0}.$$

Let us drop the asterisks in the last equations and suppose the conditions

$$\theta_{nn} = \omega_n, \quad \theta_{00} = \omega_{n0} \tag{48}$$

be already verified. The most general change of frame (42) leaving the conditions (48) unaltered is of the form

$$\left.\begin{aligned} A_0^* &= a_0 A_0, \quad A_n^* = a_n A_n, \\ B_0^* &= \frac{1}{a_n}(a_0^2 B_0 - m A_0), \\ B_n^* &= \frac{1}{a_0}(a_n^2 B_n - l A_n), \end{aligned}\right\} \tag{49}$$

where l and m arbitrary and where

$$a_0^4 = a_n^4.$$

Now, by forming the exterior derivative of any one of the equations (48), we we find that φ_{0n} is a linear combination of ω_α, $\theta_{0\sigma}$. When the change of frame (49) is applied, we have

$$c^* \varphi^*_{0n} = \frac{c}{a_0 a_n}(a_0^2 a_n^2 \varphi_{0n} - m a_n^2 \omega_n - l a_0^2 \theta_{00}).$$

Hence it is possible to choose l and m such that φ_{0n}^{*} is a linear combination of ω_λ, θ_λ ($\lambda=1, \ldots, n-1$). But, as it can easily be verified, the relative components ω_λ, θ_λ ($\lambda=1, \ldots, n-1$) undergo a non-singular linear transformation by the change of frame (49). It follows that we can change the frame attached to A_0, without disturbing the conditions (48), such that φ_{0n} is a linear combination of ω_λ, θ_λ ($\lambda=1, \ldots, n-1$) only. The change of frame leaving all these conditions unaltered is given by

$$A_0^*=kA_0, \quad A_n^*=kA_n, \quad B_0^*=kB_0, \quad B_n^*=kB_n, \tag{50}$$

or

$$A_0^*=kA_0, \quad A_n^*=-kA_n, \quad B_0^*=-kB_0, \quad B_n^*=kB_n. \tag{50a}$$

We shall call such frames attached to A_0 the *fundamental frames*. The family of fundamental frames about a fixed point on the hypersurface depends on $(n-1)(2n-1)$ paramenters.

By summarizing the foregoing results, we get the theorem:

There exists at each point of a hypersurface an $(n-1)(2n-1)$-parameter family of frames such that within the family of all these frames attached to different points of the hypersurface the following conditions hold:

$$\left.\begin{aligned} &\theta_n=0,\\ &\theta_{nn}=\omega_n, \quad \theta_{00}=\omega_{n0},\\ &\varphi_{0n}=\text{linear combination of } \omega_\lambda, \theta_\lambda, \lambda=1, \ldots, n-1. \end{aligned}\right\} \tag{51}$$

The change of frame leaving the conditions (51) unaltered is given by (50) or (50a). We say that the analytic points A_0, A_n, B_0, B_n are determined up to a common factor and an "orientation".

When the conditions (51) are satisfied, the Pfaffian forms

$$\omega_{n\alpha}-\delta_{\alpha n}\,\omega_{00}, \ \omega_{\alpha 0}, \ \omega_{\lambda n}, \ \theta_{\alpha n}, \ \varphi_{0\varrho}, \ \varphi_{\alpha n},$$
$$\pi_{00}-\omega_{00} \qquad \lambda=1, \ldots, n-1,$$

are linear combinations of ω_α, $\theta_{0\varrho}$. Moreover, the forms

$$\omega_n, \ \theta_{00}, \ \omega_{nn}-\omega_{00}, \ \pi_{00}-\omega_{00}, \ \varphi_{00}, \ \varphi_{0n}, \ \varphi_{nn}$$

are invariant.

REFERENCES

1. G. Noth, Leipziger Berichte **56**, 19, (1904).
2. Cf., for example, E. Bertini, *Projektive Geomtrie mehrdimensionaler Räume*, Wien p. 112 (1924).
3. We assume, throughout this paper, the following ranges for the indices used: i, j, k from 0 to n; α, β, γ from 1 to n; ϱ, σ, τ from 0 to n—1. The ranges for other indices are to be indicated separately.
4. The relative components occur in the following discussions only in their differences.
5. Cf. E. Kähler, *Einführung in die Theorie der Systeme von Differentialgleichungen*, Leipzing pp. 54-55, (1934) or E. Cartan, "Sur les variétés de courbure constante d'un espace euclidien ou non-euclidien", Chap. III, Bull. Soc. math. de France, **47** (1919), **48** (1920).

Reprinted from
Science Reports Nat. Tsing Hua Univ.
4 (1947) 453–477

Reprinted from *Science Record*
Vol. 2, No. 2, (1948) pp. 137-139

NOTE ON PROJECTIVE DIFFERENTIAL LINE GEOMETRY

By SHIING-SHEN CHERN (陳省身)

Institute of Mathematics, Academia Sinica

(Received November 20, 1947)

This note consists of a number of definitions, theorems, and remarks, which result from an attempt to start some work in a field that might be designated by the name "projective differential geometry in the large".

We begin by trying to define the notion of a ruled surface in the three-dimensional real projective space P. The best way to give such a definition is to take into consideration the four-dimensional Grassmann manifold H of all the lines of P. A *ruled surface* is then defined to be the set of points which belong to the lines of the topological image of a circle S^1 into H. Let the topological mapping be denoted by $t: S^1 \to H$. If S^1 can be referred to a finite number of local parameters, which are related by differentiable transformations of class $k \geqq 0$ in neighborhoods where any two of them are valid, such that the mapping $t(S^1)$ is of class k in the local parameters, the ruled surface is said to be of *class* k. The ruled surface is called *imbedded* or *immersed*, according as the following condition is satisfied or not: no two lines of $t(S^1)$ intersect.

Since each projective line is topologically a circle, it follows from the theory of sphere bundles that a ruled surface is a sphere bundle[1)], whose base space is a circle and whose fibres are circles. By the classification theorem there are two classes of such sphere bundles[2)], and we are led to the theorem:

Theorem 1. *A ruled surface imbedded in the three-dimensional real projective space is either homeomorphic to the torus or to the Klein bottle. In other words, such a ruled surface is topologically determined by its orientability or non-orientability.*

Without relying on the classification theorem a direct proof can also be given of this theorem in a fairly simple way. It is also to be remarked that for ruled surfaces immersed in P the orientability or non-orientability will also have a sense, provided that the points common to two or more lines of $t(S^1)$ are counted properly.

When a ruled surface is defined this way, one of the immediate questions to be settled is whether such ruled surfaces do exist. A natural way to generate the ruled surfaces is to take a linear line congruence and consider the topological images of a circle into the linear congruence. It is well-known that a linear line congruence has in general two directrices and is called non-special if the directrices are distinct. If the directrices are conjugate imaginary, any topological mapping of S^1 into the linear congruence defines a ruled surface imbedded in P. For through a point of P there passes exactly one line of the congruence, so that two distinct lines of $t(S^1)$ will never meet. If the directrices, say d and d', are real and distinct, we denote by $f: d \rightarrow d'$ a topological mapping. Then the lines joining p and $f(p)$, $p \varepsilon d$, generate a ruled surface imbedded in P. No two distinct lines $p\,f(p)$ will intersect, as otherwise d and d' would not be skew.

Let $p_{12}, p_{13}, p_{14}, p_{34}, p_{42}, p_{23}$ be the Plückerian line coordinates and let new line coordinates be introduced by setting

$$x_1 - x_4 = p_{12}, \quad x_2 - x_5 = p_{13}, \quad x_3 - x_6 = p_{14},$$
$$x_1 + x_4 = p_{34}, \quad x_2 + x_5 = p_{42}, \quad x_3 + x_6 = p_{23}.$$

Then the fundamental identity is

$$x_1^2 + x_2^2 + x_3^2 - x_4^2 - x_5^2 - x_6^2 = 0.$$

A linear line congruence with real and distinct directrices can be defined by

$$x_3 = x_6 = 0.$$

and a linear line congruence with conjugate imaginary directrices can be defined by

$$x_5 = x_6 = 0.$$

It follows that the former is homeomorphic to the torus and the latter to the sphere.

The simplest ruled surface is a ruled quadric, which is homeomorphic to the torus and is therefore orientable. The ruled quadric is contained in every non-special linear congruence, which therefore always contains an orientable ruled surface imbedded in P. We have, moreover, the following theorem:

Theorem 2. *Every ruled surface contained in a linear congruence with conjugate imaginary directrices is orientable.*

In fact, let S^2 denote the space of the lines of a linear congruence with conjugate imaginary directrices, which is, as we have remarked, topologically a two-sphere. Let S^1 be a circle. Then the ruled quadric is defined by a mapping $f_0: S^1 \rightarrow S^2$, and any other ruled surface contained in the congruence by a mapping $f: S^1 \rightarrow S^2$. Since S^2 is simply connected, the mappings f_0 and f are homotopic. It follows that the ruled surface f is also orientable.

It would therefore be interesting to see whether there do exist non-orientable ruled surfaces. For ruled surfaces immersed in P the answer is well-known. In fact, Severi[3] proved that only a few exceptional cases of the cubic ruled surfaces are orientable. The most notable among the cubic ruled surfaces is perhaps Cayley's cubic surface[4]. In terms of the homogeneous projective coordinates ξ_1, ξ_2, ξ_3, ξ_4 in P the surface consists of the lines which intersect the lines $\xi_1 = \xi_2 = 0$, $\xi_3 = \xi_4 = 0$ and the conic

$$2\,\xi_1\,\xi_3 + \xi_2^2 - 2_4^2 = 0,\ \xi_2 - \xi_4 = 0.$$

Its equation is easily found to be

$$2\,\xi_1\,\xi_3\,\xi_4 + \xi_2\,\xi_3^2 - \xi_2\,\xi_4^2 = 0.$$

This surface is known to be non-orientable.

Added April, 1948.

In a recent letter I was informed by Wen-tsun Wu that he succeeded in proving the following very interesting theorem: Every ruled surface imbedded in the three-dimensional real projective space is orientable. This greatly sharpens the theorems of this note and adds interest to the notions here introduced. His proof will be published subsequently.

References:

(1) N. E. Steenrod, *Ann. of Math.*, **45**(1944) 294-311.
(2) Steenrod, *loc. cit.*, 299.
(3) F. Severi, *Atti. Ist. Ven.*, **62**(1903) 863.
(4) A. Cayley, *Phil. Trans.*, **154**(1864) 559 = Coll. M. P., **5**(1892) 201.

Local Equivalence and Euclidean Connections in Finsler Spaces

SHIING-SHEN CHERN (陳省身)

(*Received September 22, 1948*)

Contents

Introduction

By a Finsler space of n dimensions is meant a space with the local coordinates x^{i}[1], relative to which is given an integral

$$S = \int_{t_0}^{t_1} F\left(x^1, \ldots, x^n; \frac{dx^1}{dt}, \ldots, \frac{dx^n}{dt}\right) dt. \tag{1}$$

The function F in the integrand is supposed to satisfy, besides certain regularity conditions, the condition that it be positively homogeneous of the first degree in the last n arguments

$$F(x^1, \ldots, x^n; \lambda y^1, \ldots, \lambda y^n) = \lambda F(x^1, \ldots, x^n; y^1, \ldots, y^n), \lambda > 0. \tag{2}$$

Under the assumption (2) the value of the integral (1) between two points of a curve will be independent of the choice of parameter, and is defined to be the arc length. A Finsler space is therefore a space in which the arc lengths of curves are defined. For this reason it is sometimes called a general metric space in differential geometry.

One of the fundamental problems in Finslerian Geometry is to define, on the basis of the integral (1), an Euclidean connection in the space. Such definitions were given by J. L. Synge,[2] J. H. Taylor,[3] L. Berwald,[4] and E. Cartan.[5] It has been agreed that Cartan's definition is the most satisfactory one. The purpose of this paper is to solve the problem of local equivalence of Finslerian Geometries and to deduce from the solution of the problem of equivalence a family of Euclidean connections in the space, which all have the property that the equivalence of the connections is a necessary and sufficient condition for the equivalence of the Finslerian Geometries. These connections include as particular cases the ones defined by previous authors. This study also shows the exceptional rôle played by Cartan's connection.

Characteristic to this method of treatment is the algebraic nature of the solution. With the exception of some discussions at the beginning a large number of deductions are algebraic consequences of earlier equations and no further properties of the problem are used. We shall also present our discussion in such a way that it will be useful for the study of differential geometry in the large of Finslerian manifolds, that is, differentiable manifolds which carry a Finsler metric.

Some of the results in this paper have been announced in an earlier Note.[6)]

§1. Formulation of the Problem of Equivalence

To the integral (1) there is intrinsically associated the Pfaffian form[7)]

$$\omega = \frac{\partial F}{\partial y^i} dx^i, \tag{3}$$

where we have put

$$y^i = \frac{dx^i}{dt}. \tag{4}$$

In fact, to (1) the Pfaffian forms $F\,dt + \lambda_i(dx^i - y^i\,dt)$, where λ_i are arbitrary, are intrinsically associated and among these forms ω is characterized by the intrinsic condition

$$d\omega \equiv 0 \qquad \text{mod } dx^i - y^i\,dt. \tag{5}$$

We shall formulate the equivalence of two Finslerian Geometries in terms of the equivalence of two systems of Pfaffian forms. For this purpose it is remarked that to a Pfaffian form $v_k dx^k$ the condition $v_k y^k = 0$ is intrinsic, that is, under a point transformation this condition remains satisfied. It is therefore advisable to introduce $n-1$ Pfaffian forms $v_k^\alpha\,dx^k$, whose coefficients v_k^α are auxiliary variables satisfying the conditions

$$v_k^\alpha y^k = 0. \tag{6}$$

We also put

$$v_k^n = \frac{\partial F}{\partial y^k} \tag{7}$$

and

$$\omega^i = v_k^i\,dx^k, \tag{8}$$

so that

$$\omega^n = \omega \tag{9}$$

Suppose there be a second Finsler space whose quantities we denote by the same symbols proceded by dashes. Two such Finsler spaces are called locally equivalent, if there exists an analytic homeomorphism between two neighborhoods such that under the homeomorphism corresponding arcs have the same length. If the analytic homeomorphism is written as

$$\bar{x}^k = \bar{x}^k(x^i) \tag{10}$$

with a non-vanishing Jacobian in the neighborhood, it can be adjoined by a transformation

$$\begin{aligned} \bar{y}^k &= \bar{y}^k(x^i, y^i), \\ \bar{v}_k^\alpha &= \bar{v}_k^\alpha(x^i, y^i, v_k^\alpha), \end{aligned} \tag{11}$$

such that under (10) and (11) we have

$$\bar{\omega}^i = \omega^i. \tag{12}$$

Conversely, if a transformation of the form

$$\begin{aligned} \bar{x}^k &= \bar{x}^k(x^i, y^i, v_k^\alpha), \\ \bar{y}^k &= \bar{y}^k(x^i, y^i, v_k^\alpha), \\ \bar{v}_k^\alpha &= \bar{v}_k^\alpha(x^i, y^i, v_k^\alpha), \end{aligned} \tag{13}$$

exists under which (12) holds, then the functions $\bar{x}^k$ will not involve y^i, v_k^α and define a local homeomorphism which establishes the local equivalence of the two Finslerian Geometries. It follows that the problem of equivalence is reduced to that of two systems of n linearly independent Pfaffian forms in the two sets of $n(n+1)$ variables each, namely, x^i, y^i, v_k^α and $\bar{x}^i, \bar{y}^i, \bar{v}_k^\alpha$. Such an equivalence problem can be solved by a process given by Cartan.

Before proceeding to the solution let us introduce the variables u_k^i defined by

$$u_j^i v_k^j = v_j^i u_k^j = \delta_k^i. \tag{14}$$

Since

$$v_k^\alpha y^k/F = 0, \qquad v_k^n y^k/F = 1,$$

we get

$$u_n^k = y^k/F. \tag{15}$$

§2. Normalizations Involving Derivatives of the Second Order

To apply Cartan's method we first calculate $d\,\omega^n$. We find

$$\begin{aligned} d\omega^n &= \frac{\partial^2 F}{\partial x^i \partial y^k} dx^i dx^k + \frac{\partial^2 F}{\partial y^i \partial y^k} dy^i dx^k \\ &= \frac{\partial^2 F}{\partial x^i \partial y^k} u_p^i u_b^k \omega^p \omega^b + \frac{\partial^2 F}{\partial y^j \partial y^k} u_b^k dy^j \omega^b. \end{aligned}$$

In this expression the coefficient of $dy^j\omega^n$ is

$$\frac{\partial^2 F}{\partial y^j \partial y^k}\frac{1}{F}y^k,$$

which is equal to zero, since $\partial F/\partial y^j$ is a homogeneous function of the zeroth degree in y^k. It follows that we can choose $n-1$ Pfaffian forms ω_α^n, with which we can write

$$d\omega^n = \omega^\alpha \omega_\alpha^n. \tag{16}$$

Developing the terms in $d\omega^n$, we find that the most general expression for ω_α^n is

$$\omega_\alpha^n = -u_\alpha^k \frac{\partial^2 F}{\partial y^j \partial y^k} dy^j + \frac{1}{F} u_\alpha^j \left(\frac{\partial F}{\partial x^j} - \frac{\partial^2 F}{\partial x^k \partial y^j} y^k\right)\omega^n$$
$$+ u_\alpha^j u_\beta^k \frac{\partial^2 F}{\partial x^j \partial y^k}\omega^\beta + \lambda_{\alpha\beta}\omega^\beta, \qquad \lambda_{\alpha\beta} = \lambda_{\beta\alpha}, \tag{17}$$

where the $\frac{1}{2}n(n-1)$ quantities $\lambda_{\alpha\beta}$ are arbitrary.

Up to this point we are naturally led to impose a condition which, in the language of the calculus of variations, means that the integral (1) leads to a *regular* problem. We put

$$F_i = \frac{\partial F}{\partial y^i}, \qquad F_{ik} = \frac{\partial^2 F}{\partial y^i \partial y^k}, \text{ etc.} \tag{18}$$

Then Euler's theorem on homogeneous functions gives

$$y^k F_k = F, \qquad y^k F_{ik} = 0. \tag{19}$$

From the second set of equations it follows that

$$|F_{ik}| = 0. \tag{20}$$

To simplify further discussions the following assumption on our Finsler metric should be made:

Regularity Assumption: The matrix (F_{ik}) is of rank $n-1$.

This assumption will be transformed in a new form as follows: Let f^{ij} be the cofactor of F_{ij} in the determinant $|F_{ij}|$. Then from (19_2) we can write

$$f^{jk} = \lambda^j y^k$$

In fact, if (F_{ik}) is of rank $n-1$, this follows from the fact that for each j, f^{jk} and y^k satisfy the same system of n linear homogeneous equations of rank $n-1$, while if (F_{jk}) is of rank $<n-1$ we only have to set $\lambda^j = 0$. Since $f^{jk} = f^{kj}$, it follows that λ^k and y^k are proportional. Hence there exists a function F_1 such that

$$\lambda^k = y^k F_1,$$

whence

$$f^{jk} = F_1 y^j y^k. \tag{21}$$

To find an explicit expression for F_1 we multiply this equation by F_jF_k or by y^jy^k and sum with respect to j, k. This gives respectively

$$F_1 = \frac{1}{F^2} f^{jk} F_j F_k = -\frac{1}{F^2} \begin{vmatrix} F_{jk} & F_j \\ F_k & 0 \end{vmatrix} \tag{22}$$

$$F_1 = \frac{1}{(\sum y^{i2})^2} f^{jk} y^j y^k = \frac{-1}{(\sum y^{i2})^2} \begin{vmatrix} F_{jk} & y^j \\ y^k & 0 \end{vmatrix}. \tag{23}$$

Clearly *an equivalent form of our regularity assumption is the condition* $F_1 \neq 0$.

We shall show that under the regularity assumption *the Pfaffian forms* ω^i, ω_α^n *are linearly independent*. Suppose in fact that

$$\mu^\alpha \omega_\alpha^n + v_i \omega^i = 0.$$

Then

$$\mu^\alpha u_\alpha^k F_{jk} dy^j = 0.$$

Since

$$y^k F_{jk} dy^j = 0.$$

and $n - 1$ of the forms $F_{jk} dy^j$ are linearly independent, we can write

$$\mu^\alpha u_\alpha^k = \lambda y^k.$$

Multiplying this equation by v_k^n and summing, we get

$$\lambda y^k F_k = \lambda F = \mu^\alpha u_\alpha^k v_k^n = 0,$$

whence $\lambda = 0$. It follows that

$$\mu^\alpha u_\alpha^k = 0.$$

The matrix (u_α^k) being certainly of rank $n - 1$, we get

$$\mu^\alpha = 0.$$

Then

$$v_i \omega^i = 0,$$

which holds only when $v_i = 0$. This proves our assertion.

We shall next form the exterior derivative $d\omega^\alpha$. It follows from (14) by differentiation that

$$du_k^j v_j^i + u_k^j dv_j^i = 0$$

or

$$dv_j^i = -v_l^i du_k^l v_j^k. \tag{24}$$

When we make use of this relation, we have

$$d\omega^\alpha = dv_k^\alpha dx^k = -v_k^\alpha du_i^k \omega^i = -v_k^\alpha du_\beta^k \omega^\beta - v_k^\alpha d\left(\frac{y^k}{F}\right)\omega^n$$

$$= -v_k^\alpha du_\beta^k \omega^\beta - \frac{1}{F} v_k^\alpha dy^k \omega^n.$$

It follows that we can write

$$d\omega^\alpha \equiv \omega^n \omega_n^\alpha, \qquad \text{mod } \omega^\beta, \tag{25}$$

where

$$\omega_n^\alpha \equiv \frac{1}{F} v_k^\alpha dy^k, \quad \text{mod } \omega^i. \tag{26}$$

It is seen that the auxiliary variables u_j^i and the forms ω_n^α, ω_α^n can be so chosen that

$$\omega_\alpha^n + \delta_{\alpha\beta}\omega_n^\beta = 0, \tag{27}$$

if the coefficients of dy^k are zero in the left-hand member. This leads to the condition

$$u_\alpha^j G_{jk} = \delta_{\alpha\beta} v_k^\beta, \tag{28}$$

where

$$G_{jk} = FF_{jk} \tag{29}$$

Notice that

$$v_i^\beta u_\alpha^j u_\beta^k G_{jk} = v_i^l u_l^k u_\alpha^j G_{jk} - v_i^n u_n^k u_\alpha^j G_{jk} = u_\alpha^j G_{ji},$$

since

$$u_n^k G_{jk} = \frac{1}{F} y^k (FF_{jk}) = 0.$$

Hence the condition (28) is equivalent to the following more symmetrical one:

$$u_\alpha^j u_\beta^k G_{jk} = \delta_{\alpha\beta}. \tag{30}$$

We shall show that u_i^j or v_i^j can be so chosen that the conditions (28) or (30) are satisfied. It is clearly sufficient to show that the conditions imposed on u_α^k will not make the determinant $|v_i^j|$ or $|u_i^j|$ vanish. We have

$$v_i^\alpha u_\alpha^j G_{jk} = v_i^l u_l^j G_{jk} - v_i^n u_n^j G_{jk} = G_{ik},$$

so that the conditions (28) or (30) are equivalent to the conditions

$$\delta_{\alpha\beta} v_i^\alpha v_j^\beta = G_{ij}. \tag{31}$$

From (31) we derive

$$|v_i^j|^2 = |\delta_{kl} v_i^k v_j^l| = |FF_{ij} + F_i F_j|,$$

of which the last determinant can be transformed as follows:

$$|FF_{ij} + F_i F_j| = \begin{vmatrix} FF_{ij} + F_i F_j & F_i \\ 0 & 1 \end{vmatrix} = \begin{vmatrix} FF_{ij} & F_i \\ -F_j & 1 \end{vmatrix}$$

$$= -F^{n-1} \begin{vmatrix} F_{ij} & F_i \\ F_j & -F \end{vmatrix} = -F^{n-1} \begin{vmatrix} F_{ij} & F_i \\ F_j & 0 \end{vmatrix} = F^{n+1} F_1,$$

by (22). It follows that under the regularity assumption the determinant $|v_i^j| \neq 0$ and that the conditions (30) can be imposed.

Suppose ω_n^α be so chosen as to satisfy (27). Making use of its expression in (17), we get

$$d\omega^\alpha = \omega^n \omega_n^\alpha + \omega^\beta \left(v_k^\alpha du_\beta^k - \delta^{\alpha\gamma} u_\gamma^j u_\beta^k \frac{\partial^2 F}{\partial x^j \partial y^k} \omega^n - \delta^{\alpha\gamma} \lambda_{\beta\gamma} \omega^n \right).$$

By introducing $(n-1)^2$ Pfaffian forms ω_β^α, we can write the above equation in the form

$$d\omega^\alpha = \omega^n \omega_n^\alpha + \omega^\beta \omega_\beta^\alpha, \tag{32}$$

where the most general expression for ω_β^α is

$$\omega_\beta^\alpha = v_k^\alpha du_\beta^k - \delta^{\alpha\gamma} \left(u_\gamma^j u_\beta^k \frac{\partial^2 F}{\partial x^j \partial y^k} + \lambda_{\beta\gamma} \right) \omega^n + \mu_{\beta\gamma}^\alpha \omega^\gamma, \tag{33}$$

the $\mu_{\beta\gamma}^\alpha$ being symmetric in the lower indices and otherwise arbitrary.

§3. Normalizations Involving Derivatives of the Third Order

The auxiliary variables $\lambda_{\alpha\beta}$, $\mu_{\beta\gamma}^\alpha$ will now be determined by making use of the conditions (28), (30) or (31). In fact, differentiation of (31) gives

$$\delta_{\alpha\beta} v_j^\beta dv_i^\alpha + \delta_{\alpha\beta} v_i^\alpha dv_j^\beta = dG_{ij}.$$

or

$$\delta_{\alpha\sigma} u_\varrho^i dv_i^\alpha + \delta_{\beta\varrho} u_\sigma^j dv_j^\beta = u_\sigma^j u_\varrho^i dG_{ij}.$$

Substituting into this equation the expression for ω_β^α, we get

$$\delta_{\alpha\sigma} \left\{ -\omega_\varrho^\alpha - \delta^{\alpha\gamma} \left(u_\gamma^j u_\varrho^k \frac{\partial^2 F}{\partial x^j \partial y^k} + \lambda_{\varrho\gamma} \right) \omega^n + \mu_{\varrho\gamma}^\alpha \omega^\gamma \right\}$$
$$+ \delta_{\alpha\varrho} \left\{ -\omega_\sigma^\alpha - \delta^{\alpha\gamma} \left(u_\gamma^j u_\sigma^k \frac{\partial^2 F}{\partial x^j \partial y^k} + \lambda_{\sigma\gamma} \right) \omega^n + \mu_{\sigma\gamma}^\alpha \omega^\gamma \right\} = dG_{ij} u_\varrho^i u_\sigma^j$$

or

$$\delta_{\alpha\sigma} \omega_\varrho^\alpha + \delta_{\alpha\varrho} \omega_\sigma^\alpha = -dG_{ij} u_\varrho^i u_\sigma^j - u_\varrho^i u_\sigma^j \left(\frac{\partial^2 F}{\partial x^i \partial y^j} + \frac{\partial^2 F}{\partial x^j \partial y^i} \right) \omega^n$$
$$- 2\lambda_{\varrho\sigma} \omega^n + (\delta_{\alpha\sigma} \mu_{\varrho\gamma}^\alpha + \delta_{\alpha\varrho} \mu_{\sigma\gamma}^\alpha) \omega^\gamma. \tag{34}$$

Now the forms ω_α^n have the property that if v_k is the coefficient of dy^k in any of their linear combination, then $v_k y^k = 0$. Since the ω_α^n are $n-1$ linearly independent forms with respect to dy^k, it follows conversely that a form $v_k dy^k$ satisfying the condition $v_k y^k = 0$ is congruent to a linear combination of ω_α^n mod ω^i. As G_{ij} is homogeneous of the zeroth degree in y^k, dG_{ij} has this property, and we can write

$$dG_{ij} = G_{ij}^\alpha \omega_\alpha^n + G_{ijk} \omega^k. \tag{35}$$

We choose $\lambda_{\varrho\sigma}$, $\mu_{\varrho\sigma}^\alpha$ so that the following equation holds:

$$\delta_{\alpha\sigma} \omega_\varrho^\alpha + \delta_{\alpha\varrho} \omega_\sigma^\alpha = H_{\varrho\sigma}^\alpha \omega_\alpha^n. \tag{36}$$

From (34) we see that this is possible if we take

$$\lambda_{\varrho\sigma} = -\frac{1}{2} u^i_\varrho u^j_\sigma \left(G_{ijn} + \frac{\partial^2 F}{\partial x^i \partial y^j} + \frac{\partial^2 F}{\partial x^j \partial y^i} \right).$$
$$\mu^\alpha_{\varrho\sigma} = \frac{1}{2}(\delta^{\alpha\beta}\xi_{\varrho\beta\sigma} + \delta^{\alpha\beta}\xi_{\sigma\beta\varrho} - \delta^{\alpha\beta}\xi_{\varrho\sigma\beta}), \tag{37}$$

where

$$\xi_{\varrho\sigma\beta} = G_{ij\beta} u^i_\varrho u^j_\sigma. \tag{38}$$

Thus the form of the equations (36) completely determines the auxiliary variables $\lambda_{\varrho\sigma}$, $\mu^\alpha_{\varrho\sigma}$.

The discussion in §, §2, 3 can be summarized as follows: Besides the local coordinates x^i there have been introduced the variables: 1) y^i, of which only the ratios are essential; 2) $n(n-1)$ variables v^α_k, which are connected by $n-1$ relations (6) and $\frac{1}{2}n(n-1)$ relations (30), so that there remain only $\frac{1}{2}(n-1)(n-2)$ essential ones. The total number of essential variables is therefore

$$n + (n-1) + \tfrac{1}{2}(n-1)(n-2)$$

The same number of Pfaffian form has been completely determined namely

$$\omega^i, \omega^n_\alpha, \omega^\beta_\alpha, \alpha < \beta, \tag{39}$$

and we see that the process is intrinsic, that is, the same for spaces which are locally equivalent. The problem of local equivalence of Finsler spaces is solved in the sense that it is reduced to a problem of equivalence of systems of Pfaffian forms, which involve the same number of forms as the number of variables. It follows from the general theory that differential invariants of the Finsler space are obtained by forming the exterior derivatives of the forms (39).

In a modern terminology the space of all the variables x^i, y^i, v^α_k is a fibre space with the given Finsler space as its base space. This interpretation is natural and is useful for the study of differential geometry in the large of Finslerian manifolds. We see in this case that the definition of the fibre space depends very much on the metric. It involves in fact derivatives up to the third order inclusive of the function F. It is the consideration of this fibre space that will play an important rôle in Finslerian Geometry, both for the local theory and the global theory.

§4. The Invariants $H^\alpha_{\varrho\sigma}$; Riemann Spaces

The first set of invariants which we have obtained in the above discussion are $H^\alpha_{\varrho\sigma}$. Their properties and geometrical significance will be studied in this section.

For our later discussions it will be convenient to introduce the process of "ascending and lowering of indices" with respect to the fundamental tensor δ_{ij} or $\delta_{\alpha\beta}$. Explicitly they are defined as follows:

$$\delta^{ij} H_{jk.}{}^l = H^i{}_{.k.}{}^l, \qquad \delta_{ij} H^j{}_{.k.}{}^l = H_{ik.}{}^l,$$
$$\delta_{ij}\omega_{k.}{}^j = \omega_{ki}, \qquad \text{etc}, \tag{40}$$

Now, by (34), (35), (36), we have

$$H_{\varrho\sigma\cdot}{}^{\alpha} = -G^{\alpha}_{ij}u^i_{\varrho}u^j_{\sigma}. \tag{41}$$

Comparing the coefficients of dy^l in dG_{ij}, we get

$$G^{\alpha}_{ij}u^k_{\alpha}F_{kl} = -FF_{ijl} - F_lF_{ij},$$

from which we derive

$$(G^{\alpha}_{ij}y^j)u^k_{\alpha}F_{kl} = FF_{il}$$

or

$$(G^{\alpha}_{ij}y^ju^k_{\alpha} - \delta^k_iF)F_{kl} = 0.$$

Since the matrix (F_{kl}) is of rank $n-1$, this holds only when

$$G^{\alpha}_{ij}y^ju^k_{\alpha} - \delta^k_iF = p_iy^k.$$

Multiplication of this equation by F_k and subsequent summation give

$$p_i = -F_i.$$

It follows that

$$G^{\alpha}_{ij}y^ju^k_{\alpha} = -F_iy^k + \delta^k_iF.$$

and

$$G^{\alpha}_{ij}y^j = v^{\alpha}_iF. \tag{42}$$

With the help of this formula we find

$$v^{\varrho}_lv^{\sigma}_mH_{\varrho\sigma\cdot}{}^{\alpha} = -G^{\alpha}_{ij}\left(\delta^i_l - \frac{y^i}{F}F_l\right)\left(\delta^j_m - \frac{y^j}{F}F_m\right) = -G^{\alpha}_{lm} + v^{\alpha}_lF_m + v^{\alpha}_mF_l.$$

An easy transformation of this equation will express $H_{\varrho\sigma\cdot}{}^{\alpha}$ in terms of the function F. In fact, we have

$$v^{\varrho}_lv^{\sigma}_mH_{\varrho\sigma\cdot}{}^{\alpha}u^k_{\alpha}F_{kj} = FF_{lmj} + F_jF_{lm} + F_mF_{lj} + F_lF_{mj} = (\tfrac{1}{2}F^2)_{lmj}.$$

Multiplying this equation by u^j_{β}, we get, by making use of (30),

$$v^{\varrho}_lv^{\sigma}_mH_{\varrho\sigma\beta} = F(\tfrac{1}{2}F^2)_{lmj}u^j_{\beta},$$

which gives

$$H_{\varrho\sigma\alpha} = F(\tfrac{1}{2}F^2)_{ijk}u^i_{\varrho}u^k_{\sigma}u^j_{\alpha} \tag{43}$$

and

$$F(\tfrac{1}{2}F^2)_{ijk} = H_{\varrho\sigma\alpha}v^{\varrho}_iv^{\sigma}_kv^{\alpha}_j. \tag{44}$$

From (43) we see that $H_{\varrho\sigma\alpha}$ is symmetric in all its indices.

Equations (43) and (44) give the following important theorem:

A necessary and sufficient condition for a Finsler space to be a Riemann space is $H_{\varrho\sigma\alpha} = 0$.

This theorem shows the exceptional position that the Riemann metric is occupying among Finsler metrics.

§5. Equations of Structure

By the equations of structure we shall mean the equations which give the exterior derivatives of the Pfaffian forms ω^i, ω^n_α, ω^α_β. As we shall see later, these equations give all the local properties of the Finsler space.

As our interest will only be the form of these equations, we shall apply a method which will deal with the algebra of exterior forms. It seems that in this way lengthy calculations are saved.

Let us first write the equations (16), (32) and (27), (36) in the form

$$\begin{aligned} d\omega^i &= \omega^j \omega_{j.}^{\ i}, \\ \delta_{jk}\omega_{i.}^{\ j} + \delta_{ji}\omega_{k.}^{\ j} &= H_{ik.}^{\ \ j}\omega_j^n. \end{aligned} \tag{45}$$

with the convention that $H_{ik.}^{\ \ j}$ is zero, whenever one of its indices is equal to n. To (45_1) we apply Poincaré's theorem that the exterior derivative of the exterior derivative of a Pfaffian form is zero. We get

$$\omega^k(\omega_{k.}^{\ j}\omega_{j.}^{\ i} - d\omega_{k.}^{\ i}) = 0.$$

This leads us to put

$$\Omega_{k.}^{\ i} = d\omega_{k.}^{\ i} - \omega_{k.}^{\ j}\omega_{j.}^{\ i}, \tag{46}$$

so that

$$\omega^k \Omega_{k.}^{\ i} = 0. \tag{47}$$

It follows that $\Omega_{k.}^{\ i}$, being quadratic, must contain an ω^j in each of its terms, and we can write

$$\Omega_{k.}^{\ i} = \pi_{k.\ j}^{\ \ i}\omega^j.$$

Each $\pi_{k.\ j}^{\ \ i}$ can be decomposed into two summands, as follows:

$$\pi_{k.\ j}^{\ \ i} = R_{k.\ jl}^{\ \ i}\omega^l + \theta_{k.\ j}^{\ \ i},$$

$\theta_{k.\ j}^{\ \ i}$ containing no ω^l. Substituting, we get

$$\Omega_{k.}^{\ i} = R_{k.\ jl}^{\ \ i}\omega^l\omega^j + \theta_{k.\ j}^{\ \ i}\omega^j, \tag{48}$$

from which we see that we can assume

$$R_{k.\ jl}^{\ \ i} + R_{k.\ lj}^{\ \ i} = 0. \tag{49}$$

The relation (47) then gives

$$\begin{aligned} \theta_{k.\ j}^{\ \ i} &= \theta_{j.\ k}^{\ \ i}, \\ R_{k.\ jl}^{\ \ i} + R_{j.\ lk}^{\ \ i} + R_{l.\ kj}^{\ \ i} &= 0 \end{aligned} \tag{50}$$

It remains to find the expression for $\theta_{k.\ j}^{\ \ i}$. Forming the exterior derivative of the second equation of (45) and making use of (46), we get

$$\Omega_{ik} + \Omega_{ki} = H_{ik.}^{\ \ \alpha}\Omega_{\alpha.}^{\ n} + (dH_{ik.}^{\ \ \alpha} + H_{ik.}^{\ \ \beta}\omega_{\beta.}^{\ \alpha} + H_{.k.}^{\gamma\ \alpha}\omega_{\gamma i} + H_{.i.}^{\gamma\ \alpha}\omega_{\gamma k})\omega_{\alpha.}^{\ n}. \tag{51}$$

This relation is symmetric in i,k. It implies a number of consequences which we shall draw presently.

Putting $k = n$, we get

$$\Omega_{in} + \Omega_{ni} = H_{.i.}^{\gamma\ \alpha}\omega_{\gamma.}^{\ n}\omega_{\alpha.}^{\ n}.$$

From the form of $\Omega_{i.}^{\ n}$ and $\Omega_{n.}^{\ j}$ this equation holds only when the right-hand member vanishes. For $i \neq n$ this gives nothing new. For $i = n$ we get

$$H_{\beta..}^{\ \gamma\alpha} = H_{\beta..}^{\ \alpha\gamma},$$

which follows from the property that $H_{\varrho\sigma\alpha}$ is symmetric in all its indices. The relation in question thus becomes

$$\Omega_{in} + \Omega_{ni} = 0. \tag{52}$$

We now multiply (51) by the product

$$P = \omega_{1.}^{\ n} \cdots \omega_{n-1.}^{\ \ \ \ n}.$$

We find

$$\Omega_{ik}P + \Omega_{ki}P = H_{ik.}^{\ \ \alpha}\Omega_{\alpha.}^{\ n}P.$$

On putting

$$\Pi_{k.j}^{\ i} = \theta_{k.j}^{\ i}P,$$

it follows from the last relation that

$$\Pi_{ikm} + \Pi_{kim} = H_{ik.}^{\ \ \alpha}\Pi_{\alpha.m}^{\ n}.$$

Noticing that Π_{ikm} is symmetric in i,m, we can solve this set of equations for Π_{ikm}, getting

$$2\Pi_{ikm} = H_{ik.}^{\ \ \alpha}\Pi_{\alpha.m}^{\ n} + H_{mk.}^{\ \ \alpha}\Pi_{\alpha.i}^{\ n} - H_{im.}^{\ \ \alpha}\Pi_{\alpha.k}^{\ n}.$$

From this relation it is easy to derive that all $\Pi_{ikm} = 0$. In fact, for $i = m = n$, we have

$$\Pi_{njn} = 0$$

that is,

$$\Pi_{n.n}^{\ j} = 0.$$

Putting $i = \beta$, $m = n$, we get

$$2\Pi_{\beta kn} = H_{\beta k.}^{\ \ \alpha}\Pi_{\alpha.n}^{\ n},$$

from which we get, for $k = n$,

$$\Pi_{\beta.n}^{\ n} = 0,$$

so that

$$\Pi_{\beta\cdot n}^{\;j} = 0,$$

Lastly, when $i = \beta$, $m = \gamma$, we have

$$2\Pi_{\beta k\gamma} = H_{\beta k\cdot}^{\;\;\alpha}\Pi_{\alpha\cdot\gamma}^{\;\;n} + H_{\gamma k\cdot}^{\;\;\alpha}\Pi_{\alpha\cdot\beta}^{\;\;n} - H_{\beta\gamma\cdot}^{\;\;\alpha}\Pi_{\alpha\cdot k}^{\;\;n}.$$

This gives, for $k = n$,

$$\Pi_{\beta n\gamma} = 0,$$

whence

$$\Pi_{\beta\cdot\gamma}^{\;j} = 0.$$

We have therefore proved that all $\Pi_{ikm} = 0$.

Since $\omega_{\alpha\cdot}^{\;n}$ are linearly independent, it follows that $\theta_{i\cdot m}^{\;j}$ is a linear combination of $\omega_{\alpha\cdot}^{\;n}$, and we can set

$$\theta_{i\cdot m}^{\;j} = P_{i\cdot m\cdot}^{\;j\;\alpha}\omega_{\alpha\cdot}^{\;n}. \tag{53}$$

where $P_{i\cdot m\cdot}^{\;j\;\alpha}$ is symmetric in its lower indices.

$$P_{i\cdot m\cdot}^{\;j\;\alpha} = P_{m\cdot i\cdot}^{\;j\;\alpha}. \tag{54}$$

When the expressions from (53) are substituted into (48), we get

$$\Omega_{k\cdot}^{\;i} = R_{k\cdot jl}^{\;i}\omega^l\omega^j + P_{k\cdot j\cdot}^{\;i\;\alpha}\omega_{\alpha\cdot}^{\;n}\omega^j. \tag{55}$$

We shall show that $P_{k\cdot j\cdot}^{\;i\;\alpha}$ can be expressed in terms of $\Pi_{\alpha\beta\cdot}^{\;\;\gamma}$ and their "covariant derivatives".

In fact, taking account of the form of $\Omega_{k\cdot}^{\;i}$, it follows from (51) that we can write

$$dH_{ik\cdot}^{\;\;\alpha} + H_{ik\cdot}^{\;\;\beta}\omega_{\beta\cdot}^{\;\alpha} + H_{\cdot k\cdot}^{\gamma\;\alpha}\omega_{\gamma i} + H_{\cdot i\cdot}^{\gamma\;\alpha}\omega_{\gamma k} = \bar{H}_{ik\cdot\cdot}^{\;\;\alpha\beta}\omega_{\beta\cdot}^{\;n} + H_{ik\cdot l}^{\;\;\alpha}\omega^l \tag{56}$$

Substituting into (51) and taking account of (55), we get, on equating corresponding coefficients of $\omega^m\omega^l$, $\omega_\alpha^{\;n}\omega_\beta^{\;n}$, $\omega^l\omega_\alpha^{\;n}$ respectively,

$$\begin{aligned} R_{iklm} + R_{kilm} &= H_{ik\cdot}^{\;\;\beta}R_{\beta\cdot lm}^{\;\;n}, \\ \bar{H}_{ik\cdot\cdot}^{\;\;\alpha\beta} &= \bar{H}_{ik\cdot\cdot}^{\;\;\beta\alpha}, \\ P_{ikl\cdot}^{\;\;\;\alpha} + P_{kil\cdot}^{\;\;\;\alpha} &= H_{ik\cdot}^{\;\;\beta}P_{\beta\cdot l\cdot}^{\;n\;\alpha} - H_{ik\cdot l}^{\;\;\alpha}. \end{aligned} \tag{57}$$

The third set of relations can be solved for $P_{ikl\cdot}^{\alpha}$, with the result

$$P_{klj\cdot}^{\;\;\;\alpha} = \tfrac{1}{4}(H_{kl\cdot}^{\;\;\beta}H_{\beta j\cdot n}^{\;\;\alpha} + H_{jl\cdot}^{\;\;\beta}H_{\beta k\cdot n}^{\;\;\alpha} - H_{kj\cdot}^{\;\;\beta}H_{\beta l\cdot n}^{\;\;\alpha} - \tfrac{1}{2}(H_{kl\cdot j}^{\;\;\alpha} + H_{jl\cdot k}^{\;\;\alpha} - H_{jk\cdot l}^{\;\;\alpha}), \tag{58}$$

so that $P_{klj\cdot}^{\;\;\;\alpha}$ can be expressed in terms of $H_{ik\cdot}^{\;\;\alpha}$, $H_{ik\cdot l}^{\;\;\alpha}$.

The equations (46), (51), (52), (55) are called the *equations of structure* of the Finsler space.

§6. Solution of the Problem of Local Equivalence

E. Cartan has solved the following problem:[8)]

Given two sets of r linearly independent Pfaffian forms θ_A, $\bar{\theta}_A$ each, in the variables x_A, $\bar{x}_A$ respectively. To determine the conditions that there is a transformation

$$\bar{x}_A = \bar{x}_A(x_B), \tag{59}$$

such that the equations

$$\bar{\theta}_A = \theta_A \tag{60}$$

hold under this transformation.

Cartan proved that this problem can be solved by a finite algorithm, and gave a process for its solution. We shall understand that our problem of local equivalence of Finsler spaces is solved, if it can be reduced to a problem of the above type. According to our count at the end of §3 we have introduced a number of new variables into our problem and intrinsically defined the linearly independent Pfaffian forms (39), so that the number of Pfaffian forms is equal to the total number of essential variables. Thus the problem of equivalence is actually reduced to one of the above type. It follows from the general theory that $H_{\alpha\beta}{}^{\gamma}_{.}$, $R_{k}{}^{i}_{.jl}$ constitute the first set of invariants of the space.

For geometrical interpretation it has been found desirable to modify the forms ω_i^j. We set

$$\pi_{i.}^{\,j} = \omega_i^j + \gamma_{i..}^{\,j\alpha}\omega_\alpha^n, \tag{61}$$

where we suppose

$$\gamma_{ki.}^{\;\alpha} + \gamma_{ik.}^{\;\alpha} = -H_{ik.}^{\;\alpha}, \tag{62}$$

or

$$\gamma_{ki\alpha} + \gamma_{ik\alpha} = -H_{ik\alpha}. \tag{62a}$$

Since $H_{ik\alpha} = 0$ when i or k is equal to n, it is natural to suppose

$$\gamma_{in\alpha} = \gamma_{ni\alpha} = 0. \tag{63}$$

The relations (62) do not suffice to determine $\gamma_{\varrho\sigma\alpha}$. With the introduction of a set of quantities $\lambda_{\alpha\varrho\sigma}$, skew-symmetric in the last two indices, we find that the most general expression for $\gamma_{\varrho\sigma\alpha}$ is

$$2\gamma_{\varrho\sigma\alpha} = \lambda_{\varrho\sigma\alpha} - \lambda_{\alpha\rho\sigma} - \lambda_{\sigma\varrho\alpha} - H_{\varrho\alpha\sigma}. \tag{64}$$

It is of course to be assumed that $d\lambda_{\varrho\sigma.}^{\;\alpha}$ should have the same form as $dH_{\varrho\sigma.}^{\;\alpha}$, as given in (56). A logical choice will be $\lambda_{\varrho\sigma\alpha} = 0$, which, as we shall see later, leads to Cartan's connection in Finsler space.

From the definition in (61) we have

$$\pi_{ij} + \pi_{ji} = 0. \tag{65}$$

Obviously, the sets of Pfaffian forms $\omega_{i.}^{\,j}$ and $\pi_{i.}^{\,j}$ determine each other, so that the following statement is true: A necessary and sufficient condition for two Finsler spaces to be locally equivalent is the equivalence of the sets of Pfaffian forms ω^i, π_i^j and $\bar{\omega}^i$, $\bar{\pi}_i^j$.

To find the first set of invariants in terms of these new forms, we write π^i for ω^i and set

$$\begin{aligned} \Pi^i &= d\pi^i - \pi^j\pi_j^i, \\ \Pi_j^i &= d\pi_j^i - \pi_j^k\pi_k^i. \end{aligned} \tag{66}$$

Then we find

$$\Pi^i = \Omega^i - \gamma_{j..}^{i\alpha}\omega^j\omega_\alpha^n,$$

$$\Pi_j^i = \Omega_j^i + \gamma_{j..}^{i\alpha}\Omega_\alpha^n - \gamma_{j..}^{\alpha\beta}\gamma_{\alpha..}^{i\gamma}\omega_{\beta.}^{\;n}\omega_{\gamma.}^{\;n} + (d\gamma_{j..}^{i\alpha} + \gamma_{j..}^{i\beta}\omega_{\beta.}^{\;\alpha} - \gamma_{\beta..}^{i\alpha}\omega_{j.}^{\;\beta} + \gamma_{j..}^{\beta\alpha}\omega_{\beta.}^{\;i})\omega_\alpha^n.$$

We assert that

$$d\gamma_{j..}^{i\alpha} + \gamma_{j..}^{i\beta}\omega_\beta^\alpha - \gamma_{\beta..}^{i\alpha}\omega_j^\beta + \gamma_{j..}^{\beta\alpha}\omega_\beta^i \equiv 0, \quad \operatorname{mod} \omega^k, \omega_\alpha^n,$$

In fact, for $i = n$ or $j = n$, this condition is obviously fulfilled. For $i = \sigma$, $j = \varrho$ it is equivalent to the following more symmetrical condition

$$d\gamma_{\varrho\sigma\alpha} + \gamma_{\varrho\sigma\beta}\omega^\beta_{\cdot\alpha} + \gamma_{\beta\sigma\alpha}\omega^\beta_{\cdot\varrho} + \gamma_{\varrho\beta\alpha}\omega^\beta_{\cdot\sigma} \equiv 0, \quad \operatorname{mod} \omega^k, \omega_\alpha^n. \tag{67}$$

This condition is satisfied by $H_{\varrho\sigma\alpha}$, and hence by our assumption, also by $\gamma_{\varrho\sigma\alpha}$.

It follows that we can write

$$\begin{aligned} \Pi^i &= -\gamma_{\beta..}^{i\alpha}\omega^\beta\omega_{\alpha.}^{\;n}, \\ \Pi_{j.}^{\;i} &= S_{j...}^{i\alpha\beta}\omega_{\alpha.}^{\;n}\omega_{\beta.}^{\;n} + Q_{j.k.}^{\;i\;\alpha}\omega^k\omega_{\alpha.}^{\;n} + T_{j.kl}^{\;i}\omega^k\omega^l, \end{aligned} \tag{68}$$

where we can suppose

$$\begin{aligned} -S_{j...}^{i\alpha\beta} &= S_{j...}^{i\beta\alpha}, \\ -T_{j.kl}^{\;i} &= T_{j.lk}^{\;i} \end{aligned} \tag{69}$$

The coefficients in Π^i, $\Pi_{j.}^{\;i}$ constitute the first set of invariants derived from the Pfaffian forms π^i, $\pi_{j.}^{\;i}$.

§7. Geometrical Interpretation

From the above results we are going to derive a geometrical interpretation. It is in this way that we are led most naturally to the Euclidean connection in the Finsler space.

Each of the Pfaffian forms ω^i defines a covariant vector and the n covariant vectors defined by them are linearly independent. Corresponding to ω^i there exists one and only one set of contravariant vectors $\mathfrak{e}_i$ such that $\omega^i\mathfrak{e}_i$ is the tensor δ_j^i. This tensor is essentially what Carten denoted by dM. To define a Euclidean connection in the space is to give a fundamental quadratic form and the equations of infinitesimal displacement of the frames $M\mathfrak{e}_1 \ldots \mathfrak{e}_n$, which are of the form

$$\begin{aligned} dM &= \theta^i\mathfrak{e}_i, \\ d\mathfrak{e}_i &= \theta_i^j\mathfrak{e}_j. \end{aligned} \tag{70}$$

It is natural to take

$$\theta^i = \omega^i = \pi^i, \tag{71}$$

as the vectors $\mathfrak{e}_i$ are determined by the first set of the equations (70), with (71) substituted.

To explain this relationship more clearly, let us take the local coordinates x^i and let $\mathfrak{a}_i$ denote the vector which is tangent to the coordinate curve

$$x^1 = \text{const.}, \ldots, x^{i-1} = \text{const.}, x^{i+1} = \text{const.}, \ldots, x^n = \text{const.},$$

having x^i as parameter. Then we have

$$dM = dx^i \mathfrak{a}_i \tag{72}$$

Comparing the two expressions for dM, we get

$$\mathfrak{a}_i = v_i^k \mathfrak{e}_k \tag{73}$$

or

$$\mathfrak{e}_i = u_i^k \mathfrak{a}_k \tag{73a}$$

These equations give the components of $\mathfrak{e}_i$ in terms of the local coordinate system and shows in particular that $\mathfrak{e}_n$ is along the direction y^i.

We define the fundamental quadratic form by the equation[9)]

$$dM^2 = (\omega^1)^2 + \cdots + (\omega^n)^2, \tag{74}$$

which means that scalar products of vectors are defined such that

$$\mathfrak{e}_i \mathfrak{e}_j = \delta_{ij}. \tag{75}$$

In order words, the vectors $\mathfrak{e}_i$ are mutually perpendicular unit vectors in terms of the fundamental quadratic form (74).

Utilizing (7) and (31), we find

$$dM^2 = \tfrac{1}{2} F_{ik}(dx^i\, dx^k).$$

This formula has served as the starting-point in the study of Finslerian Geometry to various authors.

In order that the Euclidean connection (70) preserves the fundamental quadratic form, that is, the relations (75), we must have

$$\theta_{ij} + \theta_{ji} = 0.$$

This necessary condition is satisfied, if we put

$$\theta_i^j = \pi_i^j.$$

We therefore arrive at the following conclusion:

It is possible to define in a Finsler space an Euclidean connection, whose metric is

$$dM^2 = (\omega^1)^2 + \cdots + (\omega^n)^2 = \tfrac{1}{2} F_{ik}(dx^i\, dx^k), \tag{76}$$

and whose equations of infinitesimal displacement are

$$\begin{aligned} dM &= \pi^i \mathfrak{e}_i, \\ d\mathfrak{e}_i &= \pi_i^j \mathfrak{e}_j. \end{aligned} \tag{77}$$

The vectors $\mathfrak{e}_i$ are determined by the first set of the equations (77) and are mutually perpendicular unit vectors in terms of the fundamental quadratic form (76).

It is to be remarked that the connection still depends on the choice of the quantities $\gamma_{\alpha..}^{\varrho\sigma}$ or $\lambda_{\alpha..}^{\varrho\sigma}$. For Riemann spaces we have $H_{\alpha..}^{\varrho\sigma} = 0$ and we choose $\gamma_{\alpha..}^{\varrho\sigma} = \lambda_{\alpha..}^{\varrho\sigma} = 0$. This leads to the well-known parallel displacement of Levi-Civita.

§8. Formulas for the Connection in Terms of Local Coordinates

Thus far the Euclidean connections in the Finsler space are expressed in terms of the Pfaffian forms introduced in the solution of the problem of equivalence. We are naturally led to the notions of the absolute differential calculus, when we proceed to give the connection in terms of the local coordinates. Formulas for this purpose will be derived in the present section.

Given a coordinate system x^i, we first introduce the vectors $\mathfrak{a}^i$ according to (72). By (76) the scalar products $\mathfrak{a}_i \mathfrak{a}_j$ are given by

$$\mathfrak{a}_i \mathfrak{a}_j = \tfrac{1}{2}(F^2)_{ij} \equiv g_{ij}(\text{say}). \tag{78}$$

It is sufficient now to determine the Pfaffian forms θ_i^j in the equations

$$d\mathfrak{a}_i = \theta_i^j \mathfrak{a}_j. \tag{79}$$

From

$$\pi^i \mathfrak{e}_i = dx^i \mathfrak{a}_i,$$

we get, by exterior differentiation,

$$\Pi^i \mathfrak{e}_i = d\mathfrak{a}_i \, dx^i$$

or

$$(d\mathfrak{a}_i - \gamma_{\beta..}^{k\alpha} v_i^\beta u_k^j \omega_\alpha^n \mathfrak{a}_j)\, dx^i = 0.$$

This allows us to put

$$d\mathfrak{a}_i = (\gamma_{\beta..}^{k\alpha} v_i^\beta u_k^j \omega_\alpha^n + \Gamma_{i.k}^{j*}\, dx^k)\mathfrak{a}_j, \tag{80}$$

where

$$\Gamma_{i.k}^{j*} = \Gamma_{k.i}^{j*}. \tag{81}$$

It follows that we can write

$$\theta_i^j = C_{i.h}^{j}\, dy^h + \Gamma_{i.h}^{j}\, dx^h. \tag{82}$$

The coefficients $C_{i.h}^{j}$ are easily determined. In fact, by comparing the coefficients of dy^h in (80), (82) and making use of (28), (29), we get

$$C_{i.h}^{j} = -\frac{1}{F}\gamma_{\alpha.\varrho}^{\beta} v_i^\alpha v_h^\varrho u_\beta^j. \tag{83}$$

To determine the coefficients $\Gamma_{i.k}^{j}$ we shall first transform the equations

$$\omega_\alpha^n = 0 \tag{84}$$

to a more symmetrical form. From (17) and (37) we see that these conditions are equivalent to the conditions

$$F_{kj}\, dy^j + \frac{1}{2}\left(G_{kjn} - \frac{\partial F_j}{\partial x^k} + \frac{\partial F_k}{\partial x^j}\right) dx^j = F_k \theta_1,$$

or

$$g_{kj}\,dy^j + \frac{1}{2}F\left(G_{kjn} - \frac{\partial F_j}{\partial x^k} + \frac{\partial F_k}{\partial x^j}\right)dx^j = F_k\theta_2, \tag{85}$$

where θ_1 and θ_2 are two Pfaffian forms. To make a further reduction of these equations we need the expression for G_{kjn}, which is defined in (35). From (35) we have at first

$$G^\alpha_{ij} u^k_\alpha g_{kl} = -F^2 F_{ijl} - F F_l F_{ij}. \tag{86}$$

For simplicity of writing we shall agree from now on that the ascending and lowering of *Latin* indices are taken with respect to the fundamental tensor g_{ij}. Then, by making use of the relations

$$g_{ik} y^k = F_i F, \qquad g^{ik} F_k = y^i/F, \tag{87}$$

we get

$$G^\alpha_{ij} u^k_\alpha = -F^2 F_{ij.}{}^k - y^k F_{ij}. \tag{88}$$

It follows again from (35) that

$$G_{ijn} = F F_{ij.}{}^k\left(\frac{\partial F}{\partial x^k} - \frac{\partial F_k}{\partial x^l} y^l\right) + \frac{1}{F} F_{ij} y^k \frac{\partial F}{\partial x^k} + y^k \frac{\partial F_{ij}}{\partial x^k}. \tag{89}$$

For simplicity of notation we put

$$2G_m = \frac{\partial^2 G}{\partial y^m\, \partial x^k} y^k - \frac{\partial G}{\partial x^m}, \qquad G = \tfrac{1}{2}F^2. \tag{90}$$

Then we find

$$\frac{\partial G_j}{\partial y^h} = \frac{1}{F}F_h G_j + F_j\left(\frac{1}{F}G_h - \frac{F_h}{F}\frac{\partial F}{\partial x^k}y^k + \frac{\partial F}{\partial x^h}\right) + \frac{1}{2}F_{jh}\frac{\partial F}{\partial x^k}y^k$$
$$+ \frac{1}{2}F\left(y^k \frac{\partial F_{hj}}{\partial x^k} - \frac{\partial F_h}{\partial x^j} + \frac{\partial F_j}{\partial x^h}\right).$$

Comparing this expression with (89), we get

$$\frac{1}{2}F\left(G_{kjn} - \frac{\partial F_j}{\partial x^k} + \frac{\partial F_k}{\partial x^j}\right) = \frac{\partial G_k}{\partial x^j} - F F_{kj.}{}^l G_l - \frac{1}{F}F_j G_k - \frac{1}{2}F_{kj}\frac{\partial F}{\partial x^l}y^l$$
$$+ F_k\left(-\frac{1}{F}G_j + \frac{1}{F}F_j\frac{\partial F}{\partial x^l}y^l - \frac{\partial F}{\partial x^j}\right).$$

A further reduction will show that (85) is equivalent to the system

$$dy^i + \frac{\partial G^i}{\partial y^h}dx^h = y^i\varphi, \tag{91}$$

where

$$\varphi = \frac{dF}{F}. \tag{92}$$

Suppose now that the line element (x^i, y^i) is displaced parallel to itself. Then the system (84) or (91) is satisfied. From (80) we have

$$\theta_i^j = \Gamma_{i.k}^{j*}\, dx^k,$$

while from (82), (83), (91), we have

$$\theta_i^j = \left(-C_{i.h}^{j}\frac{\partial G^h}{\partial y^k} + \Gamma_{i.k}^{j}\right) dx^k.$$

Comparing the coefficients of dx^k, we get

$$\Gamma_{i.k}^{j*} = -C_{i.h}^{j}\frac{\partial G^h}{\partial y^k} + \Gamma_{i.k}^{j}. \tag{93}$$

By (81), $\Gamma_{i.k}^{j*}$ or Γ_{ijk}^{*} are symmetric in the indices i, k, so that we have

$$\Gamma_{ikh} - \Gamma_{hki} = C_{ikj}\frac{\partial G^j}{\partial y^h} - C_{hkj}\frac{\partial G^j}{\partial y^i}. \tag{94}$$

On the other hand, by differentiating the relation

$$\mathfrak{a}_i \mathfrak{a}_j = g_{ij},$$

we get

$$dg_{ij} = \theta_{ji} + \theta_{ij},$$

from which it follows in particular that

$$\Gamma_{ikh} + \Gamma_{kih} = \frac{\partial g_{ik}}{\partial x^h}. \tag{95}$$

The equations (94), (95) can be solved for Γ_{ikh}. The result is

$$2\Gamma_{ikh} = \frac{\partial g_{ik}}{\partial x^h} + \frac{\partial g_{kh}}{\partial x^i} - \frac{\partial g_{ih}}{\partial x^k} + (C_{ikr} - C_{kir})\frac{\partial G^r}{\partial y^h} + (C_{ihr} + C_{hir})\frac{\partial G^r}{\partial y^k}$$
$$- (C_{hkr} + C_{khr})\frac{\partial G^r}{\partial y^i}. \tag{96}$$

We have therefore determined the coefficients in θ_i^j in terms of the coordinates x^i.

§9. Various Connections in the Space

From the discussions in the last two sections we see that the Euclidean connection in the Finsler space is completely determined, provided that we make a choice of $\gamma_{\alpha\beta\varrho}$.

The simplest choice is $\lambda_{\varrho\sigma\alpha} = 0$, so that

$$\gamma_{\varrho\sigma\alpha} = -\tfrac{1}{2}H_{\varrho\sigma\alpha}. \tag{97}$$

Then we have, from (83),

$$C_{ijh} = \frac{1}{2F}H_{\alpha\varrho\sigma}v_i^{\alpha}v_j^{\varrho}v_h^{\sigma} = \tfrac{1}{2}(\tfrac{1}{2}F^2)_{ijh} = \frac{1}{2}\frac{\partial g_{il}}{\partial y^h}. \tag{98}$$

This equation also shows that C_{ijh} is symmetric in all its indices. From (96) we then have

$$\Gamma_{ikh} = \frac{1}{2}\left(\frac{\partial g_{ik}}{\partial x^h} + \frac{\partial g_{kh}}{\partial x^i} - \frac{\partial g_{ih}}{\partial x^k}\right) + C_{ihr}\frac{\partial G^r}{\partial y^k} - C_{hkr}\frac{\partial G^r}{\partial y^i}. \tag{99}$$

Formulas (98) and (99) show that the connection so obtained is the one defined by Cartan.

It is clear that other choices of $\gamma_{\varrho\sigma\alpha}$ can be made, but Cartan's choice (97) is by far the simplest.

In the definitions of the parallelism by Synge, Taylor, and Berwald, the direction y^i is always along the curve of propagation, that is, (84) is always supposed to hold. We shall omit here the details of the comparison, referring to [5] for the relation between Cartan's connection and Berwald's.

Institute of Mathematics
Academia Sinica
Nanking

RFERENCES

1. Throughout this paper we shall agree that small Latin indices run from 1 to n, small Greek indices from 1 to $n-1$, and capital Latin indices from 1 to r. We also use the convention that repeated indices imply summation.
2. J. L. Synge, A generalization of the Riemannian line element. Transactions of the American Mathematical Society **27**, 61–67 (1925).
3. J. H. Taylor, A generalization of Levi-Civita's parallelism and the Frenet formulas. Transactions of the American Mathematical Society **27**, 246–264 (1925).
4. L. Berwald, Parallelübertragung in allgemeinen Räumen, Atti Congresso intern. Matem. Bologna. 1928, IV, 263–270.
5. E. Cartan, Les Espaces de Finsler, Paris 1934.
6. Shiing-shen Chern, On the Euclidean connections in a Finsler space, Proceedings of the National Academy of Sciences, U.S.A. **29**, 33–37 (1943).
7. Compare also [5], p. 8.
8. E. Cartan, Les problèmes d'équivalence. Selecta de M. Elie Cartan, Paris 1939, 113–136.
9. Products with parentheses denote ordinary products of differential forms.

Reset from
Science Reports Nat. Tsing Hua Univ. **5** (1948) 95–121

THE IMBEDDING THEOREM FOR FIBRE BUNDLES[1]

BY

SHIING-SHEN CHERN AND YI-FONE SUN

Introduction. In the theory of sphere bundles the imbedding theorem of Whitney-Steenrod[2] has played an important rôle, as it reduces the problem of the classification of sphere bundles to that of the homotopy classification of the mappings of the base space into a Grassmann manifold. With this theorem the *characteristic ring* (relative to a coefficient ring) of the sphere bundle can be defined as the image under the dual homomorphism of the cohomology ring of the Grassmann manifold. It is natural to ask whether an analogous theorem holds for any fibre bundle. The main purpose of this paper is to establish such a theorem, and to give some of its generalizations and extensions.

The paper is divided into five sections. §1 gives the definitions of various notions concerning fibre bundles. The imbedding theorem and its proof, for the case that the base space is a finite polyhedron, are given in §2. Its extension to the case of metric compact ANR (=absolute neighborhood retract) is given in §3. In §4 we extend the notion of the product of two sphere bundles in the sense of Whitney[3] to general fibre bundles and prove a simultaneous imbedding theorem for the product of fibre bundles. A treatment is given in §5 of the cases where the reference groups are the classical groups, namely, the orthogonal, the properly orthogonal, the general linear, the unitary, and the symplectic groups. As is well known, the former two cases give the sphere bundles.

1. **Definitions and notations.** The notion of a fibre bundle arises in a sense from problems which are concerned with the applications of topology, and is therefore somewhat complicated in abstract formulation. We give in this section the definition of various concepts connected with it. A novel feature consists in the definition of the topology of the fibre bundle in terms of the coordinate functions, which simplifies the treatment somewhat.

1. *Fibre bundle.* A *fibre bundle*, to be denoted by $\mathfrak{F}$ or $\{F, G; X, B; \psi, \phi_U\}$, consists of:

(1) A space F, called the *director space*, which is transformed by a topo-

Presented to the Society, November 26, 1949; received by the editors December 10, 1948.

(1) After the paper had been submitted for publication, Professor N. E. Steenrod informed us that the main result of this paper was also proved by him and will be included in his forthcoming monograph on fibre bundles. To him are also due some suggestions for improvements of our treatment.

(2) Whitney [5]; Steenrod [1]. Numbers in brackets refer to the bibliography at the end of the paper.

(3) Whitney [5], [6].

logical group of homeomorphisms G, called the *reference group*, such that the map $G\times F\to F$ defined by the operations of G on F is continuous.

(2) A set X, which will be given a natural topology in the course of definition;

(3) A space B, called the *base space*;

(4) A transformation ψ, called the *projection*, of X onto B;

(5) A system of neighborhoods $\mathfrak{U}=\{U\}$ which cover B, such that to each U there exists a one-one transformation

$$\phi_U\colon U\times F\to\psi^{-1}(U)$$

satisfying the condition $\psi\phi_U(b, y)=b$, $b\in U$, $y\in F$.

These entities are supposed to satisfy the following:

PASTE CONDITION. For given U, b with $b\in U$ denote by $\phi_{U,b}$ the one-one transformation of F onto $\psi^{-1}(b)$ defined by $\phi_{U,b}(y)=\phi_U(b, y)$, $y\in F$. If b belongs to the neighborhoods U and V, then $\phi_{V,b}^{-1}\phi_{U,b}\in G$ and depends continuously on $b\in U\cap V$.

We shall define a topology in X. Let $\{N\}$ be a base in F, and W an open set contained in a neighborhood U of $\mathfrak{U}$. X is topologized by the condition that the sets $\phi_U(W\times N)$ form a base. With this topology X is a space, ψ is a mapping (that is, a continuous transformation), and ϕ_U are homeomorphisms. We shall call X the *total space*, the homeomorphisms ϕ_U the *coordinate functions*, and the neighborhoods of $\mathfrak{U}$ the *coordinate neighborhoods*. For a given $b\in B$ the set $\psi^{-1}(b)$ is called the *fibre* at b. For simplicity we shall also say that X is a fibre bundle *over* B.

Suppose B' be a subspace of B. As a covering of B' we take $\mathfrak{U}'=\{U\cap B' \mid U\in\mathfrak{U}\}$. Define

$$X' = \bigcup_{b\in B'} \psi^{-1}(b)\subset X$$

and the coordinate functions

$$\phi'_{U\cap B'}(b, y) = \phi_U(b, y), \qquad b\in U\cap B',\ y\in F.$$

Then $\mathfrak{F}'=\{F, G; X', B'; \psi, \phi'_{U\cap B'}\}$ is a fibre bundle with the base space B', which may be called the *part of* $\mathfrak{F}$ *over* B' and denoted by $\mathfrak{F}\mid B'$.

Equally naturally we may define an extension of $\mathfrak{F}$. Let I be the unit interval $0\leqq t\leqq 1$. We consider the Cartesian product $B\times I$ and take $\{U\times I\mid U\in\mathfrak{U}\}$ to be its covering. Put

$$X^* = \bigcup_{b\in B}\psi^{-1}(b)\times I$$

and define

$$\psi^*(\psi^{-1}(b)\times t) = b\times t, \qquad \phi^*_{U\times I}(b\times t, y) = \phi_U(b, y)\times t.$$

Then $\{F, G; X^*, B\times I; \psi^*, \phi^*_{U\times I}\}$ is a fibre bundle, to be denoted by $\mathfrak{F}\times I$. For a given value $t\in I$ we shall denote by $\mathfrak{F}\times t$ the contraction $\mathfrak{F}\times I|B\times t$. It is easy to verify that $\mathfrak{F}\times t$, $t\in I$, is equivalent to $\mathfrak{F}\times 0$.

The fibre bundle is called a *sphere bundle* if F is a sphere and G the group of orthogonal transformations of F. It is called a *vector bundle* if F is a vector space and G the group of linear transformations in F. The sphere bundle and the vector bundle are said to be *oriented* if G is the group of proper orthogonal transformations and the group of linear transformations of positive determinant respectively.

2. *Equivalence.* Equivalence is here defined for two fibre bundles with the same F, G, B. Two fibre bundles $\{F, G; X, B; \psi, \phi_U\}$ and $\{F, G; X^*, B; \psi^*, \phi^*_{U^*}\}$ are called *equivalent* if there exists a homeomorphism h of X onto X^*, which satisfies the conditions:

(1) For each $b\in B$, $h(\psi^{-1}(b))=\psi^{*-1}(b)$.

(2) To each $b\in B$ and any two neighborhoods U, U^* containing b of the coverings $\mathfrak{U}$, $\mathfrak{U}^*$, we have

$$\phi^{*-1}_{U^*,b} h \phi_{U,b} \in G$$

and depends continuously on $b\in U\cap U^*$.

Clearly this equivalence relation is reflexive, symmetric, and transitive. It therefore enables us to divide the fibre bundles with given F, G, B into mutually disjoint equivalence classes. We shall use the notation $\equiv$ to denote equivalence.

3. *Mapping and induced bundle.* Given a fibre bundle $\mathfrak{F}=\{F, G; X, B; \psi, \phi_U\}$, and a mapping $f: A\rightarrow B$. We shall define a fibre bundle $\{F, G; X^*, A; \psi^*, \phi^*_{U^*}\}$, called the *induced bundle* and to be denoted by $(\mathfrak{F}; f: A)$ or $A(f)$, as follows: X^* is the union $\bigcup_{a\in A} a\times\psi^{-1}(f(a))$, and

$$\psi^*(a\times\psi^{-1}(f(a))) = a.$$

The neighborhoods U^* are defined to be the open sets $f^{-1}(U)$ so that $\{U^*\}$ is a covering of A. Then we define

$$\phi^*_{U^*}(a, y) = a\times\phi_U(f(a), y).$$

It is easy to verify that the Paste Condition is satisfied.

4. *Admissible mapping of fibre bundles.* Let

$$\mathfrak{F} = \{F, G; X, B; \psi, \phi_U\}, \qquad \mathfrak{F}^* = \{F, G; X^*, B^*; \psi^*, \phi^*_{U^*}\}$$

be two fibre bundles with the same F, G. We take the points $b\in B$, $b^*\in B^*$, and consider the fibres $\psi^{-1}(b)$, $\psi^{*-1}(b^*)$. *A* mapping k: $\psi^{-1}(b)\rightarrow\psi^{*-1}(b^*)$ is called *admissible* if b, b^* have respectively the coordinate neighborhoods U, U^*, such that $\phi^{*-1}_{U^*,b^*}k\phi_{U,b}\in G$. Notice that the condition is independent of the choice of U, U^*. This definition also applies to the case that $\mathfrak{F}$ and $\mathfrak{F}^*$ are identical.

A mapping

$$h: X \to X^*$$

is called *admissible* if the following conditions are satisfied:

(1) For each $b \in B$, $h(\psi^{-1}(b)) = \psi^{*-1}(b^*)$, where $b^* \in B^*$. It follows that h induces a transformation h': $B \to B^*$ defined by $h'(b) = b^*$. Because of our definition of the topologies in X and X^* the transformation h' is continuous.

(2) For each $b \in B$ the partial mapping $h | \psi^{-1}(b)$ is admissible. Moreover, if U and U^* are coordinate neighborhoods which contain b and $b^* = h'(b)$ respectively, the homeomorphism $\phi^{*-1}_{U^*,b^*} h \phi_{U,b} \in G$ depends continuously on $b \in U \cap h'^{-1}(U^*)$.

For simplicity h is said to define an *admissible mapping* h: $\mathfrak{F} \to \mathfrak{F}^*$. Admissible mapping of fibre bundles generalizes the notion of equivalence.

The following theorem is easily verified:

THEOREM 1.1. *If* h: $\mathfrak{F} \to \mathfrak{F}^*$ *is an admissible mapping, the induced bundle* $(\mathfrak{F}^*; h': B)$ *is equivalent to* $\mathfrak{F}$.

5. *Principal fibre bundles; the operations* τ *and* τ_F^{-1}. Given a fibre bundle $\{F, G; X, B; \psi, \phi_U\}$, we shall, following Ehresmann([4]), define its *principal fibre bundle* $\{G, G; X^*, B; \psi^*, \phi^*_U\}$, as follows:

Let $b \in U \subset B$. We denote by G_b the set of functions of the form $\phi_{U,b}g$ for all $g \in G$. G_b depends only on b; for, if $b \in V$, then $\phi_{V,b} = \phi_{U,b}g_0$, $g_0 \in G$, and the set $\phi_{V,b}G$ is identical with $\phi_{U,b}G$. We put $X^* = \bigcup_{b \in B} G_b$ and define the projection ψ^* to be $\psi^*(G_b) = b$. The director space is G, operated on by G as the group of left translations.

The coordinate functions ϕ^*_U are defined by

$$\phi^*_U(b, g) = \phi_{U,b}g, \qquad g \in G.$$

If $b \in U$, V, and $g \in G$, we have $\phi_{V,b} = \phi_{U,b}g_0$, $g_0 \in G$, and

$$\phi^*_{U,b}(g) = \phi_{U,b}g, \qquad \phi^*_{V,b}(g) = \phi_{V,b}g = \phi_{U,b}g_0g,$$

so that $\phi^{*-1}_{V,b}\phi^*_{U,b}$ is the mapping $g \to g_0^{-1}g$ in G and the Paste Condition is satisfied.

We shall write $\{G, G; X^*, B; \psi^*, \phi^*_U\} = \tau\{F, G; X, B; \psi, \phi_U\}$.

To the operation τ so defined there is an inverse operation. In fact, let $\mathfrak{F}^* = \{G, G; X^*, B; \psi^*, \phi^*_U\}$ be a principal fibre bundle, and let F be a space operated on by the group G such that the transformation $F \times G \to F$ defined by the group operation is continuous. Define

$$X = \bigcup_{b \in B} \phi^*_{U,b}F, \qquad \psi(\phi^*_{U,b}F) = b, \qquad \phi_U(b, y) = \phi^*_{U,b}y, \qquad y \in F.$$

([4]) Ehresmann [1].

Then $\{F, G; X, B; \psi, \phi_U\}$ is a fibre bundle, to be denoted by $\tau_F^{-1}\mathfrak{F}^*$. It follows from definition that the following relations hold:

$$\tau_F^{-1}\tau\mathfrak{F} \equiv \mathfrak{F}, \qquad \tau\tau_F^{-1}\mathfrak{F}^* \equiv \mathfrak{F}^*.$$

Two fibre bundles $\mathfrak{F}_1$ and $\mathfrak{F}_2$ are called *associated* if $\tau\mathfrak{F}_1 \equiv \tau\mathfrak{F}_2$. From the above relations we see that two associated bundles are equivalent if they have the same director space.

For later use we state here the following theorems, whose proofs are immediate:

THEOREM 1.2 (EHRESMANN). *Two fibre bundles are equivalent if and only if their principal bundles are equivalent.*

It follows that the operations τ and τ_F^{-1} are defined for equivalence classes of fibre bundles.

THEOREM 1.3. *Let $\mathfrak{F}$ be a fibre bundle and A a space which is mapped by f into the base space of $\mathfrak{F}$. Then*

$$(\tau\mathfrak{F}; f\colon A) \equiv \tau(\mathfrak{F}; f\colon A).$$

If $\mathfrak{F}$ is a principal fibre bundle, then

$$(\tau_F^{-1}\mathfrak{F}; f\colon A) \equiv \tau_F^{-1}(\mathfrak{F}; f\colon A).$$

In other words, the induction of fibre bundles commutes with the operations τ and τ_F^{-1}.

In a principal fibre bundle it is possible to define a group of transformations, which will be of importance in several connections. In fact, let $\mathfrak{F} = \{G, G; X, B; \psi, \phi_U\}$ be a principal bundle. Let $g_0 \in G$. Then g_0 acts on X as a right translation as follows:

$$x \cdot g_0 = \phi_{U,b}(\phi_{U,b}^{-1}(x)g_0),$$

where $b = \psi(x)$ and U is a neighborhood containing b. Clearly, the point in the right-hand side is independent of the choice of U. The mapping $x \cdot g_0$ induces a mapping on each fibre, and is in general not admissible. For any fixed x_0 the mapping of G onto the fibre through x_0 given by $g \to x_0 \cdot g$ is admissible.

6. *Universal fibre bundle.* Consider the fibre bundles with given F, G, B. A fibre bundle $\{\mathring{F}, G; X, A; \psi, \phi_U\}$ is called *universal* (relative to F, G, B) if the equivalence classes of fibre bundles over B are in one-one correspondence with the homotopy classes of mappings $B \to A$, or, more precisely, if the following properties hold:

(1) Every bundle with the same F, G, B is equivalent to the bundle induced by a mapping $B \to A$.

(2) The bundles induced by the mappings f, g: $B \to A$ are equivalent if and only if f and g are homotopic.

It will be our main purpose to prove theorems on the existence of universal fibre bundles and to draw consequences therefrom.

7. *Simple bundles.* As an example we consider a class of fibre bundles called *simple.* There is a natural way to define from the product space $B\times F$ a bundle whose reference group G_0 consists of the identity. If $\mathfrak{F}$ is any bundle over B with the director space F and reference group G, then $G_0\subset G$, and $B\times F$ determines an equivalence class of bundles relative to F, G, B, of which $B\times F$ is a member. Each such bundle is said to be equivalent to the product bundle or simply a *product bundle.*

A fibre bundle $\mathfrak{F}=\{F, G; X, B; \psi, \phi_U\}$ is called *parallelisable* if it admits a cross section, that is, if there exists a mapping λ: $B\rightarrow X$ such that $\psi\lambda$ is the identity. $\mathfrak{F}$ is called *simple* if its principal bundle is parallelisable.

Suppose now that $\mathfrak{F}^*=\{G, G; X^*, B; \psi^*, \phi_U^*\}$ is a principal bundle over B which is parallelisable with the cross section λ: $B\rightarrow X^*$. Define f: $B\times G\rightarrow X^*$ by $f(b, g)=\lambda(b)\cdot g$. Then f provides an equivalence of $B\times G$ and $\mathfrak{F}^*$. Thus a principal bundle is a product bundle if and only if it is parallelisable.

Now let $\mathfrak{F}$ be a simple bundle. Then $\mathfrak{F}^*=\tau\mathfrak{F}$ is parallelisable and is hence a product bundle. It follows from Theorem 1.2 that $\mathfrak{F}$ is also a product bundle. Thus simple bundle means the same thing as product bundle.

If $\mathfrak{F}^*$ is a fibre bundle over B^* and f_0 maps B into a point of B^*, then the induced bundle is a product bundle. Hence, if $\mathfrak{F}^*$ is a universal bundle relative to F, G, B, the induced bundle $(\mathfrak{F}^*; f: B)$ is a product bundle if and only if f is homotopic to a constant.

2. The imbedding theorem for finite polyhedra.

The imbedding theorem is concerned with sufficient conditions for the existence of universal fibre bundles. An answer will be given in this section for the case that the base space is a finite polyhedron. This generalizes the imbedding theorem of Whitney-Steenrod for sphere bundles.

1. *Fibre bundles induced by homotopic mappings.* Let $\mathfrak{F}$ be a fibre bundle with the base space B, A a compact space, and f_0, f_1: $A\rightarrow B$ two mappings of A into B. We shall, following a procedure due essentially to Steenrod(5), prove the theorem that $A(f_0)$ and $A(f_1)$ are equivalent if f_0 and f_1 are homotopic. By the use of the notion of admissible mapping of fibre bundles the result can be put in a slightly more general form, which includes both the last mentioned result and the covering homotopy theorem as particular cases. This theorem can be stated as follows:

THEOREM 2.1. *Let $\mathfrak{F}$ and $\mathfrak{F}'$ be two fibre bundles with the base spaces B and B' respectively, of which B' is compact. Let f: $\mathfrak{F}'\times 0\rightarrow\mathfrak{F}$ be an admissible mapping and f': $B'\times I\rightarrow B$ an extension of its induced mapping f_0': $B'\times 0\rightarrow B$. Then there exists an admissible mapping f: $\mathfrak{F}'\times I\rightarrow\mathfrak{F}$ which coincides with f_0 on $\mathfrak{F}'\times 0$ and whose induced mapping is f'.*

(5) Steenrod [4, pp. 302–303].

Proof. Let $\mathfrak{F}=\{F, G; X, B; \psi, \phi_U\}$, $\mathfrak{F}'=\{F, G; X', B'; \psi', \phi'_{U'}\}$. For each $b\in B$ we select a pair of neighborhoods V, W of b such that $\overline{V}\subset W$ and that $\overline{W}$ is contained in some coordinate neighborhood U. Since $B'\times I$ is compact, there exists a $\delta>0$ such that, for any point $b'\in B'$, the image $f'(b'\times I')$ is contained in some member of the family $\{V\}$, provided that $I'\subset I$ is an interval of length less than δ. Let $0=t_0<t_1<\cdots<t_n=1$ be a division of I, with $t_{k+1}-t_k<\delta$, and let I_k denote the interval $t_k\leq t\leq t_{k+1}$.

Since, for each b', there exists a V containing the image of $b'\times I_k$ under f', it follows by continuity that there is a neighborhood N of b' such that f' maps $N\times I_k$ into some V. Since B' is compact, we may select a finite covering by these neighborhoods: $N_{k,1}, \cdots, N_{k,m}$. Let $V_{k,i}\supset f'(N_{k,i}\times I_k)$ and let $W_{k,i}$ form a pair with $V_{k,i}$.

By the Urysohn lemma, there exists a continuous real-valued function $u_{k,i}(b)$ on B such that $0\leq u_{k,i}(b)\leq 1$ and

$$u_{k,i}(b) = 1, \qquad b\in \overline{V}_{k,i}, \qquad u_{k,i}(b) = 0, \qquad b\in B - W_{k,i}.$$

Let

$$\tau_{k,j}(b') = t_k + (t_{k+1} - t_k) \max_{i\leq j} \left\{ \min_{t_k\leq t\leq t_{k+1}} u_{k,i}(f'(b', t))\right\}.$$

Clearly $\tau_{k,j}$ is continuous, $t_k\leq\tau_{k,j}\leq t_{k+1}$, $\tau_{k,j}(b')\leq\tau_{k,j+1}(b')$, and $\tau_{k,m}(b')=t_{k+1}$. We define by convention $\tau_{k,0}(b')=t_k$.

The proof of the theorem hinges on the definition of a mapping

$$f\colon\ X'\times I\to X$$

with the desired properties. This will be achieved by double induction on k and j. It is clearly sufficient to define

$$f\colon\ X'\times I_k\to X$$

for $(x', t)\in X'\times I_k$ such that $\tau_{k,j}(b')<t\leq\tau_{k,j+1}(b')$, $b'=\psi'(x')$, where $f(x', \tau_{k,j}(b'))$ is given.

Denote by T the set of points $(b', t)\in B'\times I$ such that $\tau_{k,j}(b')<t\leq\tau_{k,j+1}(b')$. Then $\min_{t_k\leq t\leq t_{k+1}} u_{k,j+1}(f'(b', t))\neq 0$, and we have $u_{k,j+1}(f'(b', t))\neq 0$ or $f'(b', t)\in W_{k,j+1}$ for $(b', t)\in T$. It follows that $f'(\overline{T})\subset\overline{W}_{k,j+1}$, where the latter belongs to a coordinate neighborhood, say U, of B. Let ϕ be the coordinate function relative to U. Since ϕ establishes a homeomorphism between $U\times F$ and $\psi^{-1}(U)$, there exists a mapping $\zeta\colon \psi^{-1}(U)\to F$ such that $\zeta\phi(b, y)=y$ for $b\in U$, $y\in F$. Writing $b'=\psi'(x')$, we define

$$f(x', t) = \phi(f'(b'), \zeta f(x', \tau_{k,j}(b'))).$$

This completes the induction. It is easy to verify that f defines an admissible mapping $f\colon \mathfrak{F}'\times I\to\mathfrak{F}$.

REMARK. The theorem remains true if B' is the union of a countable number of compact spaces.

THEOREM 2.2. *Let $\mathfrak{F}$ be a fibre bundle with the base space B, A a compact space, and f_0, f_1: $A \to B$ two mappings of A into B. If f_0 and f_1 are homotopic, then $A(f_0)$ and $A(f_1)$ are equivalent.*

Proof. It follows from Theorems 1.1 and 2.1 that

$$\mathfrak{F} \times i \equiv A(f_i), \qquad i = 0, 1.$$

Since $\mathfrak{F}\times 0 \equiv \mathfrak{F}\times 1$, we get $A(f_0) \equiv A(f_1)$.

For the sake of completeness we state here the following useful theorem:

THEOREM 2.3 (COVERING HOMOTOPY THEOREM). *If X is the total space of a fibre bundle over B, A a compact space, f a continuous map $A \to X$, and $h(a, t)$ a homotopy of the map ψf of A into B, then there exists a homotopy $g(a, t)$ of f which covers $h(a, t)$.*

2. *Equivalent fibre bundles.* We shall first reduce the problem of universal fibre bundles to that of universal principal fibre bundles by means of the following theorem:

THEOREM 2.4. *Let G operate on F such that the induced transformation $G \times F \to F$ is continuous. If $\mathfrak{F}_0$ is a universal principal fibre bundle relative to G, G, B, then $\tau_F^{-1}\mathfrak{F}_0$ is a universal fibre bundle relative to F, G, B.*

Proof. Denote by B_0 the base space of $\mathfrak{F}_0$. Let $\mathfrak{F} = \{F, G; X, B; \psi, \phi_U\}$ be a fibre bundle. Then $\tau\mathfrak{F}$ is equivalent to the bundle $(\mathfrak{F}_0; f: B)$ induced by a mapping f: $B \to A$. It follows that

$$\mathfrak{F} \equiv \overset{-1}{\tau_F}\, \tau\mathfrak{F} \equiv \overset{-1}{\tau_F}(\mathfrak{F}_0; f: B) \equiv (\overset{-1}{\tau_F}\mathfrak{F}_0; f: B),$$

by Theorem 1.3.

Consider next two induced bundles $(\tau_F^{-1}\mathfrak{F}_0; f: B)$ and $(\tau_F^{-1}\mathfrak{F}_0; g: B)$. They are equivalent if and only if $\tau_F^{-1}(\mathfrak{F}_0; f: B)$ and $\tau_F^{-1}(\mathfrak{F}_0; g: B)$ are equivalent, and hence if and only if $(\mathfrak{F}_0; f: B)$ and $(\mathfrak{F}_0; g: B)$ are equivalent. $\mathfrak{F}_0$ being a universal principal fibre bundle, a necessary and sufficient condition for the latter property is that f and g are homotopic.

THEOREM 2.5. *Let $\mathfrak{F} = \{G, G; X, B: \psi, \phi_U\}$ be a principal fibre bundle having as base space B a polyhedron of dimension n. Let $\mathfrak{F}^* = \{G, G; X^*, B^*; \psi^*, \phi^*_{U^*}\}$ be a principal fibre bundle such that $\pi_i(X^*) = 0$, $0 \leq i \leq n-1$. Denote by B_0 a subpolyhedron of B and by $\mathfrak{F}_0$ the part of $\mathfrak{F}$ over B_0. Then every admissible mapping f_0: $\mathfrak{F}_0 \to \mathfrak{F}^*$ can be extended to an admissible mapping f: $\mathfrak{F} \to \mathfrak{F}^*$.*

Proof. We take a simplicial decomposition of B which is so fine that each simplex belongs to a coordinate neighborhood. By hypothesis there exists a mapping f_0: $\psi^{-1}(B_0)\to X^*$ such that $f_0(\psi^{-1}(b))=\psi^{*-1}(b^*)$ for $b\in B_0$ and that the partial mapping $f_0|\psi^{-1}(b)$ is admissible. It is sufficient to define an extension f of f_0 over $\psi^{-1}(B)$, with the desired properties. This extension f will be defined on $\psi^{-1}(B_0\cup B^r)$, by induction on r, where B^r denotes the r-dimensional skeleton of B.

For $r=0$ the definition of $f|\psi^{-1}(B_0\cup B^0)$ is obvious, it being only necessary to take $f|\psi^{-1}(\sigma^0)$ to be an admissible mapping into a fibre of X^* for any 0-dimensional simplex $\sigma^0\notin B_0$. Suppose $f|\psi^{-1}(B_0\cup B^{r-1})$ be defined and let σ^r be an r-simplex not belonging to B_0. Take a coordinate neighborhood U containing σ^r, and denote by e the identity of G. Then the map $f\phi_U(\partial\sigma^r\times e)$ is the map of an $(r-1)$-sphere into X^* and is contractible. It follows that there is an extension $f\phi_U(\sigma^r\times e)$ of $f\phi_U(\partial\sigma^r\times e)$. Define then

$$f\phi_U(b, g) = (f\phi_U(b, e))\cdot g, \qquad b\in\sigma^r,\ g\in G.$$

Since ϕ_U: $\sigma^r\times G\to\psi^{-1}(\sigma^r)$ is a homeomorphism, f is defined for $\psi^{-1}(\sigma^r)$ and the induction is complete. It can be verified that f defines an admissible mapping of $\mathfrak{F}$ into $\mathfrak{F}^*$.

From Theorem 2.5 we derive the following theorems which give sufficient conditions for a universal principal fibre bundle.

THEOREM 2.6. *Let $\mathfrak{F}=\{G, G; X, B; \psi, \phi_U\}$ be a principal fibre bundle having as base space B a polyhedron of dimension n. Let $\{G, G; X^*, B^*; \psi^*, \phi^*_{U^*}\}$ be a principal fibre bundle such that $\pi_i(X^*)=0$, $0\leq i\leq n-1$*(6)*. There exists a mapping f: $B\to B^*$ such that $\mathfrak{F}\equiv B(f)$.*

Proof. Take B_0 to be empty and apply Theorem 2.5.

THEOREM 2.7. *Let $\{G, G; X^*, B^*; \psi^*, \phi^*_{U^*}\}$ be a principal fibre bundle such that $\pi_i(X^*)=0$, $0\leq i\leq n$. If B is a polyhedron of dimension n and f, g: $B\to B^*$ are mappings which induce equivalent fibre bundles $B(f)\equiv B(g)$, then the mappings f and g are homotopic.*

Proof. Replacing, in Theorem 2.5, $\mathfrak{F}$ by $\mathfrak{F}\times I$, B by $B\times I$, and B_0 by $(B\times 0)\cup(B\times 1)$, we get the theorem.

3. *Existence of universal fibre bundles.* From Theorems 2.2, 2.6, and 2.7, it is now easy to prove, by an explicit construction, the existence of universal fibre bundles for the case that the base space is a finite polyhedron and that the reference group is a linear group. The assumption on the reference group is reasonable in view of applications, as the most important case of compact Lie groups is included.

We denote by A_n the general linear group in n variables, and A_n^+ the subgroup of all linear transformations of positive determinant of A_n. Let I_n denote the identical linear transformation in n variables. Then we can imbed $I_m\times A_n\subset A_{m+n}$, $I_m\times A_n^+\subset A_{m+n}^+$, by assuming that I_m operates on the first m

(6) $\pi_0(X^*)=0$ means by definition that X^* is arcwise connected.

and A_n or A_n^+ on the last n variables. Using these notations, we have the theorem:

THEOREM 2.8. *The homogeneous spaces*

$$A_{m+n}/I_m \times A_n, \quad \overset{+}{A}_{m+n}/I_m \times \overset{+}{A}_n, \qquad m, n \geqq 1,$$

are arcwise connected and have their homotopy groups $\pi_i = 0$ *for* $1 \leqq i \leqq n-1$.

Proof. The two homogeneous spaces in question are homeomorphic. In fact, for $a \in A_{m+n}$, define the mapping

$$a(I_m \times A_n) \to \eta a(I_m \times \overset{+}{A}_n), \qquad \eta = I_{m+n-1} \times \operatorname{sgn} \det a.$$

This induces a mapping

$$A_{m+n}/I_m \times A_n \to \overset{+}{A}_{m+n}/I_m \times \overset{+}{A}_n,$$

which is easily proved to be a homeomorphism. It is therefore sufficient to consider the second space.

We have

$$\pi_i(\overset{+}{A}_{m+n}/I_m \times \overset{+}{A}_n) \approx \pi_i(\overset{+}{A}_{m+n}, I_m \times \overset{+}{A}_n).$$

An element of the latter group is represented by the mapping of an i-cell into A_{m+n}^+ with its boundary mapped into $I_m \times A_n^+$. By the covering homotopy theorem(7) we know that such a map is contractible into $I_m \times A_n^+$ with its boundary fixed, if $i \leqq n-1$. The covering homotopy theorem also proves that the space $A_{m+n}^+/I_m \times A_n^+$ is arcwise connected. Hence the theorem is proved.

Let G be a linear group in m variables. Then $A_{m+n+1}/I_m \times A_{n+1}$ is a fibre bundle over $A_{m+n+1}/(G \times A_{n+1})$ with director space G subject to the same group G as left translations. By Theorems 2.2, 2.6, 2.7, 2.8, it follows that this fibre bundle is universal relative to G, G, B, if B is a finite polyhedron of dimension n. Applying the operation τ_F^{-1} to this universal principal fibre bundle, we get a universal fibre bundle relative to F, G, B. This result is now stated in the following theorem:

THEOREM 2.9. *For fibre bundles whose base space* B *is a finite polyhedron and whose reference group* G *is a linear group, universal fibre bundles exist relative to* F, G, B.

3. The imbedding theorem for compact metric ANR. We shall extend in this section the theorem on the existence of universal fibre bundles to cover the case that the base space is a compact metric ANR (absolute neighborhood retract). For this purpose it is convenient to make use of the notion of a *bridge* introduced by Hu(8) in his study of mappings. We begin by recalling its definition and basic properties.

(7) Hurewicz-Steenrod [3, p. 64].

(8) Hu [2].

1. *Résumé of some theorems of Hu.* Let X be a compact metric ANR, X_0 a closed subset of X, and Y a connected ANR. We shall denote a finite open covering of X by the notation $\alpha=\{a_1, \cdots, a_r\}$. Suppose the nerve of α be geometrically realized, and denote by A the geometrical complex which realizes the nerve. The sets $X_0 \cap a_i$, $i=1, \cdots, r$, form an open covering of X_0, and its realization A_0 is a subcomplex of A. Without danger of confusion we use the symbol a_i to denote at the same time a set of the covering and a vertex of A. A mapping ϕ_α: $X \rightarrow A$ is called a *canonical mapping* of α if for each point $x \in X$, $\phi_\alpha(x)$ is contained in the closure of the simplex $a_{i_0}a_{i_1} \cdots a_{i_m}$ of A, where $a_{i_0}, \cdots, a_{i_m}$ are the members of α containing x.

Let f: $X_0 \rightarrow Y$ be a given mapping and α a covering of X. A mapping ψ_α: $A_0 \rightarrow Y$ is called a *bridge mapping* for f if the partial mapping $\psi_\alpha\phi_\alpha | X_0$ is homotopic to f for each canonical mapping ϕ_α: $X \rightarrow A$ of the covering α. If such a bridge mapping ψ_α exists, α is called a *bridge* for the mapping.

Concerning this notion of bridge Hu has established the following three basic theorems:

(1) Bridge refinement theorem. *For a given mapping f: $X_0 \rightarrow Y$, any refinement β of a bridge α is a bridge.*

(2) Bridge existence theorem. *Every mapping f: $X_0 \rightarrow Y$ has a bridge α.*

(3) Bridge homotopy theorem. *If α, β are two bridges for a given mapping f: $X_0 \rightarrow Y$, with the bridge mappings ψ_α: $A \rightarrow Y$, ψ_β: $B \rightarrow Y$, there exists a common refinement γ of α and β such that $\psi_\alpha p_{\gamma\alpha} | C_0$ and $\psi_\beta p_{\gamma\beta} | C_0$ are homotopic where $p_{\gamma\alpha}$: $C \rightarrow A$, $p_{\gamma\beta}$: $C \rightarrow B$ are arbitrary simplicial projections.*

2. *The imbedding theorem.*

Theorem 3.1. *Let B be a compact metric ANR of dimension n, and let F be the director space and G the reference group. Let $\mathfrak{F}_0$ be a universal fibre bundle relative of F, G, B', where B' is a finite polyhedron of dimension n. Then $\mathfrak{F}_0$ is also a universal fibre bundle relative to F, G, B.*

Proof. Consider the identity mapping ι: $B \rightarrow B$. By the bridge existence theorem and the bridge refinement theorem there exist a bridge α, a geometrical realization as an n-dimensional complex A of the nerve of α, and a bridge mapping g: $A \rightarrow B$ such that gh is homotopic to ι for each canonical mapping h: $B \rightarrow A$. The mapping g induces a bundle $A(g)$ over A. Since A is an n-dimensional complex, there exists a mapping s: $A \rightarrow B_0$, where B_0 is the base space of $\mathfrak{F}_0$, such that $A(s) \equiv A(g)$. It follows that $B(sh) \equiv B(gh)$. Since gh is homotopic to ι, we have $B(gh) \equiv \mathfrak{F}$. Therefore $B(sh) \equiv \mathfrak{F}$, and the mapping sh: $B \rightarrow B_0$ induces a bundle over B, which is equivalent to the given bundle $\mathfrak{F}$.

Suppose next that f_0, f_1: $B \rightarrow B_0$ are mappings such that $B(f_0) \equiv B(f_1)$. We shall prove that then $f_0 \simeq f_1$. In fact, introducing the bridge mapping g and the canonical mapping h as above, we have $A(f_0 g) \equiv A(f_1 g)$. Since A is a finite complex, it follows from Theorem 2.7 that $f_0 g \simeq f_1 g$. Then $f_0 gh \simeq f_1 gh$. Since $gh \simeq \iota$, we have $f_i gh \simeq f_i$, $i=0, 1$. Consequently, we get $f_0 \simeq f_1$, which is to be proved.

The above theorem reduces the question on the existence of universal fibre bundles for the case that the base space is a compact metric ANR to the case that the base space is a finite polyhedron.

3. *The characteristic ring.* The imbedding theorem permits us to define an important invariant of a fibre bundle, its *characteristic ring.* Let $\mathfrak{F}_0$, with the base space B_0, be a universal fibre bundle relative to F, G, B. Let R be a commutative ring, and $H(B_0, R)$ the cohomology ring of B_0 with the coefficient ring R. The classes of equivalent fibre bundles with the same F, G, B being in one-one correspondence with the homotopy classes of mappings $B \rightarrow B_0$, to each equivalence class of fibre bundles corresponds a definite ring homomorphism $H(B_0, R) \rightarrow H(B, R)$. We shall call the image of this ring homomorphism the *characteristic ring* relative to the coefficient ring R and denote it by $C(B, R)$. The cohomology classes of $C(B, R)$ are called the *characteristic cohomology classes.*

The characteristic ring depends by definition on the choice of the universal fibre bundle $\mathfrak{F}_0$, which is by no means unique. It is very likely that the ring $C(B, R)$ as an abstract ring is independent of the choice of $\mathfrak{F}_0$, but we are not able to prove it(9).

4. Product of fibre bundles. In studying the problem of position of one sphere bundle in another we are naturally led to the notion of the product of sphere bundles. This section will be devoted to a discussion of the product of fibre bundles. A simultaneous imbedding theorem will be established, which is useful for the description of the position of one fibre bundle in another.

1. *Definitions.* Let

$$\mathfrak{F}' = \{F', G'; X', B; \psi', \phi'_U\}, \qquad \mathfrak{F}'' = \{F'', G''; X'', B; \psi'', \phi''_U\}$$

be two fibre bundles, with the same base space B and the same family of coordinate neighborhoods $\{U\}$. We shall define a fibre bundle, their *product,*

$$\mathfrak{F}' \times \mathfrak{F}'' = \{F' \times F'', G' \times G''; X, B; \psi, \phi_U\}$$

as follows: The director space is $F' \times F''$ and is transformed by $G' \times G''$ according to the formula $(g' \times g'')(y' \times y'') = g'(y') \times g''(y'')$, $g' \in G'$, $g'' \in G''$, $y' \in F'$, $y'' \in F''$. The total space is

$$X = \bigcup_{b \in B} \phi'_{U,b}(F') \times \phi''_{U,b}(F''),$$

(9) In the case of sphere bundles this can be proved by interpreting the generators of the characteristic ring as certain obstructions.

while projection ψ is defined by

$$\psi\{\phi'_{U,b}(F') \times \phi''_{U,b}(F'')\} = b.$$

The coordinate functions are

$$\phi_U(b, y' \times y'') = \phi'_{U,b}(y') \times \phi''_{U,b}(y'').$$

It is easily verified that the Paste Condition is satisfied.

This definition of the product bundle does not include the case of sphere bundles as defined by Whitney. We shall, however, show in the next section how, with the help of a relationship between sphere bundles and vector bundles, the product of vector bundles in the sense just defined will lead to the product of sphere bundles in the sense of Whitney.

2. *Change of the reference group.* In order to derive from our product of fibre bundles Whitney's product of sphere bundles, we shall study a relationship between fibre bundles with different reference groups. In fact, let H be a subgroup of G. A fibre bundle with the reference group H can be considered as a bundle with the group G. The converse question is solved by the following result of Ehresmann(10).

THEOREM 4.1. *Let $\mathfrak{F}$ be a fibre bundle with the reference group G and let H be a subgroup of G. Construct from the principal bundle $\tau\mathfrak{F}$ the bundle $\tau_{G/H}^{-1}\tau\mathfrak{F}$ whose director space is the homogeneous space G/H of left cosets of G relative to H. $\mathfrak{F}$ is equivalent to a bundle with the reference group H if and only if $\tau_{G/H}^{-1}\tau\mathfrak{F}$ is parallelisable.*

Suppose the base space B be given. Denote by (B, G) an equivalence class of principal fibre bundles relative to G, G, B. If H is a subgroup of G, then there is a natural correspondence λ: $(B, H)\rightarrow(B, G)$. From Theorem 4.1 it follows that the correspondence is onto if and only if the bundles of the class $\tau_{G/H}^{-1}(B, G)$ is parallelisable. If B is a polyhedron of dimension n, a sufficient condition for this is $\pi_i(G/H)=0$, $0\leqq i\leqq n-1$.

Suppose the bundles relative to G, G, B have a universal bundle $\mathfrak{F}_0$, whose base space we denote by B_0. Two induced bundles $B(f_0)$ and $B(f_1)$ are equivalent if and only if f_0 and f_1 are homotopic, that is, if and only if there is a mapping F: $B\times I\rightarrow B_0$, with $F(B\times 0)=f_0$, $F(B\times 1)=f_1$. If we write $B'=B\times I$, the mapping F induces a bundle $B'(F)$ over B', and the bundles $B(f_0)$ and $B(f_1)$ are equivalent relative to H if $B'(F)$ is equivalent relative to G to a bundle with the reference group H. A necessary and sufficient condition for the latter property is that $\tau_{G/H}^{-1}B'(F)$ is parallelisable. This condition is satisfied if $\pi_i(G/H)=0$, $0\leqq i\leqq n$.

It follows, when B is a polyhedron of dimension n, that the correspondence λ is one-one and onto if $\pi_i(G/H)=0$, $0\leqq i\leqq n$.

(10) Ehresmann [1].

When $G=A^+(n)$, $H=0^+(n)$, the homogeneous space G/H is a Euclidean space and is hence contractible. Therefore there is a one-one correspondence between classes of sphere bundles and classes of vector bundles.

3. *The simultaneous imbedding theorem.* It is important to remark that the above "product" admits an inverse operation. In fact, let $\{F'\times F'', G'\times G''; X, B; \psi, \phi_U\}$ be a fibre bundle, where the reference group $G'\times G''$ operates on the director space $F'\times F''$ as in the last paragraph. We shall define two fibre bundles whose product is a fibre bundle equivalent to the given one. Write a point of $F'\times F''$ in the form $y'\times y''$, $y'\in F'$, $y''\in F''$, and define the projections $\zeta_1(y'\times y'')=y'$, $\zeta_2(y'\times y'')=y''$. Let $b\in U\subset B$. Two points z, z' $\in\psi^{-1}(b)$ are called equivalent if $\zeta_i(\phi_{U,b}^{-1}(z))=\zeta_i(\phi_{U,b}^{-1}(z'))$, $i=1, 2$. This equivalence relation is independent of the choice of the neighborhood U. The equivalence classes thus obtained form a decomposition space $Z_i(b)$, $i=1, 2$. Denote by $(z)_i$ the class which contains z. Put

$$X' = \bigcup_{b\in B} Z_1(b), \qquad X'' = \bigcup_{b\in B} Z_2(b),$$

and define

$$\psi'(Z_1(b)) = b, \qquad \psi''(Z_2(b)) = b,$$

$$\psi'_U(b\times y') = (\phi_U(b\times(y'\times y'')))_1,$$

$$\phi''_U(b\times y'') = (\phi_U(b\times(y'\times y'')))_2.$$

According to these definitions we get the fibre bundles

$$\mathfrak{F}' = \{F', G'; X', B; \psi', \phi'_U\}, \qquad \mathfrak{F}'' = \{F'', G''; X'', B; \psi'', \phi''_U\}.$$

It is easily verified that $\mathfrak{F}'\times\mathfrak{F}''$ is equivalent to the given fibre bundle.

Consider two fibre bundles

$$\mathfrak{F}' = \{F', G'; X', B; \psi', \phi'_U\}, \qquad \mathfrak{F}'' = \{F'', G''; X'', B; \psi'', \phi''_U\},$$

with the same base space B and their product bundle $\mathfrak{F}'\times\mathfrak{F}''$. Such a triple of bundles we shall call a *triad.* We consider the triads with given F', F'', G', G'', B. A triad $\{\mathfrak{C}', \mathfrak{C}'', \mathfrak{C}'\times\mathfrak{C}''\}$, with the director spaces F', F'', the reference groups G', G'', and the base space A, is called a ***universal triad of fibre bundles*** relative to F', F'', G', G'', B, if the following conditions are satisfied:

(1) To every triad $\mathfrak{F}'$, $\mathfrak{F}''$, $\mathfrak{F}'\times\mathfrak{F}''$ there exists a mapping $f: B\to A$ such that $(\mathfrak{C}'; f: B)\equiv\mathfrak{F}'$, $(\mathfrak{C}''; f: B)\equiv\mathfrak{F}''$, $(\mathfrak{C}'\times\mathfrak{C}''; f: B)\equiv\mathfrak{F}'\times\mathfrak{F}''$.

(2) The three fibre bundles in the triads induced by the mappings $f_0, f_1: B\to A$ are respectively equivalent if and only if $f_0\simeq f_1$.

Relative to F', F'', G', G'', B the problem on the existence of a universal triad of fibre bundles is solved by the following theorem:

THEOREM 4.2. *The triad* $\{\mathfrak{E}', \mathfrak{E}''; \mathfrak{E}'\times\mathfrak{E}''\}$ *is a universal triad of fibre bundles, relative to* F', F'', G', G'', B, *if and only if the fibre bundle* $\mathfrak{E}'\times\mathfrak{E}''$ *is a universal fibre bundle relative to* $F'\times F'', G'\times G'', B$.

The proof of this theorem is straightforward, and we shall omit it here.

5. The classical groups. We shall devote this section to exhibit explicitly the universal fibre bundles for the cases that the reference groups are the classical groups, namely, the general linear group $GL(n, R)$ with real coefficients, its subgroup $GL^+(n, R)$ of the linear transformations with positive determinant, the orthogonal group $O(n, R)$, the proper orthogonal group $SO(n, R)$, the unitary group $U(n, C)$, and the unitary symplectic group $Sp(n)$.

1. *The universal fibre bundles for the classical groups.* The universal fibre bundles in question will be constructed according to the following general process: Let R be a Lie group, and K, H closed subgroups of R, $H\subset K$. Define the projection of R/H onto R/K as the mapping which maps the coset rH into the coset rK, $r\in R$. With coordinate functions defined by the canonical coordinates, Steenrod([11]) proved that a fibre bundle can be defined with the director space K/H, operated on by the group K as left translations, and with the total space R/H and the base space R/K. We shall denote such a fibre bundle by

$$\{K/H, K; R/H, R/K\},$$

omitting the projection and the coordinate functions, or simply by $\{R/H, R/K\}$, indicating only the total space and the base space.

To construct universal fibre bundles for the classical groups the general idea is to consider the same classical groups with more variables and construct their coset spaces. We begin with the following notations:

$GL(n)$ denotes the group of all n-rowed real matrices with nonzero determinant, $GL^+(n)$ the subgroup of $GL(n)$ consisting of all matrices of positive determinant. $H^+(n, N)$ and $H(n, N)$ denote respectively the subgroups of $GL^+(n, N)$ consisting of all matrices of the form

$$\begin{pmatrix} GL^+(n) & 0 \\ * & GL^+(N) \end{pmatrix}, \quad \begin{pmatrix} GL(n) & 0 \\ * & GL(N) \end{pmatrix},$$

where the elements in the upper right corners are zero. $O(n)$ denotes the group of all n-rowed real orthogonal matrices and $SO(n)$ the subgroup of $O(n)$ consisting of all the matrices of determinant $+1$. $U(n)$ denotes the group of all n-rowed unitary matrices, that is, matrices σ with complex elements satisfying $\bar{\sigma}=\sigma'^{-1}$. For a quaternion $q=a_0+a_1i+a_2j+a_3k$ let $\bar{q}=a_0-a_1i-a_2j-a_3k$ be its conjugate quaternion. Then $Sp(n)$ denotes the group of all n-rowed matrices τ with quaternion elements satisfying $\bar{\tau}=\tau'^{-1}$.

([11]) Steenrod [4, pp. 300–302].

When there are several groups in different numbers of variables, it is possible to imbed one in another. We agree by convention that the imbeddings $SO(N) \subset SO(n+N)$, $SO(n) \times SO(N) \subset SO(n+N)$ are such that

$$\begin{pmatrix} I_n & 0 \\ 0 & SO(N) \end{pmatrix}, \quad \begin{pmatrix} SO(n) & 0 \\ 0 & SO(N) \end{pmatrix},$$

and similarly for the other groups.

With all these definitions and conventions we consider the following fibre bundles:

$$\begin{aligned} \mathfrak{F}_{GL^+}(n, N) &= \{GL^+(n+N)/GL^+(N), GL^+(n+N)/H^+(n, N)\}, \\ \mathfrak{F}_{SO}(n, N) &= \{SO(n+N)/SO(N), SO(n+N)/SO(n) \times SO(N)\}, \\ \mathfrak{F}_{U}(n, N) &= \{U(n+N)/U(N), U(n+N)/U(n) \times U(N)\}, \\ \mathfrak{F}_{Sp}(n, N) &= \{Sp(n+N)/Sp(N), Sp(n+N)/Sp(n) \times Sp(N)\}. \end{aligned}$$

Take, for instance, $\mathfrak{F}_{SO}(n, N)$. Its director space is $SO(n) \times SO(N)/I_n \times SO(N)$, operated on by $SO(n) \times SO(N)$ as left translations. By a natural homeomorphism we can take $SO(n)$ for director space, which is then operated on by $SO(n)$ as left translations. It follows that $\mathfrak{F}_{SO}(n, N)$ is a principal fibre bundle. Similarly, we see that $\mathfrak{F}_{GL^+}(n, N)$, $\mathfrak{F}_{U}(n, N)$ and $\mathfrak{F}_{Sp}(n, N)$ are principal fibre bundles. These fibre bundles will serve as universal principal fibre bundles for the cases that the reference groups are the classical groups, as given by the following theorem:

THEOREM 5.1. *Let B be a finite polyhedron of dimension k, and $\mathfrak{P}$ a universal principal fibre bundle relative to G, G, B. When G is a classical group, a corresponding $\mathfrak{P}$ is given by the following table.*

G	$\mathfrak{P}$	
$SO(n)$	$\mathfrak{F}_{SO}(n, N)$	$k \leqq N-1$
$GL^+(n)$	$\mathfrak{F}_{GL^+}(n, N)$	$k \leqq N-1$
$U(n)$	$\mathfrak{F}_{U}(n, N)$	$k \leqq 2N$
$Sp(n)$	$\mathfrak{F}_{Sp}(n, N)$	$k \leqq 4N+2$

Proof. It suffices to prove that the homotopy groups π_i, $0 \leqq i \leqq k$, of the total spaces of $\mathfrak{P}$ are zero. As in the proof of Theorem 2.5, this follows from successive applications of the covering homotopy theorem. The desired deformations can be carried out, because the spaces $SO(N+1)/SO(N)$, $U(N+1)/U(N)$, $Sp(N+1)/Sp(N)$ are homeomorphic to spheres of dimensions N, $2N+1$, $4N+3$, respectively. This completes the proof of the theorem.

When the reference group is $GL(n)$ or $O(n)$, the situation differs slightly from the preceding ones, as these groups are not connected. In order that the foregoing manipulations prevail, we have to consider the fibre bundles

$$\mathfrak{F}_O(n, N) = \{SO(n + N)/SO(N), SO(n + N)/SO(n) \otimes SO(N)\},$$
$$\mathfrak{F}_{GL}(n, N) = \{GL^+(n + N)/GL^+(N), GL^+(n + N)/H(n, N)\},$$

where

$$SO(n) \otimes SO(N) = SO(n + N) \cap (O(n) \times O(N)) \text{ in } O(n + N).$$

This makes the total spaces connected, and the proof of Theorem 5.1 can be applied. We state this result in the theorem:

THEOREM 5.2. *Let B be a finite polyhedron of dimension k. If* $k \leqq N-1$; $\mathfrak{F}_O(n, N)$ *and* $\mathfrak{F}_{GL}(n, N)$ *are universal principal fibre bundles relative to* $O(n)$, B *and* $GL(n)$, B, *respectively.*

By the process τ_F^{-1} universal fibre bundles can be constructed whenever the reference group is one of the classical groups. We notice that the process τ_F^{-1} does not affect the base space. For the description of the characteristic ring it will be useful to know the base space, and in particular its cohomology ring.

We consider the linear vector spaces $V(n+N, R)$, $V(n+N, C)$, $V(n+N, Q)$, of dimension $n+N$, over the real field, the complex field, and the quaternion field, respectively. Denote by $G(n, N, R)$, $G(n, N, C)$, $G(n, N, Q)$ respectively the manifolds of the linear spaces of dimension n through the origin of these vector spaces. They are known as the *Grassmann manifolds.* For the case of the real field we may also consider the Grassmann manifold $\tilde{G}(n, N, R)$ of the oriented linear spaces of dimension n through the origin. Each of these Grassmann manifolds is operated on transitively by the corresponding classical group in the vector space of dimension $n+N$, and we easily identify it with a space of cosets. In this way we deduce the theorem:

THEOREM 5.3. *The base spaces of the universal fibre bundles for the classical groups are given by the following table.*

Fibre Bundle	*Base Space*
$\mathfrak{F}_{GL^+}(n, N)$	$\tilde{G}(n, N, R)$
$\mathfrak{F}_{SO}(n, N)$	$\tilde{G}(n, N, R)$
$\mathfrak{F}_U(n, N)$	$G(n, N, C)$
$\mathfrak{F}_{Sp}(n, N)$	$G(n, N, Q)$
$\mathfrak{F}_O(n, N)$	$G(n, N, R)$
$\mathfrak{F}_{GL}(n, N)$	$G(n, N, R)$

2. *Whitney's product for sphere bundles and a universal triad.* From Theorem 5.3 we see that the universal sphere bundles and the universal vector bundles, whether oriented or non-oriented, have the same base space,

namely, $\tilde{G}(n, N, R)$ and $G(n, N, R)$, respectively. It follows from this and also from the considerations of §4.2 that to each equivalence class of sphere bundles over a base space B is associated an equivalence class of vector bundles, and vice versa. Denote these operations by w and w^{-1}, respectively. Let $\mathfrak{F}_1$ and $\mathfrak{F}_2$ be two sphere bundles over the same base space B. The bundle $w^{-1}(w\mathfrak{F}_1 \times w\mathfrak{F}_2)$ is then Whitney's product of the sphere bundles $\mathfrak{F}_1$ and $\mathfrak{F}_2$.

To construct a universal triad for the product $w\mathfrak{F}_1 \times w\mathfrak{F}_2$ suppose m and n be the dimensions of the director spaces of $w\mathfrak{F}_1$ and $w\mathfrak{F}_2$, respectively. Consider in $V(m+n+N, R)$ the manifold whose elements consist of a linear subspace of dimension m and a linear subspace of dimension n in general position through the origin. Call this manifold an *E-manifold* and denote it by $E(m, n, N, R)$(12). A mapping $B \rightarrow E(m, n, N, R)$ then induces two vector bundles and their product over B. From Theorem 4.1 it follows that $E(m, n, N, R)$ is the base space of a universal triad, provided that $\dim B \leqq N-1$. The study of the homology properties, and in particular of the cohomology ring, of $E(m, n, N, R)$ has therefore an important significance for the description of the position of one sphere bundle in another.

Bibliography

1. C. Ehresmann, C. R. Acad. Sci. Paris vol. 213 (1941) pp. 762–764; vol. 214 (1942) pp. 144–147; vol 216 (1943) pp. 628–630.

2. S. T. Hu, *Mappings of a normal space in an absolute neighborhood retract*, Trans. Amer. Math. Soc. vol. 64 (1948) pp. 336–358.

3. W. Hurewicz, and N. Steenrod, *Homotopy relations in fibre spaces*, Proc. Nat. Acad. Sci. U.S.A. vol. 27 (1941) pp. 61–64.

4. N. E. Steenrod, *The classification of sphere bundles*, Ann. of Math. vol. 45 (1944) pp. 294–311.

5. H. Whitney, *Topological properties of differentiable manifolds*, Bull. Amer. Math. Soc. vol. 43 (1937) pp. 785–805.

6. ———, *On the topology of differentiable manifolds*, Lectures in Topology, University of Michigan Press 1941, pp. 101–141.

Academia Sinica,
Nanking, China.

(12) The homology properties of the E-manifolds have been studied by Ehresmann; cf. C. Ehresmann, Journal de Mathématiques vol. 16 (1937) pp. 69–100.

Reprinted from
Trans. Amer. Math. Soc. 67 (1949) 286–303

Reprinted from the Proceedings of the NATIONAL ACADEMY OF SCIENCES
Vol. 36, No. 4, pp. 248–255. April, 1950

THE HOMOLOGY STRUCTURE OF SPHERE BUNDLES

BY SHIING-SHEN CHERN AND E. SPANIER

DEPARTMENT OF MATHEMATICS, UNIVERSITY OF CHICAGO

Communicated by S. MacLane, February 25, 1950

1. Since a fiber bundle is a generalization of a product space, it is natural to expect some relations between the homology groups of the bundle, the base space, and the fiber. The simplest relation is between the Euler characteristics, the characteristic of the bundle being the product of the characteristics of the base space and the fiber. For sphere bundles the first comprehensive result was obtained by Gysin.[1] Recently, Steenrod[2] gave a new derivation of the Gysin results. On the other hand, results have been announced by Hirsch[3] and Leray[4] for more general types of bundles.

In the works of both Gysin and Steenrod the base space is assumed to be an orientable manifold. The main purpose of our work is to extend this result to the case that the base space is a complex and to furnish a simpler and more direct method of approach which can be applied to more general fiber bundles. The principal theorem is a statement that a certain sequence of homomorphisms of cohomology (or homology) groups is exact. It seems to us that this theorem includes all known results about the cohomology structure of sphere bundles. As a basis of our treatment we make use of the axiomatic homology theory of Eilenberg and Steenrod.[5]

In §2 we define the type of bundle to be studied. In §3 we state the main results. Some applications of these results are given in §4, including a proof of Gysin's main theorem and a derivation of some relations between the Betti numbers of the bundle, the base space and the fiber. An indication of the methods used is given in §5.

2. Let K denote a finite connected simplicial complex of dimension n. Let $B = |K|$ be the space of K, and let X be a topological space. A continuous map $f:X \to B$ is called a *fibering of X into d-spheres over B* if for each simplex s of K there exists a homeomorphism

$$\varphi_s:|s| \times S^d \approx f^{-1}(|s|),$$

where S^d denotes the d-dimensional sphere, such that:

(*a*) $f\varphi_s(y, z) = y \qquad \text{for } y \in |s|,\ z \in S^d.$

(*b*) If s is a face of s', then

$$g_{ss'}(y):S^d \approx S^d$$

defined by

$$\varphi_s(y, g_{ss'}(y)(z)) = \varphi_{s'}(y, z), \qquad \text{for } y \in |s|,\ z \in S^d$$

is a continuous mapping of $|s|$ into the group of homeomorphisms of S^d of degree $+1$.

X is called the *bundle*, and B is called the *base space* of the bundle. The mapping f is called the *projection* of the bundle. The projection induces homomorphisms

$$f^*:H^p(B) \to H^p(X)$$

of the cohomology groups of B into those of X over any coefficient group. In the following the coefficient group is arbitrary but fixed for all cohomology groups under consideration.

3. Using the bundle structure two homomorphisms will be defined. The first homomorphism maps the cohomology groups of X into those of B and lowers dimension bv d. It is denoted by

$$\Phi:H^p(X) \to H^{p-d}(B).$$

This homomorphism has been considered by Gysin and Lichnerowicz[6] in the case when the base space is an orientable manifold.

The second homomorphism maps cohomology groups of B into cohomology groups of B and raises dimension by $d + 1$. It will be denoted by

$$\Psi: H^p(B) \to H^{p+d+1}(B).$$

These two homomorphisms together with f^* form an infinite sequence

$$\ldots \xrightarrow{\Psi} H^p(B) \xrightarrow{f^*} H^p(X) \xrightarrow{\Phi} H^{p-d}(B) \xrightarrow{\Psi} H^{p+1}(B) \xrightarrow{f^*} \ldots$$

called the *cohomology sequence of the bundle.*

The main result is embodied in the following theorem:

THEOREM 3.1. *The cohomology sequence of a sphere bundle is exact.*

This theorem shows that if the homomorphism Ψ can be determined in B then the cohomology groups of X are group extensions of groups defined in B by other groups defined in B. This shows the significance of characterizing the homomorphism Ψ, which can be described as follows:

Any sphere bundle gives rise in a natural way to a unique $(d + 1)$-dimensional cohomology class of B with integer coefficients.[7] This class is called the *characteristic class* of the bundle and will be denoted by Ω. It is the obstruction to the construction of a cross-section over the $(d + 1)$-dimensional skeleton of B. Using the natural pairing of the integers and the abelian coefficient group to the coefficient group, the cup products $u \cup \Omega$ and $\Omega \cup u$ are defined for $u \in H^p(B)$ and are elements of $H^{p+d+1}(B)$. The order of multiplication is immaterial. For odd d this follows simply from the commutation rule of the cup product. For even d it can be shown that $2\Omega = 0$, so that the two products are again equal. Then we have the theorem:

THEOREM 3.2. *For any* $u \in H^p(B)$,

$$\Psi(u) = u \cup \Omega = \Omega \cup u.$$

Naturally there exist dual theorems for the homology groups. In particular, we define homomorphisms

$$f_*: H_p(X) \to H_p(B),$$
$$\Phi': H_p(B) \to H_{p+d}(X),$$
$$\Psi': H_p(B) \to H_{p-d-1}(B).$$

These homomorphisms form an infinite sequence,

$$\ldots \xleftarrow{\Psi'} H_p(B) \xleftarrow{f_*} H_p(X) \xleftarrow{\Phi'} H_{p-d}(B) \xleftarrow{\Psi'} H_{p+1}(B) \xleftarrow{f_*} \ldots.$$

called the *homology sequence* of the bundle. The results for the case of homology are summarized in the following theorem:

THEOREM 3.3. *The homology sequence of a sphere bundle is exact. Moreover, if Ω denotes the characteristic cohomology class of the bundle, the homomorphism*

$$\Psi' : H_p(B) \to H_{p-d-1}(B)$$

satisfies the equation

$$\Psi'(z) = \Omega \cap z, \qquad \textit{for } z \in H_p(B),$$

4. Gysin's isomorphism theorem is contained in Theorem 3.3, for let $K_p(B)$ be the subgroup of $H_p(B)$ consisting of the kernel of Φ'. Then exactness of the sequence implies that Ψ' induces an isomorphism

$$\bar{\Psi} : H_{p+d+1}(B)/f_* H_{p+d+1}(X) \approx K_p(B).$$

The inverse of this isomorphism is the isomorphism denoted by H by Gysin.

A slightly different isomorphism is obtained for cohomology if we define $K^p(B)$ to be the subgroup of $H^p(B)$ consisting of the kernel of Ψ. Then Φ induces isomorphisms

$$\bar{\Phi} : H^p(X)/f^* H^p(B) \approx K^{p-d}(B).$$

The inverse of $\bar{\Phi}$ is Steenrod's functional cup product.[2] More precisely, if $u \in H^{p-d}(B)$ such that $u \cup \Omega = 0$, then $u \underset{f}{\cup} \Omega$ is defined and belongs to $H^p(X)/f^* H^p(B)$. Then we have

$$\bar{\Phi}\,(u \underset{f}{\cup} \Omega) = u.$$

As a second application we study the influence of the relative position of the fiber in the bundle on the homology structure of the bundle. If S^d denotes a fiber, the inclusion map $i : S^d \subset X$ induces homomorphisms

$$i_* : H_d(S^d) \to H_d(X)$$

$$i^* : H^d(X) \to H^d(S^d).$$

The fiber is said to *bound in the bundle* ($S^d \sim 0$) relative to a coefficient group G if

$$i_* H_d(S^d) = 0$$

where $H_d(S^d)$ is taken over G. In terms of cohomology this condition is known to be equivalent to

$$i^* H^d(X) = 0,$$

$H^d(X)$ being taken over a group dual to G.

To avoid cumbersome statements we assume in the rest of this section that the coefficient group for homology and cohomology is a field.

The groups $H^0(B)$ and $H^d(S^d)$ are both isomorphic to the coefficient field and hence to each other. An isomorphism

$$\nu : H^0(B) \approx H^d(S^d)$$

between them can be found such that

$$\nu\Phi = i^*.$$

Therefore, if $S^d \sim 0$, then $K^0(B) = 0$. This in turn implies that $\Omega \neq 0$ and $k\ \Omega \neq 0$ for any integer k. In particular, if d is even, then $2\ \Omega = 0$ so S^d does not bound in X.

If, conversely, S^d does not bound in X, the argument can be reversed, so we see that $k\ \Omega = 0$ for some integer k. It follows then from the cohomology sequence that there is an isomorphism between $H^p(X)$ and $H^p(B \times S^d)$.

Finally, we shall derive some relations between the Betti numbers of X and B, which were obtained by Gysin, Hirsch, and Leray. Let $\rho^p(X)$ and $\rho^p(B)$ be the pth Betti numbers of X and B and $\rho_0^p(B)$ the dimension of the kernel of Ψ in $H^p(B)$. Then we have, from the cohomology sequence of the bundle,

$$\rho^p(X) = \rho^p(B) - \rho^{p-d-1}(B) + \rho_0^{p-d-1}(B) + \rho_0^{p-d}(B). \qquad (4.1)$$

Let $X(t)$, $B(t)$, $S^d(t)$ be the Poincaré polynomials of X, B, S^d, respectively, and let

$$P(t) = \sum_{p \geqq d} (\rho^{p-d}(B) - \rho_0^{p-d}(B))t^p,$$

so that $P(t)$ has non-negative coefficients. It is easily verified by using (4.1) that

$$X(t) = S^d(t)B(t) - (1 + t)P(t),$$

and that

$$[1 - t(S^d(t) - 1)]B(t) \leqq X(t) \leqq S^d(t)B(t),$$

where an inequality between polynomials means a set of inequalities between their corresponding coefficients.

5. We shall indicate in this section the method used in proving the theorems stated in §3.

Let $B^p = |K^p|$ be the space of the p-dimensional skeleton of K. Let $X_p = f^{-1}(B^p)$. Then

$$X = X_n \supset \ldots \supset X_p \supset X_{p-1} \supset \ldots \supset X_0 \supset X_{-1} = 0.$$

Using the fact that $f^{-1}(|s|)$ is homeomorphic to $|s| \times S^d$ for each simplex s of K, it is easy to prove the following

LEMMA 5.1. *Let $f_p:(X_p, X_{p-1}) \to (B^p, B^{p-1})$ be the map defined by f. Then*

$$f_p^*:H^p(B^p, B^{p-1}) \approx H^p(X_p, X_{p-1})$$

is an isomorphism onto. Also there is an isomorphism

$$\lambda_p:H^{p-d}(B^{p-d}, B^{p-d-1}) \approx H^p(X_{p-d}, X_{p-d-1}),$$

while

$$H^q(X_p, X_{p-1}) = \{0\}, \textit{ if } q \neq p, p + d.$$

The group of cochains of K, $C^p(K)$, is defined by

$$C^p(K) = H^p(B^p, B^{p-1}),$$

so the lemma shows that $H^q(X_p, X_{p-1})$ is either the trivial group or is isomorphic to a group of cochains of K.

The following diagram will be referred to as the main diagram

$$\begin{array}{ccccc}
 & \{0\} \xrightarrow{\alpha_p} H^p(X) & & \{0\} \xrightarrow{\alpha_{p+1}} H^{p+1}(X) & \\
 & \downarrow{\scriptstyle \beta_p} & & \downarrow{\scriptstyle \beta_{p+1}} & \\
\cdots \xrightarrow{\delta_{p-1}} H^p(X_p, X_{p-1}) \xrightarrow{\rho_p} & H^p(X_p) \xrightarrow{\delta_p} & & & \\
 & \downarrow{\scriptstyle \gamma_p} & H^{p+1}(X_{p+1}, X_p) \xrightarrow{\rho_{p+1}} & H^{p+1}(X_{p+1}) & \xrightarrow{\delta_{p+1}} \cdots \\
\cdots \xrightarrow{\bar{\delta}_{p-1}} H^p(X_{p-d}, X_{p-d-1}) \xrightarrow{\sigma_p} & H^p(X_{p-d}) & & & \\
 & \downarrow{\scriptstyle \tau_p} \quad \downarrow{\scriptstyle \bar{\delta}_p} & & \downarrow{\scriptstyle \gamma_{p+1}} & \\
 & \{0\} & & & \\
 & & H^{p+1}(X_{p-d+1}, X_{p-d}) \xrightarrow{\sigma_{p+1}} & H^{p+1}(X_{p-d+1}) & \xrightarrow{\bar{\delta}_{p+1}} \cdots \\
 & & & \downarrow{\scriptstyle \tau_{p+1}} & \\
 & & & \{0\} &
\end{array}$$

In the main diagram δ_p and $\bar{\delta}_p$ are coboundary operators and α_p, β_p, γ_p, ρ_p, σ_p, τ_p are induced by inclusion mappings. Using the exactness axiom of cohomology theory and (5.4) the following can be proved.

LEMMA 5.5. *For any $p \geqq 0$, the sequence of homomorphisms α_p, β_p, δ_p, ρ_{p+1}, γ_{p+1}, $\bar{\delta}_{p+1}$, σ_{p+2}, τ_{p+2} is exact.*

The fact that in defining the bundle, $g_{ss'}(y)$ is required to be a homeomorphism of S^d of degree $+1$ furnishes the following results.

LEMMA 5.6. *Commutativity holds in the diagram*

$$\begin{array}{ccc} H^p(X_p, X_{p-1}) & \xrightarrow{\delta_p \rho_p} & H^{p+1}(X_{p+1}, X_p) \\ f_p{}^* \uparrow & & \uparrow f_{p+1}{}^* \\ C^p(K) & \xrightarrow{\delta} & C^{p+1}(K). \end{array}$$

LEMMA 5.7. *Commutativity holds in the diagram*

$$\begin{array}{ccc} H^p(X_{p-d}, X_{p-d-1}) & \xrightarrow{\bar{\delta}_p \sigma_p} & H^{p+1}(X_{p-d+1}, X_{p-d}) \\ \lambda_p \uparrow & & \uparrow \lambda_{p+1} \\ C^{p-d}(K) & \xrightarrow{\delta} & C^{p-d+1}(K) \end{array}$$

We now proceed to construct the homomorphisms used in forming the cohomology sequence of the bundle from the main diagram.

To compute

$$f^*: H^p(B) \rightarrow H^p(K)$$

from the main diagram, let $u \in H^p(B)$. Let $c \in C^p(K)$ represent u. Then $\delta c = 0$, so that $\delta_p \rho_p f_p{}^*(c) = 0$. Hence there exists $v \in H^p(X)$ such that

$$\beta_p(v) = \rho_p f_p{}^*(c).$$

It is easy to see that $f^*(u) = v$.

To define the homomorphism

$$\Phi: H^p(X) \rightarrow H^{p-d}(B),$$

let $u \in H^p(X)$. Then $\gamma_p \beta_p(u) \in H^p(X_{p-d})$. Since $\tau_p(\gamma_p \beta_p(u)) = 0$, there is $v \in H^p(X_{p-d}, X_{p-d-1})$ such that

$$\sigma_p(v) = \gamma_p \beta_p(u).$$

By (5.3) there is $c \in C^{p-d}(K)$ such that

$$\lambda_p(c) = v.$$

Then

$$\lambda_{p+1} \delta(c) = \bar{\delta}_p \sigma_p \lambda_p(c) = \bar{\delta}_p \sigma_p(v) = \bar{\delta}_p \gamma_p \beta_p(u) = 0,$$

so $\delta(c) = 0$. It follows that c is a cocycle of K and determines a cohomology class of B. This class depends only on u and is defined to be $\Phi(u)$.

To define

$$\Psi: H^{p-d}(B) \rightarrow H^{p+1}(B),$$

let $u \in H^{p-d}(B)$. Choose $c \in C^{p-d}(K)$ to represent u. Then $\delta\, c = 0$, so $\delta_p \sigma_p \lambda_p(c) = 0$. Hence there is $v \in H^p(X_p)$ such that

$$\gamma_p(v) = \sigma_p \lambda_p(c).$$

Now $\delta_p(v) = \bar{v} \in H^{p+1}(X_{p+1}, X_p)$. Let $\bar{c} \in C^{p+1}(K)$ such that $f_{p+1}^*(\bar{c}) = \bar{v}$. Since

$$\delta_{p+1}\rho_{p+1}(\bar{v}) = \delta_{p+1}\rho_{p+1}\delta_p(v) = 0,$$

it follows that $\delta(\bar{c}) = 0$. The cohomology class in B to which c belongs depends only on u and is defined to be $\Psi(u)$.

Having defined the homomorphisms f^*, Φ, Ψ, Theorem 3.1 follows immediately from Lemma 5.5.

If the integers are used as coefficients, let $\omega \in H^0(B)$ be the unit cohomology class in B. It is easy to show that the characteristic class Ω satisfies the equation

$$\Omega = \Psi(\omega).$$

The following is a brief sketch of the proof of Theorem 3.2. We consider $X \times B$ as a bundle over $B \times B$. Let Ψ_1 be the homomorphism Ψ defined for this bundle. Then it can be shown that

$$\Psi_1(\alpha \times \beta) = \Psi(\alpha) \times \beta, \qquad \text{for } \alpha \in H^p(B),\ \beta \in H^q(B).$$

Let $d:B \rightarrow B \times B$ be the diagonal map. The induced bundle by d over B is equivalent to the given bundle. It follows from this that

$$d^*\Psi_1 = \Psi d^*.$$

Then

$$d^*\Psi_1(\omega \times u) = \Psi(\omega \cup u) = \Psi(u).$$

But

$$d^*\Psi_1(\omega \times u) = d^*(\Psi(\omega) \times u) = \Psi(\omega) \cup u = \Omega \cup u.$$

Added in proof March 27, 1950. While this was in press an article by Thom appeared in the *C. R. Paris* of January 30, 1950 which essentially contains our theorems 3.1 and 3.2.

[1] Gysin, W., *Comm. Math. Helv.*, **14**, 61–122 (1942).
[2] Steenrod, N. E., *Ann. Math.*, **50**, 954–988 (1949).
[3] Hirsch, G., *C. R. Paris*, **227**, 1328–1330 (1948).
[4] Leray, J., *ibid.*, **223**, 395–397 (1946).
[5] Eilenberg, S., and Steenrod, N. E., these PROCEEDINGS, **31**, 117–120 (1945).
[6] Lichnerowicz, A., *Comm. Math. Helv.*, **22**, 271–301 (1949).
[7] Whitney, H., these PROCEEDINGS, **26**, 148–153 (1940).

Reprinted from Vol. II, Proceedings of the International Congress of Mathematicians, 1950
Printed in U.S.A.

DIFFERENTIAL GEOMETRY OF FIBER BUNDLES

Shiing-shen Chern

The aim of this lecture is to give a discussion of the main results and ideas concerning a certain aspect of the so-called differential geometry in the large which has made some progress in recent years. Differential geometry in the large in its vaguest sense is concerned with relations between global and local properties of a differential-geometric object. In order that the methods of differential calculus may be applicable, the spaces under consideration are not only topological spaces but are differentiable manifolds. The existence of such a differentiable structure allows the introduction of notions as tangent vector, tangent space, differential forms, etc. In problems of differential geometry there is usually an additional structure such as: (1) a Riemann metric, that is, a positive definite symmetric covariant tensor field of the second order; (2) a system of paths with the property that through every point and tangent to every direction through the point there passes exactly one path of the system; (3) a system of cones of directions, one through each point, which correspond to the light cones in general relativity theory, etc. Among such so-called geometric objects the Riemann metric is perhaps the most important, both in view of its rôle in problems of analysis, mechanics, and geometry, and its richness in results. In 1917 Levi-Civita discovered his celebrated parallelism which is an infinitesimal transportation of tangent vectors preserving the scalar product and is the first example of a connection. The salient fact about the Levi-Civita parallelism is the result that it is the parallelism, and not the Riemann metric, which accounts for most of the properties concerning curvature.

The Levi-Civita parallelism can be regarded as an infinitesimal motion between two infinitely near tangent spaces of the Riemann manifold. It was Elie Cartan who recognized that this notion admits an important generalization, that the spaces for which the infinitesimal motion is defined need not be the tangent spaces of a Riemann manifold, and that the group which operates in the space plays a dominant rôle. In his theory of generalized spaces (*Espaces généralisés*) Cartan carried out in all essential aspects the local theory of what we shall call connections [1; 2]. With the development of the theory of fiber bundles in topology, begun by Whitney for the case of sphere bundles and developed by Ehresmann, Steenrod, Pontrjagin, and others, [8; 19], it is now possible to give a modern version of Cartan's theory of connections, as was first carried out by Ehresmann and Weil [7; 22].

Let F be a space acted on by a topological group G of homeomorphisms. A fiber bundle with the director space F and structural group G consists of topological spaces B, X and a mapping ψ of B onto X, together with the following: (1) X is covered by a family of neighborhoods $\{U_\alpha\}$, called the coordinate neighborhoods, and to each U_α there is a homeomorphism (a coordinate function) φ_α: $U_\alpha \times F \to \psi^{-1}(U_\alpha)$, with $\psi\varphi_\alpha(x, y) = x, x \in U_\alpha, y \in F$.

(2) As a consequence of (1), a point of $\psi^{-1}(U_\alpha)$ has the coordinates (x, y), and a point of $\psi^{-1}(U_\alpha \cap U_\beta)$ has two sets of coordinates (x, y) and (x, y'), satisfying $\varphi_\alpha(x, y) = \varphi_\beta(x, y')$. It is required that $g_{\alpha\beta}(x): y' \to y$ be a continuous mapping of $U_\alpha \cap U_\beta$ into G.

The spaces X and B are called the base space and the bundle respectively. Each subset $\psi^{-1}(x) \subset B$ is called a fiber.

This definition of a fiber bundle is too narrow in the sense that the coordinate neighborhoods and coordinate functions form a part of the definition. An equivalence relation has thus to be introduced. Two bundles (B, X), (B', X) with the same base space X and the same F, G are called equivalent if, $\{U_\alpha, \varphi_\alpha\}$, $\{V_\beta, \theta_\beta\}$ being respectively their coordinate neighborhoods and coordinate functions, there is a fiber-preserving homeomorphism T: $B \to B'$ such that the mapping $h_{\alpha\beta}(x)$: $y \to y'$ defined by $\theta_\beta(x, y) = T\varphi_\alpha(x, y')$ is a continuous mapping of $U_\alpha \cap V_\beta$ into G.

An important operation on fiber bundles is the construction from a given bundle of other bundles with the same structural group, in particular, the principal fiber bundle which has G as director space acted upon by G itself as the group of left translations. The notion of the principal fiber bundle has been at the core of Cartan's method of moving frames, although its modern version was first introduced by Ehresmann. It can be defined as follows: For $x \in X$, let G_x be the totality of all maps $\varphi_{\alpha,g}(x)$: $F \to \psi^{-1}(x)$ defined by $y \to \varphi_\alpha(x, g(y))$, $y \in F$, $g \in G$, relative to a coordinate neighborhood U_α containing x. G_x depends only on x. Let $B^* = \bigcup_{x \in X} G_x$ and define the mapping ψ^*: $B^* \to X$ by $\psi^*(G_x) = x$ and the coordinate functions $\overset{*}{\varphi}_\alpha(x, g) = \varphi_{\alpha,g}(x)$. Topologize B^* such that the $\overset{*}{\varphi}_\alpha$'s define homeomorphisms of $U_\alpha \times G$ into B^*. The bundle (B^*, X) so obtained is called a principal fiber bundle. This construction is an operation on the equivalence classes of bundles in the sense that two fiber bundles are equivalent if and only if their principal fiber bundles are equivalent. Similarly, an inverse operation can be defined, which will permit us to construct bundles with a given principal bundle and having as director space a given space acted upon by the structural group G. An important property of the principal fiber bundle is that B^* is acted upon by G as right translations.

For the purpose of differential geometry we shall assume that all spaces under consideration are differentiable manifolds and that our mappings are differentiable with Jacobian matrices of the highest rank everywhere. In particular, the structural group G will be assumed to be a connected Lie group. For simplicity we suppose our base space X to be compact, although a large part of our discussions holds without this assumption.

The implications of these assumptions are very far-reaching indeed. First of all we can draw into consideration the Lie algebra $L(G)$ of G. $L(G)$ is invariant under the left translations of G, while the right translations and the inner automorphisms of G induce on $L(G)$ a group of linear endomorphisms ad(G), called the adjoint group of G. Relative to a base of $L(G)$ there are the left-invariant linear differential forms ω^i and the right-invariant linear differential forms π^i, each set consisting of linearly independent forms whose number is equal to the dimension of G. A fundamental theorem on Lie groups asserts that their exterior derivatives are given by

$$(1)\qquad \begin{aligned} d\omega^i &= -\frac{1}{2}\sum_{j,k} c^i_{jk}\omega^j \wedge \omega^k, \\ d\pi^i &= +\frac{1}{2}\sum_{j,k} c^i_{jk}\pi^j \wedge \pi^k, \qquad i, j, k = 1, \cdots, \dim G, \end{aligned}$$

where c^i_{jk} are the so-called constants of structure which are antisymmetric in the lower indices and which satisfy the well-known Jacobi relations.

Returning to our fiber bundle, the dual mapping of the mapping $g_{\alpha\beta}$: $U_\alpha \cap U_\beta \to G$ carries ω^i and π^i into linear differential forms in $U_\alpha \cap U_\beta$, which we shall denote by $\omega^i_{\alpha\beta}$ and $\pi^i_{\alpha\beta}$ respectively. Since $g_{\alpha\gamma} = g_{\alpha\beta}g_{\beta\gamma}$ in $U_\alpha \cap U_\beta \cap U_\gamma$, we have

$$(2)\qquad \omega^i_{\alpha\gamma} = \sum_j \operatorname{ad}(g_{\beta\gamma})^i_j \omega^j_{\alpha\beta} + \omega^i_{\beta\gamma}.$$

We can also interpret $\omega^i_{\alpha\beta}$ as a vector-valued linear differential form in $U_\alpha \cap U_\beta$, with values in $L(G)$, and shall denote it simply by $\omega_{\alpha\beta}$ when so interpreted.

The generalization of the notion of a tensor field in classical differential geometry leads to the following situation: Let E be a vector space acted on by a representation $M(G)$ of G. A tensorial differential form of degree r and type $M(G)$ is an exterior differential form u_α of degree r in each coordinate neighborhood U_α, with values in E, such that, in $U_\alpha \cap U_\beta$, $u_\alpha = M(g_{\alpha\beta})u_\beta$. The exterior derivative du_α of u_α is in general not a tensorial differential form. It is in order to preserve the tensorial character of the derivative that an additional structure, a connection, is introduced into the fiber bundle.

A connection in the fiber bundle is a set of linear differential forms θ_α in U_α, with values in $L(G)$, such that

$$(3)\qquad \omega_{\alpha\beta} = -\operatorname{ad}(g_{\alpha\beta})\theta_\alpha + \theta_\beta, \qquad \text{in } U_\alpha \cap U_\beta.$$

It follows from (2) that such relations are consistent in $U_\alpha \cap U_\beta \cap U_\gamma$. As can be verified without difficulty, a connection defines in the principal fiber bundle a field of tangent subspaces transversal to the fibers, that is, tangent subspaces which, together with the tangent space of the fiber, span at every point the tangent space of the principal bundle. It follows from elementary extension theorems that in every fiber bundle there can be defined a connection. As there is great freedom in the choice of the connection, the question of deciding the relationship between the properties of the bundle and those of the connection will be our main concern in this paper.

Let us first define the process of so-called absolute differentiation. Let $\bar{M}(X)$, $X \in L(G)$, be the representation of the Lie algebra $L(G)$ induced by the representation $M(G)$ of G. Then we have

$$dM(g_{\alpha\beta}) = M(g_{\alpha\beta})\bar{M}(\theta_\beta) - \bar{M}(\theta_\alpha)M(g_{\alpha\beta}). \tag{4}$$

It follows that if we put for our tensorial differential form u_α of degree r and type $M(G)$

$$Du_\alpha = du_\alpha + \bar{M}(\theta_\alpha) \wedge u_\alpha, \tag{5}$$

the form Du_α will be a tensorial differential form of degree $r + 1$ and the same type $M(G)$.

To study the local properties of the connection we again make use of a base of the Lie algebra, relative to which the form θ_α has the components θ_α^i. We put

$$\Theta_\alpha^i = d\theta_\alpha^i + \frac{1}{2}\sum_{j,k} c_{jk}^i \theta_\alpha^j \wedge \theta_\alpha^k, \qquad \text{in } U_\alpha. \tag{6}$$

The form Θ_α, whose components relative to the base are Θ_α^i, is then an exterior quadratic differential form of degree 2, with values in $L(G)$. It is easy to verify that $\Theta_\alpha = \mathrm{ad}(g_{\alpha\beta})\Theta_\beta$ in $U_\alpha \cap U_\beta$. The Θ_α's therefore define a tensorial differential form of degree 2 and type $\mathrm{ad}(G)$, called the curvature tensor of the connection. In a manner which we shall not attempt to describe here, the curvature tensor and tensors obtained from it by successive absolute differentiations give all the local properties of the connection. In particular, the condition $\Theta_\alpha = 0$ is a necessary and sufficient condition for the connection to be flat, that is, to be such that $\theta_\alpha = 0$ by a proper choice of the coordinate functions.

The following formulas for absolute differentiation can easily be verified:

$$\begin{aligned} \bar{M}(\Theta_\alpha) &= d\bar{M}(\theta_\alpha) + \bar{M}(\theta_\alpha)^2, \\ D\Theta_\alpha &= 0, \\ D^2u_\alpha &= \bar{M}(\Theta_\alpha)u_\alpha. \end{aligned} \tag{7}$$

Such relations are known in classical cases, the second as the Bianchi identity.

We now consider real-valued symmetric multilinear functions $P(Y_1, \cdots, Y_k)$, $Y_i \in L(G)$, $i = 1, \cdots, k$, which are invariant, that is, which are such that $P(\mathrm{ad}(a)Y_1, \cdots, \mathrm{ad}(a)Y_k) = P(Y_1, \cdots, Y_k)$ for all $a \in G$. For simplicity we shall call such a function an invariant polynomial, k being its degree. By the definition of addition,

$$(P + Q)(Y_1, \cdots, Y_k) = P(Y_1, \cdots, Y_k) + Q(Y_1, \cdots, Y_k), \tag{8}$$

all invariant polynomials of degree k form an abelian group. Let $I(G)$ be the direct sum of these abelian groups for all $k \geqq 0$. If P and Q are invariant polynomials of degrees k and l respectively, we define their product PQ to be an invariant polynomial of degree $k + l$ given by

$$(PQ)(Y_1, \cdots, Y_{k+l}) = \frac{1}{N}\sum P(Y_{i_1}, \cdots, Y_{i_k})Q(Y_{i_{k+1}}, \cdots, Y_{i_{k+l}}), \tag{9}$$

where the summation is extended over all permutations of the vectors Y_i, and N is the number of such permutations. This definition of multiplication, together with the distributive law, makes $I(G)$ into a commutative ring, the ring of invariant polynomials of G.

Let $P \in I(G)$, with degree k. For Y_i we substitute the curvature tensor Θ. Then $P(\Theta) = P(\Theta, \cdots, \Theta)$ is an exterior differential form of degree $2k$, which, because of the invariance property of P, is defined everywhere in the base space X. From the Bianchi identity (7_2) it follows that $P(\Theta)$ is closed. Therefore, by the de Rham theory, $P(\Theta)$ determines an element of the cohomology ring $H(X)$ of X having as coefficient ring the field of real numbers. This mapping is a ring homomorphism

$$h: I(G) \rightarrow H(X) \tag{10}$$

of the ring of invariant polynomials of G into the cohomology ring of X. It is defined with the help of a connection in the bundle.

Our first main result is the following theorem of Weil: h is independent of the choice of the connection [22]. In other words, two different connections in the fiber bundle give rise to the same homomorphism h. To prove this we notice that if θ_α and θ'_α are the linear differential forms defining these connections, their difference $u_\alpha = \theta'_\alpha - \theta_\alpha$ is a linear differential form of type ad(G), with values in $L(G)$. With the help of u_α Weil constructs a differential form whose exterior derivative is equal to the difference $P(\Theta') - P(\Theta)$, for a given invariant polynomial P. Another proof has been given recently by H. Cartan, by means of an invariant definition of the homomorphism h.

Our next step consists in setting up a relationship between this homomorphism h and a homomorphism which is defined in a purely topological manner. This requires the concepts of an induced fiber bundle and a universal fiber bundle.

Let a mapping $f: Y \rightarrow X$ be given. The neighborhoods $\{f^{-1}(U_\alpha)\}$ then form a covering of Y and coordinate functions $\varphi'_\alpha: f^{-1}(U_\alpha) \times F \rightarrow f^{-1}(U_\alpha) \times \psi^{-1}(U_\alpha)$ can be defined by $\varphi'_\alpha(\eta, y) = \eta \times \varphi_\alpha(f(\eta), y)$. This defines a fiber bundle $Y \times \psi^{-1}(f(Y))$ over Y, with the same director space F and the same group G. The new bundle is said to be induced by the mapping f. If the original bundle has a connection given by the differential form θ_α in U_α, the dual mapping f^* of f carries θ_α into $f^*\theta_\alpha$ in $f^{-1}(U_\alpha)$ for which the relation corresponding to (3) is valid. The forms $f^*\theta_\alpha$ therefore define an induced connection in the induced bundle.

This method of generating new fiber bundles from a given bundle is very useful. Its value is based on the fact that it provides a way for the enumeration of fiber bundles. In fact, let the director space and the structural group G be given and fixed for our present considerations. A bundle with the base space X_0 is called universal relative to a space X if every bundle over X is equivalent to a bundle induced by a mapping $X \rightarrow X_0$ and if two such induced bundles are equivalent when and only when the mappings are homotopic. If, for a space X, there exists a universal bundle with the base space X_0, then the classes of bundles over X are in one-one correspondence with the homotopy classes of mappings $X \rightarrow X_0$, so that the enumeration of the bundles over X reduces to a homotopy classification problem.

It is therefore of interest to know the circumstances under which a universal bundle exists. A sufficient condition for the bundle over X_0 to be universal for all compact spaces X of dimension less than or equal to n is that the bundle B_0 of its principal fiber bundle have vanishing homotopy groups up to dimension n inclusive: $\pi_i(B_0) = 0, 0 \leqq i \leqq n$, where the condition $\pi_0 = 0$ means connectedness.

Under our assumptions that X is compact and that G is a connected Lie group, bundles can be found such that these conditions are fulfilled. First of all, according to a theorem due to E. Cartan, Malcev, Iwasawa, and Mostow, [12; 14; 15], G contains a maximal compact subgroup G_1, and the homogeneous space G/G_1 is homeomorphic to a Euclidean space. This makes it possible to reduce problems of equivalence, classification, etc. of bundles with the group G to the corresponding problems for G_1. Since G_1 is a compact Lie group, it has a faithful orthogonal representation and can be considered as a subgroup of the rotation group $R(m)$ operating in an m-dimensional Euclidean space E^m. Imbed E^m in an $(m + n + 1)$-dimensional Euclidean space E^{m+n+1} and consider the homogeneous space $\tilde{B} = R(m + n + 1)/(I_m \times R(n + 1))$ as a bundle over $X_0 = R(m + n + 1)/(G_1 \times R(n + 1))$, where I_m is the identical automorphism of E^m, and $R(n + 1)$ is the rotation group of the space E^{n+1} perpendicular to E^m in E^{m+n+1}. This is a principal bundle with G_1 as its structural group. By the covering homotopy theorem we can prove that $\pi_i(\tilde{B}) = 0, 0 \leqq i \leqq n$. In this way the existence of a universal bundle is proved by an explicit construction.

Suppose that a universal bundle exists, with the base space X_0. Let $H(X, R)$ be the cohomology ring of X, relative to the coefficient ring R. Since the classes of bundles over X are in one-one correspondence with the homotopy classes of mappings $X \to X_0$, the homomorphism $h' : H(X_0, R) \to H(X, R)$ is completely determined by the bundle. h' will be called the characteristic homomorphism, its image $h'(H(X_0, R)) \subset H(X, R)$ the characteristic ring, and an element of the characteristic ring a characteristic cohomology class. It will be understood that the coefficient ring R will be the field of real numbers whenever it is dropped in the notation.

The universal bundle is of course not unique. However, given any two bundles which are universal for compact base-spaces of dimension less than or equal to n, it is possible to establish between their base spaces X_0 and X_0' a chain transformation of the singular chains of dimension less than or equal to n which gives rise to a chain equivalence. From this it follows that up to the dimension n inclusive, the cohomology rings of X_0 and X_0' are in a natural isomorphism. The characteristic homomorphism is therefore independent of the choice of the universal bundle. Although this conclusion serves our purpose, it may be remarked that, in terms of homotopy theory, a stronger result holds between X_0 and X_0', namely, they have the same homotopy-n-type. From this the above assertion follows as a consequence.

A knowledge of $H(X_0, R)$ would be necessary for the description of the characteristic homomorphism. Since elements of dimension greater than n ($= \dim X$) of $H(X_0, R)$ are mapped into zero by dimensional considerations, $H(X_0, R)$ can be replaced by any ring which is isomorphic to it up to dimension n inclusive. On the other hand, it follows from the discussions of the last section that the choice of the universal bundle is immaterial, so that we can take the one whose base space is $X_0 = R(m + n + 1)/(G_1 \times R(n + 1))$. Using a connection in this universal bundle, we can, according to a process given above, define a homomorphism $h_0 = I(G_1) \rightarrow H(X_0)$ of the ring of invariant polynomials of G_1 into $H(X_0)$. X_0 being a homogeneous space, its cohomology ring $H(X_0)$ with real coefficients can be studied algebraically by methods initiated by E. Cartan and recently developed with success by H. Cartan, Chevalley, Kozsul, Leray, and Weil [13]. Thus it has been shown that, up to dimension n, h_0 is a one-one isomorphism. We may therefore replace $H(X_0)$ by $I(G_1)$ in the homomorphism h' and write the characteristic homomorphism as

$$h': I(G_1) \rightarrow H(X). \tag{11}$$

This homomorphism h' is defined by the topological properties of the fiber bundle.

On the other hand, the homomorphism $h: I(G) \rightarrow H(X)$ defined above can be split into a product of two homomorphisms. Since an invariant polynomial under G is an invariant polynomial under G_1, there is a natural homomorphism

$$\sigma: I(G) \rightarrow I(G_1). \tag{12}$$

Since G_1 can be taken to be the structural group, the homomorphism

$$h_1: I(G_1) \rightarrow H(X) \tag{13}$$

is defined. Now, a connection with the group G_1 can be considered as a connection with the group G. Using such a connection, we can easily prove

$$h = h_1\sigma. \tag{14}$$

Our main result which seems to include practically all our present knowledge on the subject consists in the statement:

$$h' = h_1. \tag{15}$$

Notice that h' is defined by the topological properties of the bundle and h_1 by the help of a connection, so that our theorem gives a relationship between a bundle and a connection defined in it, which is restrictive in one way or the other. In particular, when the structural group G is compact, we have $G_1 = G$ and σ is the identity, and the characteristic homomorphism is in a sense determined by the connection. For instance, it follows that the characteristic ring of the bundle has to be zero when a connection can be defined such that $h(I(G)) = 0$.

A proof for this theorem is obtained by first establishing it for the universal bundle. Under the mapping $f: X \rightarrow X_0$ it is then true for the induced bundle and the induced connection. Using the theorem of Weil that h is independent of the choice of the connection, we see that the relation is true for any connection in the bundle.

A great deal can be said about the rings of invariant polynomials $I(G)$, $I(G_1)$ and the homomorphism σ. When the structural group is compact, such statements can usually be proved more simply by topological considerations. In the other case we have to make use of the cohomology theory of Lie algebras. As we do not wish to discuss this, we shall restrict ourselves to the explanation of the corresponding topological notions. For this purpose we shall first discuss compact groups, that is, we begin by confining our attention to G_1.

We first recall some results on compact group manifolds. All the maximal abelian subgroups are conjugate and are isomorphic to a torus whose dimension is called the rank of the group. By an idea due essentially to Pontrjagin [16] we can define an operation of the homology classes of G_1 on the cohomology classes of G_1. In fact, $m: G_1 \times G_1 \rightarrow G_1$ being defined by the group multiplication, the image $m^*\gamma^k$ of a cohomology class of dimension k of G under the dual homomorphism m^* can be written $m^*\gamma^k = \sum u_i^r \times v_i^{k-r}$. The operation of a homology class c^s of dimension $s \leqq k$ on γ^k is then defined as $i(c^s)\gamma^k = \sum_i KI(c^s, u_i^s)v_i^{k-s}$. We call this operation an interior product. A cohomology class γ^k of G_1 is called primitive if its interior product by any homology class of dimension s, $1 \leqq s \leqq k - 1$, is zero. The homology structure of compact group manifolds (with real coefficients) has a description given by the following theorem of Hopf and Samelson [11; 18]: (1) all primitive cohomology classes are of odd dimension; (2) the vector space of the primitive classes has as dimension the rank of G; (3) the cohomology ring of G_1 is isomorphic to the Grassmann algebra of the space of primitive classes.

The primitive classes play a rôle in the study of the universal principal fiber bundle $\psi: B_0 \rightarrow X_0$. Identify a fiber $\psi^{-1}(x)$ $(x \in X_0)$ with G_1, and let i be the inclusion mapping of G_1 into B_0. If γ^k is a cocycle of X_0, $\psi^*\gamma^k$ is a cocycle of B_0. Since B_0 is homologically trivial, there exists a cochain β^{k-1} having $\psi^*\gamma^k$ as coboundary. Then $i^*\beta^{k-1}$ is a cocycle in G_1 whose cohomology class depends only on that of γ^k. The resulting mapping of the cohomology classes is called a transgression. It is an additive homomorphism of the ring of invariant polynomials of G_1 into the cohomology ring of G_1 and it carries an invariant polynomial of degree k into a cohomology class of dimension $2k - 1$. Chevalley and Weil proved that the image is precisely the space of the primitive classes.

When the group G is noncompact, the consideration of its Lie algebra allows us to generalize the above notions, at least under the assumption that G is semi-simple. H. Cartan, Chevalley, and Koszul have developed a very comprehensive theory dealing with the situation, which can be considered in a sense as the algebraic counterpart of the above treatment. Among their consequences we mention the following which is interesting for our present purpose: The ring of invariant polynomials under G has a set of generators equal to the rank of G; these can be so chosen that their images under transgression span the space of primitive classes of G.

Using the fact that the cohomology theory of Lie algebras and transgression can be defined algebraically, and therefore for G, we have the following diagram

$$\begin{array}{ccc} H(G) & \xrightarrow{i^*} & H(G_1) \\ t\uparrow & & \uparrow t_1 \\ I(G) & \xrightarrow{\sigma} & I(G_1). \end{array}$$

It is not difficult to prove that commutativity holds in this diagram. Hence the image under σ depends on the image under i^* of $H(G)$, that is, on the "homological position" of G_1 in G. In general, $\sigma[I(G)] \neq I(G_1)$.

There are relations between the characteristic cohomology classes in our definition and the classes carrying the same name in the topological method of obstructions but we cannot discuss them in detail. The latter come into being when one attempts to define a cross-section in the fiber bundle (that is, a mapping f of X into B, such that ψf is the identity) by extension over the successive skeletons; they are cohomology classes over groups of coefficients which are the homotopy groups of the director space. As we shall see from examples, it is sometimes possible to identify them by identifying the coefficient groups. In general, however, our characteristic classes are based on homological considerations, while those of obstruction theory are based on homotopy considerations. Their rôles are complementary.

We shall devote the rest of this lecture to the consideration of examples. Although the main results will follow from the general theorems, special problems arise in individual cases which can be of considerable interest. To begin with, take for G the rotation group in m variables, and suppose that a connection is given in the bundle. This includes in particular the case of orientable Riemann manifolds with a positive definite metric, the bundle being the tangent bundle of the manifold and the connection being given by the parallelism of Levi-Civita; it also includes, among other things, the theory of orientable submanifolds imbedded in an orientable Riemann manifold.

By a proper choice of a base of the Lie algebra of $G = R(m)$, the space of the Lie algebra can be identified with the space of skew-symmetric matrices of order m. The connection can therefore be defined, in every coordinate neighborhood, by a skew-symmetric matrix of linear differential forms $\theta = (\theta_{ij})$, and its curvature tensor by a skew-symmetric matrix of quadratic differential forms $\Theta = (\Theta_{ij})$. The effect of the adjoint group is given by $\mathrm{ad}(a)\Theta = A\Theta\, {}^tA$, where A is a proper orthogonal matrix and ${}^tA$ is its transpose.

The first question is of course to determine a set of generators for the ring of invariant polynomials; using the fundamental theorem on invariants, it is easy to do this explicitly [23]. Instead of the invariant polynomials we write the corresponding differential forms:

$$\begin{aligned} &\Delta_s = \Theta_{i_1 i_2} \cdots \Theta_{i_s i_1}, &&s = 2, 4, \cdots, m+1, &&m \text{ odd} \\ (16)\quad &\Delta_s = \Theta_{i_1 i_2} \cdots \Theta_{i_s i_1}, &&s = 2, 4, \cdots, m-2, &&m \text{ even} \\ &\Delta_0 = \epsilon_{i_1 \cdots i_m} \Theta_{i_1 i_2} \cdots \Theta_{i_{m-1} i_m}, && &&m \text{ even,} \end{aligned}$$

where repeated indices imply summation and where $\epsilon_{i_1 \cdots i_m}$ is the Kronecker tensor, equal to $+1$ or -1 according as $i_1, \cdots, i_m$ form an even or odd permutation of $1, \cdots, m$ and otherwise to 0. Since the rank of $R(m)$ is $(m+1)/2$ or $m/2$ according as m is odd or even, we verify here that the number of the above generators is equal to the rank. They form a complete set of generators, because they are obviously independent.

It follows that the cohomology classes determined by these differential forms or by polynomials in these differential forms depend only on the bundle and not on the connection. As a consequence, if all these differential forms are zero, the characteristic ring is trivial. The differential forms in (16) were first given by Pontrjagin [17].

For geometric applications it is useful to have a more explicit description of the base space of a universal bundle. This is all the more significant, since it would then allow us to study the characteristic homomorphisms with coefficient rings other than the field of real numbers. Our general theory gives as such a base space the Grassmann manifold

$$X_0 = R(m+n+1)/(R(m) \times R(n+1)),$$

which can be identified with the space of all oriented m-dimensional linear spaces through a point 0 of an $(m+n+1)$-dimensional Euclidean space E^{m+n+1}.

The homology structure of Grassmann manifolds has been studied by Ehresmann [9, 10]. A cellular decomposition can be constructed by the following process: Take a sequence of linear spaces

$$0 \subset E^1 \subset E^2 \subset \cdots \subset E^{m+n} \subset E^{m+n+1}.$$

Corresponding to a set of integers

$$0 \leqq a_1 \leqq a_2 \leqq \cdots \leqq a_m \leqq n+1,$$

denote by $(a_1 \cdots a_m)$ the set of all m-dimensional linear spaces $\xi \in X_0$ such that

$$\dim (\xi \cap E^{a_i + i}) \geqq i, \qquad i = 1, \cdots, m.$$

The interior points of $(a_1 \cdots a_m)$ form two open cells of dimension $a_1 + \cdots + a_m$. These open cells constitute a cellular decomposition of X_0, whose incidence relations can be determined. From this we can determine the homology and cohomology groups of X_0. In particular, it follows that the symbol $(a_1 \cdots a_m)^{\pm}$ can be used to denote a cochain, namely, the one which has the value $+1$ for the corresponding open cells and has otherwise the value zero. The characteristic homomorphism can then be described as a homomorphism of combinations of such symbols into the cohomology ring $H(X, R)$ of X. When R is the field of real numbers, the result is particularly simple. In fact, a base for the cohomology groups of dimensions less than or equal to n consists of cocycles having as symbols those for which all a_i are even, together with the cocycle $(1 \cdots 1)$ when m is even.

This new description of the characteristic homomorphism allows us to give a geometric meaning to individual characteristic classes. In this respect the class $h'((1 \cdots 1))$, which exists only when m is even, deserves special attention. In fact, the bundle with the director space $S^{m-1} = R(m)/R(m-1)$ constructed from the principal bundle is a bundle of $(m-1)$-spheres in the sense of Whitney. For such a sphere bundle, Whitney introduced a characteristic cohomology class W^m with integer coefficients. It can be proved that W^m, when reduced to real coefficients, is precisely the class $h'((1 \cdots 1))$. On the other hand, the latter can be identified on the universal bundle with a numerical multiple of the class defined by the differential form Δ_0. Taking the values of these classes for the fundamental cycle of the base manifold, we can write the result in an integral formula

$$W^m \cdot X = c \int_X \Delta_0, \tag{17}$$

where c is a numerical factor and X denotes a fundamental cycle of the base manifold. For a Riemann manifold, $W^m \cdot X$ is equal to the Euler-Poincaré characteristic of X and our formula reduces to the Gauss-Bonnet formula [3].

We introduce the notations

$$\begin{aligned} P^{4k} &= h'(0 \cdots 0\underbrace{2 \cdots 2}_{2k \text{ times}}) \\ \bar{P}^{4k} &= h'(0 \cdots 0\ 2k\ 2k) \\ \chi^m &= h'(1 \cdots 1), \qquad m \text{ even}, \end{aligned} \tag{18}$$

where the symbols denote also the cohomology classes to which the respective cocycles belong. By studying the multiplicative structure of the cohomology ring of X_0, we can prove that the characteristic homomorphism is determined by the classes P^{4k}, χ^m, $4k \leqq \dim X$ or the classes $\bar{P}^{4k}$, χ^m, $4k \leqq \dim X$.

We shall mention an application of the classes $\bar{P}^{4k}$. Restricting ourselves for simplicity to the tangent bundle of a compact differentiable manifold, the conditions $\bar{P}^{4k} = 0$, $2k \geqq n + 2$, are necessary for the manifold to be imbeddable into a Euclidean space of dimension $m + n + 1$. We get thus criteria on the impossibility of imbedding which can be expressed in terms of the curvature tensor of a Riemann metric on the manifold.

The second example we shall take up is the case that G is the unitary group. Such bundles occur as tangent bundles of complex analytic manifolds, and the introduction of an Hermitian metric in the manifold would give rise to a connection in the bundle.

The space of the Lie algebra of the unitary group $U(m)$ in m variables can be identified with the space of $m \times m$ Hermitian matrices A (${}^t\bar{A} = A$). A connection is therefore defined in each coordinate neighborhood by an Hermitian matrix of linear differential forms $\theta = (\theta_{ij})$ and its curvature tensor by an Hermitian matrix of quadratic differential forms $\Theta = (\Theta_{ij})$. Under the adjoint group the curvature tensor is transformed according to $\mathrm{ad}(a)\Theta = A\Theta{}^t\bar{A}$, A being a unitary matrix. Using this representation of the adjoint group, a set of invariant polynomials can be easily exhibited. We give their corresponding differential forms as

$$\Lambda_k = \Theta_{i_1 i_2} \cdots \Theta_{i_k i_1}, \qquad k = 1, \cdots, m. \tag{19}$$

Since they are clearly independent and their number is equal to the rank m of $U(m)$, they form a complete set of generators in the ring of invariant polynomials.

As in the case of the rotation group the complex Grassmann manifold $X_0 = U(m+n)/(U(m) \times U(n))$ is the base space of a universal bundle, whose study would be useful for some geometric problems. The results are simpler than the real case, but we shall not describe them here. A distinctive feature of the complex case is that a set of generators can be chosen in the ring of invariant polynomials whose corresponding differential forms are

$$\Psi_r = \frac{1}{(2\pi(-1)^{1/2})^{m-r+1}(m-r+1)!} \sum \delta(i_1 \cdots i_{m-r+1}; j_1 \cdots j_{m-r+1}) \cdot \Theta_{i_1 j_1} \cdots \Theta_{i_{m-r+1} j_{m-r+1}}, \qquad r = 1, \cdots, m, \tag{20}$$

where $\delta(i_1 \cdots i_{m-r+1}; j_1 \cdots j_{m-r+1})$ is zero except when $j_1, \cdots, j_{m-r+1}$ form a permutation of $i_1, \cdots, i_{m-r+1}$, in which case it is $+1$ or -1 according as the permutation is even or odd, and where the summation is extended over all indices $i_1, \cdots, i_{m-r+1}$ from 1 to m. This set of generators has the advantage that the cohomology classes determined by the differential forms have a simple geometrical meaning. In fact, they are the classes, analogous to the Stiefel-Whitney classes, for the bundle with the director space $U(m)/U(m-r)$. As such they are primary obstructions to the definition of a cross-section and are therefore more easily dealt with [4]. Substantially the same classes have been introduced by M. Eger and J. A. Todd in algebraic geometry, even before they first made their appearance in differential geometry [6; 20].

The situation is different for bundles with the rotation group, since the Stiefel-Whitney classes, except the highest-dimensional one, are essentially classes mod 2 and therefore do not enter into our picture. However, there is a close relationship between bundles with the group $R(m)$ and bundles with the group $U(m)$. In fact, given a bundle with the group $R(m)$, we can take its Whitney product with itself, which is a bundle with the same base space and the group $R(m) \times R(m)$. The latter can be imbedded into $U(m)$, so that we get a bundle with the group $U(m)$. Such a process is frequently useful in reducing problems on bundles with the rotation group to those on bundles with the unitary group.

We shall take as last example the case that the group is the component of the identity of the general linear group $GL(m)$ in m variables. A connection in the bundle is called an affine connection. An essential difference from the two previous examples is that the group is here noncompact.

The Lie algebra of the group $GL(m)$ can be identified with the space of all m-rowed square matrices, so that the curvature tensor in each coordinate neighborhood is given by such a matrix of exterior quadratic differential forms: $\Theta = (\Theta^j_i)$. The effect of the adjoint group being defined by $\mathrm{ad}(a)\Theta = A\,\Theta\,A^{-1}$, $a \in GL(m)$, it is easily seen that a set of generators of the ring of invariant polynomials can be so chosen that the corresponding differential forms are

$$M_s = \Theta^{i_2}_{i_1} \cdots \Theta^{i_1}_{i_s}, \qquad s = 1, \cdots, m-1. \tag{21}$$

According to the general theory it remains to determine the homomorphism of the ring of invariant polynomials under $GL(m)$ into the ring of invariant polynomials under its maximal compact subgroup, which is in this case the rotation group $R(m)$. It is seen that M_s, for even s, is mapped into Δ_s, and, for odd s, is mapped into zero. The class defined by Δ_0 does not belong to the image of the homomorphism. This fact leads to the interesting explanation that a formula analogous to the Gauss-Bonnet formula does not exist for an affine connection.

Perhaps the most important of the bundles is the tangent bundle of a differentiable manifold. We mentioned above the identification of a certain characteristic class with the Euler-Poincaré characteristic of the manifold, at least for the case that the manifold is orientable and of even dimension. Beyond this very little is known on the relations between topological invariants of the manifold and the characteristic homomorphism of its tangent bundle. Recently, contributions have been made by Thom and Wu which bear on this question [21; 25]. Although it is not known whether a topological manifold always has a differentiable structure, nor whether it can have two essentially different differentiable structures, Thom and Wu proved that the characteristic homomorphisms of the tangent bundle, with coefficients mod 2 and with coefficients mod 3, are independent of the choice of the differentiable structure, provided one exists. Briefly speaking, this means that such characteristic homomorphisms are topological invariants of differentiable manifolds. The proof for coefficients mod 3 is considerably more difficult than the case mod 2.

For bundles with other groups such questions have scarcely been asked. The next case of interest is perhaps the theory of projective connections derived from the geometry of paths. In this case the bundle with the projective group depends both on the tangent bundle and the family of paths. It would be of interest to know whether or what part of the characteristic homomorphism is a topological invariant of the manifold.

Before concluding we shall mention a concept which has no close relation with the above discussion, but which should be of importance in the theory of connections, namely, the notion of the group of holonomy. It can be defined as follows: if ω is the left-invariant differential form in G, with values in $L(G)$, and if θ_α defines a connection, the equation

$$\theta_\alpha + \omega = 0 \tag{22}$$

is independent of the coordinate neighborhood. When a parametrized curve is given in the base manifold, this differential equation defines a family of integral curves in G invariant under left translations of the group. Let $x \in X$ and consider all closed parametrized curves in X having x as the initial point. To every such curve C let $a(C)$ be the endpoint of the integral curve which begins at the unit element e of G. All such points $a(C)$ form a subgroup H of G, the group of holonomy of the connection.

Added in proof: The details of some of the discussions in this article can be found in mimeographed notes of the author, *Topics in differential geometry*, Institute for Advanced Study, Princeton, 1951.

Bibliography

1. E. Cartan, *Les groupes d'holonomie des espaces généralisés*, Acta Math. vol. 48 (1926) pp. 1–42.

2. ———, *L'extension du calcul tensoriel aux géométries non-affines*, Ann. of Math. (2) vol. 38 (1947) pp. 1–13.

3. S. S. Chern, *A simple intrinsic proof of the Gauss-Bonnet formula for closed Riemannian manifolds*, Ann. of Math. (2) vol. 45 (1944) pp. 747–752.

4. ———, *Characteristic classes of Hermitian manifolds*, Ann. of Math. (2) vol. 47 (1946) pp. 85–121.

5. ———, *Some new viewpoints in differential geometry in the large*, Bull. Amer. Math. Soc. vol. 52 (1946) pp. 1–30.

6. M. Eger, *Sur les systèmes canoniques d'une variété algébrique*, C. R. Acad. Sci. Paris vol. 204 (1937) pp. 92–94, 217–219.

7. C. Ehresmann, *Les connexions infinitésimales dans un espace fibré différentiable*, Seminaire Bourbaki, March, 1950. This paper contains references to earlier notes of the author on the subject in C. R. Acad. Sci. Paris.

8. ———, *Sur la théorie des espaces fibrés*, Colloque de Topologie Algébrique, Paris, 1947, pp. 3–15.

9. ———, *Sur la topologie de certains espaces homogènes*, Ann. of Math. (2) vol. 35 (1934) pp. 396–443.

10. ———, *Sur la topologie des certaines variétés algébriques réelles*, Journal de Mathématiques vol. 104 (1939) pp. 69–100.

11. H. Hopf, *Über die Topologie der Gruppenmannigfaltigkeiten und ihre Verallgemeinerungen*, Ann. of Math. (2) vol. 42 (1941) pp. 22–52.

12. K. Iwasawa, *On some types of topological groups*, Ann. of Math. (2) vol. 50 (1949) pp. 507–558.

13. J. L. Koszul, *Homologie et cohomologie des algèbres de Lie*, Bull. Soc. Math. France vol. 78 (1950) pp. 65–127.

14. A. Malcev, *On the theory of Lie groups in the large*, Rec. Math. (Mat. Sbornik) N. S. vol. 16 (1945) pp. 163–189.

15. G. D. Mostow, *A new proof of E. Cartan's theorem on the topology of semi-simple groups*, Bull. Amer. Math. Soc. vol. 55 (1949) pp. 969–980.

16. L. Pontrjagin, *Homologies in compact Lie groups*, Rec. Math. (Mat. Sbornik) N. S. vol. 48 (1939) pp. 389–422.

17. ———, *On some topological invariants of Riemannian manifolds*, C. R. (Doklady) Acad. Sci. URSS vol. 43 (1944) pp. 91–94.

18. H. Samelson, *Beiträge zur Topologie der Gruppen-Mannigfaltigkeiten*, Ann. of Math. (2) vol. 42 (1941) pp. 1091–1137.

19. N. E. Steenrod, *Topology of fiber bundles*, Princeton, 1951. With few exceptions this will be our standard reference on fiber bundles.

20. J. A. Todd, *The geometrical invariants of algebraic loci*, Proc. London Math. Soc. (2) vol. 43 (1937) pp. 127–138.

21. R. Thom, *Variétés plongées et i-carrés*, C. R. Acad. Sci. Paris vol. 230 (1950) pp. 507–508.

22. A. Weil, *Géométrie différentielle des espaces fibrés*, unpublished.

23. H. Weyl, *Classical groups*, Princeton, 1939.

24. H. Whitney, *On the topology of differentiable manifolds*, Lectures on Topology, Michigan, 1941, pp. 101–141.

25. Wen-tsün Wu, *Topological invariance of Pontrjagin classes of differentiable manifolds*, to be published.

University of Chicago,
Chicago, Ill., U. S. A.

Reprinted from *Amer. J. Math.* 74 (1952) 227–236

ON THE KINEMATIC FORMULA IN THE EUCLIDEAN SPACE OF *N* DIMENSIONS.*

By SHIING-SHEN CHERN.

Introduction. The idea of considering the kinematic density in problems of geometrical probability was originated by Poincaré. It was further exploited by L. A. Santalò and W. Blaschke in their work on integral geometry [1], culminating in the following theorem:

Let Σ_0, Σ_1 be two closed surfaces in space, which are twice differentiable, and let D_0, D_1 be the domains bounded by them. Let V_i, $\chi_i = K_i/4\pi$ be the volume and Euler characteristic of D_i and let A_i, M_i be the area and the integral of mean curvature of Σ_i, $i = 0, 1$. Suppose Σ_0 fixed and Σ_1 moving. Then the integral of $K(D_0 \cdot D_1) = 4\pi\chi(D_0 \cdot D_1)$ over the kinematic density of Σ_1 is given by the formula

$$\int K(D_0 \cdot D_1)\dot{\Sigma}_1 = 8\pi^2(V_0K_1 + A_0M_1 + M_0A_1 + K_0V_1). \tag{1}$$

This formula includes most formulas in Euclidean integral geometry as special or limiting cases. The purpose of this paper is to apply E. Cartan's method of moving frames and to derive the generalization of this formula in an Euclidean space of n dimensions. By doing this, we hope that some insight can be gained on integral geometry in a general homogeneous space. Moreover, one of the ideas introduced, the consideration of measures in spaces which are now called fiber bundles, will most likely find further applications. The main procedures of our proof have been given in a previous note [2].

We consider a compact orientable hypersurface Σ, twice differentiably imbedded in an Euclidean space E of n ($\geqq 2$) dimensions. At a point P of Σ there are $n-1$ principal curvatures κ_α, $\alpha = 1, \cdots, n-1$, whose i-th elementary symmetric function we shall denote by S_i, $i = 0, \cdots, n-1$, where $S_0 = 1$ by definition. Let dA be the element of area of Σ, and let

$$M_i = \int_\Sigma S_i dA \Big/ \binom{n-1}{i}, \qquad i = 0, 1, \cdots, n-1. \tag{2}$$

These M_i are integro-differential invariants of Σ. In particular, M_0 is the

* Received May 10, 1951.

area and M_{n-1} is a numerical multiple of the degree of mapping of Σ into the unit hypersphere defined by the field of normals.

Take now two such hypersurfaces Σ_0, Σ_1, whose invariants we distinguish by superscripts. The volume of the domain D_i bounded by Σ_i we denote by V_i, $i = 0, 1$. Let Σ_0 be fixed and Σ_1 be moving, and let $\dot{\Sigma}_1$ be the kinematic density of Σ_1. We suppose our hypersurfaces to be such that for all positions of Σ_1 the intersection $D_0 \cdot D_1$ has a finite number of components. Then the Euler-Poincaré characteristic $\chi(D_0 \cdot D_1)$ is well defined. If I_{n-1} denotes the area of the unit hypersphere in E and if

$$J_n = I_1 I_2 \cdots I_{n-1}, \tag{3}$$

the *kinematic formula* in E is

$$\begin{aligned} &\int K(D_0 \cdot D_1)\dot{\Sigma}_1 \\ &\quad = J_n\{M_{n-1}^{(0)}V_1 + M_{n-1}^{(1)}V_0 + \frac{1}{n}\sum_{k=0}^{n-2}\binom{n}{k+1} M_k^{(0)}M_{n-2-k}^{(1)}\}, \end{aligned} \tag{4}$$

where

$$K(D_0 \cdot D_1) = I_{n-1}\chi(D_0 \cdot D_1). \tag{5}$$

For $n = 3$ this reduces to the formula (1). The formula for $n = 4$ is

$$\begin{aligned} &\int K(D_0 \cdot D_1)\dot{\Sigma}_1 \\ &\quad = 16\pi^4(M_3^{(0)}V_1 + M_3^{(1)}V_0 + M_0^{(0)}M_2^{(1)} + M_0^{(1)}M_2^{(0)} + \tfrac{3}{2}M_1^{(0)}M_1^{(1)}). \end{aligned} \tag{6}$$

1. Measures in spaces associated with a Riemann manifold. We shall first review a few notions in Riemannian geometry, in a form which will be useful for our later purpose.

Let M be an orientable Riemann manifold of class $\geqq 3$ and dimension n. Associated with M are the spaces B_h $(h = 1, \cdots, n)$ formed by the elements $Pe_1 \cdots e_h$, each of which consists of a point P of M and an ordered set of h mutually perpendicular tangent unit vectors $e_1, \cdots, e_h$ at P. When $h = n$, such an element will be called a frame. In the current terminology B_n is a principal fiber bundle over M with the rotation group as structural group and B_h are the associated bundles [3]. We shall introduce a measure in B_h. Since B_h is clearly an orientable differentiable manifold, this can be done by defining an exterior differential form of degree $\frac{1}{2}(h+1)(2n-h)$(= dim of B_h).

There is a natural mapping $\psi_h\colon B_n \to B_h$ defined by taking as the image of $Pe_1 \cdots e_n$ the element $Pe_1 \cdots e_h$. It induces a dual homomorphism of the differential forms of B_h into those of B_n. This process has in a sense a converse. In fact, let

$$e_r = \sum_s u_{rs} e^*_s, \qquad h+1 \leqq r, s \leqq n \tag{7}$$

be a rotation of the last $n-h$ vectors. A differential form of B_n which is invariant under the action of (7) can be regarded as a form of B_n.

The well-known parallelism of Levi-Civita can be interpreted as defining a set of $n(n+1)/2$ linearly independent Pfaffian forms in B_n, which we shall denote by ω_i, $\omega_{ij}(=-\omega_{ji})$, $1 \leqq i, j \leqq n$. To give it a brief description [4] we start from the following useful lemma on exterior forms: *Let ω_i be linearly independent Pfaffian forms, and let $\pi_{ij}=-\pi_{ji}$ be Pfaffian forms such that* [1]

$$(8) \qquad \sum_j \omega_j \wedge \pi_{ji} = 0.$$

Then $\pi_{ij}=0$. In fact, it follows from (8) that

$$\pi_{ji} = \sum_{k=1}^{n} a_{jik}\omega_k.$$

Then a_{jik} is skew-symmetric in its first two indices, because the π_{ji} are, and is symmetric in its last two indices, on account of (8). Therefore $a_{jik}=0$ or $\pi_{ji}=0$.

For geometric reasons we denote by dP the identity mapping in the tangent space at P, which maps every tangent vector into itself. Then dP can be written in the form

$$(9) \qquad dP = \sum_i \omega_i \otimes e_i,$$

where the multiplication is tensor product, and the ω_i are Pfaffian forms in B_n and are linearly independent. The fundamental theorem on local Riemannian geometry asserts that there exists a uniquely determined set of Pfaffian forms ω_i, ω_{ij} in B_n, linearly independent, which satisfy (9) and

$$(10) \qquad d\omega_i = \sum_j \omega_j \wedge \omega_{ji}.$$

In fact, the uniqueness follows from the above lemma.

For our purpose we shall study the effect of the rotation (7) on these forms. Denote the new forms by the same symbols with asterisks. Clearly we have

$$(11) \qquad \omega_\alpha{}^* = \omega_\alpha, \qquad \omega_r{}^* = \sum_s u_{sr}\omega_s, \qquad 1 \leqq \alpha \leqq h,\ h+1 \leqq r, s \leqq n.$$

Taking the exterior derivatives of both sides of these equations and making use of (10), we get

---

[1] We shall, following Bourbaki, use wedge product to denote exterior multiplication. It will sometimes be dropped, when the meaning is clear. Parentheses will be used to denote ordinary products of differential forms.

$$(12)\quad \begin{aligned} &\sum_\beta \omega_\beta \wedge (\omega_{\beta\alpha}{}^* - \omega_{\beta\alpha}) + \sum_r \omega_r{}^* \wedge (\omega_{r\alpha}{}^* - \sum_s u_{sr}\omega_{s\alpha}) = 0, \\ &\sum_\alpha \omega_\alpha \wedge (\omega_{\alpha r}{}^* - \sum_s u_{sr}\omega_{\alpha s}) + \sum_s \omega_s{}^* \wedge \phi_{sr}{}^* = 0, \end{aligned}$$

where $\phi_{sr}{}^*$ are Pfaffian forms skew-symmetric in the indices s, r. The system of equations (12) is of the same form as (8), and the above lemma is then applicable. It follows that

$$(13)\qquad \omega_{\beta\alpha}{}^* = \omega_{\beta\alpha}, \qquad \omega_{\alpha r}{}^* = \sum_s u_{sr}\omega_{\alpha s}.$$

If we put

$$(14)\qquad \Omega_\alpha = \prod_r \omega_{\alpha r},$$

we see from (13) that Ω_α is invariant under the action of (7). The same is therefore true of the form

$$(15)\qquad L_{n,h} = \prod_\alpha \Omega_\alpha \prod_{\alpha<\beta} \omega_{\alpha\beta} \prod_i \omega_i.$$

This form is clearly not identically zero, and we define it to be the density in B_h. It gives rise to a measure in B_h.

2. Differential geometry of a submanifold in Euclidean space. As a further preparation we need some notions on the geometry of a hypersurface in Euclidean space. As no additional complication is involved, we develop them for a submanifold V of p dimensions, which is twice differentiably imbedded in E. We agree in this section on the following ranges of indices:

$$(16)\qquad 1 \leqq \alpha, \beta, \gamma \leqq p, \qquad p+1 \leqq r, s, t \leqq n, \qquad 1 \leqq i, j, k \leqq n.$$

Since E is a Riemann manifold, the discussions of the last section are valid. In this case B_n is naturally homeomorphic to the group of proper motions in E. To study V we consider the submanifold of B_n characterized by the conditions that $P \varepsilon V$ and that the $\mathfrak{e}_\alpha$ are tangent vectors to V at P. If we denote by the same notation the forms on this submanifold induced by the identity mapping, we have

$$(17)\qquad \omega_r = 0.$$

From (10) it follows that

$$d\omega_r = \sum_\alpha \omega_\alpha \wedge \omega_{\alpha r} = 0.$$

Since the ω_α are linearly independent, we have

$$(18)\qquad \omega_{r\alpha} = \sum_\beta A_{r\alpha\beta}\omega_\beta$$

where the $A_{r\alpha\beta}$ are symmetric in α, β:

$$A_{r\alpha\beta} = A_{r\beta\alpha}. \tag{19}$$

From these Pfaffian forms it is possible to construct some significant "ordinary" quadratic differential forms. The first is a set

$$\Phi_r = \sum_{\alpha} (\omega_{r\alpha}\omega_\alpha) = \sum_{\alpha,\beta} A_{r\alpha\beta}(\omega_\alpha\omega_\beta), \tag{20}$$

which generalizes the second fundamental form in ordinary surface theory. The second is

$$\Psi = \sum_{r,\alpha} (\omega_{r\alpha})^2 = \sum_{r,\alpha,\beta,\gamma} A_{r\alpha\beta}A_{r\alpha\gamma}(\omega_\beta\omega_\gamma), \tag{21}$$

generalizing the third fundamental form. The latter seems to deserve some attention. However, so far as the writer is aware, it has not been considered in the literature.

For a hypersurface we have $p = n - 1$, and we shall write Φ, $A_{\alpha\beta}$ for Φ_n, $A_{n\alpha\beta}$ respectively. The $n - 1$ roots of the characteristic equation

$$| A_{\alpha\beta} - \kappa\delta_{\alpha\beta} | = 0 \tag{22}$$

are called the principal curvatures.

In the case of the Euclidean space E we can also write ω_i, ω_{ij} as scalar products, thus:

$$\omega_i = dP \cdot e_i, \qquad \omega_{ij} = de_i \cdot e_j. \tag{23}$$

3. A formula on densities. The situation we are going to consider consists of two hypersurfaces Σ_0, Σ_1 in E, with Σ_0 fixed and Σ_1 moving, which intersect in a manifold V^{n-2} of dimension $n - 2$, such that at a point of V^{n-2} the normals to Σ_0, Σ_1 never coincide. We denote by ϕ, $\phi \neq 0, \pi$, the angle between these normals and by $\dot{\Sigma}_1$ the kinematic density of Σ_1. An $(n-2)$-frame on V^{n-2} has a density on each of V^{n-2}, Σ_0, Σ_1, to be denoted by L_V, $\dot{L}_0$, L_1 respectively. Our formula to be proved can be written

$$\dot{L}_V\dot{\Sigma}_1 = \sin^{n-1}\phi \dot{L}_0\dot{L}_1\dot{\phi}. \tag{24}$$

Throughout this section we shall agree on the following ranges of indices:

$$1 \leqq i, j, k \leqq n, \qquad 1 \leqq \alpha, \beta \leqq n-2, \qquad 1 \leqq A, B \leqq n-1. \tag{25}$$

Let $O\mathfrak{a}_1 \cdots \mathfrak{a}_n$ be the fixed frame and $O'\mathfrak{a}'_1 \cdots \mathfrak{a}'_n$ the moving frame. For a given relative position between $O\mathfrak{a}_1 \cdots \mathfrak{a}_n$ and $O'\mathfrak{a}'_1 \cdots \mathfrak{a}'_n$ let $Pe_1 \cdots e_{n-2}$ be an $(n-2)$-frame on V^{n-2}. We complement this into a frame $Pe_1 \cdots e_n$ such that e_n is normal to Σ_0 and also into a frame $P'e'_1 \cdots e'_n$

such that e'_n is normal to Σ_1 at P, and $P' = P$, $e'_\alpha = e_\alpha$. Between e_{n-1}, e_n, e'_{n-1}, e'_n we have then the relations

$$e'_{n-1} = \cos\phi e_{n-1} + \sin\phi e_n, \qquad e'_n = -\sin\phi e_{n-1} + \cos\phi e_n. \tag{26}$$

From this we derive the following useful relation

$$de'_{n-1} \cdot e'_n = d\phi + de_{n-1} \cdot e_n. \tag{27}$$

Let us now express the relations between the frames so introduced by the equations

$$\begin{aligned} P &= O + \sum_i x_i a_i, & e_i &= \sum_k u_{ik} a_k, \\ P' &= O' + \sum_i x'_i a'_i, & e'_i &= \sum_k u'_{ik} a'_k. \end{aligned} \tag{28}$$

We shall denote the differentiation by d' when $O'a'_1 \cdots a'_n$ is regarded as fixed. In other words, d' is differentiation relative to the moving frame. Then we have, from (28),

$$dO' = dP - d'P - \sum_i x'_i da'_i. \tag{29}$$

It follows that, on neglecting terms in da'_i,

$$\begin{aligned} \prod_\alpha (dP \cdot e_\alpha) \prod_i (dO' \cdot a'_i) &\equiv \prod_\alpha (dP \cdot e_\alpha) \prod_i (dP \cdot a'_i - d'P \cdot a'_i) \\ &\equiv \prod_\alpha (dP \cdot e_\alpha) \prod_i (dP \cdot e'_i - d'P \cdot e'_i) \\ &\equiv \pm \prod_\alpha (dP \cdot e_\alpha)(d'P \cdot e'_\alpha)(dP \cdot e'_{n-1} - d'P \cdot e'_{n-1})(dP \cdot e'_n - d'P \cdot e'_n) \\ &\equiv \pm \sin\phi \prod_A (dP \cdot e_A)(d'P \cdot e'_A). \end{aligned} \tag{30}$$

These are to be taken as congruences mod da'_i. In particular, the last step follows from the fact that e'_n is normal to Σ_1 at P and that the product of n factors involving dP is zero, because the locus of P is a hypersurface Σ_0.

In order to get a further reduction of the left-hand side of (24) we start from the formula

$$de'_i - d'e'_i = \sum_k u'_{ik} da'_k. \tag{31}$$

From the invariance of the kinematic density under a rotation it follows that

$$\prod_{i<j} (da'_i \cdot a'_j) = \prod_{i<j} ((de'_i - d'e'_i) \cdot e'_j).$$

Then we have

$$\begin{aligned}
&\prod_{\alpha<\beta}(de_\alpha\cdot e_\beta)\prod_{i<j}(d\alpha'_i\cdot\alpha'_j)=\prod_{\alpha<\beta}(de_\alpha\cdot e_\beta)\prod_{i<j}(de'_i-d'e'_i)\cdot e'_j\\
&\quad=\prod_{\alpha<\beta}(de_\alpha\cdot e_\beta)\prod_{\alpha<i}((de_\alpha-d'e_\alpha)\cdot e'_i)\,((de'_{n-1}-d'e'_{n-1})\cdot e'_n)\\
&\quad=\prod_{\alpha<\beta}(de_\alpha\cdot e_\beta)\prod_{\alpha<\beta}((de_\alpha-d'e_\alpha)\cdot e'_\beta)\\
&\qquad\qquad\wedge\prod_\alpha\{(de_\alpha\cdot e'_{n-1}-d'e_\alpha\cdot e'_{n-1})(de_\alpha\cdot e'_n-d'e_\alpha\cdot e'_n)\}\\
(32)\qquad&\qquad\qquad\wedge(de'_{n-1}\cdot e'_n-d'e'_{n-1}\cdot e'_n)\\
&\quad\equiv\pm\prod_{\alpha<\beta}(de_\alpha\cdot e_\beta)(d'e_\alpha\cdot e_\beta)\prod_\alpha(de_\alpha\cdot e'_{n-1}-d'e_\alpha\cdot e'_{n-1})\\
&\qquad\qquad\wedge(de_\alpha\cdot e'_n)(de'_{n-1}\cdot e'_n)\\
&\quad\equiv\pm\sin^{n-2}\phi\prod_{\alpha<A}\{(de_\alpha\cdot e_A)(d'e_\alpha\cdot e'_A)\}d\phi.
\end{aligned}$$

Here the congruences are to be understood mod $dP\cdot e_A$, $d'P\cdot e'_A$. The step next to the last follows from the relations

$$d'e'_A\cdot e'_n=-d'e'_n\cdot e'_A\equiv 0,\quad \mathrm{mod}\ d'P\cdot e'_A,$$

which in turn are consequences of (18). In the reduction of the last step we make use of the relations (26), (27), and

$$de_A\cdot e_n=-de_n\cdot e_A\equiv 0,\quad \mathrm{mod}\ dP\cdot e_A.$$

If we notice that

$$\dot{\Sigma}=\prod_i(dO'\cdot\alpha'_i)\prod_{i<j}(d\alpha'_i\cdot\alpha'_j),$$

and recall the expressions for $\dot{L}_V$, $\dot{L}_0$, $\dot{L}_1$, then (30) and (32) together give the formula (24).

4. Total curvature and Euler characteristic. The success of our procedure depends on the possibility of expressing the Euler-Poincaré characteristic of a domain bounded by a hypersurface Σ by an integral over Σ, a result known as the Gauss-Bonnet formula. Let Λ be the volume element of the unit hypersphere in E, and N^+ the field of outward normals of Σ. By means of N^+ we define the normal mapping of Σ. The Gauss-Bonnet formula in this particular case can be written

$$(33)\qquad \int_{N^+}\Lambda=\chi(D)I_{n-1},$$

where D is the domain bounded by Σ, and $\chi(D)$ is its Euler-Poincaré characteristic. The left-hand side of this equation is sometimes called the total curvature of the domain.

In our later application the domain D will not be bounded by a smooth hypersurface but will be such that its boundary consists of a finite number of hypersurfaces which intersect in a number of submanifolds V^{n-2} of dimension $n-2$. To the integral of Λ over the outward normals we must then add the integral over the vectors belonging to the angle subtended by the outward normals of the two hypersurfaces. To express the latter analytically let us use the notation of the last section, together with the ranges of indices (25). In addition we denote by $\mathfrak{v}_A$, $\mathfrak{v}'_A$ the unit vectors in the principal directions of Σ_0, Σ_1 respectively. For a differentiation on Σ_0 we can then write

$$\theta_A = dP \cdot \mathfrak{v}_A, \qquad \kappa_A\theta_A = d\mathfrak{e}_n \cdot \mathfrak{v}_A, \tag{34}$$

where κ_A are the principal curvatures. Similarly, for a differentiation on Σ_1 we have

$$\theta'_A = dP \cdot \mathfrak{v}'_A, \qquad \kappa'_A\theta'_A = d\mathfrak{e}'_n \cdot \mathfrak{v}'_A, \tag{35}$$

κ'_A being the principal curvatures of Σ_1. Since the $\mathfrak{e}_\alpha$ lie in the intersection of the tangent hyperplanes, we have relations of the form

$$\mathfrak{e}_\alpha = \sum_A c_{\alpha A}\mathfrak{v}_A = \sum_A c'_{\alpha A}\mathfrak{v}'_A \tag{36}$$

To simplify notation we introduce the unit vectors $\mathfrak{v}$, $\mathfrak{w}$ in the directions of the angle bisectors of $\mathfrak{e}_n$, $\mathfrak{e}'_n$. Then we have

$$\mathfrak{e}_n = (\cos\tfrac{1}{2}\phi)\mathfrak{v} - (\sin\tfrac{1}{2}\phi)\mathfrak{w}, \qquad \mathfrak{e}'_n = (\cos\tfrac{1}{2}\phi)\mathfrak{v} + (\sin\tfrac{1}{2}\phi)\mathfrak{w}, \tag{37}$$

or

$$\mathfrak{e}_n + \mathfrak{e}'_n = 2(\cos\tfrac{1}{2}\phi)\mathfrak{v}, \qquad -\mathfrak{e}_n + \mathfrak{e}'_n = 2(\sin\tfrac{1}{2}\phi)\mathfrak{w}. \tag{38}$$

Let $\mathfrak{x}$ be a unit vector between $\mathfrak{e}_n$. $\mathfrak{e}'_n$, and $\mathfrak{y}$ the unit vector perpendicular to $\mathfrak{x}$ and in the plane of $\mathfrak{e}_n$, $\mathfrak{e}'_n$. We can then write

$$\mathfrak{x} = \cos\sigma\mathfrak{v} + \sin\sigma\mathfrak{w}, \qquad \mathfrak{y} = -\sin\sigma\mathfrak{v} + \cos\sigma\mathfrak{w}; \qquad -\tfrac{1}{2}\phi \leqq \sigma \leqq \tfrac{1}{2}\phi. \tag{39}$$

It follows that the total curvature, i. e., I_{n-1} times the Euler-Poincaré characteristic of D, is given by

$$K = \int_{N^+} \Lambda + \int_{-\frac{1}{2}\phi}^{\frac{1}{2}\phi} d\sigma \int_V \prod_\alpha \{\cos\sigma(d\mathfrak{v}\cdot\mathfrak{e}_\alpha) + \sin\sigma(d\mathfrak{w}\cdot\mathfrak{e}_\alpha)\}. \tag{40}$$

The product in the second integral admits some further simplification. In fact, using (38), we have

$$\prod_\alpha \{\cos\sigma(d\mathfrak{v}\cdot\mathfrak{e}_\alpha) + \sin\sigma(d\mathfrak{w}\cdot\mathfrak{e}_\alpha)\}$$
$$= \prod_\alpha \{\sin(\tfrac{1}{2}\phi - \sigma)(d\mathfrak{e}_n\cdot\mathfrak{e}_\alpha) + \sin(\tfrac{1}{2}\phi + \sigma)(d\mathfrak{e}'_n\cdot\mathfrak{e}_\alpha)\}/\sin^{n-2}\phi.$$

By (36), we get

$$de_n \cdot e_\alpha = \sum_A \kappa_A c_{\alpha A} \theta_A = \sum_{A,\beta} \kappa_A c_{\alpha A} c_{\beta A} (dP \cdot e_\beta),$$

$$de'_n \cdot e_\alpha = \sum_{A,\beta} \kappa'_A c'_{\alpha A} c'_{\beta A} (dP \cdot e_\beta).$$

It follows that

$$\prod_\alpha \{\cos \sigma (d\mathfrak{v} \cdot e_\alpha) + \sin \sigma (d\mathfrak{w} \cdot e_\alpha)\} = D\dot{V}/\sin^{n-2} \phi, \tag{41}$$

where $\dot{V}$ is the volume element of V^{n-2}, and where

$$D = |\sin(\tfrac{1}{2}\phi - \sigma) \sum_A \kappa_A c_{\alpha A} c_{\beta A} + \sin(\tfrac{1}{2}\phi + \sigma) \sum_A \kappa'_A c'_{\alpha A} c'_{\beta A}|. \tag{42}$$

The determinant D can be expanded in the form

$$D = \sum_{p=0}^{n-2} H_p \sin^{n-2-p}(\tfrac{1}{2}\phi - \sigma) \sin^p(\tfrac{1}{2}\phi + \sigma), \tag{43}$$

where

$$H_p = \sum \begin{vmatrix} c_{1A_1} & \cdots & c_{1A_q} & c'_{1B_1} & \cdots & c'_{1B_p} \\ & \cdots & & & \cdots & \\ c_{n-2,A_1} & \cdots & c_{n-2,A_q} & c'_{n-2,B_1} & \cdots & c'_{n-2,B_p} \end{vmatrix}^2 \kappa_{A_1} \cdots \kappa_{A_q} \kappa'_{B_1} \cdots \kappa'_{B_p}, \quad p + q = n - 2, \tag{44}$$

the summation being extended over all independent combinations $A_1, \cdots, A_q$ and $B_1, \cdots, B_p$ of $1, \cdots, n-1$. To prove this we observe that the expansion of D is of the above form and that the question is only to determine the coefficient of $\kappa_{A_1} \cdots \kappa_{A_q} \kappa'_{B_1} \cdots \kappa'_{B_p}$ in H_p. This coefficient is, up to the factor $\sin^q(\tfrac{1}{2}\phi - \sigma)\sin^p(\tfrac{1}{2}\phi + \sigma)$, the value of D, when we set

$$\kappa_{A_1} = \cdots = \kappa_{A_q} = 1, \qquad \kappa'_{B_1} = \cdots = \kappa'_{B_p} = 1,$$

and equal to zero otherwise. Writing

$$\bar{c}_{\alpha A} = \{\sin(\tfrac{1}{2}\phi - \sigma)\}^{\frac{1}{2}} c_{\alpha A}, \qquad \bar{c}'_{\alpha A} = \{\sin(\tfrac{1}{2}\phi + \sigma)\}^{\frac{1}{2}} c'_{\alpha A},$$

we have

$$D = |\sum_{s=A_1,\ldots,A_q} \bar{c}_{\alpha s} \bar{c}_{\beta s} + \sum_{t=B_1,\ldots,B_p} \bar{c}'_{\alpha t} \bar{c}'_{\beta t}| = \begin{vmatrix} \bar{c}_{1A_1} & \cdots & \bar{c}_{1A_q} & \bar{c}'_{1B_1} & \cdots & \bar{c}'_{1B_p} \\ & \cdots & & & \cdots & \\ \bar{c}_{n-2,A_1} & \cdots & \bar{c}_{n-2,A_q} & \bar{c}'_{n-2,B_1} & \cdots & \bar{c}'_{n-2,B_p} \end{vmatrix}^2.$$

This shows that the coefficient is actually the one asserted in (43), (44).

5. Proof of the kinematic formula. Let Σ_0, Σ_1 be two hypersurfaces twice differentiably imbedded in E, with Σ_0 fixed and Σ_1 moving. We denote by D_i the domain bounded by Σ_i, $i = 0, 1$, and suppose that the intersection $D_0 \cdot D_1$ consists of a finite number of components F_s. The boundary of ΣF_s consists of the sets $\Sigma_1 \cdot D_0$, $\Sigma_0 \cdot D_1$, $\Sigma_0 \cdot \Sigma_1$, so that we can write

$$(45)\qquad \int K(D_0\cdot D_1)\dot{\Sigma}_1=\int K(\Sigma F_s)\dot{\Sigma}_1$$
$$=\int K(\Sigma_1\cdot D_0)\dot{\Sigma}_1+\int K(\Sigma_0\cdot D_1)\dot{\Sigma}_1+\int K(\Sigma_0\cdot\Sigma_1)\dot{\Sigma}_1.$$

The first two integrals are easily evaluated. Take, for instance, the second integral. For every position of Σ_1 the integrand $K(\Sigma_0\cdot D_1)$ is the integral of Λ over the outward normals to Σ_0 at points of $\Sigma_0\cdot D_1$. This domain of integration can be decomposed in a different way by first fixing a common point of D_1 and Σ_0, rotating D_1 about this point, and then letting this point vary over D_1 and Σ_0 respectively. The result of this iterated integration is

$$(46)\qquad \int K(\Sigma_0\cdot D_1)\Sigma_1=J_nK_0V_1=J_nM_{n-1}{}^{(0)}V_1.$$

Similarly, using the fact that the kinematic density is invariant under the "inversion" of a motion, we have

$$(47)\qquad \int K(\Sigma_1\cdot D_0)\dot{\Sigma}_1=J_nK_1V_0=J_nM_{n-1}{}^{(1)}V_0.$$

To evaluate the third integral in (45) we use the density formula (24), and the formulas (40)-(44) for the total curvature arising from $\Sigma_0\cdot\Sigma_1$. We get

$$\int K(\Sigma_0\cdot\Sigma_1)\dot{\Sigma}_1=\int(D/\sin^{n-2}\phi)d\sigma\dot{V}\dot{\Sigma}_1=(1/J_{n-2})\int(D/\sin^{n-2}\phi)d\sigma\dot{L}_V\dot{\Sigma}_1$$
$$=(1/J_{n-2})\int(\sin\phi)Dd\sigma d\phi\dot{L}_0\dot{L}_1$$
$$=b_{n-2}\int H_0\dot{L}_0\dot{L}_1+\cdots+b_0\int H_{n-2}\dot{L}_0\dot{L}_1$$
$$=a_{n-2}M_{n-2}{}^{(0)}M_0{}^{(1)}+\cdots+a_0M_0{}^{(0)}M_{n-2}{}^{(1)}$$

where the a's and b's are numerical constants. These constants can be determined if we take Σ_0, Σ_1 to be two hyperspheres of radii 1 and h respectively. This completes the proof of the kinematic formula.

UNIVERSITY OF CHICAGO.

REFERENCES.

[1] W. Blaschke, *Integralgeometrie*, Hamburg, 1936 and 1937.

[2] S. Chern and C. T. Yen, "Sulla formula principale cinematica dello spazio ad n dimensioni," *Bolletino della Unione Matematica Italiana* (2), vol. 2 (1940), pp. 434-437.

[3] N. Steenrod, *Fibre Bundles*, Princeton, 1951.

[4] Cf., for instance, Chern, *Topics in Differential Geometry* (Mimeographed notes), Institute for Advanced Study, Princeton, 1951.

Reprinted from *Bull. Amer. Math. Soc.* **58** (1952) 217–250

ÉLIE CARTAN AND HIS MATHEMATICAL WORK

SHIING-SHEN CHERN AND CLAUDE CHEVALLEY

After a long illness Élie Cartan died on May 6, 1951, in Paris. His death came at a time when his reputation and the influence of his ideas were in full ascent. Undoubtedly one of the greatest mathematicians of this century, his career was characterized by a rare harmony of genius and modesty.

Élie Cartan was born on April 9, 1869 in Dolomieu (Isère), a village in the south of France. His father was a blacksmith. Cartan's elementary education was made possible by one of the state stipends for gifted children. In 1888 he entered the "École Normale Supérieure," where he learned higher mathematics from such masters as Tannery, Picard, Darboux, and Hermite. His research work started with his famous thesis on continuous groups, a subject suggested to him by his fellow student Tresse, recently returned from studying with Sophus Lie in Leipzig. Cartan's first teaching position was at Montpellier, where he was "maitre de conférences"; he then went successively to Lyon, to Nancy, and finally in 1909 to Paris. He was made a professor at the Sorbonne in 1912. The report on his work which was the basis for this promotion was written by Poincaré;[1] this was one of the circumstances in his career of which he seemed to have been genuinely proud. He remained at the Sorbonne until his retirement in 1940.

Cartan was an excellent teacher; his lectures were gratifying intellectual experiences, which left the student with a generally mistaken idea that he had grasped all there was on the subject. It is therefore the more surprising that for a long time his ideas did not exert the influence they so richly deserved to have on young mathematicians. This was perhaps partly due to Cartan's extreme modesty. Unlike Poincaré, he did not try to avoid having students work under his direction. However, he had too much of a sense of humor to organize around himself the kind of enthusiastic fanaticism which helps to form a mathematical school. On the other hand, the bulk of the mathematical research which was accomplished at the beginning of this century in France centered around the theory of analytic functions; this subject, made glamorous by the achievement represented

[1] This report was in part published in Acta Math. vol. 38 (1921) pp. 137–145. It should be of considerable historic interest to have now a complete version of this report.

by Picard's theorem, offered many not too difficult problems for a young mathematician to tackle. In the minds of inexperienced beginners in mathematics, Cartan's teaching, mostly on geometry, was sometimes very wrongly mistaken for a remnant of the earlier Darboux tradition of rather hollow geometric elegance. When, largely under the influence of A. Weil, a breeze of fresh air from the outside came to blow on French mathematics, it was a great temptation to concentrate entirely on the then ultra-modern fields of topology or modern algebra, and the ideas of Cartan once more, though for other reasons, partially failed to attract the amount of attention which was their due. This regrettable situation was partly corrected when Cartan's work was taken (at the suggestion of A. Weil) in 1936 to be the central theme of the seminar of mathematics organized by Julia. In 1939, at the celebration of Cartan's scientific jubilee, J. Dieudonné could rightly say to him: " . . . vous êtes un "jeune," et vous comprenez les jeunes"—it was then beginning to be true that the young understood Cartan.

In foreign countries, particularly in Germany, his recognition as a great mathematician came earlier. It was perhaps H. Weyl's fundamental papers on group representations published around 1925 that established Cartan's reputation among mathematicians not in his own field. Meanwhile, the development of abstract algebra naturally helped to attract attention to his work on Lie algebra. However, the reception of his contributions to differential geometry was varied. This was partly due to his approach which, though leading more to the heart of the problem, was unconventional, and partly due to inadequate exposition. Thus Weyl, in reviewing one of Cartan's books [**41**],[2] wrote in 1938:[3] "Cartan is undoubtedly the greatest living master in differential geometry. . . . I must admit that I found the book, like most of Cartan's papers, hard reading. . . ." This sentiment was shared by many geometers.

Cartan was elected to the French Academy in 1931. In his later years he received several other honors. Thus he was a foreign member of the National Academy of Sciences, U.S.A., and a foreign Fellow of the Royal Society. In 1936 he was awarded an honorary degree by Harvard University.

Closely interwoven with Cartan's life as a scientist and teacher has been his family life, which was filled with an atmosphere of happiness and serenity. He had four children, three sons, Henri, Jean,

---

[2] Numbers in brackets refer to the bibliography at the end of the paper.

[3] Bull. Amer. Math. Soc. vol. 44 (1938) p. 601.

and Louis, and a daughter, Hélène. Jean Cartan oriented himself towards music, and already appeared to be one of the most gifted composers of his generation when he was cruelly taken by death. Louis Cartan was a physicist; arrested by the Germans at the beginning of the Résistance, he was murdered by them after a long period of detention. There is no need to say here that Henri Cartan followed in the footsteps of his father to become a mathematician.

Cartan's mathematical work can be roughly classified under three main headings: group theory, systems of differential equations, and geometry. These themes are, however, constantly interwoven with each other in his work. Almost everything Cartan did is more or less connected with the theory of Lie groups.

S. Lie introduced the groups which were named after him as groups of transformations, i.e., as systems of analytic transformations on n variables such that the product of any two transformations of the system still belongs to the system and each transformation of the system has an inverse in the system. The idea of considering the abstract group which underlies a given group of transformations came only later; it is more or less implicit in Killing's work and appears quite explicitly already in the first paper by Cartan. Whereas, for Lie, the problem of classification consisted in finding all possible transformation groups on a given number of variables—a far too difficult problem in the present stage of mathematics as soon as the number of variables is not very small—for Killing and Cartan, the problem was to find all possible abstract structures of continuous groups; and their combined efforts solved the problem completely for simple groups. Once the structures of all simple groups were known, it became possible to look for all possible realizations of any one of these structures by transformations of a specified nature, and, in particular, for their realizations as groups of linear transformations. This is the problem of the determination of the representations of a given group; it was solved completely by Cartan for simple groups. The solution led in particular to the discovery, as early as 1913, of the spinors, which were to be re-discovered later in a special case by the physicists.

Cartan also investigated the infinite Lie groups, i.e., the groups of transformations whose operations depend not on a finite number of continuous parameters, but on arbitrary functions. In that case, one does not have the notion of the abstract underlying group. Cartan and Vessiot found, at about the same time and independently of each other, a substitute for this notion of the abstract group which consists in defining when two infinite Lie groups are to be considered as isomorphic. Cartan then proceeded to classify all possible types of non-isomorphic infinite Lie groups.

Cartan paid also much attention to the study of topological properties of groups considered in the large. He showed how many of these topological problems may be reduced to purely algebraic questions; by so doing, he discovered the very remarkable fact that many properties of the group in the large may be read from the infinitesimal structure of the group, i.e., are already determined when some arbitrarily small piece of the group is given. His work along these lines resembles that of the paleontologist reconstructing the shape of a prehistoric animal from the peculiarities of some small bone.

The idea of studying the abstract structure of mathematical objects which hides itself beneath the analytical clothing under which they appear at first was also the mainspring of Cartan's theory of differential systems. He insisted on having a theory of differential equations which is invariant under arbitrary changes of variables. Only in this way can the theory uncover the specific properties of the objects one studies by means of the differential equations they satisfy, in contradistinction to what depends only on the particular representation of these objects by numbers or sets of numbers. In order to achieve such an invariant theory, Cartan made a systematic use of the notion of the exterior differential of a differential form, a notion which he helped to create and which has just the required property of being invariant with respect to any change of variables.

Raised in the French geometrical tradition, Cartan had a constant interest in differential geometry. He had the unusual combination of a vast knowledge of Lie groups, a theory of differential systems whose invariant character was particularly suited for geometrical investigations, and, most important of all, a remarkable geometrical intuition. As a result, he was able to see the geometrical content of very complicated calculations, and even to substitute geometrical arguments for some of the computations. The latter practice has often been baffling to his readers. But it is an art whose presence is usually identical with the vigor of a geometrical thinker.

In the 1920's the general theory of relativity gave a new impulse to differential geometry. This gave rise to a feverish search of spaces with a suitable local structure. The most notable example of such a local structure is a Riemann metric. It can be generalized in various ways, by modifying the form of the integral which defines the arc length in Riemannian geometry (Finsler geometry), by studying only those properties pertaining to the geodesics or paths (geometry of paths of Eisenhart, Veblen, and T. Y. Thomas), by studying the properties of a family of Riemann metrics whose fundamental forms differ from each other by a common factor (conformal geometry), etc. While in all these directions the definition of a parallel displacement is considered to be the major concern, the approach of Cartan to these problems is most original and satisfactory. Again the notion of group plays the central rôle. Roughly speaking, a generalized space (espace généralisé) in the sense of Cartan is a space of tangent spaces such that two infinitely near tangent spaces are related by an infinitesimal transformation of a given Lie group. Such a structure is known as a connection. The tangent spaces may not be the spaces of tangent vectors. This generality, which is absolutely necessary, gave rise to misinterpretation among differential geometers. As we shall show below, it is now possible to express these concepts in a more satisfactory way, by making use of the modern notion of fiber bundles.

We can perhaps conclude from the above brief description that Cartan's mathematical work, unlike that of Poincaré or Hadamard, centers around a few major concepts. This is partly due to the richness of the field, in which his pioneering work has opened avenues where much further development is undoubtedly possible. While many of Cartan's ideas have received clarification in recent years, the difficulties of conceiving the proper concepts at the early stage of development can hardly be overestimated. Thus in writing on the psychology of mathematical thinking, Hadamard had to admit "the insuperable difficulty in mastering more than a rather elementary and superficial knowledge of Lie groups."[4] Thanks to the development of modern mathematics, such difficulties are now eased.

Besides several books Cartan published about 200 mathematical papers. It is earnestly to be hoped that the publication of his collected works may be initiated in the near future. Not only do they fully deserve to find their place on the bookshelves of our libraries at the side of those of other great mathematicians of the past, but they will be, for a long time to come, a most indispensable tool for all those who will attempt to proceed further in the same directions.

We now proceed to give a more detailed review of some of the most important of Cartan's mathematical contributions.

I. Group theory

Cartan's papers on group theory fall into two categories, distinguished from each other both by the nature of the questions

[4] J. Hadamard, *The psychology of invention in the mathematical field*, Princeton, 1945, p. 115.

treated and by the time at which they were written. The papers of the first cycle are purely algebraic in character; they are more concerned with what are now called Lie algebras than with group theory proper. In his thesis [3], Cartan gives the complete classification of all simple Lie algebras over the field of complex numbers. They fall into four general classes (which are the Lie algebras of the unimodular groups, of the orthogonal groups in even or odd numbers of variables, and of the symplectic groups) and a system of five "exceptional" algebras, of dimensions 14, 52, 78, 133, and 248. Killing had already discovered the fact that, outside the four general classes, there can exist only these five exceptional Lie algebras; but his proofs were incorrect at several important points, and, as to the exceptional algebras, it is not clear from his paper whether he ever proved that they actually existed. Moreover, in his work, the algebra of dimension 52 appears under two different forms, whose equivalence he did not recognize. Cartan gave rigorous proofs that the classification into four general classes and five exceptional algebras is complete, and constructed explicitly the exceptional algebras.

Let $\mathfrak{g}$ be any Lie algebra; to every element X of $\mathfrak{g}$ there is associated a linear transformation, the adjoint ad X of X, operating on the space $\mathfrak{g}$, which transforms any element Y of $\mathfrak{g}$ into $[X, Y]$. Because of the relation $[X, X]=0$, this linear transformation always admits 0 as a characteristic root; those elements X of $\mathfrak{g}$ for which 0 is a characteristic root of least possible multiplicity of ad X are called regular elements. Let H be a regular element; then those elements of $\mathfrak{g}$ which are mapped into 0 by powers of ad H are seen to form a certain subalgebra $\mathfrak{h}$ of $\mathfrak{g}$, and this subalgebra is always nilpotent (which means that, in the adjoint representation of such an algebra, every element has 0 as its *only* characteristic roots). A subalgebra such as $\mathfrak{h}$ has been called a Cartan subalgebra of $\mathfrak{g}$. It is a kind of inner core of the algebra $\mathfrak{g}$, and many properties of the big algebra $\mathfrak{g}$ are reflected in properties of this subalgebra $\mathfrak{h}$. In the case where $\mathfrak{g}$ is semi-simple, $\mathfrak{h}$ is always abelian (which means, for a Lie algebra, that $[X, Y]$ is always 0 for any X and Y in the algebra). Moreover, $\mathfrak{g}$ has a base which is composed of elements which are eigenvectors simultaneously for all adjoint operations of elements of $\mathfrak{h}$. The factors by which these elements are multiplied when bracketed with elements of $\mathfrak{h}$ are called the *roots* of the Lie algebra; it is the study of the properties of these roots which leads to the classification of simple Lie algebras. In establishing these properties, Cartan made a systematic use of the "fundamental quadratic form" of $\mathfrak{g}$, whose value at an element X is the trace of the square of ad X (if $\mathfrak{g}$ is semi-simple, or more generally if it coincides with its derived algebra, then the trace of ad X itself is always zero). One of the most important results of Cartan's thesis is that a necessary and sufficient condition for $\mathfrak{g}$ to be semi-simple is that its fundamental quadratic form be nondegenerate (i.e. that its rank be equal to the dimension of $\mathfrak{g}$). Incidentally, Cartan also applied similar methods to the study of systems of hypercomplex numbers (cf. [**4**]) and obtained in this manner the main structure theorems for associative algebras over the fields of real and of complex numbers; however, these results were superseded by the work of Wedderburn, which applies to algebras over arbitrary basic fields. By studying those algebras which have only one integrable (or, as we say now, solvable) ideal, Cartan also laid the foundations in his thesis for his subsequent study of linear representations of simple Lie algebras; in particular, he determined, for each class of simple groups, the linear representation of smallest possible degree.

The general theory of linear representations is the object of the paper [**5**]. As above let $\mathfrak{g}$ be any semi-simple Lie algebra over the field of complex numbers (any algebraically closed field of characteristic 0 would do just as well); a linear representation of $\mathfrak{g}$ is a law which assigns to every X in $\mathfrak{g}$ a linear transformation $\rho(X)$ on some finite-dimensional space; $\rho(X)$ depends linearly on X, and is such that $\rho([X, Y]) = \rho(X)\rho(Y) - \rho(Y)\rho(X)$ for any X and Y in $\mathfrak{g}$. Let $\mathfrak{h}$ be a Cartan subalgebra of $\mathfrak{g}$. Then it turns out that the matrices which represent the elements of $\mathfrak{h}$ may all be reduced simultaneously to the diagonal form; the diagonal coefficients which occur in these matrices, considered as linear functions of the element which is represented, are called the *weights* of the representation. The roots of $\mathfrak{g}$ are the weights of a particular linear representation, viz. the adjoint representation. Cartan proved that all relations between weights of one or several representations are consequences of certain linear relations with rational coefficients between these weights, a fact which can now be explained in two different manners: it reflects the properties of characters of compact abelian groups, and also the properties of algebraic groups of linear transformations. Cartan then introduces an order relation in the system of all weights and roots, and proves that any irreducible representation is uniquely determined by its highest weight for this order relation. The problem of finding all irreducible linear representations of $\mathfrak{g}$ is thereby reduced to that of finding all possible highest weights of representations. The sum of the highest weights of two irreducible representations is again the highest weight of an irreducible representation, which is contained in the tensor product (or rather, sum, if we speak of representations of Lie algebras and not of groups) of the two given representations. If r is the rank of the Lie algebra $\mathfrak{g}$ (i.e. the dimension of any Cartan subalgebra of $\mathfrak{g}$), Cartan established that all possible highest weights of irreducible representations may be written as linear combinations with non-negative integral coefficients of r particular linear functions which depend only on the structure of $\mathfrak{g}$ and the order relation in the system of roots. Considering one by one the various types of simple Lie algebras, he established that every one of these r basic functions is the highest weight of some irreducible representation; this led to a complete classification of all irreducible linear representations of simple Lie algebras. This theory of linear representations was later completed in an important point by H. Weyl, who established by transcendental methods that every representation of a semi-simple Lie algebra is completely reducible, and who expressed the degree of an irreducible representation in terms of its highest weight. It has also recently become possible to give a direct proof of the existence of irreducible representations corresponding to the possible highest weights predicted by Cartan's theory, a proof which applies not only to simple but also to semi-simple algebras (Lie algebras behave differently from associative algebras in this respect that algebras which are not simple may have faithful irreducible representations; this happens for instance for the Lie algebra of the orthogonal group in 4 variables). In the process of classifying all possible linear representations, Cartan discovered the spin representations of the orthogonal Lie algebras, which later played such an important rôle in physics. In a book published later (*Leçons sur la théorie des spineurs*, Hermann, Paris, 1938), Cartan developed the theory of spinors from a geometric point of view.

In [6] Cartan classifies all simple Lie algebras over the field of real numbers instead of that of complex numbers. The method is to study the "complexification" of the Lie algebra under consideration; this complexification is either simple or the sum of two simple Lie algebras, and there is defined in it an operation of passing to the imaginary conjugate which admits the elements of the given simple Lie algebra as fixed elements. Starting with a suitable complex algebra, Cartan determines all possible operations of conjugation in it and arrives in this manner at a complete classification of simple real Lie algebras (the method was later simplified by himself and by others, making use of the compact real forms and replacing the determination of all possible conjugations by that of all classes of involutive automorphisms of the compact form).

It turned out that, but for one exception, the structure of a simple real Lie algebra is characterized by that of its complexification and by its *character*, i.e., by the index of inertia of its fundamental quadratic form. Cartan noticed that, for every complex form, there is a unique real form whose fundamental quadratic form is negative definite; this real form is the Lie algebra of a compact Lie group. This fact was to play a very important rôle in the subsequent theory of Lie groups, because it establishes a one-to-one correspondence between semi-simple connected complex Lie groups and compact semi-simple Lie groups. While the former are more readily amenable to an algebraic study, because of the algebraically closed character of the basic field, the latter lend themselves more easily to study by transcendental methods because the volume of the whole group (in the sense of Haar measure) is finite.

The last paper of the first cycle is [7] in which Cartan determined all real linear representations of simple real Lie algebras.

The work of Cartan's second group-theoretic period is concerned with the groups themselves, and not with their Lie algebras, and in general with the global aspect of the group. This period opens with a paper [8] which contains a study of group manifolds from a local differential geometric point of view. A group G may operate on itself in three different ways: by the left translations, by the right translations, and by transformation (we call here transformation a mapping $t \to sts^{-1}$, where s is a fixed element of the group). In relation to this, Cartan shows that there are three affine connections which are intrinsically defined on G. Two of these connections (those which correspond to the left and right translations) are without curvature but in general have torsion; the third one has no torsion but has in general a curvature. The geodesic lines are the same for all three connections: they are the cosets with respect to the one-parameter subgroups. Cartan determines also the totally geodesic varieties on G (varieties such that any geodesic which is tangent to it is entirely contained in it); they are of two different kinds. The varieties of the first kind are the subgroups of G and their cosets. The varieties of the second kind are determined by what have since been called the Lie triple systems contained in the Lie algebra $\mathfrak{g}$ of G, i.e., the linear subspaces of $\mathfrak{g}$ which, together with three elements X, Y, and Z, contain the element $[[X, Y], Z]$.

After the paper we have just mentioned, Cartan's interest orients itself very definitely towards the topological study of Lie groups in the large.

This period begins around 1925, at the time when H. Weyl had just published his fundamental papers on the theory of compact Lie groups. It is difficult to appreciate to what extent Cartan was influenced by Weyl's methods and results; at any rate his book *Leçons sur la Géométrie des espaces de Riemann* shows clearly that, even before Weyl's paper, Cartan was already getting more and more interested in topological questions. Whereas Weyl's line of attack was, if we may say so, brutally global, depending essentially on the method of integration on the whole group, the work of Cartan puts the emphasis on the connection between the local and the global. This essential difference has a great bearing on the nature of the results it is possible to expect from these two methods. Weyl's methods are not bound to the differentiable structure of the group under consideration; as soon as the possibility of integrating on any locally compact group was established Weyl's results could be extended to all compact topological groups. However, the assumption of compactness is essential (it insures the convergence of the integrals on the group), and Weyl's methods give nothing on noncompact groups, whereas those of Cartan, applied in the domain of Lie groups, have led to a very complete knowledge of the topology of these groups, whether compact or not.

In his paper [9] Cartan studies the topology of compact semi-simple Lie groups and of their complexifications. Let G be a compact semi-simple Lie group, $\mathfrak{g}$ the Lie algebra of G, and $\mathfrak{h}$ a Cartan subalgebra of $\mathfrak{g}$. Cartan establishes that every element of G belongs to a one-parameter subgroup (at least), and that every infinitesimal transformation may be transformed by an operation of the adjoint group into an element of $\mathfrak{h}$. Every element H of $\mathfrak{h}$ gives rise to a one-parameter subgroup with a definite parameter on it; let us denote by $\exp H$ the point of parameter 1 on this subgroup. Then every element of G is conjugate to some element of the form $\exp H$; the next question is to find out under which condition two elements $\exp H$ and $\exp H'$ of this form are conjugate to each other. Cartan shows that a necessary and sufficient condition for this to happen is that H' can be transformed into H by an operation of a certain discontinuous group S operating on the space $\mathfrak{h}$. This group admits a fundamental domain which is a polyhedron P with a finite number of vertices. The points of P, with suitable identification of faces, will represent in a one-to-one manner the classes of conjugate elements of G. The inner points of P correspond to the regular elements of the group, i.e., to those elements which may be represented in the form $\exp H$, where H is a regular element of $\mathfrak{h}$. It follows that any closed path in G which does not meet the set of singular (i.e., not regular) elements will be represented by a continuous path in P, not necessarily closed. Now, it turns out that the singular elements of G form a set of dimension at least 3 less than the dimension of the whole group; as a consequence they may be entirely disregarded in the determination of the fundamental group. This allows one to proceed to the determination of this fundamental group π by the mere consideration of the polyhedron P itself; the order of π is the number of vertices H of P such that exp H is the unit element of G. From this follows Weyl's theorem to the effect that the fundamental group of a compact semi-simple Lie group is finite. This implies that, if the fundamental quadratic form of a Lie group G is definite negative, then not only does there exists at least one compact group which is locally isomorphic with G, but G itself is compact. Moreover Cartan's methods allow one to determine, for every semi-simple infinitesimal structure with a negative definite fundamental quadratic form, the number of times that the simply-connected group with this structure covers the adjoint group, and to study the various types of closed one-parameter groups (or geodesics) in the space of a compact group. The last part of the paper, devoted to the study of simple complex groups, is a prelude to the future theory of noncompact simple groups; it is proved in particular that, for any simple Lie algebra over the complex numbers, there always exists a simply-connected complex *linear* group having this Lie algebra.

After the determination of the fundamental group was accomplished, the next step was to determine the higher-dimensional Betti numbers of compact Lie groups. The method of accomplishing this was indicated in Cartan's paper [11]. Cartan considers a homogeneous space E whose group of transformations G is compact; the space may then be considered to be the space of cosets of G modulo some closed subgroup g. Let us say that two exact differential forms are equivalent when their difference is the differential of a form of degree $p-1$, and let b_p be the maximal number of forms of degree p no linear combination of which is equivalent to 0. It was conjectured, but not yet proved at the time, that b_p is equal to the pth Betti number of the space E (this was established soon afterwards by de Rham). Let ω be any exact form of degree p. Any operation s of the group G transforms into a new exact form $s\omega$, which Cartan proves to be equivalent to ω. He then constructs the average of the form $s\omega$, s running over all elements of the group G; this new form is still exact, is equivalent to ω, and is furthermore invariant under the group G. Moreover, Cartan proves by a similar argument that, if ω is invariant and equivalent to 0, then ω is the differential of a form which is itself invariant. These theorems reduce the determination of the number b_p to that of the integral invariants of the space E. Cartan then shows that the latter problem may be reduced to a purely algebraic problem depending only on the Lie algebra of the group G and the subalgebra corresponding to the subgroup g (this, at any rate, in the case where the group g is connected). This algebraic problem has been solved since for the case where E is either the group G itself or a symmetric Riemann space with G as its group of isometric transformations.

We now come to the results obtained by Cartan in the study of Lie groups which are not compact. The proof he gave of the converse of Lie's third theorem (i.e., every Lie algebra over the field of real numbers is the Lie algebra of some group) implied that every simply-connected Lie group is topologically the product of a Euclidean space by the space of a simply connected semi-simple group. The problem was therefore reduced to the special case in which the group G under consideration is semi-simple. Let then $\mathfrak{g}$ be its Lie algebra, and let $\mathfrak{g}'$ be the complexification of $\mathfrak{g}$. Since $\mathfrak{g}'$ is semi-simple, it has a compact real form $\mathfrak{g}_c$; i.e., it may be considered as the complexification of a Lie algebra $\mathfrak{g}_c$ which is the algebra of a compact semi-simple group. Cartan proves (in his paper [10]) that there exists an involutory automorphism a of $\mathfrak{g}_c$ such that $\mathfrak{g}$ is spanned by those elements of $\mathfrak{g}_c$ which are invariant under a and by the products by i of those elements of $\mathfrak{g}_c$ which are changed into their opposites by a. The subalgebra $\mathfrak{g}\cap\mathfrak{g}_c$ is the Lie algebra of a maximal compact subgroup g of G. Cartan then considers the space R whose points are the conjugates of g in G; the essential fact is that the existence of the automorphism a implies that R is a symmetric Riemannian space, having the adjoint group Γ of G as its group of isometries (more precisely, Γ is the component of the identity in the group of isometries of R). This opened the way to the application of the theory of symmetric Riemannian spaces which Cartan had already developed for its own merits (cf. the part of this article which is concerned with the geometric aspects of Cartan's work). Assume that G is Γ itself. Then every operation of G may be decomposed uniquely into a rotation around the origin (the origin in R being the point which represents the group g itself) and a transvection. The transvections are the operations of G which result from the integration of those infinitesimal transformations which are changed into their negatives by the operation a. They form a subvariety of G, homeomorphic to R, and it turns out that this variety has the topological structure of a Euclidean space. This establishes that the adjoint group of any connected semi-simple Lie group is topologically the product of the space of a compact group by a Euclidean space. Moreover, making use of the fact that any compact group of isometric transformations of the space R admits a fixed point, Cartan proves that every compact subgroup of G is conjugate to a subgroup of g. These results relative to the adjoint groups may be extended without any difficulty to the simply-connected groups with the same infinitesimal structure; they are also true for all intermediary groups, as follows from the recent work of Iwasawa.

II. Systems of differential equations

The principal paper of Cartan on the theory of differential systems is [15]. The reader will find a very clear exposition of the theory of Pfaffian systems in Cartan's book: *Les systèmes différentiels extérieurs et leurs applications géométriques*, Paris, Hermann, 1945.

A Pfaffian system is a system consisting of a certain number of equations of the form $A_1dx_1 + \cdots + A_ndx_n = 0$, where $A_1, \cdots, A_n$ are functions of $x_1, \cdots, x_n$, and possibly of certain equations of the form $F(x_1, \cdots, x_n) = 0$. A parametric r-dimensional manifold, given by $x_i = f_i(t_1, \cdots, t_r)$ $(1 \leqq i \leqq n)$ is a solution of the system if the equations of the system become identically satisfied when one replaces the variables x and their differentials by their expressions in terms of the variables t and their differentials.

It is first necessary to indicate how any system of differential equations may be reduced to a Pfaffian system. If the system contains equations of order higher than one, we may first reduce it to the order one by introducing new unknown functions which represent certain derivatives of the original ones. This being done, we obtain a system of equations of the form $F_i(x_1, \cdots, x_m; z_1, \cdots, z_p; \cdots, \partial z_r/\partial x_s, \cdots) = 0$ where the z's are the unknown and the x's the independent variables. If we set $\partial z_r/\partial x_s = t_{rs'}$, the original system of partial differential equations may be replaced by the Pfaffian system composed of the equations $F_i(x; z; t) = 0$, $dz_r - \sum_s t_{rs}dx_s = 0$. The solutions of the original system correspond to those solutions of the Pfaffian system which are manifolds of dimension r on which the variables $x_1, \cdots, x_n$ are independent.

The essential originality of Cartan consists in having introduced, besides the Pfaffian forms, the exterior differential forms of higher degree. The algebra of exterior forms had been developed by Grassmann for geometric purposes; before it could be used in the theory of differential systems, it was necessary to introduce the operation of exterior differentiation. An exterior differential form is an expression of the form $\sum A_{i_1 \cdots i_p} dx_{i_1} \cdots dx_{i_p}$, where the coefficients $A_{i_1 \cdots i_p}$ are functions of the variables x; such expressions may be multiplied with each other with the convention that $dx_i dx_j = -dx_j dx_i$ (in particular, $(dx_i)^2 = 0$). The exterior differential of the form written above is $\sum dA_{i_1 \cdots i_p} dx_{i_1} \cdots dx_{i_p}$ where it is understood that the differentials $dA_{i_1 \cdots i_p}$ are expressed as linear combinations of $dx_1, \cdots, dx_n$. The fundamental property of the operation of differentiation is that it is invariant with respect to any change of variables. Now, let us consider any Pfaffian system $\omega_1 = 0, \cdots, \omega_h = 0$, $F_1 = 0, \cdots, F_m = 0$, where $F_1, \cdots, F_m$ are functions and $\omega_1, \cdots, \omega_h$ Pfaffian forms; then we see immediately that any solution of the system will also be a solution of the system obtained by adjoining to the original one the equations $dF_1 = 0, \cdots, dF_m = 0$, $d\omega_1 = 0, \cdots, d\omega_h = 0$. More generally, let I be the smallest set of differential forms containing $F_1, \cdots, F_m, \omega_1, \cdots, \omega_h$ and such that, whenever ω and ω' are in I, then $\omega + \omega'$ is in I, the product of ω by an arbitrary differential form is in I, and $d\omega$ is in I; I is called the differential ideal generated by $F_1, \cdots, F_m, \omega_1, \cdots, \omega_h$. Then any solution of the original system will be a solution of the system obtained by equating to 0 all forms in I. The operation of adjoining to a system the exterior differentials of its forms is an invariant counterpart of the method of obtaining conditions of compatibility by writing that certain higher derivatives which may be computed in two different ways by means of the equations of the system have the same value.

By a contact element E_p of dimension p in the Cartesian n-space R^n is meant a pair (M, P) formed by a point M of R^n and a p-dimensional linear subspace P of R^n going through M. It is sometimes necessary to generalize this notion to the case of an arbitrary n-dimensional manifold V instead of R^n; M is then any point of V, and P any p-dimensional vector subspace of the n-dimensional tangent vector space to V at M. The totality of all p-dimensional contact elements of a given manifold V is itself a manifold V_1, the so-called *first prolonged manifold* of V. If $x_1, \cdots, x_n$ are local coordinates at the origin of the contact element (M, P), P may be represented parametrically by equations of the form $dx_i = L_i(v_1, \cdots, v_p)$, where the L_i's are linear forms in p parameters $v_1, \cdots, v_p$. Let ω be a differential form; if we substitute the coordinates of the origin of a contact element E_p for the variables $x_1, \cdots, x_n$ in the coefficients of ω, and the linear forms L_i for the differentials dx_i which occur in ω, we obtain an exterior form in the variables v. If this form is 0, then we say that ω is 0 at E_p. If we are given a differential ideal I of differential forms, and if every form of I is 0 at E_p, then we say that E_p is *an integral element* of I.

Let W be a p-dimensional submanifold of V. If M is a point of W, we may represent W locally around M by equations of the form $x_i = f_i(u_1, \cdots, u_p)$, the u_i's being parameters. The contact element (M, P) formed by M and by the p-dimensional tangent space P to W at N is called the tangential element of W at M; the space P may be represented by the equations $dx_i = df_i$, the du_i's taking the place of the parameters v_i considered above. A solution (or integral manifold) of the system obtained by equating to 0 the forms of a differential ideal I is a manifold W whose tangential contact elements are integral elements of I. The problem of looking for such solutions may be decomposed into two parts: the determination of all integral elements, which is an algebraic problem, and the determination of the ways of grouping these integral elements together in such a way that they may be the tangential elements of some manifold.

The ideal I may contain forms of degree 0, i.e., functions of the variables. Assume that these functions, equated to 0, represent an irreducible analytic manifold V_0. Assume that any point of V_0 is the origin of ∞^{r_1} integral elements of dimension 1. There may be points of V which are the origins of more than ∞^{r_1} integral elements of dimension 1, but they form lower-dimensional submanifolds of V. A point of V which does not lie on any one of these submanifolds is called an *ordinary* point. The smallest manifold (in the space of integral elements of dimension 1) which contains the integral elements whose origins are ordinary points is called the manifold of *general integral elements* of dimension 1; the origin of a general integral element is not necessarily ordinary. Assume now that any general integral element of dimension 1 is contained in ∞^{r_2} integral elements of dimension 2; then those integral elements of dimension 1 which are not contained in more than ∞^{r_2} integral elements of dimension 2 are called ordinary; proceeding as above, we define the notion of a general integral element of dimension 2. We may continue in the same manner, and define inductively the integers $r_1, r_2, \cdots, r_n$. For a certain dimension n, the r_{n+1} will be 0, which means that not every general integral element of dimension n will be contained in an integral element of dimension $n+1$. The number n is called the *genus* of the system, and the system is said to be *in involution* for every dimension not greater than n. A p-dimensional general integral element E_p is called *regular* if there exists a chain $E_0 \subset E_1 \subset \cdots$

$\subset E_{p-1} \subset E_p$ where, for each $i<p$, E_i is an ordinary *general* integral element of dimension i. A p-dimensional *general* solution of I is a p-dimensional manifold whose tangential contact elements are general integral elements of I, at least one of these integral elements being regular. In this manner, Cartan succeeded for the first time in giving a precise definition of the notion of the general solution of any differential system. The existence theorem for general solutions states that any regular p-dimensional integral element E_p is a tangential contact element of some manifold which is a solution of I. More precisely, if E_{p-1} is a regular element of dimension $p-1$ contained in E_p and is a tangential element of a $(p-1)$-dimensional integral manifold V^{p-1} of I, then V^{p-1} is contained in at least one integral manifold V^p tangent to E_p. This general theorem allowed Cartan to determine exactly the degree of indetermination of the general solution (i.e. on how many arbitrary constants, arbitrary functions of 1, 2, $\cdots$ arguments, it depends). Its application is however limited to the consideration of *analytic* differential systems and to the determination of *analytic* solutions.

The next step was to try to determine the *singular solutions* of the system, i.e. the solutions which are not in the general solution (for instance, for a differential equation in the plane, the envelope of the general solutions). Here the idea of Cartan was to construct from the given differential system new systems which are obtained from it by *a method of prolongation*, in such a manner that any singular solution of the original system should become a general solution of one of these new systems. The method consists, generally speaking, in adjoining new unknowns which are the coordinates of integral elements which are not general, and constructing *a priori* the finite and differential equations these new unknowns must satisfy. However, an exact description of the method would be a little too long to be given here. In every concrete case in which it was applied, Cartan's method led to the complete determination of all singular solutions. But a general proof that it always does so is still missing; this is a theme of research which would richly deserve to attract the attention of ambitious young mathematicians.

One of the main applications of Cartan's theory of systems of differential equations is his theory of infinite groups of transformations (cf. [16; 17; 18; 19]). We touch here a branch of mathematics which is very rich in results but which very badly needs clarification of its foundations. For the infinite Lie groups, in spite of their name, are probably no groups at all, in the precise sense the word has received in modern algebra; what they really are is not clear yet. Lie defined them as follows: he considers a set of analytic transformations $x'_i = F_i(x_1, \cdots, x_n)$ on n variables which is closed with respect to the ordinary operations of forming the product of two transformations and taking the inverse of a transformation of the set, and which has furthermore the property of being composed of all transformations of the form indicated above for which the functions F_i satisfy a certain system of partial differential equations. The hitch is of course that nothing is said about the domains in which the transformations are to be defined and invertible, and that this domain may apparently vary from one transformation to another. Cartan establishes that any Lie group, whether finite or infinite, may be defined (after possible adjunction of new variables, which transform in a suitable way when the original variables are transformed by an operation of the group) to be the group of all transformations which leave invariant a certain number of functions and Pfaffian forms. A simple but not typical example is the group of transformations of the form $x' = F(x)$, $y' = G(y)$ on two variables x and y (where F and G are arbitrary analytic functions). It may be considered to be the group which leaves invariant the two Pfaffian forms udx and vdy, where u and v are new variables, which are transformed as follows: we have $u' = u(dF/dx)^{-1}$, $v' = v(dG/dx)^{-1}$. Having written a group in the form we have just indicated, Cartan was able to extend to infinite groups the structure theory which Lie had developed for finite groups. Assume that we have a group G which is defined by the conditions that some of the variables, say $x_{r+1}, \cdots, x_n$, are invariant and that some Pfaffian forms $\omega_1, \cdots, \omega_h$ are invariant, the ω_i's containing only the differentials of the variables $x_1, \cdots, x_n$ but their coefficients involving possibly certain other variables u. Then we may write

$$d\omega_i = \sum c_{ijk}\omega_j\omega_k + \sum a_{ijk}\omega_j\eta_k$$

where the η_k's are certain linear combinations of the differentials of the auxiliary variables u. Cartan shows that it is always possible to assume that the coefficients c_{ijk}, a_{ijk} depend only on the invariants $x_{r+1}, \cdots, x_n$ (if the group may be put into a form where it is transitive, which always happens for finite-dimensional groups, then the c_{ijk}'s and a_{ijk}'s are constants). These coefficients define the *structure* of the group. Just as in the case of finite-dimensional groups, they cannot be taken arbitrarily; Cartan gave the conditions they must satisfy in order to define a group, thus generalizing to infinite groups Lie's third fundamental theorem.

The operation of adjoining new variables to those which are transformed by a group is called the prolongation of the group. Cartan says that two groups are isomorphic to each other if they admit prolongations which are similar, i.e., which can be deduced from each other by a change of independent variables. He showed how it is possible to recognize whether two infinite groups whose structures are known are isomorphic or not. He applied this method to the problem of the classification of simple infinite groups, and found that they fall into 8 general types.

Cartan's theory of infinite groups had its origin in the study he made of equivalence problems. The general problem can be formulated as follows: Let G be a linear group acting in a space of n dimensions. Let $\theta_1, \cdots, \theta_n$ and $\bar{\theta}_1, \cdots, \bar{\theta}_n$ be two sets of linearly independent Pfaffian forms in the variables $x_1, \cdots, x_n$ and $\bar{x}_1, \cdots, \bar{x}_n$, respectively. Determine whether there exists an admissible transformation of coordinates

$$\bar{x}_i = \bar{x}_i(x_1, \cdots, x_n), \qquad i = 1, \cdots, n,$$

such that

$$\bar{\theta}_i = \sum_{j=1}^{n} a_{ij}(x)\theta_j, \qquad i = 1, \cdots, n,$$

where the linear transformation belongs to G. To treat this problem let $u_1, \cdots, u_m$ be the parameters of G and let us introduce the Pfaffian forms

$$\omega_i = \sum_{j=1}^{n} a_{ij}(u)\theta_j,$$

$$\bar{\omega}_i = \sum_{j=1}^{n} a_{ij}(\bar{u})\bar{\theta}_j, \qquad i = 1, \cdots, n,$$

in which we regard the u's and the $\bar{u}$'s as auxiliary variables. The sets of forms θ_i and $\bar{\theta}_i$ are equivalent in the above sense if and only if $\bar{x}^i$, $\bar{u}^r$ can be determined as functions of x^j, u^s ($i, j=1, \cdots, n$; $r, s=1, \cdots, m$) so that

$$\bar{\omega}_i = \omega_i \qquad i = 1, \cdots, n.$$

Such a system may be discussed by the general methods for dealing with Pfaffian systems. The first step is, of course, to adjoin to the system the equations $d\bar{\omega}_i = d\omega_i$; if we express the forms $d\omega_i$ by means of the forms ω_i themselves and of the differentials of the auxiliary variables, the coefficients of these expressions, when they contain the variables x only, will yield invariants $I_k(x)$, and we may enlarge the original system by adjoining the equations $I_k(\bar{x}) = I_k(x)$, together with the equations which are obtained from them by differentiation.

Cartan shows that the continuation of this procedure eventually leads to a complete system of invariants which can be obtained by operations of differentiation only. However, when the systems one is led to consider are not in involution for the dimension n, the completeness of the system of invariants depends on the theorem that all singular solutions of a differential system may be obtained by the method of prolongation, a theorem which is not completely proved as yet (cf. above).

Among other applications which Cartan made of his theory of differential equations we mention the following: (1) Various applications to differential-geometric problems; (2) Principle of integral invariants in analytical dynamics; (3) Theory of general relativity.

As a matter of fact, the study of differential equations arising from problems of differential geometry had always interested him, and his papers on this subject run through most of his scientific career. The numerous examples given in [23] show quite decisively the advantages of using differential forms. One of the notable results is his proof of a conjecture of Schläfli to the effect that every Riemann metric of n dimensions can be imbedded locally into a Euclidean space of dimension $n(n+1)/2$ [22]. This theorem played no small part in Levi-Civita's original definition of his parallelism and has attracted the attention of differential geometers.

Also his work on integral invariants in analytical dynamics can be considered as an application of the theory of differential equations [21]. Mathematically the problem is to determine the trajectories, which are to be solutions of a differential system of the type

$$\frac{dx_i}{dt} = X_i(x_1, \cdots, x_n, t), \qquad i = 1, \cdots, n.$$

The standard way is by means of Hamilton's principle, which defines the trajectories as the extremals of a certain variational problem. Unfortunately the integrand of the latter does not have a simple physical interpretation. An alternative way was suggested by Poincaré. He called a multiple integral

$$\int \cdots \int_p \sum_{i_1, \cdots, i_p} a_{i_1 \cdots i_p}(x_1, \cdots, x_n, t) dx_{i_1} \cdots dx_{i_p}$$

invariant when its value over a domain covered by the trajectories is invariant under the motion. In fact, it is called absolute if the domain is arbitrary and relative if the invariance is true only for closed domains. If p_i, q_i, $i=1, \cdots, n$, are the canonical variables of a dynamical system with n degrees of freedom, Poincaré's principle asserts that the trajectories can be characterized as the curves admitting the relative integral invariant

$$\sum_{i=1}^{n} p_i dq_i.$$

Cartan's principle is a modification of Poincaré's. He derived his ideas from his theory of differential systems. The differential system of the trajectories has $2n-2$ functionally independent first integrals. Cartan observed that the property of an exterior differential form to be invariant and thus to depend only on the trajectories is that it is a form in these first integrals. Expressed in terms of the original variables, it may involve the independent variable t. Omitting from this the terms involving dt, we obtain the integrand of an invariant integral in the sense of Poincaré. Thus the latter is the truncated form of an invariant differential form (i.e., a form in the first integrals of the trajectories). Conversely, it can be proved that, given the integrand of an invariant integral of Poincaré, terms involving dt can be added to it so as to obtain an invariant differential form. Cartan's principle characterizes the trajectories as admitting an invariant differential form. Moreover, the latter has a simple physical interpretation. The work therefore furnishes an interesting complement to formal dynamics.

In connection with the general theory of relativity and the unified field theory Cartan studied on several occasions the question of the possible forms of the equations of gravitation and of the unified gravitational and electro-magnetic field. He made a very detailed analysis and determined all possible forms of such differential systems. He was also the first one to introduce the notion of Riemann spaces without curvature and with torsion, which later served as the basis of Einstein's unified field theory. Apparently these studies are not of the same importance as his studies on pure mathematics.

III. Geometry

Although the theory of Lie groups has an intimate relationship with differential geometry, Cartan did not begin his substantial work on differential geometry until a relatively late stage. His first series of papers on differential geometry was concerned with the problem of deformation [27; 28]. It is clear that he had then all the essential ideas of the method of moving frames, one of his favorite subjects in later years, which has not been fully exploited even now.

The method was not new. It is a generalization, to an arbitrary homogeneous space, of the method of moving trihedrals, so successfully used by Darboux, Ribaucour, and others [39; 41]. Even in the most general case some of its essential ideas had been given by Emile Cotton. It is also closely related to Cesaro's "intrinsic method" in differential geometry, as later developed by Kowalewski. To Cartan the attraction was not the method, but the geometrical results to which it so effectively leads. It is interesting to notice in his book [41] how he took pleasure in studying numerous examples and did not care to discuss the generalities, except in very sketchy outlines.

We attempt to give a description of this method in modern terminology. The problem is the local theory of a p-dimensional submanifold M in a homogeneous space E of dimension n, acted on by a Lie group G of dimension r. Let O be a point of E and H the subgroup of G leaving O fixed. Then the set of all transformations of G carrying O to a point P of E is a left coset gH of G relative to H, and E can be identified with the space of left cosets G/H. Under this identification the action of G on E is represented by left multiplication. This process depends on the choice of O. If we replace O by O' and if g_O is a transformation of G carrying O' to O, the subgroup of G leaving O' fixed will be $g_O^{-1}Hg_O$ and the set of transformations of G carrying O' to P will be gHg_O. In other words, the latter is defined up to multiplication by a fixed element to the right.

The method of moving frames is a method for the determination of differential invariants of M under G, and in fact for determining enough of them to enable us to decide whether two given submanifolds differ from each other only by a transformation of G. Its main idea is that of passing from the homogeneous space E to the group space G. In fact, denote by $\psi: G \to G/H$ the natural projection which assigns to an element $g \in G$ the coset gH. From M we get the submanifold $F_O = \psi^{-1}(M) \subset G$, determined up to multiplication by a fixed element on the right (depending on the choice of the point O). F_O is in general a manifold of dimension higher than p and will be called the manifold of frames of order 0 of M. Now the Lie algebra $\mathfrak{h}$ of H is a subalgebra of the Lie algebra $\mathfrak{g}$ of G. There is therefore in the dual space $\mathfrak{g}^*$ of $\mathfrak{g}$, whose elements are the so-called Maurer-Cartan forms, a linear subspace $\mathfrak{n}^*(G, H)$ of dimension n consisting of all elements of $\mathfrak{g}^*$ orthogonal to $\mathfrak{h}$. The dual mapping of the identity mapping $i_0: F_0 \to G$ maps the elements of $\mathfrak{n}^*(G, H)$ into Pfaffian forms on F_0, called by Cartan the principal components of order 0. Among them there are exactly p linearly independent ones, the others being their linear combinations. The coefficients of such linear combinations play an important rôle in the method of moving frames. From them it is sometimes possible to derive differential invariants of M by elimination.

In order to get more information about M, we have to pass to the elements of contact of higher order. The general principle is to extend the above considerations to them. Let M be defined locally by the equations

$$x_i = f_i(u_1, \cdots, u_p), \qquad i = 1, \cdots, n,$$

where the functions f_i possess sufficiently many continuous partial derivatives or even are analytic. The values, at a certain point, of the functions x_i and their partial derivatives up to the order s inclusive, subject to the usual laws of transformation when either the x's or the u's undergo an admissible coordinate transformation, constitute an element of contact C_s of order s. Thus an element of contact of order 0 is the point itself. Moreover, an element of contact of order $s>0$ determines uniquely an element of contact of order $s-1$, obtained by ignoring the derivatives of order s.

The totality of the elements of contact of order s, for all submanifolds of dimension p in E, is a space E_s on which G acts. The elements of contact of order s of M constitute a submanifold M_s of E_s. The case for general s differs from the case $s=0$ in two essential aspects: (1) The group G does not necessarily act in a transitive manner on E_s, so that the latter decomposes into domains of transitivity; (2) The subgroup H_s of G leaving fixed a given C_s may not be connected, as for instance in the case when E is the Euclidean space with the group of rigid motions, and $p=1$, $s=1$. In this example M is a curve and C_1 can be identified with the tangent direction; the motions leaving a line fixed have two connected components.

This phenomenon shows that the same C_s may carry several "oriented" elements of contact of order s, obtained by replacing H_s by the component of the unit element in H_s. The latter process involves an arbitrary choice, because only the class of conjugate subgroups of H_s is defined by C_s. On the other hand, the first fact, together with generality assumptions, allows us to coordinatize the generic oriented elements of contact of order s by a finite set of numbers. These are the differential invariants of order not greater than s.

Denote by T a domain of transitivity of E_s under G, by $H_s(T)$ the subgroup of G leaving one of its points fixed, and by $H_s'(T)$ the connected component of the unit element of $H_s(T)$. Then the space of oriented elements of contact of order s can be identified with the union $\cup_T G/H_s'(T)$, and those of M can be considered as a submanifold M_s of $\bigcup_T G/H_s'(T)$. If $\psi_{s,T}: G \to G/H_s'(T)$ is the natural projection, the submanifold

$$F_s = \bigcup_{C_s \in M_s} \psi_{s,T}^{-1}(C_s)$$

is called the manifold of frames of order s of M. Generalizing the situation on frames of order 0, let $\mathfrak{h}_s$ denote the Lie algebra of $H_s(T)$ and $\mathfrak{n}^*(G, H_s(T))$ the linear subspace of the dual space $\mathfrak{g}^*$ of $\mathfrak{g}$ which consists of all elements of $\mathfrak{g}^*$ orthogonal to $\mathfrak{h}_s$. The dual mapping of the identity mapping $i_s: F_s \to G$ maps the elements of $\mathfrak{n}^*(G, H_s(T))$ into Pfaffian forms on F_s, called the principal components of order not greater than s. The main feature of the method of moving frames is the result that the study of the manifold F_s in G gives some of the most important of the local geometric properties of M in E. The determination of the manifolds of frames of different orders is achieved by induction on s.

Cartan had a more geometrical picture of the frames. To him they are configurations in E such that there exists exactly one transformation of G carrying one such configuration into another. In Euclidean space with the group of rigid motions we can take as frames sets of points (P, U_1, U_2, U_3) with the properties: (1) the points U_1, U_2, U_3 are at a distance 1 from P; (2) any two of the lines PU_1, PU_2, PU_3 are perpendicular; (3) the vectors PU_1, PU_2, PU_3 form a right-handed system. For a surface in Euclidean space the frames of order 0 are those with P on the surface. The frames of order one satisfy the further condition that PU_3 is normal to the surface. Here an orientation has to be made according as U_3 is along one sense of the normal or the other. If P is not an umbilic, the frame of order two is uniquely determined at each point by the condition that PU_1, PU_2 are in the principal directions. The two principal curvatures are invariants of order two.

So far we have restricted our discussion to the generic elements of contact. Among the most interesting properties of differential geometry are perhaps those concerning the nongeneric ones. Thus the four-vertex theorem for closed plane curves and the theorem on the existence of umbilics on closed surfaces of genus $\neq 1$ are statements on the existence of certain types of elements of contact on a closed submanifold. No general result along this direction is known, and it is hoped that the method of moving frames will give some clues.

From this viewpoint the problem of deformation naturally presents itself. Two submanifolds M^p and M^{*p} in E are said to be applicable or deformable of order s relative to G if there is a transformation of G which carries the elements of contact of order s of M^p into those of

M^{*p}, that is, if there is a one-one mapping between the submanifolds under which the invariants and the principal components of order not greater than s are equal. When G is the group of motions in Euclidean space and $p=2$, $s=1$, this notion of applicability reduces to the classical one studied by Gauss, Minding, Darboux, and others. When G is the group of projective collineations in a real projective space and $p=2$, $s=2$, the problem was known as the projective deformation of surfaces and was studied at great length by Fubini and Čech.

When the order s is sufficiently large, the method gives a solution of the fundamental problem of local differential geometry in a homogeneous space, namely, that of deciding whether two submanifolds differ from each other by a transformation of the group G.

Actually it may be quite complicated to carry out the method (i.e., to determine the invariants and frames of different orders) in concrete cases, particularly when p is large. Moreover, the generality assumptions may soon become unrealistic. Cartan developed various ways of simplifying the computations and adapting the method to special cases. As frequently happens in mathematics, the generality of the viewpoint helps also to treat the special cases in a more effective way.

Some of the applications he made are: the conformal deformation of hypersurfaces [**28**], the projective deformation of surfaces [**27**], and the theory of submanifolds of constant curvature in a Euclidean or non-Euclidean space [**25**; **26**]. The second was a problem which had received considerable attention from the Italian geometers. Cartan proved that, except for a class of surfaces which depends on six arbitrary functions of one variable, a surface is not deformable in a nontrivial way. Moreover, if a surface is projectively deformable (i.e., in a nontrivial way), the surfaces to which it is deformable depend at most on three arbitrary constants. This does not settle the problem. It perhaps makes the study of projectively deformable surfaces even more interesting. To cite an example, the question whether there exist surfaces projectively deformable to ∞^2 projectively inequivalent surfaces is not yet solved.

The study of the submanifolds of constant curvature in a Euclidean or non-Euclidean space generalizes the classical treatment of developable surfaces [**25**; **26**]. Cartan made an exhaustive study and determined the degree of generality of such submanifolds. They do not exist under all circumstances. For instance, if the curvature of the submanifold of dimension p is less than the curvature of the space, the latter must have dimension not less than $2p-1$. His emphasis was on the necessity of investigating the existence of submanifolds with prescribed properties, and their degree of generality, before their study can be undertaken with sense. For this purpose his theory of differential systems in involution was applied to the best advantage.

Most remarkable among his works along this line are the results on the isoparametric families of hypersurfaces in a spherical space [42; 43]. It started with a problem of Levi-Civita: to study the scalar functions in a Riemannian space which are functionally dependent on both their first and second Beltrami differential parameters. The hypersurfaces obtained by equating such a scalar function to a constant are said to form an isoparametric family. When the Riemannian space is of zero or constant negative curvature, the determination of its isoparametric families of hypersurfaces does not cause much difficulty. This is due to the fact that every such hypersurface has at most two distinct principal curvatures. When the Riemannian space is of constant positive curvature, the situation is very complicated, but also most interesting. Cartan proved that in this case there do exist isoparametric families of hypersurfaces having three distinct principal curvatures but that this can happen only when the dimension of the space is 4, 7, 13, or 25. The last family admits the exceptional simple Lie group in 52 parameters; this was the first time that this group was geometrically realized. Similar results hold for isoparametric families of hypersurfaces in spherical space with four distinct principal curvatures. These exist only in spaces of 5 and 9 dimensions. This is one of the few instances of geometrical problems where the dimension of the ambient space plays an essential rôle.

Einstein's theory of general relativity gave a new impetus to differential geometry. In their efforts to find an appropriate model of the universe geometers have broadened their horizon from the study of submanifolds in classical spaces (Euclidean, noneuclidean, projective, conformal, etc.) to that of more general spaces intrinsically defined. The result is an extension of the work of Gauss and Riemann on Riemannian geometry to spaces with a connection, which may be an affine connection, a Weyl connection, a projective connection, of a conformal connection. In these generalizations, sometimes called non-Riemannian geometry, an important tool is the absolute differential calculus of Ricci and Levi-Civita. The results achieved are of considerable geometric interest. For instance, in the theory of projective connections, developed independently by Cartan, Veblen, Eisenhart, and Thomas, it is shown that when the space has a system of paths defined by a system of differential equations of the second order, a generalized projective geometry can be defined in the space which reduces to ordinary projective geometry when the differential system is that of the straight lines. Numerous other examples can be cited. The problem at this stage is twofold: (1) to give a definition of "geometry" which will include most of the existing spaces of interest; (2) to develop analytic methods for the treatment of the new geometries, it being increasingly clear that the absolute differential calculus is inadequate.

For this purpose Cartan developed what seems to be the most comprehensive and satisfactory program and demonstrated its advantages in a decisive way [**31**; **39**]. This contribution clearly illustrates his geometric insight and we consider it to be the most important among his works on differential geometry. It can be best explained by means of the modern notion of a fiber bundle. Let $p: B \to X$ be a fiber bundle with fiber Y and structural group G. We assume X to be a differentiable manifold and G a Lie group. Let $U, V, W, \cdots$ be a covering of X by coordinate neighborhoods. A point of B belonging to $p^{-1}(U \cap V)$ has, with respect to U and V, respectively, the coordinates (x, y) and $(x, g_{UV}(x)y)$, $x \in U \cap V$, $y \in Y$, where we denote the action of G on Y by multiplication to the left. The function $g_{UV}(x)$ is defined for $x \in U \cap V$, with values in G. Its dual mapping maps the Maurer-Cartan forms of G into forms in $U \cap V$, to be denoted by ω^i_{UV}, $i = 1, \cdots, r$. Let $(a^i_j(g))$, $g \in G$, denote the adjoint representation of G in the space of the Maurer-Cartan forms. A connection in the bundle is defined in each coordinate neighborhood U by a set of Pfaffian forms θ^i_U, $i = 1, \cdots, r$, such that in $U \cap V$,

$$\overset{i}{\theta}_U = \overset{i}{\omega}_{UV} + \sum_{j=1}^{r} a^i_j(g_{UV})\overset{j}{\theta}_V, \qquad i = 1, \cdots, r.$$

It is easy to verify that this condition is coherent in the intersection of any three coordinate neighborhoods U, V, W.

The curvature tensor is given by the exterior quadratic differential forms

$$\overset{i}{\Theta}_U = d\overset{i}{\theta}_U - \frac{1}{2} \sum_{j,k=1}^{r} c^i_{jk}\overset{j}{\theta}_U \wedge \overset{k}{\theta}_U, \qquad i = 1, \cdots, r$$

where the c^i_{jk} are the constants of structure of G. Under a change of the coordinate neighborhood they are transformed according to the adjoint representation.

Actually Cartan proceeded in a different way. Guided by the classical notion of parallelism he laid more emphasis on the possibility of "developing" the fiber along a parametrized curve. In the present formulation this possibility arises from the fact that the differential system

$$\sum_{j=1}^{r} a_j^i(g)\theta^j + \omega^i(g) = 0, \quad i = 1, \cdots, r; g \in G,$$

is independent of the choice of the coordinate neighborhood. Corresponding to a parametrized curve in X there is a uniquely determined integral curve $g(t)$ of the differential system through the unit element of G. The curve $g(t)$ gives rise to a one-parameter family of transformations in Y. Cartan called this process the development along a parametrized curve, and took it as the definition of a connection.

Without the notion and terminology of fiber bundles it was difficult to explain these concepts in a satisfactory way. The situation was further complicated by the fact that Cartan called tangent space what is now known as fiber while the base space X, being a differentiable manifold, has a tangent space from its differentiable structure. But he saw clearly that the geometrical situation demands the introduction of fiber bundles with rather general fibers. Attempts by several other mathematicians to tie up the fiber with the differentiable structure of the base space were suggested by their experience from affinely connected spaces and led to complicated computations which have nothing to do with the geometrical problem.

When one takes the notion of a connection as the guiding principle in differential geometry, the fundamental problem is to define the fiber bundle and the connection in every geometrical problem. This is not at all a routine matter, and Cartan carried it out in various cases. Some of the more important ones are: (1) the projective connection of the geometry of paths [29]; (2) the conformal connection in the conformal theory of Riemann spaces; (3) the metrical connection in Finsler spaces; (4) the metrical connection in spaces based on the notion of area of hypersurfaces, now known as Cartan spaces [38]; (5) the geometry of the integral $\int F(x, y, y', y'')dx$, which is a generalization of plane Finsler geometry.

Results from these particular examples tend to substantiate the belief that the notion of a connection is a guiding principle in differential geometry. For instance, in Finsler geometry, the spaces of interest are those defining a "regular" problem in the calculus of variations. This fact manifests itself clearly when one tries to define a connection in the space.

In order that the connection in a space may be useful it must have a further property: it should give all the geometrical properties of the space. This can be made precise by the requirement that two spaces are to be equivalent under admissible transformations of the coordinates if and only if the connections are equivalent. This naturally leads to an equivalence problem whose analytical aspects have been discussed in Part II. It suffices to remark here that, while in simpler geometrical problems (such as Riemannian geometry) the introduction of a connection in the space automatically solves the equivalence problem, it is advisable in more general cases to go the other way around by first solving the equivalence problem and then interpreting the solution geometrically. Cartan's treatment of the equivalence problem furnishes a method which is particularly suitable for such geometrical problems.

Cartan, first in his Mémorial volume and later in his *Leçons*, applied his general ideas on connections to the case of Riemannian geometry [**30**; **35**]. Although he himself never refrained from computations, he did not hide his distaste for the computational work on differential geometry which was then very fashionable and much of which was of little geometrical interest. He stated his aim in the preface of [**35**] as that of bringing out the simple geometrical facts which have often been hidden under a debauch of indices. The result is a very original account of Riemannian geometry, still the standard book in the field.

His most important work on Riemannian geometry is undoubtedly the theory of symmetric Riemann spaces [**33**; **34**; **37**]. It is well known that the local properties of a Riemann metric are given by Riemann-Christoffel curvature tensor and its successive covariant derivatives. Besides the locally Euclidean spaces the simplest Riemann spaces are therefore the ones for which the covariant derivative of the Riemann-Christoffel tensor is zero. These spaces, which include the Riemann spaces of constant curvature, are called symmetric by Cartan, for an obvious reason which will be brought out below. He published papers on the subject during the period from 1927 to 1935. Perhaps because of their extensiveness the subject did not receive the attention it deserves. Its significance for the determination of the real forms of complex simple Lie groups has been discussed above. We proceed to give a very brief survey of the geometrical aspects of the theory, together with its relations to classical geometries, the theory of analytical functions of several complex variables, number theory, and topology.

Cartan soon discovered that the definition can be put in a more geometrical form. A symmetric Riemann space can be defined either as one for which the Levi-Civita parallelism preserves the sectional Riemannian curvature or as one for which the symmetry about a point is an isometry. From the second equivalent definition it follows immediately that the space admits a transitive group of isometries and that the connected component of the subgroup of isometries leaving fixed a point of the space is compact. This result brings the symmetric Riemann spaces into relation with homogeneous spaces.

The enumeration of all the symmetric Riemann spaces is not a simple problem. Cartan first observed that if a symmetric Riemann metric can be decomposed (locally) into a sum of two lower-dimensional Riemann metrics, each of the latter is symmetric. The problem is thus reduced to the irreducible case in which such a decomposition is not possible. Cartan then applied two different methods to the problem.

The first method consists in the determination of the subgroups of the orthogonal groups which can be the groups of holonomy of an irreducible symmetric Riemann space. Such a subgroup leaves invariant a form

$$\sum R_{ij,kl}x^i y^j x^k y^l,$$

where R_{ijkl} is the Riemann-Christoffel curvature tensor. It is thus not a most general subgroup of the orthogonal group, and this limitation makes it possible to carry out the program to the end. Unfortunately the method leads to very complicated computations.

It is the second method that opens up entirely unexpected views. Denote by G the connected component of the group of all isometries, and by H the connected component of the subgroup of G leaving a point O fixed. Then H is compact. If σ denotes the symmetry about O, the mapping which sends $g \in G$ into $\sigma g \sigma \in G$ is an involutory automorphism of G. Under this automorphism all elements of H remain fixed. Conversely, when a connected Lie group G has an involutory automorphism such that the connected component of the set of fixed elements is compact, the homogeneous space G/H can be given a symmetric Riemann metric. Now choose a base in the Lie algebra of G such that the endomorphism induced by the involutory automorphism changes the signs of some of the base vectors and leaves the remaining ones fixed. This normalization allows one to draw far-reaching conclusions on the infinitesimal structure of G. In fact, it follows that if the space, which we can now denote by G/H, is irreducible and is not locally Euclidean (that is, its Riemannian-Christoffel tensor is not 0), the group G is simple or is the direct product of two isomorphic compact simple groups.

In the latter case the elements of G can be written as (a, b), $a, b \in H$, H being a simple group. Then the involutory automorphism has to be $(a, b) \rightarrow (b, a)$, and the space can be identified with the space of H. In other words, this case reduces to the geometry of the space of a compact simple Lie group.

More interesting is the case in which G is simple. If G is a complex simple Lie group and H its compact real form, the space G/H is homeomorphic to a Euclidean space and is the only symmetric Riemann space with G as its group of isometries. Cartan called it the fundamental Riemann space of G. When G is a noncompact simple real Lie group, we consider its corresponding complex group G and the fundamental Riemann space E of G. The involutory automorphism in G, which maps every element into its complex conjugate, induces a symmetry in E leaving invariant a totally geodesic manifold of E. The latter is homeomorphic to a Euclidean space and is the only symmetric Riemann space with the group G. The situation is more complicated when G is a compact simple real Lie group, as then the symmetric Riemann space with G as the group of isometries is not unique. Thus there is, from the point of view of the infinitesimal structure, one and only one symmetric Riemann space belonging to a given noncompact simple group G: it is that of the homogeneous space G/H, where H is a maximal compact subgroup of G (which turns out to be uniquely determined up to an inner automorphism of G). For instance, the symmetric Riemann space belonging to the unimodular real linear group $GL_n(R)$ is the space of positive definite quadratic forms in n variables; to the unimodular complex linear group $GL_n(C)$ there belongs similarly the space of positive definite Hermitian forms in n variables. This undoubtedly accounts for the rôle played by those forms both in classical and in modern number theory.

The study of symmetric Riemann spaces also throws considerable light on the relations between Riemannian geometry and the classical geometries, and helps to unify and explain some of the phenomena in classical geometries. Cartan carried out this idea for the case of complex projective geometry in his book [36]. It is known, for instance, that there is a correspondence between the geometry on the complex projective line and non-Euclidean hyperbolic geometry in space. In the present terminology the hyperbolic space is the fundamental Riemann space of the group of projective collineations on the complex projective line.

Actually Cartan's interest in symmetric Riemann spaces was aroused by a related but different problem. It is the study of Riemann spaces which admit an absolute parallelism, whose autoparallel curves are geodesics. The guiding example is the Clifford parallelism in non-Euclidean elliptic space. The first result in this direction was achieved jointly by Cartan and J. A. Schouten. They found that the irreducible spaces with absolute parallelism are exactly the spaces of compact simple Lie groups with one exception, which is the 7-dimensional elliptic space. The existence of the latter is related to properties of Cayley numbers.

Another application of the theory of symmetric Riemann spaces is to functions of several complex variables. Henri Cartan studied the group of all pseudo-conformal transformations which leave invariant a bounded domain in a space of several complex variables and proved that it is a Lie group. Using this result, Cartan studied the domains which are homogeneous, that is, which admit a transitive group of pseudo-conformal transformations. He did not succeed in determining all such domains, because there are perhaps too many. However, he did determine all those which are also symmetric, that is, which have the further property that to every point O of the domain there exists an involutory pseudo-conformal transformation of the domain onto itself which admits O as an isolated invariant point. This is due to the fact that the group is then semi-simple. The irreducible bounded symmetric homogeneous domains form four large classes and two exceptional cases, of dimensions 16 and 27 respectively. These domains have recently been found to play an important rôle in Siegel's work on automorphic functions and analytic number theory, where the discontinuous subgroups of these groups are studied. No bounded homogeneous domain is known which is not symmetric.

The notion of a symmetric space can be extended to the case which Cartan called non-Riemannian. It is a homogeneous space G/H such that there is an involutory automorphism σ of G with H as the connected component of the subgroup of invariant elements of σ, where H is not necessarily compact. His main contribution to general symmetric homogeneous spaces is concerned with their Betti numbers, of which an account was given above in connection with the Betti numbers of compact Lie groups [11]. The results given there are valid for any compact symmetric homogeneous space, so that the determination of the Betti numbers of such a space can be reduced to a purely algebraic problem. Non-Riemannian symmetric spaces have otherwise hardly been studied.

Besides the importance of integral invariants in studying the topological properties of a space as a consequence of de Rham's theorems, they play a rôle in another field of geometry, now known as integral geometry. There again Cartan's exterior differential forms can be applied to the best advantage. In 1898 he devoted a paper to multiple integrals in the spaces of lines and planes of Euclidean space, which are invariant under the group of motions. The paper marks an important step in integral geometry, a subject founded by the English mathematician Crofton and later developed by Blaschke and his school. It is quite curious that, although Cartan laid much emphasis on the idea of defining a group as the set of transformations leaving invariant a set of linear differential forms and took this to be the starting point of his theory of infinite groups, he did not come back to invariant differential forms of higher degree, except in the paper discussed above. His exterior differential forms have now become an indispensable tool in integral geometry.

Bibliography

The following is a partial list of Cartan's mathematical papers. It is complete in the papers published after 1939. A complete list of his papers up to 1939 can be found in [2], while [1] contains a complete list up to 1931.

I. Sources

1. *Notice sur les travaux scientifiques de M. Élie Cartan*, Paris, Gauthier-Villars, 1931.

2. *Selecta; Jubilé scientifique de M. Élie Cartan*, Paris, Gauthier-Villars, 1939.

II. Group theory

3. *Sur la structure des groupes de transformations finis et continus*, Thèse, Paris, 1894; 2d ed., Paris, Vuibert, 1933.

4. *Les groupes bilinéaires et les systèmes de nombres complexes*, Ann. Fac. Sci. Univ. Toulouse vol. 12B (1898) pp. 1–99.

5. *Les groupes projectifs qui ne laissent invariante aucune multiplicité plane*, Bull. Soc. Math. France vol. 41 (1913) pp. 53–96.

6. *Les groupes réels simples finis et continus*, Ann. École Norm. vol. 31 (1914) pp. 263–355.

7. *Les groupes projectifs continus réels qui ne laissent invariante aucune multiplicité plane*, J. Math. Pures Appl. (6) vol. 10 (1914) pp. 149–186.

8. *La géométrie des groupes de transformations*, J. Math. Pures Appl. (9) vol. 6 (1927) pp. 1–119.

9. *La géométrie des groupes simples*, Ann. di Mat. vol. 4 (1927) pp. 209–256.

10. *Groupes simples clos et ouverts et géométrie riemannienne*, J. Math. Pures Appl. (9) vol. 8 (1929) pp. 1–33.

11. *Sur les invariants intégraux de certains espaces homogènes clos et les propriétés topologiques de ces espaces*, Annales de la Société Polonaise de Mathématiques vol. 8 (1929) pp. 181–225.

12. *La théorie des groupes finis et continus et l'analysis situs*, Mémorial des Sciences Mathématiques vol. 42, 1930.

13. *Les représentations linéaires des groupes de Lie*, J. Math. Pures Appl. (9) vol. 17 (1938) pp. 1–12.

14. *Leçons sur la théorie des spineurs* I, II, Actualités Scientifiques et Industrielles, nos. 643, 701, 1938.

III. Differential equations

15. *Sur l'intégration des systèmes d'équations aux différentielles totales*, Ann. École Norm. vol. 18 (1901) pp. 241–311.

16. *Sur la structure des groupes infinis de transformations*, Ann. École Norm. vol. 21 (1904) pp. 153–206.

17. *Sur la structure des groupes infinis de transformations*, Ann. École Norm. vol. 22 (1905) pp. 219–308.

18. *Les sous-groupes des groupes continus de transformations*, Ann. École Norm. vol. 25 (1908) pp. 57–194.

19. *Les groupes de transformations continus, infinis, simples*, Ann. École Norm. vol. 26 (1909) pp. 93–161.

20. *Sur les équations de la gravitation d'Einstein*, J. Math. Pures Appl. (9) vol. 1 (1922) pp. 141–203.

21. *Leçons sur les invariants intégraux*, Paris, Hermann, 1922.

22. *Sur la possibilité de plonger un espace riemannien donné dans un espace euclidien*, Annales de la Société Polonaise de Mathématiques vol. 6 (1927) pp. 1–7.

23. *Les systèmes différentiels extérieurs et leurs applications géométriques*, Paris, Hermann, 1945.

IV. Geometry

24. *Le principe de dualité et certaines intégrales multiples de l'espace tangentiel et d l'espace réglé*, Bull. Soc. Math. France vol. 24 (1896) pp. 140–176.

25. *Sur les variétés de courbure constante d'un espace euclidien ou non euclidien*, Bull. Soc. Math. France vol. 47 (1919) pp. 125–160.

26. *Sur les variétés de courbure constante d'un espace euclidien ou non euclidien*, Bull. Soc. Math. France vol. 48 (1920) pp. 132–208.

27. *Sur la déformation projective des surfaces*, Ann. École Norm. vol. 37 (1920) pp. 259–356.

28. *Sur le problème général de la déformation*, C.R. Congrès Strasbourg, 1920, pp. 397–406.

29. *Sur les variétés à connexion projective*, Bull. Soc. Math. France vol. 52 (1924) pp. 205–241.

30. *La géométrie des espaces de Riemann*, Mémorial des Sciences Mathématiques, vol. 9, 1925.

31. *Les groupes d'holonomie des espaces généralisés*, Acta Math. vol. 48 (1926) pp. 1–42.

32. (with J. A. Schouten), *On the Riemannian geometries admitting an absolute parallelism*, Neder. Akad. Wetensch. vol. 29 (1926) pp. 933–946.

33. *Sur une classe remarquable d'espaces de Riemann*, Bull. Soc. Math. France vol. 54 (1926) pp. 214–264.

34. *Sur une classe remarquable d'espaces de Riemann*, Bull. Soc. Math. France vol. 55 (1927) pp. 114–134.

35. *Leçons sur la géométrie des espaces de Riemann*, Paris, Gauthier-Villars, 1928; 2d ed., 1946.

36. *Leçons sur la géométrie projective complexe*, Paris, Gauthier-Villars, 1931.

37. *Les espaces riemanniens symétriques*, Verh. Int. Math. Kong. Zurich, vol. I, 1932, pp. 152–161.

38. *Les espaces métriques fondés sur la notion d'aire*, Actualités Scientifiques et Industrielles, no. 72, 1933.

39. *La méthode du repère mobile, la théorie des groupes continus, et les espaces généralisés*, Actualités Scientifiques et Industrielles no. 194, 1935.

40. *Sur les domains bornés homogènes de l'espace de n variables complexes*, Abh. Math. Sem. Hamburgischen Univ. vol. 11 (1935) pp. 116–162.

41. *La théorie des groupes finis et continus et la géométrie différentielle*, Paris, Gauthier-Villars, 1937.

42. *Famille de surfaces isoparamétriques dans les espaces à courbure constante*, Ann. di Mat. vol. 17 (1938) pp. 177–191.

43. *Sur des familles remarquables d'hypersurfaces isoparamétriques dans les espaces sphériques*, Math. Zeit. vol. 45 (1939) pp. 335–367.

V. PAPERS PUBLISHED AFTER 1939

44. *Sur quelques familles remarquables d'hypersurfaces*, C.R. Congrès Sci. Math. Liège, 1939, pp. 30–41.

45. *Sur les groupes linéaires quaternioniens*, Vierteljahrsschrift der Naturforschenden Gesellschaft in Zurich vol. 85 (1940) pp. 191–203.

46. *Sur des familles d'hypersurfaces isoparamétriques des espaces sphériques à 5 et à 9 dimensions*, Universidad Nacional de Tucumán, Revista vol. A 1 (1940) pp. 5–22.

47. *Sur un théorème de J. A. Schouten et W. van der Kulk*, C.R. Acad. Sci. Paris vol. 211 (1940) pp. 21–24.

48. *Sur une classe de surfaces apparentées aux surfaces R et aux surfaces de Jonas*, Mem. Vol. Dedicated to D. A. Grave, Moscow, 1940, pp. 72–78.

49. *La geometria de las ecuaciones diferenciales de tercer orden*, Revista Matemática Hispano-Americana vol. 1 (1941) pp. 1–31.

50. *Sur les surfaces admettant une seconde forme fondamentale donnée*, C.R. Acad. Sci. Paris vol. 212 (1941) pp. 825–828.

51. *La notion d'orientation dans les différentes géométries*, Bull. Soc. Math. France vol. 69 (1941) pp. 47–70.

52. *Sur les couples de surfaces applicables avec conservation des courbures principales*, Bull. Sci. Math. vol. 66 (1942) pp. 55–72, 74–85.

53. *Notice sur M. Tullio Levi-Civita*, C.R. Acad. Sci. Paris vol. 215 (1942) pp. 233–235.

54. *Les surfaces qui admettent une seconde forme fondamentale donnée*, Bull. Sci. Math. vol. 67 (1943) pp. 8–32.

55. *Sur une classe d'espaces de Weyl*, Ann. École Norm. vol. 60 (1943) pp. 1–16.

56. *Sur une classe de surfaces apparentées aux surfaces R et aux surfaces de Jonas*, Bull. Sci. Math. vol. 68 (1944) pp. 41–50.

57. *Sur un problème de géométrie différentielle projective*, Ann. École Norm. vol. 62 (1945) pp. 205–231.

58. *Quelques remarques sur les 28 bitangentes d'une quartique plane et des 27 droites d'une surface cubique*, Bull. Sci. Math. vol. 70 (1946) pp. 42–45.

59. *L'oeuvre scientifique de M. E. Vessiot*, Bull. Soc. Math. France vol. 75 (1947) pp. 1–8.

60. *Sur l'espace annalagmatique réel à n dimensions*, Annales de la Société Polonaise de Mathématiques vol. 20 (1948) pp. 266–278.

ANNALS OF MATHEMATICS
Vol. 56, No. 3, November, 1952
Printed in U.S.A.

SOME THEOREMS ON THE ISOMETRIC IMBEDDING OF COMPACT RIEMANN MANIFOLDS IN EUCLIDEAN SPACE

SHIING-SHEN CHERN AND NICOLAAS H. KUIPER

(Received November 2, 1951)

Introduction

This paper will be concerned with some estimates on the lower bound of the dimension of the Euclidean space in which a compact Riemann manifold can be imbedded isometrically, if its curvatures satisfy certain conditions. Our basic geometrical idea is a very simple one. Denote by M a compact Riemann manifold of dimension n in an Euclidean space E of dimension $n + N$, the Riemann metric on M being induced by the imbedding. Let O be a fixed point of E. The distance OP, $P \in M$, is a continuous function in M and attains a maximum at a point $P_0 \in M$, since M is compact. It is intuitively clear that M will be "concave toward O" at P_0, so that there will be some restrictions on the Riemann curvature of M at P_0. If M is given abstractly, the imbedding will not be possible, if these restrictions are not fulfilled by the given Riemann metric at any of the points of M. Actually, however, if the difference N of the dimensions of M and E is greater than one, the implication of this geometrical fact on the Riemann curvature of M is not very simple. The question leads to algebraic problems which probably do not have simple answers. We propose to give in this paper a few conclusions which can be drawn. It should be mentioned that the above geometrical idea has been used by Tompkins[1] to prove that a locally flat compact Riemann manifold of dimension n cannot be isometrically imbedded in an Euclidean space of dimension $2n - 1$. Among other things this theorem will be generalized and the invariants entering in the problem geometrically interpreted. As for the differentiability assumptions we suppose our manifold and the imbedding to be of class $\geqq 4$.

1. Submanifolds in Euclidean space

Let E be the Euclidean space of dimension $n + N$ and M a compact submanifold of dimension n in E. To avoid confusion we agree to use the following ranges of indices throughout the paper:

$$\begin{aligned} &1 \leqq i, j, k, l \leqq n, \\ (1) \qquad &n + 1 \leqq r, s, t \leqq n + N, \\ &1 \leqq A, B, C \leqq n + N. \end{aligned}$$

We shall also follow the convention that repeated indices imply summation.

To study the geometry in E we introduce its right-handed rectangular frames $Pe_1 \cdots e_{n+N}$, where P is a point of E and $e_1, \cdots, e_{n+N}$ an ordered set of mutually

[1] TOMPKINS, C., *Isometric embedding of flat manifolds in Euclidean space*, Duke Math. J. 5, (1939), pp. 58–61.

perpendicular unit vectors with its determinant equal to $+1$. We use the symbol P also to denote the position vector of P. Define the Pfaffian forms ω_A, ω_{AB} according to the equations

$$(2) \qquad \begin{aligned} dP &= \omega_A e_A, \\ de_A &= \omega_{AB} e_B, \end{aligned}$$

so that

$$(3) \qquad \omega_{AB} + \omega_{BA} = 0.$$

Since the exterior derivative of an exact differential is zero:

$$(4) \qquad d(dP) = 0, \qquad d(de_A) = 0,$$

exterior differentiation of (2) gives

$$(5) \qquad \begin{aligned} d\omega_A &= \omega_B \wedge \omega_{BA}, \\ d\omega_{AB} &= \omega_{AC} \wedge \omega_{CB}. \end{aligned}$$

Equations (5) are known as the equations of structure of the group of proper motions in E.

When M is given in E, we consider the sub-family of frames $Pe_1 \cdots e_{n+N}$ satisfying the conditions: (1) $P \in M$; (2) $e_1, \cdots, e_n$ are tangent vectors to M at P. Then we have

$$(6) \qquad \omega_r = 0,$$

and the first equation of (5) gives

$$(7) \qquad \omega_i \wedge \omega_{ir} = 0.$$

By a lemma of Cartan[2] on exterior algebra, this implies

$$(8) \qquad \omega_{ir} = A_{rij}\, \omega_j,$$

with

$$(9) \qquad A_{rij} = A_{rji}.$$

For the sub-family of frames under consideration, equations (5) give

$$(10) \qquad \begin{aligned} d\omega_i &= \omega_j \wedge \omega_{ji}, \\ d\omega_{ij} &= \omega_{ik} \wedge \omega_{kj} + \Omega_{ij}, \end{aligned}$$

where

$$(11) \qquad \Omega_{ij} = -\omega_{ir} \wedge \omega_{jr} = -A_{rik} A_{rjl} \omega_k \wedge \omega_l.$$

The exterior quadratic differential forms Ω_{ij} therefore give the Riemann curvature of the induced metric on M. We observe that they involve quadratic expressions in the quantities A_{rij}.

[2] Cf., for instance, E. Cartan, *Leçons sur la théorie des espaces à connexion projective*, Paris 1937, p. 117.

On the other hand, the A_{rij} themselves have geometrical significance in terms of properties of M in E. In fact, introduce the ordinary quadratic differential forms

$$\Phi_r = A_{rij}\omega_i\omega_j, \tag{12}$$

to be called the *second fundamental forms* of M. To a tangent direction PT to M at P consider a curve of M tangent to PT and a unit normal vector

$$\xi = \lambda_r e_r. \tag{13}$$

If s is the arc length of this curve, κ its curvature at P, and θ the angle between ξ and its first normal, we have

$$\kappa \cos \theta = \frac{d^2P}{ds^2} \cdot (\lambda_r e_r).$$

Now, we have from (2)

$$\frac{d^2P}{ds^2} = (\cdots)e_i + \frac{\Phi_r}{ds^2}e_r,$$

so that

$$\kappa \cos \theta = \frac{d^2P}{ds^2}(\lambda_r e_r) = \lambda_r \frac{\Phi_r}{ds^2}. \tag{14}$$

This gives a geometrical interpretation of the forms Φ_r and therefore of the quantities A_{rij}.

A tangent direction through P is called *asymptotic*, if it annihilates the Φ_r, i.e., if

$$\Phi_r = 0. \tag{15}$$

We now consider the function

$$f = f(P) = \overline{OP}^2, \tag{16}$$

where O is the origin of E. Its first and second differentials are given by

$$\begin{aligned} \tfrac{1}{2}\, df &= PdP, \\ \tfrac{1}{2}\, d^2f &= dP^2 + Pd^2P = \omega_i\omega_i + Pd^2P. \end{aligned} \tag{17}$$

These equations justify the introduction of the quantities

$$y_A = Pe_A. \tag{18}$$

Then we can write

$$\begin{aligned} \tfrac{1}{2}\, df &= y_i\omega_i, \\ \tfrac{1}{2}\, d^2f &= \omega_i\omega_i + (\cdots)\, y_j + \Phi_r y_r. \end{aligned}$$

At a point $P_0 \in M$ where f attains a maximum we must have

$$df = 0, \qquad d^2f \leqq 0,$$

which gives

$$y_i = 0, \qquad y_r\Phi_r < 0 \qquad \text{(for } \omega_j \text{ not all zero).} \tag{19}$$

These considerations imply the theorem:

THEOREM 1. *A compact submanifold in an Euclidean space has at least one point at which there is no real asymptotic direction.*

This last property, however, pertains to M as a submanifold of E and is not a property of the induced Riemann metric of M. Our purpose is therefore to find properties of the Riemann metric of M which will imply that M has real asymptotic directions.

To arrive at such a property we introduce the integer $\nu(P)$ such that $n - \nu(P)$ is the minimum number of linearly independent linear differential forms in terms of which the Φ_r can be expressed. The integer $n - \nu(P)$ is consequently the number of linearly independent equations in the system

$$\omega_{ir} = A_{rij}\omega_j = 0. \tag{20}$$

The latter define a linear subspace of dimension $\nu(P)$ of the tangent space to M at P. Although this linear subspace can be given a simple geometrical interpretation, we shall not bother with it here. It will, however, be convenient to give a name to the integer $\nu(P)$. We propose to call it the *index of relative nullity.*

2. An inequality between the indices of nullity and relative nullity

A similar integer can be introduced, which depends only on the Riemann metric of M. It is the integer $\mu(P)$ such that $n - \mu(P)$ is the minimum number of linearly independent linear differential forms in terms of which Ω_{ij} can be expressed. $\mu(P)$ will be called the *index of nullity* at P.

THEOREM 2. *Between the indices of nullity and relative nullity the following inequalities hold:*

$$\nu(P) \leqq \mu(P) \leqq N + \nu(P). \tag{21}$$

Since $n - \nu(P)$ is the number of linearly independent equations in (20), we have, from (11),

$$n - \mu(P) \leqq n - \nu(P),$$

which is precisely the first inequality in (21).

To prove the second inequality let us first agree on the following ranges of indices throughout the proof:

$$\begin{aligned} &1 \leqq \alpha \leqq n - \nu(P), \\ &1 \leqq \rho \leqq n - \mu(P), \\ &n - \mu(P) + 1 \leqq u, \quad v \leqq n - \nu(P). \end{aligned} \tag{22}$$

For simplicity of writing we also put $h = n - \mu(P)$, $m = \mu(P) - \nu(P)$. Let θ_α be linear differential forms such that ω_{ir} are linear combinations of θ_α and that Ω_{ij} are exterior quadratic differential forms in θ_ρ. Writing

$$\omega_{ir} = l_{ri\alpha}\theta_\alpha, \tag{23}$$

we consider $(l_{n+1i\alpha}, \cdots, l_{n+Ni\alpha})$ as the components of a vector $l_{i\alpha}$ in an auxiliary vector space V of N dimensions. We introduce in V the usual scalar product which, for two vectors

$$(24)\qquad \begin{aligned} l &= (l_{n+1}, \cdots, l_{n+N}), \\ l' &= (l'_{n+1}, \cdots, l'_{n+N}), \end{aligned}$$

is given by

$$(25)\qquad (ll') = l_r l'_r .$$

With these preparations consider the matrix

$$(26)\qquad L = (l_{iu}),$$

whose elements are vectors, the indices i and u being the row and column indices respectively. By our assumptions, ω_{ir} cannot be expressed in terms of less than $n - \nu(P)$ linear differential forms and Ω_{ij} involve only θ_ρ. These imply that the matrix L has the following properties:

(α) No column of L is a linear combination of other columns.

(β) The vectors of L satisfy the relations

$$(27)\qquad (l_{iu} l_{jv}) = (l_{iv} l_{ju}).$$

From (27) it follows that, if $l_{iu} = 0$ for fixed i, u, the vectors l_{iv} and l_{ju} are orthogonal for all j, v. Since, by (α), the column of vectors l_{ju} cannot consist entirely of zero vectors, there is a vector l_{ju} linearly independent from the vectors l_{iv}.

On the matrix L we now perform the following elementary operations:

(1) Interchange of rows (or columns);

(2) Multiplication of one row (or column) by a non-zero factor;

(3) Addition of one row (or column) to another row (or column).

Under these operations the properties (α) and (β) remain unchanged.

To establish our theorem we observe that it is sufficient to prove that the number of linearly independent vectors in L is $\geqq m$. For then it follows that $N \geqq m$, which is the second inequality in (21) to be proved.

After suitable elementary transformations we can suppose that $l_{1,h+1}, \cdots, l_{1,h+q}$ are linearly independent and that $l_{1,h+m} = \cdots = l_{1,h+q+1} = 0$. By the above remark there is a vector in the $(q + 1)$st column which is linearly independent from $l_{1,h+1}, \cdots, l_{1,h+q}$. By adding a sufficiently small multiple of the corresponding row to the first row, we do not disturb the linear independence of its first q vectors and we get at least $q + 1$ linearly independent vectors in the first row. Repetition of this process proves that the vectors of the first row can be made linearly independent by our elementary operations. This proves our statement and the theorem.

3. Conclusions in the large

Let M be a compact submanifold in E. At a point $P_0 \in M$ where the function f in (16) takes a maximum the index of relative nullity $\nu(P_0)$ must be equal to 0, by Theorem 1. From (21) it follows that

$$N \geqq \mu(P_0).$$

This gives the theorem:

THEOREM 3. *If a compact Riemann manifold of dimension n has at every point an index of nullity $\geqq \mu_0$, it cannot be isometrically imbedded in an Euclidean space of dimension $n + \mu_0 - 1$.*

Since the index of nullity of a locally flat Riemann manifold of dimension n is equal to n, this theorem contains as a corollary the following theorem of Tompkins:

THEOREM 4. *A compact locally flat Riemann manifold of dimension n cannot be isometrically imbedded in an Euclidean space of dimension $2n - 1$.*

Utilizing the same idea, we shall prove some further theorems on the isometric imbedding of Riemann manifolds. At a point $P \in M$ consider the plane element π spanned by the mutually perpendicular unit vectors

$$\begin{aligned} x &= x_i e_i , \\ y &= y_i e_i . \end{aligned} \tag{28}$$

Its sectional curvature is known to be

$$\begin{aligned} R(P, \pi) &= \tfrac{1}{2}(A_{rik}A_{rjl} - A_{ril}A_{rjk})(x_i y_j - x_j y_i)(x_k y_l - x_l y_k) \\ &= 2(A_{rik}A_{rjl} - A_{ril}A_{rjk})x_i y_j x_k y_l . \end{aligned} \tag{29}$$

It is probable that the following conjecture is true: If $R(P, \pi) \leqq 0$ for all plane elements π, the system of quadratic equations

$$\Psi_r \equiv A_{rij}x_i x_j = 0 \tag{30}$$

has a non-trivial real solution in x_i, when $N \leqq n - 1$. By Theorem 1 this conjecture of a purely algebraic nature will imply the following geometrical result: At every point of an abstract Riemann manifold M suppose there be a linear subspace of dimension q of the tangent space such that the sectional curvatures along all its plane elements are $\leqq 0$. Then M cannot be isometrically imbedded in an Euclidean space of dimension $n + q - 1$, if it is compact.

Unfortunately we are only able to prove the above statement for $n = 2, 3$. We can assume $N = n - 1$, since we do not exclude the possibility that some of the forms in (30) are identically zero. For $n = 2$ we have

$$\Psi_r = \Psi_3 = A_{3ij}x_i x_j . \tag{31}$$

The condition that the sectional curvature is non-positive gives

$$A_{311}A_{322} - A_{312}^2 \leqq 0,$$

which implies that equation (31) has a non-trivial real solution.

We proceed to give a proof of the truth of our conjecture for $n = 3$. We observe that the properties involved in the statement remain unchanged if we apply an orthogonal transformation on Ψ_r. Since $n = 3$, there are real values of λ, μ such that $\lambda\Psi_{n+1} + \mu\Psi_{n+2}$ is degenerate. Without loss of generality we can assume that Ψ_{n+2} is degenerate. By an orthogonal transformation in the space of x_i when necessary, Ψ_{n+2} can be supposed to be

$$\Psi_{n+2} = a_{11}x_1^2 + 2a_{12}x_1x_2 + a_{22}x_2^2, \tag{32}$$

while we write, for simplicity,

$$\Psi_{n+1} = A_{ij}x_ix_j. \tag{33}$$

Let (α_{ij}) be the adjoint matrix of (A_{ij}), whose elements are the cofactors of order two of the latter. Put

$$\xi_1 = x_2y_3 - x_3y_2, \qquad \xi_2 = x_3y_1 - x_1y_3, \qquad \xi_3 = x_1y_2 - x_2y_1. \tag{34}$$

Then the condition $R(P, \pi) \leqq 0$ can be written in the form

$$\alpha_{ij}\xi_i\xi_j + \Delta\xi_3^2 \leqq 0, \tag{35}$$

where

$$\Delta = a_{11}a_{22} - a_{12}^2. \tag{36}$$

Consider first the case $\Delta \leqq 0$. Under this assumption Ψ_{n+2} is a product of two real linear forms. By a rotation of axes in the plane x_1, x_2, we can assume that $a_{22} = 0$. Since, by (35), $\alpha_{11} \leqq 0$, the equations

$$x_1 = 0, \qquad A_{22}x_2^2 + 2A_{23}x_2x_3 + A_{33}x_3^2 = 0$$

have a non-trivial real solution, and the same is thus true of the system (30).

Suppose next that $\Delta > 0$. From (35) it follows that

$$\alpha_{ij}\xi_i\xi_j \leqq 0.$$

The form Ψ_{n+1} has therefore the property that the quadratic form whose coefficients are the elements of its adjoint matrix is negative semi-definite. By using a diagonal canonical form for Ψ_{n+1}, we see that, under the assumption $n = 3$, this is possible only when Ψ_{n+1} is degenerate. If Ψ_{n+1} decomposes into a product of two real linear forms, the above proof applies, by reversing the roles of Ψ_{n+1} and Ψ_{n+2}, so that our statement is true. If this is not the case, Ψ_{n+1} must be the sum of squares of two linear forms. We write, in this case,

$$\Psi_{n+1} = (a_1x_1 + a_2x_2 + a_3x_3)^2 + (b_1x_1 + b_2x_2 + b_3x_3)^2,$$

from which we find

$$\alpha_3 + \Delta = (a_1b_2 - a_2b_1)^2 + \Delta > 0.$$

But this contradicts (35). Thus the proof of our statement is complete.

We state the geometrical consequences of the above results in the following theorem:[3]

THEOREM 5. *Let M be a compact Riemann manifold with the property that at every point there is a q-dimensional linear subspace in the tangent space, $q = 2, 3$, along whose plane elements the sectional curvatures are non-positive. Then M cannot be isometrically imbedded in an Euclidean space of dimension $n + q - 1$.*

4. Some properties of the spaces of nullity

We give in this section some results on the geometrical implication of our assumptions on the index of nullity. Suppose M be an abstract Riemann manifold of dimension n, whose curvature forms are

$$\Omega_{ij} = R_{ijkl}\omega_k \wedge \omega_l, \qquad R_{ijkl} + R_{ijlk} = 0. \tag{37}$$

If $\mu(P)$ is the index of nullity at a point $P \in M$, $n - \mu(P)$ is the maximum number of linearly independent equations in the system

$$R_{ijkl}\omega_l = 0. \tag{38}$$

These equations define a linear subspace $L(P)$ of dimension $\mu(P)$ in the tangent space at P. We call $L(P)$ the *space of nullity* at P.

This space $L(P)$ has a simple geometrical interpretation. In fact, if

$$x = x_i e_i \tag{39}$$

is a tangent vector, the equations

$$R_{ijkl}x_j x_l = 0. \tag{40}$$

define all the tangent vectors x such that the sectional curvatures along plane elements containing x are zero. Each of the equations (40) defines a hypercone whose singular elements are given by

$$R_{ijkl}x_l = 0. \tag{41}$$

It follows that the space of nullity is the space of intersection of the singular elements of the hypercones (40). The existence of a non-zero space of nullity imposes therefore a restriction on the curvature of the Riemann manifold.

To derive further results let us first agree on the following ranges of indices:

$$1 \leqq \alpha, \beta, \gamma \leqq \mu(P), \qquad \mu(P) + 1 \leqq a, b, c, e, f \leqq n. \tag{42}$$

Suppose that $\mu(P)$ remains constant in a neighborhood under consideration, and let μ denote their common value. We restrict ourselves to the family of frames such that e_α span the space of nullity at P. Then the system of equations (38) is equivalent to the system

$$\omega_a = 0, \tag{43}$$

---

[3] We are indebted to the referee for pointing out that the statement in this theorem corresponding to $q = 2$ is contained in a theorem of Myers. Cf. S. B. MYERS, *Curvature of closed hypersurfaces and non-existence of closed minimal hypersurfaces*, Trans. Amer. Math. Soc., 71 (1951) p. 215.

and Ω_{ij} involve ω_a only. Now, by exterior differentiation of the first equation of (10), we have

$$\omega_i \wedge \Omega_{ij} = 0.$$

Because of the particular form of Ω_{ij} this implies that

$$\omega_\alpha \wedge \Omega_{\alpha j} = 0,$$

and hence that

$$\Omega_{\alpha j} = 0. \tag{44}$$

On the other hand, exterior differentiation of the second equation of (10) gives

$$d\Omega_{ij} + \Omega_{ik} \wedge \omega_{kj} - \omega_{ik} \wedge \Omega_{kj} = 0.$$

It follows that

$$\omega_{\alpha a} \wedge \Omega_{ab} = 0 \text{ or } R_{abce}\, \omega_{\alpha a} \wedge \omega_c \wedge \omega_e = 0.$$

For a given set of indices a, c, e, we multiply this equation by the product of ω_f, $f \neq c, e$. This gives

$$(R_{abce}\,\omega_{\alpha a}) \wedge \prod\nolimits_f \omega_f = 0.$$

Since ω_f are linearly independent, we have

$$R_{abce}\,\omega_{\alpha a} \equiv 0 \ (\omega_f).$$

For a given α exactly $n - \mu$ of these equations are linearly independent. Hence we have

$$\omega_{\alpha a} \equiv 0 \ (\omega_f). \tag{45}$$

Consider now the differential system (43). We have

$$d\omega_a = \omega_\alpha \wedge \omega_{\alpha a} + \omega_b \wedge \omega_{ba} \equiv 0 \ (\omega_f).$$

By Frobenius's Theorem it follows that the system (43) is completely integrable. Hence in a neighborhood of M there passes through every point a submanifold of dimension μ which is everywhere tangent to the space of nullity. Moreover, condition (44) implies that the induced metric on an integral manifold of (43) is flat. We summarize these results in the theorem:

THEOREM 6. *Suppose that the index of nullity has a constant value μ in a neighborhood of a Riemann manifold. Then the manifold can be locally sliced into submanifolds of dimension μ which are everywhere tangent to the spaces of nullity and are locally flat in the induced metric.*

UNIVERSITY OF CHICAGO and
LANDBOUWHOGESCHOOL WAGENINGEN, NETHERLANDS

RELATIONS BETWEEN RIEMANNIAN AND HERMITIAN GEOMETRIES

BY SHIING-SHEN CHERN

Introduction. Complex manifolds with an Hermitian or a Kählerian metric have become a topic of current interest in differential geometry. It is well-known that every complex manifold can be made Hermitian. However, a general differentiable manifold does not possess a complex structure, while the existence of a Kählerian metric imposes even more severe restrictions on the topological properties of the manifold. (For accounts of some of the recent works on the subject, we refer to papers [5], [6].) Following some questions raised to me by B. Eckmann, we study in this paper the properties of an analytic Riemannian manifold which locally behaves as an Hermitian or a Kählerian manifold. It turns out that the conditions for these properties can be expressed in a simple way in terms of the curvature tensor. We shall also show by examples that these properties are weaker than the corresponding global properties. On the other hand, we do not know whether any differentiable manifold can be given a locally Kählerian metric. In the four-dimensional case we shall prove that, if a further condition is satisfied, it does mean a restriction on the topological properties of the manifold.

By an Hermitian metric on a complex manifold is meant a positive definite Hermitian differential form which, in the local coordinates z^k, is given by

$$ds^2 = h_{jk}\, dz^j\, d\bar{z}^k \qquad (j, k = 1, \cdots, n), \tag{1}$$

where the functions h_{jk} satisfy the conditions

$$\bar{h}_{jk} = h_{kj}, \tag{2}$$

and their real and imaginary parts are real analytic in the arguments. (We agree, unless otherwise specified, that small Latin indices run from 1 to n, small Greek indices run from 1 to $2n$, and that repeated indices imply summation. Also a bar over a complex number denotes its complex conjugate.) The Hermitian metric is called Kählerian, if the corresponding exterior differential form is closed:

$$d(h_{jk}\, dz^j \wedge d\bar{z}^k) = 0. \tag{3}$$

Consider now an analytic Riemannian metric of $2n$ dimensions. We say that it is *locally Hermitian*, if in every neighborhood there exist complex-valued analytic functions z^k in the local coordinates x^α:

$$z^k = z^k(x^\alpha) \qquad (\alpha = 1, 2, \cdots, 2n) \tag{4}$$

Received November 24, 1952. The paper was written under partial support of a contract with the Office of Naval Research.

such that the metric takes the form (1) in the variables z^k. It is called *locally Kählerian* if, furthermore, the condition (3) is satisfied. A Riemannian metric of two dimensions is always locally Kählerian. For, in terms of a set of isothermal parameters x, y it can be written

$$ds^2 = g(x, y)(dx^2 + dy^2). \tag{5}$$

Putting

$$z = x + iy,$$

we have

$$ds^2 = g'(z, \bar{z})\, dz\, d\bar{z}, \tag{6}$$

which proves that the metric is locally Hermitian. It is locally Kählerian, since condition (3) is clearly satisfied. The notions introduced here can therefore be regarded as generalizations to higher dimensions of the notion of isothermal parameters. We shall show that in higher dimensions not every Riemannian metric has these properties. The study of this problem depends on Elie Cartan's theory of differential systems in involution.

To avoid misunderstanding we wish to remark that our notion of local Kählerian property has nothing in common with a concept under the same name recently introduced by A. Lichnerowicz (compare [9]).

1. **Existence theorems on differential systems in involution.** In finding the conditions for a Riemannian metric to be locally Hermitian or locally Kählerian, we shall apply Cartan's theory of differential systems in involution. We shall give in this section the statements of the existence theorems which will be used, together with some remarks. The proofs of these theorems can be found in [2] or [8].

Let $x^1, \cdots, x^N$ be the coordinates in a real Euclidean space of N dimensions. A system of a finite number of equations of the form

$$\begin{aligned}
&f_\alpha(x^1, \cdots, x^N) = 0 && (\alpha = 1, \cdots, m_0),\\
&\pi_\beta \equiv a_{\beta A}\, dx^A = 0 && (\beta = 1, \cdots, m_1),\\
&\Phi_\gamma \equiv \tfrac{1}{2} a_{\gamma AB}\, dx^A \wedge dx^B = 0 && (\gamma = 1, \cdots, m_2),\\
&\Psi_\rho \equiv \tfrac{1}{6} a_{\rho ABC}\, dx^A \wedge dx^B \wedge dx^C = 0 && (\rho = 1, \cdots, m_3),\\
&\cdots && (1 \le A, B, C \le N),
\end{aligned} \tag{7}$$

is called an exterior differential system. We suppose that the functions f_α and the coefficients $a_{\beta A}$, $a_{\gamma AB}$, $\cdots$ are real analytic functions of x^A in a neighborhood of the Euclidean space under consideration. The left-hand members of (7) generate an ideal $\mathfrak{a}$ in the ring of exterior differential forms. The system (7) is called *closed*, if the exterior derivative of the left-hand member of any of its equations belongs to $\mathfrak{a}$. By adjoining such exterior derivatives to the system, we see that there is no loss of generality in supposing that (7) is closed. We denote by $\mathfrak{a}_r$ the set of all forms of degree r in $\mathfrak{a}$.

A submanifold in the space, defined by the parametric equations

$$x^A = x^A(u^1, \cdots, u^p) \tag{8}$$

is called an integral manifold, if the functions in (8) satisfy the equations (7). Let $\theta_1, \cdots, \theta_p$ be p linearly independent Pfaffian forms. An important problem in the application of differential systems is to see whether the system (7) has an integral manifold on which

$$\theta_1 \wedge \cdots \wedge \theta_p \neq 0. \tag{9}$$

To study this problem we adjoin to $\theta_1, \cdots, \theta_p$, $N - p$ Pfaffian forms $\theta_{p+1}, \cdots, \theta_N$, such that $\theta_1 \wedge \cdots \wedge \theta_N \neq 0$. Then the differential forms in (7) can be expressed in terms of the θ's. We put

$$\theta_r = \sum_{i=1}^{p} l_{ri}\theta_i \qquad (r = p + 1, \cdots, N), \tag{10}$$

and denote by l_i the vector whose components are $l_{p+1,i}, \cdots, l_{Ni}$. Substitute (10) into the forms of $\mathfrak{a}_\nu$, $\nu = 1, \cdots, p$, and let $\mathfrak{a}_\nu(x, l_1, \cdots, l_\nu)$ denote the set of coefficients of $\theta_1 \wedge \cdots \wedge \theta_\nu$ in these forms. Clearly every equation of the set

$$\mathfrak{a}_\nu(x, l_1, \cdots, l_\nu) = 0 \qquad (\nu = 1, \cdots, p), \tag{11}$$

is linear in each set of variables l_i, $i = 1, \cdots, \nu$. On the other hand, the equations $\mathfrak{a}_0 = 0$, which are the same as

$$f_\alpha = 0 \qquad (\alpha = 1, \cdots, m_0), \tag{12}$$

define an algebroid submanifold of dimension r_0 in the space, which we shall denote by V. A point $P \in V$ is called a *simple point*, if at P the rank of the functional matrix $(\partial f_\alpha/\partial x^A)$ is equal to r_0. After these definitions our existence theorem can be stated as follows:

Let $(x^0, l_1^0, \cdots, l_p^0)$ be a system of solutions of the equations (11), (12), *such that (x^0) is a simple point on the algebroid manifold V defined by* (12). *Suppose that, in a neighborhood of $(x^0, l_1^0, \cdots, l_p^0)$, the equations* (11) *reduce to $N - r_\nu - \nu$ independent linear equations with respect to l_ν, after taking account of the equations*

$$\mathfrak{a}_{\nu-1}(x, l_1, \cdots, l_{\nu-1}) = 0. \tag{13}$$

Then, in a sufficiently small neighborhood of (x^0), the system (7) *has an integral manifold of dimension p, on which $\theta_1 \wedge \cdots \wedge \theta_p \neq 0$. If we define*

$$\delta_i = r_i - r_{i+1} - 1 \geq 0 \qquad (i = 0, 1, \cdots, p - 1),$$

the general integral manifold depends on:

s_0 arbitrary constants,

s_1 arbitrary functions in one variable,

. . .

s_{p-1} arbitrary functions in $p - 1$ variables,

r_p arbitrary functions in p variables.

A closed differential system to which this existence theorem applies will be said to be in involution.

A useful particular case of this theorem concerns with the situation in which the system (7) contains only Pfaffian equations and is closed:

$$\pi_\beta \equiv a_{\beta A}\, dx^A = 0 \qquad (\beta = 1, \cdots, m). \tag{14}$$

Such a system is usually called a completely integrable Pfaffian system. Without loss of generality we can suppose that the π_β are linearly independent. Then the conditions for (14) to be closed can be written as

$$d\pi_\beta \wedge \pi_1 \wedge \cdots \wedge \pi_m = 0 \qquad (\beta = 1, \cdots, m). \tag{15}$$

From our existence theorem we can derive that the highest dimension of an integral manifold is $N - m$ and that they depend on m constants. Moreover, there exist m functions $I_1(x), \cdots, I_m(x)$, called the first integrals, such that the differential ideal $\mathfrak{a}$ is generated by $dI_1, \cdots, dI_m$.

In our applications we shall allow the variables and functions to take complex values. By the former we simply mean that we permit linear transformations on the variables with complex coefficients, while a complex-valued function is understood to be one of the form $f + ig$, where f and g are real analytic functions. It is clear that the above existence theorems are valid under these modified circumstances.

2. Locally Hermitian metrics. We consider an analytic Riemannian metric of even dimension $2n$. In order to describe its curvature properties it is convenient to introduce the space B of all rectangular frames (the principal fiber bundle). The connection arising from the Riemannian metric gives rise in B to the Pfaffian forms ω_α, $\omega_{\alpha\beta}$, and its curvature the exterior quadratic differential forms $\Omega_{\alpha\beta}$, such that they satisfy the equations (compare, for instance, [4])

$$\begin{aligned} d\omega_\alpha &= \omega_\beta \wedge \omega_{\beta\alpha}, \\ d\omega_{\alpha\beta} &= \omega_{\alpha\gamma} \wedge \omega_{\gamma\beta} + \Omega_{\alpha\beta}, \qquad \omega_{\alpha\beta} + \omega_{\beta\alpha} = 0. \end{aligned} \tag{16}$$

By taking the exterior derivatives of (16), we get

$$\begin{aligned} \omega_\beta \wedge \Omega_{\beta\alpha} &= 0, \\ \Omega_{\alpha\gamma} \wedge \omega_{\gamma\beta} - \omega_{\alpha\gamma} \wedge \Omega_{\gamma\beta} + d\Omega_{\alpha\beta} &= 0. \end{aligned} \tag{17}$$

These equations are called the Bianchi identities. We put

$$\Omega_{\alpha\beta} = R_{\alpha\beta\rho\sigma}\omega_\rho \wedge \omega_\sigma . \tag{18}$$

Then $R_{\alpha\beta\rho\sigma}$ are functions in B satisfying the relations

$$\begin{aligned} R_{\alpha\beta\rho\sigma} &= -R_{\beta\alpha\rho\sigma} = -R_{\alpha\beta\sigma\rho} , \\ R_{\alpha\beta\rho\sigma} &= R_{\rho\sigma\alpha\beta} , \\ R_{\alpha\beta\rho\sigma} &+ R_{\beta\rho\alpha\sigma} + R_{\rho\alpha\beta\sigma} = 0. \end{aligned} \tag{19}$$

The element of arc of the space, which is an ordinary quadratic differential form, is given by

$$ds^2 = \omega_\alpha^2 . \tag{20}$$

Suppose now that complex coordinates z^k can be introduced such that the element of arc (20) takes the form (1). Since (1) is positive definite, there exist Pfaffian forms

$$\theta_j = a_{jk}\, dz^k, \tag{21}$$

in terms of which the element of arc can be written

$$ds^2 = \theta_j \bar{\theta}_j . \tag{22}$$

Separating θ_k in its real and imaginary parts, we can write

$$\theta_k = \xi_k + i\eta_k , \tag{23}$$

which gives

$$\bar{\theta}_k = \xi_k - i\eta_k . \tag{24}$$

Thus we get

$$ds^2 = \sum_k (\xi_k^2 + \eta_k^2). \tag{25}$$

It follows by a comparison of (20) and (25) that ξ_k, η_k and ω_k, ω_{n+k} differ from each other by an orthogonal transformation. The frames can therefore be so chosen that we have

$$\omega_k = \xi_k , \qquad \omega_{n+k} = \eta_k , \tag{26}$$

or

$$\theta_k = \omega_k + i\omega_{n+k} . \tag{27}$$

These considerations lead to the conclusion:

A necessary and sufficient condition that our Riemannian metric is locally Hermitian is the existence of a submanifold in B on which the Pfaffian system

$$\theta_k = \omega_k + i\omega_{n+k} = 0 \tag{28}$$

is completely integrable and $\theta_1 \wedge \cdots \wedge \theta_n \wedge \bar{\theta}_1 \wedge \cdots \wedge \bar{\theta}_n \neq 0.$

The condition is clearly necessary, in view of (21). It is also sufficient, for if z^k denote the first integrals of (28), equations (21) will hold. From (21) and (22) it follows that the metric is locally Hermitian.

To find the conditions of complete integrability of the system (28) we compute the exterior derivative $d\theta_k$. Introducing the Pfaffian forms

$$\begin{aligned} \phi_{ik} &= -\phi_{ki} = \omega_{j,n+k} - \omega_{k,n+j}, \\ \psi_{ik} &= -\psi_{kj} = -\omega_{jk} + \omega_{n+j,n+k}, \end{aligned} \tag{29}$$

we find, by using (16),

$$d\theta_k = \theta_j \wedge (\omega_{jk} + i\omega_{j,n+k}) + \omega_{n+j} \wedge (\phi_{jk} + i\psi_{jk}). \tag{30}$$

Applying the condition (15), we get

$$(\phi_{jk} + i\psi_{jk}) \wedge \omega_{n+j} \wedge \theta_1 \wedge \cdots \wedge \theta_n = 0.$$

Since the forms ω_{n+j}, θ_k are linearly independent, this condition is equivalent to

$$\phi_{jk} + i\psi_{jk} = A_{jkl}\theta_l + B_{jkl}\omega_{n+l},$$

where

$$B_{jkl} = B_{lkj}.$$

On the other hand, both A_{jkl} and B_{jkl} are anti-symmetric in their first two indices. This implies in particular that B_{jkl} is anti-symmetric in its first two indices and symmetric in its first and third indices and must therefore be zero. Consequently, the condition of complete integrability of (28) can be written in the form

$$\phi_{jk} + i\psi_{jk} = A_{jkl}\theta_l, \tag{31}$$

where

$$A_{jkl} = -A_{kjl}. \tag{32}$$

We observe that ϕ_{jk}, ψ_{jk} are real, so that

$$\phi_{jk} - i\psi_{jk} = \bar{A}_{jkl}\bar{\theta}_l,$$

which gives

$$\begin{aligned} \phi_{jk} &= \tfrac{1}{2}(A_{jkl}\theta_l + \bar{A}_{jkl}\bar{\theta}_l), \\ \psi_{jk} &= \frac{1}{2i}(A_{jkl}\theta_l - \bar{A}_{jkl}\bar{\theta}_l). \end{aligned} \tag{33}$$

Moreover, with the use of (31), equation (30) can be written in the form

$$d\theta_k = \theta_j \wedge (\omega_{jk} + i\omega_{j,n+k} - A_{lkj}\omega_{n+l}). \tag{34}$$

The condition for the Riemannian metric to be locally Hermitian is therefore equivalent to the existence of an integral manifold of the system (31) on which

$$\theta_1 \wedge \cdots \wedge \theta_n \wedge \bar{\theta}_1 \wedge \cdots \wedge \bar{\theta}_n \neq 0. \tag{35}$$

We remark that the system (31) is to be regarded as one in the Cartesian product of B and the space of the variables $A_{jkl} = -A_{kjl}$. To apply the existence theorem in §1 we first make it closed by adjoining to it the equations obtained by exterior differentiation. For this purpose we put

$$\alpha_{jk} = \phi_{jk} + i\psi_{jk} . \tag{36}$$

Then we find

$$\begin{aligned} d\alpha_{jk} = (\omega_{lk} + i\omega_{l,n+k}) \wedge \alpha_{lj} - (\omega_{lj} + i\omega_{l,n+j}) \wedge \alpha_{lk} \\ + i\alpha_{lj} \wedge \alpha_{lk} + \Phi_{jk} + i\Psi_{jk} , \end{aligned} \tag{37}$$

where

$$\begin{aligned} \Phi_{jk} &= -\Phi_{kj} = \Omega_{j,n+k} - \Omega_{k,n+j} , \\ \Psi_{jk} &= -\Psi_{kj} = -\Omega_{jk} + \Omega_{n+j,n+k} . \end{aligned} \tag{38}$$

On the other hand, we have

$$d(A_{jkl}\theta_l) = dA_{jkl} \wedge \theta_l + A_{jkm}\theta_l \wedge (\omega_{lm} + i\omega_{l,n+m} - A_{pml}\omega_{n+p}). \tag{39}$$

Equating these two expressions, we get

$$\begin{aligned} dA_{jkl} \wedge \theta_l + A_{jkm}\theta_l \wedge (\omega_{lm} + i\omega_{l,n+m} - A_{pml}\omega_{n+p}) \\ = (\omega_{pk} + i\omega_{p,n+k}) \wedge A_{pjl}\theta_l - (\omega_{pj} + i\omega_{p,n+j}) \wedge A_{pkl}\theta_l \\ + iA_{jpl}A_{kpm}\theta_l \wedge \theta_m + \Phi_{jk} + i\Psi_{jk} . \end{aligned} \tag{40}$$

From (40) we immediately derive a necessary condition for the Riemannian metric to be locally Hermitian. For $\Phi_{jk} + i\Psi_{jk}$ can be expressed as an exterior differential form in θ_l, $\bar{\theta}_m$. In order that (40) holds, it is necessary that $\Phi_{jk} + i\Psi_{jk}$ contains no terms in $\bar{\theta}_l \wedge \bar{\theta}_m$. In terms of the quantities $R_{\alpha\beta\rho\sigma}$ these conditions are found to be

$$\begin{aligned} R_{l'mjk} + R_{lm'jk} + R_{lmj'k} + R_{lmjk'} - R_{lm'j'k'} - R_{l'mj'k'} \\ - R_{l'm'j'k} - R_{l'm'jk'} = 0, \\ R_{lmj'k'} + R_{l'm'jk} + R_{lm'j'k} + R_{l'mjk'} + R_{lm'jk'} + R_{l'mj'k} \\ - R_{lmjk} - R_{l'm'j'k'} = 0, \end{aligned} \tag{41}$$

where we adopt the convention to write k' for $n + k$, *etc.*

We now study the differential system consisting of the equations (31), (40), with the conditions (41) satisfied. The system is closed, for (40) is obtained by exterior differentiation of (31), so that equations obtained by further exterior differentiations are identically satisfied. To prove that the system has an integral manifold satisfying (35), we apply the existence theorem in §1. We put

$$\alpha_{jk} = \lambda_{jkm}\theta_m + \lambda'_{jkm}\bar{\theta}_m,$$

$$dA_{jkp} = l_{jkpm}\theta_m + l'_{jkpm}\bar{\theta}_m,$$

where λ_{jkm}, λ'_{jkm}, l_{jkpm}, l'_{jkpm} are all anti-symmetric in j, k. It is easy to verify that the conditions of our existence theorem are fulfilled.

We state our result in the theorem:

In order that an analytic Riemannian metric of even dimension 2n be locally Hermitian it is necessary and sufficient that there is a family of rectangular frames relative to which the conditions (41) *are satisfied.*

For $n = 2$, that is, for a four-dimensional Riemannian metric, conditions (41) can be written more explicitly as follows:

$$\text{(42)}\qquad \begin{aligned} -R_{1223} + R_{3423} + R_{1214} - R_{3414} &= 0, \\ 2(R_{1234} - R_{1423}) - R_{1212} - R_{3434} + R_{1414} + R_{2323} &= 0. \end{aligned}$$

3. **Locally Kählerian metrics.** Using the same method, we shall derive the conditions for our Riemannian metric to be locally Kählerian. We begin by proving the following lemma:

There exists one, and only one, set of Pfaffian forms θ_{ik}, satisfying relations of the form

$$\text{(43)}\qquad \theta_{jk} + \bar{\theta}_{kj} = 0,$$

$$\text{(44)}\qquad d\theta_k = \theta_j \wedge \theta_{jk} + T_{kjl}\theta_j \wedge \theta_l, \qquad T_{kjl} + T_{klj} = 0.$$

By (34) we see that θ_{jk} are of the form

$$\theta_{jk} = \omega_{jk} + i\omega_{j,n+k} - A_{lkj}\omega_{n+l} + \lambda_{jkl}\theta_l,$$

where λ_{jkl} are to be determined. Since

$$\theta_{jk} + \bar{\theta}_{kj} = \left\{\frac{i}{2}(A_{jkl} + A_{lkj} + \bar{A}_{ljk}) + \lambda_{jkl}\right\}\theta_l$$
$$+ \left\{-\frac{i}{2}(\bar{A}_{kjl} + \bar{A}_{ljk} + A_{lkj}) + \bar{\lambda}_{kjl}\right\}\bar{\theta}_l,$$

condition (43) is satisfied if and only if

$$\text{(45)}\qquad \lambda_{jkl} = \frac{i}{2}(A_{kjl} + A_{klj} + \bar{A}_{jlk}).$$

This proves the lemma.

Using the expression (45) for λ_{ikl}, we find that the quantities T_{kjl} in (44) are given by

$$T_{kjl} = \frac{i}{2} \overline{A}_{ljk} .$$

If the forms θ_i, θ_{ik} are interpreted as defining a connection for the Hermitian metric, the exterior quadratic differential forms

$$\Theta_k = \frac{i}{2} \overline{A}_{ljk} \theta_j \wedge \theta_l \tag{46}$$

define the torsion of the connection in the sense of Cartan.

The exterior quadratic differential form associated to (22) is

$$\Theta = \theta_k \wedge \overline{\theta}_k . \tag{47}$$

Its exterior derivative is

$$d\Theta = \Theta_k \wedge \overline{\theta}_k - \theta_k \wedge \overline{\Theta}_k ,$$

which is zero if and only if

$$\Theta_k = 0, \tag{48}$$

or $A_{jkl} = 0$. It follows that a necessary and sufficient condition for our Riemannian metric to be locally Kählerian is that the differential system

$$\phi_{ik} + i\psi_{ik} = 0 \tag{49}$$

or

$$\phi_{ik} = 0, \qquad \psi_{ik} = 0 \tag{50}$$

has an integral manifold on which (35) holds.

From (37) it follows by exterior differentiation of (50) that

$$\Omega_{j,n+k} = \Omega_{k,n+j} , \qquad \Omega_{jk} = \Omega_{n+j,n+k} . \tag{51}$$

These are therefore necessary conditions. When they are fulfilled, the system (50) is in involution. Hence we have the theorem:

In order that an analytic Riemannian metric of even dimension 2n be locally Kählerian it is necessary and sufficient that there is a family of rectangular frames relative to which the conditions (51) *are satisfied. In particular, a locally flat metric is locally Kählerian.*

When $n = 2$, conditions (51) become

$$\Omega_{12} = \Omega_{34} , \qquad \Omega_{14} = \Omega_{23} . \tag{52}$$

We shall give a geometrical interpretation of conditions (51) in terms of the Riemannian curvature. For this purpose we write these conditions in the quantities $R_{\alpha\beta\rho\sigma}$, which are

$$
\begin{aligned}
R_{jklm} &= R_{jkl'm'} = R_{j'k'l'm'}, \\
R_{jklm'} &= R_{j'k'lm'}, \\
R_{jk'lm'} &= R_{kj'lm'}, \qquad (53) \\
R_{jk'lm} &= R_{kj'lm}, \\
R_{jk'l'm'} &= R_{kj'l'm'}.
\end{aligned}
$$

These relations are not independent. In particular, the last one is a consequence of the first five and the identities (19).

Before proceeding, we wish to make a remark about the extensiveness of the family of frames relative to which the conditions (51) are valid. It is known that if a frame $Pe_1 \cdots e_{2n}$ at a point P undergoes an orthogonal transformation

$$e_\alpha^* = u_{\alpha\beta} e_\beta, \tag{54}$$

the forms $\Omega_{\alpha\beta}$ undergo the transformation

$$\Omega_{\alpha\beta}^* = u_{\alpha\rho} u_{\beta\sigma} \Omega_{\rho\sigma}. \tag{55}$$

It follows from a straightforward computation that if the matrix $(u_{\alpha\beta})$ is of the form

$$(u_{\alpha\beta}) = \begin{pmatrix} A & B \\ -B & A \end{pmatrix}, \tag{56}$$

where A and B denote two $n \times n$ matrices, the conditions (51) remain satisfied. All the orthogonal matrices of the form (56) form a group. We can therefore suppose that in the neighborhood under consideration the family of frames is such that the frames with the same origin can be obtained from one of them by the transformations (54), with the matrix $(u_{\alpha\beta})$ given by (56). We note in passing that the conditions that the matrix (56) is orthogonal are

$$
\begin{aligned}
A\,'A + B\,'B &= I, \\
A\,'B - B\,'A &= 0,
\end{aligned} \tag{57}
$$

where $'A$, $'B$ are the transposes of A, B respectively.

Consider now a bivector of measure 1, with the components $p_{\alpha\beta}$ relative to the frame in question. The sectional curvature K in the plane of the bivector is given by the formula

$$-2K = R_{\alpha\beta\gamma\delta} p_{\alpha\beta} p_{\gamma\delta}. \tag{58}$$

(Compare [1; 195]; our notation is slightly different from Cartan's.) Following Bompiani, we can also consider the mixed curvature of two bivectors. If the latter have the components $p_{\alpha\beta}$, $q_{\gamma\delta}$ respectively and are both of the measure 1, their mixed curvature H is given by

$$-2H = R_{\alpha\beta\gamma\delta}p_{\alpha\beta}q_{\gamma\delta}. \tag{59}$$

Suppose the conditions (51) or (53) be satisfied. Consider the motion T_0 which maps the vector $\mathfrak{v}$ with the components $(v_k, v_{k'})$ into the vector $T_0\mathfrak{v}$ with the components $(v_{k'}, -v_k)$. Since we have

$$\begin{pmatrix} A & B \\ -B & A \end{pmatrix}\begin{pmatrix} 0 & I \\ -I & 0 \end{pmatrix}\begin{pmatrix} {}^tA & -{}^tB \\ {}^tB & {}^tA \end{pmatrix} = \begin{pmatrix} 0 & I \\ -I & 0 \end{pmatrix},$$

the mapping T_0 is independent of the choice of the frame and is well defined. Clearly T_0^2 is the negative of the identity mapping. It has n two-dimensional invariant subspaces $I_1, \cdots, I_n$, of which I_k is spanned by $\mathfrak{e}_k$ and $\mathfrak{e}_{k'}$. For two fixed and distinct indices j, k, we consider the space spanned by I_j and I_k and let p be a bivector of measure 1 in this four-dimensional space. Let q be any bivector of measure 1, and denote by $H(p, q)$, $H(T_0p, q)$ the mixed curvatures of the pairs of bivectors p, q and T_0p, q respectively. Putting

$$-S_{\beta\alpha} = S_{\alpha\beta} = R_{\alpha\beta\gamma\delta}q_{\gamma\delta}, \tag{60}$$

we have

$$-H(p, q) = S_{jk}\begin{vmatrix} v_j & v_k \\ w_j & w_k \end{vmatrix} + S_{jk'}\begin{vmatrix} v_j & v_{k'} \\ w_j & w_{k'} \end{vmatrix} + S_{kj'}\begin{vmatrix} v_k & v_{j'} \\ w_k & w_{j'} \end{vmatrix} + S_{j'k'}\begin{vmatrix} v_{j'} & v_{k'} \\ w_{j'} & w_{k'} \end{vmatrix},$$

$$-H(T_0p, q) = S_{jk}\begin{vmatrix} v_{j'} & v_{k'} \\ w_{j'} & w_{k'} \end{vmatrix} + S_{jk'}\begin{vmatrix} v_{j'} & -v_k \\ w_{j'} & -w_k \end{vmatrix} + S_{kj'}\begin{vmatrix} v_{k'} & -v_j \\ w_{k'} & -w_j \end{vmatrix} + S_{j'k'}\begin{vmatrix} v_j & v_k \\ w_j & w_k \end{vmatrix}, \tag{61}$$

$(j, j', k, k'$ not summed$)$.

It follows that

$$H(p, q) = H(T_0p, q). \tag{62}$$

We wish to prove that, conversely, these conditions are sufficient to imply the conditions (51). Our result, which may be regarded as a geometrical characterization of the local Kählerian property, can be stated in the theorem:

In order that an even-dimensional Riemannian metric be locally Kählerian it is necessary and sufficient that there is at every point a motion T_0 (varying analytically with the point) which carries a vector $\mathfrak{v}$ into the vector $T_0\mathfrak{v}$ perpendicular to $\mathfrak{v}$ and which has the further property: For any bivector p of measure 1 in the subspace spanned by any two two-dimensional invariant subspaces of T_0, the mixed curvature of p and any bivector q is equal to that of T_0p and q.

It remains to prove the sufficiency. Relative to a frame let T_0 be defined by the matrix $(t_{\alpha\beta})$. Since $T_0\mathfrak{v}$ is perpendicular to $\mathfrak{v}$ for all $\mathfrak{v}$, this matrix is skew-symmetric. There exists therefore a frame relative to which T_0 is given by the matrix

$$\begin{pmatrix} 0 & I \\ -I & 0 \end{pmatrix},$$

that is, T_0 carries the vector $(v_k, v_{k'})$ into the vector $(v_{k'}, -v_k)$. The equality (62) then implies the conditions

$$S_{jk} = S_{j'k'}, \qquad S_{jk'} = S_{kj'},$$

which are equivalent to (51). This completes the proof of the theorem.

From this theorem we can derive a conclusion concerning Riemannian metrics of constant curvature K. Let p be a bivector spanned by two unit vectors belonging to different invariant subspaces. Then p and T_0p are completely perpendicular and the mixed curvature $H(T_0p, p) = 0$. On the other hand, we have $H(p, p) = K$. It follows that *a Riemannian metric of constant curvature is locally Kählerian only if the curvature is zero.*

4. **Examples.** In spite of the above results we do not have a very clear picture of the strength of the local Kählerian property. A simple particular problem in this direction, of which we do not know the answer, is the following: Can a locally Kählerian metric be defined on the four-dimensional sphere?

On the other hand, it is easy to give an example of a compact even-dimensional Riemannian manifold with a locally Kählerian metric, which is not a Kählerian manifold. It suffices to take a three-dimensional compact Euclidean space form with first Betti number equal to zero and take its Cartesian product with a circle. On this manifold we can define a locally flat metric. But it is not a Kählerian manifold, because its first Betti number is one. According to the work of Hantzsche and Wendt, [7], there is exactly one orientable Euclidean space form with the above property. It is the fundamental domain in the three-dimensional Euclidean space of the discontinuous group generated by the translations in a unit distance along the coordinate axes and the helicoidal motions with angle π and pitch 1/2, whose axes are respectively the lines $y = 0$, $z = 0$; $x = 0$, $z = 1/4$; $x = 1/4$, $y = 1/4$.

Finally, we wish to show that if a four-dimensional compact orientable Riemannian manifold M satisfies the conditions

$$\Omega_{12} = \Omega_{34}, \qquad \Omega_{14} = \Omega_{23}, \qquad \Omega_{13} = \Omega_{42}, \tag{63}$$

there is some restriction on the topological properties of the manifold. By the Gauss-Bonnet formula, the Euler-Poincaré characteristic of M is given by

$$\chi = \frac{1}{4\pi^2} \int_M \Omega_{12}\Omega_{34} + \Omega_{13}\Omega_{42} + \Omega_{14}\Omega_{23}. \tag{64}$$

On the other hand, we have introduced a new topological invariant $\eta (=$ integer) of M given by the formula, [3; 970],

$$\eta = \frac{1}{2\pi^2} \int_M \Omega_{12}^2 + \Omega_{13}^2 + \Omega_{14}^2 + \Omega_{34}^2 + \Omega_{42}^2 + \Omega_{23}^2. \tag{65}$$

It follows that, when the conditions (63) are satisfied, we must have $\eta = 4\chi$. Since $\eta = 0$, $\chi = 2$ for the four-dimensional sphere, a Riemannian metric with the property (63) cannot exist on the sphere.

References

1. Élie Cartan, *Leçons sur la géométrie des espaces de Riemann*, 2nd edition, Paris, 1946.
2. Élie Cartan, *Les systèmes différentiels extérieurs et leurs applications géométriques*, Paris, 1945.
3. Shiing-shen Chern, *On Riemannian manifolds of four dimensions*, Bulletin of the American Mathematical Society, vol. 51(1945), pp. 964–971.
4. Shiing-shen Chern, *Topics in differential geometry* (mimeographed), Princeton, 1951.
5. Beno Eckmann, *Complex-analytic manifolds*, Proceedings of the International Congress of Mathematicians, Cambridge, Mass., 1950, vol. 2, pp. 420–427.
6. Charles Ehresmann, *Sur les variétés presque complexes*, Proceedings of the International Congress of Mathematicians, Cambridge, Mass., 1950, vol. 2, pp. 412–419.
7. W. Hantzsche and H. Wendt, *Dreidimensionale euklidische Raumformen*, Mathematische Annalen, vol. 110(1934), pp. 593–611.
8. E. Kähler, *Einführung in die Theorie der Systeme von Differentialgleichungen*, Leipzig and Berlin, 1934.
9. A. Lichnerowicz, *Généralisations de la Géométrie Kählerienne Globale, Colloque de Géométrie Différentielle*, Paris, 1951.

University of Chicago.

Reprinted from
Duke Math J. **20** (1953) 575–587

Extrait de l'*Enseignement mathém.*, T. 40, 1951-1954.

LA GÉOMÉTRIE DES SOUS-VARIÉTÉS D'UN ESPACE EUCLIDIEN A PLUSIEURS DIMENSIONS[1]

PAR

Shiing-Shen Chern (Chicago).

La géométrie des sous-variétés d'un espace euclidien de dimension quelconque contient naturellement comme cas particuliers l'étude des courbes et des surfaces de l'espace euclidien ordinaire. Cependant, malgré l'histoire très ancienne du sujet, nos renseignements dans le cas général sont assez maigres. Dans cette conférence je me propose de parler de quelques progrès qui ont été accomplis récemment.

I

1. — Soient E^{n+N} l'espace euclidien à $n + N$ dimensions et M une variété différentiable à n dimensions régulièrement plongée dans E^{n+N}. Cela signifie que tout point de M a un voisinage dans lequel la variété peut être définie en exprimant N coordonnées de E^{n+N} comme des fonctions des n autres coordonnées ayant des dérivées partielles continues d'un ordre assez

[1] Conférence faite à la séance de la Société mathématique suisse, tenue à Berne le 7 juin 1953.

élevé. Pour simplifier nous supposerons que M est compacte, bien que beaucoup de nos discussions soient valables sans cette hypothèse. Dans une formulation générale nous nous intéressons à des relations entre les courbures de la métrique riemannienne induite de M, les courbures relatives de M dans E^{n+N} et les propriétés topologiques de M elle-même et de sa position dans E^{n+N}.

La première idée féconde, remontant au moins à Gauss, consiste à étudier une application qui généralise l'application normale d'une surface dans l'espace euclidien ordinaire. Soit en effet $G(n, N)$ (resp. $\tilde{G}(n, N)$) la variété grassmannienne des variétés linéaires (resp. variétés linéaires orientées) à n dimensions passant par un point fixe O de E^{n+N}. Cette variété est de dimension nN. Pour $n = 1$ ou $N = 1$, $G(n, N)$ est homéomorphe à l'espace projectif réel et $\tilde{G}(n, N)$ à la sphère. Pour chaque point $x \in M$, on mène par l'espace linéaire $T(x)$ à n dimensions parallèle au plan tangent à M en x. Cette construction conduit à l'application tangentielle $T: M \longrightarrow G(n, N)$. D'une manière analogue on définit l'application tangentielle $\tilde{T}: M \longrightarrow \tilde{G}(n, N)$, si M est orientée. Ces applications jouent un rôle fondamental dans l'étude de la géométrie de M dans E^{n+N}.

Tout d'abord l'application T induit des homomorphismes sur les groupes d'homologie. Plus précisément, soit J un groupe de coefficients et, X étant un espace topologique, désignons par $H_r(X, J)$ (resp. $H^r(X, J)$) le groupe d'homologie (resp. de cohomologie) de dimension r de X avec le groupe de coefficients J. L'application T induit les homomorphismes suivants:

$$\begin{aligned} T_*&: H_r(M, J) \longrightarrow H_r(G(n, N), J) , \\ T^*&: H^r(G(n, N), J) \longrightarrow H^r(M, J) . \end{aligned} \tag{1}$$

De plus, si J est un anneau, la somme directe des groupes de cohomologie de différentes dimensions peut être munie d'une structure d'anneau et l'homomorphisme T^* est un homomorphisme d'anneaux. Des considérations analogues sont valables pour l'application $\tilde{T}$.

De quelle manière ces homomorphismes dépendent-ils de la variété M ? Pour étudier cette question disons que deux variétés M_0 et M_1 sont régulièrement homotopes s'il existe une famille M_t $(0 \leqq t \leqq 1)$ de variétés plongées, dépendant continuement de t et du point $x \in M$, qui contient les variétés données pour $t = 0$ et $t = 1$. Il est clair que pour deux variétés régulièrement homotopes les homomorphismes (1) sont les mêmes. D'autre part, M. Whitney a démontré que si $N \geqq n + 1$, deux variétés de dimension n dans E^{n+N} sont toujours régulièrement homotopes [1]. Dans ce cas les homomorphismes (1) dépendent de M comme variété différentiable abstraite et ne dépendent pas de sa position dans E^{n+N}. En particulier, T^* s'appellera l'homomorphisme caractéristique de M et un élément de l'image de T^* une classe caractéristique.

Il y a d'autres cas où les homomorphismes (1) ne dépendent que de M. Par exemple, si M est une hypersurface orientée $(N = 1)$, la variété $\tilde{G}(n, 1)$ est homéomorphe à une sphère de dimension n et les homomorphismes (1) sont essentiellement déterminés par le degré de l'application $\tilde{T}$. Si, de plus, n est pair, on peut démontrer que ce degré est égal à $\chi(M)/2$, où $\chi(M)$ est la caractéristique d'Euler-Poincaré de M.

Dans le cas général il sera utile d'imposer des conditions moins restrictives sur M, en admettant les cas où M peut se rencontrer elle-même de telle sorte qu'en chaque point où M se coupe elle-même il n'y a que deux branches de M et que les plans tangents à celles-ci soient transversaux l'un à l'autre. Ce sont les variétés immergées au sens de M. Whitney. L'application tangentielle et la notion d'homotopie régulière s'étendent à ces variétés. Il est encore vrai que les homomorphismes (1) sont les mêmes pour deux variétés immergées régulièrement homotopes.

On en sait plus dans le cas $n = N = 1$, c'est-à-dire le cas des courbes fermées du plan avec un nombre fini de points doubles. M. Whitney a démontré [19] qu'il est possible d'orienter une telle courbe et de donner un signe à chaque point double tel que, si ν^+ (resp. ν^-) désigne le nombre de points doubles positifs

---

[1] Ce théorème a été démontré dans [20] pour $N \geqq n + 2$; mais la méthode de démonstration et le théorème que M peut être plongée topologiquement dans E^{2n} entraînent qu'il est encore vrai pour $N \geqq n + 1$.

Les nombres entre crochets [] se réfèrent à la *Bibliographie* à la fin de cet article.

(resp. négatifs), le degré de l'application tangentielle $\tilde{T}$ est égal à $1 + \nu^+ - \nu^-$. Ce théorème contient comme cas particulier le théorème bien connu (« Umlaufssatz ») qui affirme que le degré de $\tilde{T}$ est égal à 1 pour une courbe fermée simple, convenablement orientée. De plus, M. WHITNEY a aussi démontré que, si les degrés des applications tangentielles de deux courbes fermées de ce genre sont les mêmes, les courbes sont régulièrement homotopes. En d'autres termes, les classes des courbes fermées régulièrement homotopes sont en correspondance biunivoque avec les entiers. J'ignore si un théorème analogue est valable pour le cas $n = N > 1$.

2. — Pour faire une étude plus approfondie des homomorphismes (1), le problème préliminaire est la connaissance des groupes d'homologie et de cohomologie des variétés $G(n, N)$ et $\tilde{G}(n, N)$. Ce problème a été traité par M. EHRESMANN [5], [8], [14], en utilisant les décompositions de ces variétés par les variétés de Schubert. Nous nous intéressons surtout aux cas où J est soit le corps Z_2 des entiers modulo deux soit le corps R des nombres réels. Faisons aussi l'hypothèse $n + 1 \leqq N$. Alors les éléments de dimension $\leqq n$ de l'anneau de cohomologie $H(G(n, N), Z_2)$ sont engendrés (au sens de la structure d'anneau) par des classes de cohomologie w^i de dimension i, $1 \leqq i \leqq n$. Les classes caractéristiques $W^i = T^* w^i$ dans M sont appelées les classes de Stiefel-Whitney. Dans les applications il sera commode d'introduire, avec l'indéterminée t, le polynome de Stiefel-Whitney:

$$W(t) = \sum_{i=0}^{n} W^i t^i, \quad W^0 = 1 \,. \tag{2}$$

Avec l'anneau de coefficients R les éléments de dimension $\leqq n$ de l'anneau de cohomologie de $G(n, N)$ sont engendrés par des classes p^{4k} de dimension $4k$, $4 \leqq 4k \leqq n$. Leurs images par T^*, les classes de cohomologie $P^{4k} = T^* p^{4k}$ dans M, sont appelées les classes de Pontrjagin. De même, nous introduisons le polynome de Pontrjagin:

$$\mathrm{P}(t) = \sum_{0 \leqq k \leqq \frac{n}{4}} (-1)^k \mathrm{P}^{4k} t^{4k}, \quad \mathrm{P}^0 = 1 \,.$$

Bien entendu, les classes de Stiefel-Whitney et de Pontrjagin sont des invariants de M comme une variété différentiable abstraite. La description de l'anneau de cohomologie de $G(n, N)$ dans le cas $n \geqq N$ est beaucoup plus compliquée.

D'après les théorèmes célèbres de M. DE RHAM, il correspond, à chaque classe de cohomologie de dimension r à coefficients réels, une forme différentielle extérieure fermée de degré r, définie modulo les dérivées des formes de degré $r - 1$. Pour notre variété $G(n, N)$, qui est transformée transitivement par le groupe de transformations orthogonales autour du point O, il suffit de nous limiter aux formes qui sont invariantes par rapport à ce groupe. Nous nous proposons de donner explicitement une telle forme correspondant à la classe p^{4k}.

Pour cela, considérons la famille de tous les repères rectangulaires qui consistent en $n + N$ vecteurs unitaires $e_1, \ldots, e_{n+N}$, deux à deux perpendiculaires. Supposons que l'élément de $G(n, N)$ est déterminé par les n premiers vecteurs $e_1, \ldots, e_n$. Pour éviter des répétitions, faisons les conventions suivantes sur nos indices:

$$1 \leqq i, j, k, l \leqq n \,, \quad n + 1 \leqq r, s, t \leqq n + N \,. \tag{4}$$

Cela étant, définissons les formes de Pfaff

$$\omega_{is} = de_i \cdot e_s \,, \tag{5}$$

où les produits sont des produits scalaires. Comme l'espace linéaire à n dimensions déterminé par les e_i reste fixe quand on fait sur les vecteurs e_i et e_r des transformations orthogonales indépendantes, les formes extérieures (ou ordinaires) engendrées par les ω_{is} peuvent être considérées comme des formes dans $G(n, N)$ si elles sont invariantes par ces transformations. Un exemple simple est fourni par la forme différentielle quadratique ordinaire

$$\Phi = \sum_{i,s} \omega_{is}^2 \,. \tag{6}$$

Elle définit une métrique riemannienne dans $G(n, N)$. Pour construire des formes différentielles extérieures dans $G(n, N)$, posons d'abord

$$\Omega_{ij} = - \sum_{s} \omega_{is} \wedge \omega_{js} \,, \tag{7}$$

et puis

$$\theta_{2m} = \frac{1}{(2\pi)^m m!} \Sigma \delta(i_1 \dots i_m ; j_1 \dots j_m) \Omega_{i_1 j_1} \wedge \dots \wedge \Omega_{i_m j_m}, \quad (8)$$

où le symbole $\delta(i_1 \dots i_m; j_1 \dots j_m)$ n'est pas nul si et seulement si $j_1, \dots, j_m$ constituent une permutation de $i_1, \dots, i_m$, auquel cas il est égal à $+1$ ou -1 suivant que la permutation est paire ou impaire. De plus, la sommation dans (8) est étendue à tous les $i_1, \dots, i_m$ de 1 à n. Il est facile de voir que les formes θ_{2m} restent invariantes, si l'on applique des transformations orthogonales indépendantes aux vecteurs e_i et e_r. Elles sont donc des formes différentielles extérieures de degré $2m$ dans $G(n, N)$. Comme Ω_{ij} est antisymétrique dans les indices i, j, on déduit immédiatement de la définition (8) que $\theta_{2m} = 0$, si m est impair.

Le théorème fondamental dans cet ordre d'idées est le fait que la forme θ_{4k}, $4 \leqq 4k \leqq n$, est fermée et que la classe de cohomologie qu'elle détermine est précisément p^{4k} [5], p. 82.

3. — Il faut appliquer ces considérations à la géométrie de la variété M dans E^{n+N}. Pour cela on emploie la méthode du repère mobile. On entend par repère dans E^{n+N} la figure formée d'un point x et de $n+N$ vecteurs unitaires $e_1, \dots, e_{n+N}$, deux à deux perpendiculaires, d'origine x. Nous nous limitons à la famille de tous les repères où x est un point de M et $e_1, \dots, e_n$ sont des vecteurs tangents à M en x. Alors les vecteurs e_s sont normaux à M en x. On a $e_s \,.\, dx = 0$, et on posera $\omega_i = e_i dx$, où les produits sont des produits scalaires. La variété ayant un plan tangent en chaque point, les formes de Pfaff ω_i sont toujours linéairement indépendantes. On obtient l'application tangentielle T en menant par O l'espace linéaire à n dimensions déterminé par les vecteurs e_i.

Développons en détail les formules fondamentales de la géométrie locale de M. On peut écrire, pour la famille de repères considérée,

$$\begin{aligned} dx &= \sum_i \omega_i e_i , \\ de_A &= \sum_B \omega_{AB} e_B , \quad \omega_{AB} + \omega_{BA} = 0 , \quad 1 \leqq A, B \leqq n + N . \end{aligned} \quad (9)$$

On en déduit

$$0 = e_s \cdot d(dx) = e_s \cdot \sum_i de_i \wedge \omega_i = \sum_i \omega_{is} \wedge \omega_i \ .$$

Les formes ω_i étant linéairement indépendantes, cela entraîne

$$\omega_{is} = \sum_j A_{sij}\, \omega_j \ , \tag{10}$$

où

$$A_{sij} = A_{sji} \ . \tag{11}$$

Avec ces quantités, on construit les formes différentielles quadratiques ordinaires

$$\Psi_s = \sum_{i,j} A_{sij}\, \omega_i\, \omega_j \ , \tag{12}$$

qui généralisent la seconde forme fondamentale d'une surface. Si $v = \sum_s v_s \cdot e_s$ est un vecteur unitaire normal, on a en effet

$$-dv\, dx = \sum_s v_s \Psi_s \ . \tag{13}$$

On obtient les courbures riemaniennes de la métrique riemannienne induite de M par la considération des formes

$$\begin{aligned} \Omega_{ij} &= -\sum_s \omega_{is} \wedge \omega_{js} = -\sum_{k,l,s} A_{sik} A_{sjl}\, \omega_k \wedge \omega_l \\ &= -\frac{1}{2} \sum_{k,l,s} (A_{sik} A_{sjl} - A_{sil} A_{sjk})\, \omega_k \wedge \omega_l \ . \end{aligned} \tag{14}$$

Les expressions

$$R_{ijkl} = \sum_s (A_{sik} A_{sjl} - A_{sil} A_{sjk}) \tag{15}$$

sont essentiellement les composantes du tenseur de Riemann-Christoffel. Plus précisément, si π est un élément plan passant par x déterminé par les vecteurs linéairement indépendants suivants:

$$\xi = \sum_i \xi_i e_i \ , \qquad \eta = \sum_i \eta_i e_i \ , \tag{16}$$

la courbure riemannienne en π est donnée par la formule

$$R(x,\pi) = 2\,\frac{\sum_{s,i,j,k,l}(A_{sik}A_{sjl} - A_{sil}A_{sjk})\,\xi_i\,\eta_j\,\xi_k\,\eta_l}{\sum_{i,j,k,l}(\delta_{ik}\delta_{jl} - \delta_{il}\delta_{jk})\,\xi_i\,\eta_j\,\xi_k\,\eta_l}. \tag{17}$$

D'après la définition des classes de Pontrjagin et le théorème à la fin du dernier paragraphe, il s'ensuit que les formes Θ_{4k} obtenues à partir de θ_{4k} en substituant aux Ω_{ij} les expressions (14) sont des formes différentielles fermées dans M et que Θ_{4k} détermine la classe P^{4k} de Pontrjagin au sens du théorème de M. de Rham. Cette proposition est due essentiellement à M. PONTRJAGIN [15], qui n'a du reste pas donné la correspondance exacte. On a ici une relation entre les propriétés de courbure de M et les invariants de sa structure différentiable. A ma connaissance on ne sait pas si les classes de Pontrjagin de M sont des invariants topologiques, tandis que, d'après un théorème de M. THOM [18], les classes de Stiefel-Whitney le sont.

4. — Le résultat le plus important de ce genre est peut-être la formule de GAUSS-BONNET [2], [5], [10]. Elle n'est pas exactement contenue comme cas particulier dans les considérations précédentes, mais peut-être déduite d'une manière analogue. Dans ce cas nous supposons que M est de dimension paire et orientée, de sorte que l'application tangentielle soit $\tilde{T}: M \longrightarrow \tilde{G}(n, N)$. Soient E^{n+N-1} un hyperplan passant par O et $\tilde{G}(n, N-1)$ la variété grassmannienne dans E^{n+N-1}. $\tilde{G}(n, N-1)$ est une sous-variété orientée de dimension $n(N-1)$ de $\tilde{G}(n, N)$ et définit un cycle z. A ce cycle correspond un cocycle γ de dimension n, à coefficients réels, qui est défini par la condition que, pour tout cycle z de dimension n, on a

$$\gamma \cdot z = KI(\xi, z), \tag{18}$$

où le symbole KI, à droite, désigne le nombre d'intersection des cycles entre parenthèses. Comme dans le dernier paragraphe, notre problème est de trouver une forme différentielle extérieure fermée qui est invariante par le groupe des rotations autour de O transformant $\tilde{G}(n, N)$ et qui détermine la classe de cohomologie de γ au sens du théorème de M. de Rham. On trouve qu'une telle forme est donnée par

$$\Omega = (-1)^{\frac{n}{2}} \frac{1}{2^n \pi^{\frac{n}{2}} \left(\frac{n}{2}\right)!} \sum_{i_1 \ldots i_n} \varepsilon_{i_1 \ldots i_n} \Omega_{i_1 i_2} \wedge \ldots \wedge \Omega_{i_{n-1} i_n}, \quad (19)$$

où $\varepsilon_{i_1 \ldots i_n}$ est $+1$ ou -1, suivant que $i_1, \ldots, i_n$ est une permutation paire ou impaire de $1, \ldots, n$, et est nul si deux i_k sont égaux. Quand on substitue dans (19) les expressions (14), on obtient une forme différentielle fermée qui détermine la classe de cohomologie de $\tilde{T}^* \gamma$. En d'autres termes, on a

$$\int_M \Omega = \tilde{T}^*(\gamma) \cdot M = \gamma \cdot \tilde{T}_*(M) = KI(\xi, \tilde{T}_*(M)),$$

où M désigne le cycle fondamental de la variété orientée M.

Il est possible de déterminer plus explicitement cette dernière expression. En effet, soit v_0 le vecteur unitaire perpendiculaire à E^{n+N-1}. Supposons que ξ et $\tilde{T}(M)$ n'ont qu'un nombre fini de points communs, c'est-à-dire que M n'a qu'un nombre fini de points où les plans tangents soient perpendiculaires à v_0. En projetant v_0 orthogonalement sur le plan tangent à chaque point x de M, on définit un champ continu de vecteurs sur M avec un nombre fini de points où les vecteurs sont nuls. On vérifie alors que $KI(\xi, \tilde{T}_*(M))$ est égal à la somme des indices des singularités de ce champ de vecteurs. Cette somme est, comme il est bien connu, égale à la caractéristique d'Euler-Poincaré $\chi(M)$ de M. La formule ci-dessus peut donc être écrite

$$\int_M \Omega = \chi(M). \quad (20)$$

C'est la formule de Gauss-Bonnet.

La forme sous l'intégrale dans (20) peut être écrite

$$\Omega = K(x)\, \omega_1 \wedge \ldots \wedge \omega_n = K(x)\, dV, \quad (21)$$

où dV est l'élément de volume de M et $K(x)$ est un invariant scalaire qui ne dépend que de la métrique riemannienne de M. Cet invariant n'est autre que la courbure de Lipschitz-Killing [11], [13] et possède une interprétation géométrique simple. Pour cela rappelons qu'une hypersurface a une courbure scalaire, c'est la courbure de Gauss-Kronecker qui est le quotient du déterminant de la seconde forme fondamentale par le déterminant de la première forme fondamentale. Comme la seconde forme fondamentale dépend du choix du vecteur normal, la courbure de Gauss-Kronecker est un invariant de l'hypersurface si n est pair et est déterminée au signe près si n est impair. Dans tous les cas elle est bien déterminée si l'on fixe un vecteur normal. Maintenant soient M une sous-variété générale dans E^{n+N} et $\nu = \sum_s \nu_s e_s$ un vecteur unitaire normal au point x de M. Soient $M(\nu)$ la projection orthogonale de M dans l'espace linéaire à $n+1$ dimensions déterminé par ν et le plan tangent à M en x, et $G(x, \nu)$ la courbure de Gauss-Kronecker de $M(\nu)$ au point x. Désignons par $d\sigma_{N-1}$ l'élément de volume de l'hypersphère unitaire dans l'espace normal à M en x. Son volume total est une constante donnée par

$$c_{N-1} = \frac{2\pi^{\frac{1}{2}N}}{\Gamma\left(\frac{1}{2}N\right)}. \tag{22}$$

Cela étant, on démontre que l'intégrale

$$\frac{c_n}{2c_{n+N-1}} \int G(x, \nu)\, d\sigma_{N-1} \tag{23}$$

étendue à l'hypersphère unitaire dans l'espace normal est nulle si n est impair et égale à $K(x)$ si n est pair. C'est ainsi que l'invariant $K(x)$ a été introduit par LIPSCHITZ et KILLING.

5. — Les développements ci-dessus se rapportent à l'étude de l'application tangentielle. Nous allons terminer ces discussions par des remarques sur une application analogue, l'application normale $N: M \longrightarrow G(N, n)$. Elle est définie par la condition que, pour $x \in M$, $N(x)$ est l'espace linéaire à N dimensions passant

par O et parallèle à l'espace normal à M en x. Cela conduit, d'une manière tout à fait analogue à l'application tangentielle, aux classes caractéristiques de Stiefel-Whitney et de Pontrjagin, dites normales. Nous désignons les polynomes correspondants par

$$\overline{W}(t) = \sum_{i=0}^{N} \overline{W}^i t^i, \quad \overline{W}^0 = 1, \tag{24}$$

$$\overline{P}(t) = \sum_{0 \leq k \leq 4N} (-1)^k \overline{P}^{4k} t^{4k}, \quad \overline{P}^0 = 1. \tag{25}$$

Posons

$$\overline{\Omega}_{rs} = -\frac{1}{2} \sum_{i,h,l} (A_{rik} A_{sil} - A_{ril} A_{sik}) \omega_k \wedge \omega_l, \tag{26}$$

et puis

$$\overline{\Theta}_{4k} = \frac{1}{(2\pi)^{2k}(2k)!} \Sigma \delta(r_1, \dots, r_{2k}; s_1, \dots, s_{2k}) \overline{\Omega}_{r_1 s_1} \wedge \dots \wedge \overline{\Omega}_{r_{2k} s_{2k}}. \tag{27}$$

Alors la forme différentielle $\overline{\Theta}_{4k}$ est fermée et détermine la classe normale de Pontrjagin $\overline{P}_{4k}$. Il est sous-entendu que la forme $\overline{\Theta}_{4k}$ dépend de la position de M dans E^{n+N}, et non seulement de sa métrique riemannienne, contrairement au cas des classes de Pontrjagin tangentielles.

Il y a des relations entre les classes caractéristiques tangentielles et normales qu'on peut déduire commodément par l'étude de l'homéomorphisme involutif σ: $G(n, N) \longrightarrow G(N, n)$, défini en prenant pour chaque espace linéaire à n dimensions passant par O son espace linéaire perpendiculaire. Les homomorphismes induits par σ sur les groupes d'homologie et de cohomologie ont été déterminés par Wu Wen-Tsun, au moins dans les dimensions qui nous intéressent [22]. Les relations entre les classes caractéristiques, qui en résultent, peuvent être écrites sous la forme suivante:

$$W(t)\,\overline{W}(t) = 1, \tag{28}$$

$$P(t)\,\overline{P}(t) = 1, \tag{29}$$

où les polynomes seront multipliés formellement, la multiplication des coefficients étant celle de l'anneau de cohomologie. D'après Whitney, on appelle (28), (29) les théorèmes de dualité.

Pour justifier l'étude des classes caractéristiques, il serait important de démontrer qu'elles ne sont pas triviales. On doit à M. Wu Wen-Tsun plusieurs exemples où des classes caractéristiques ne s'annulent pas [22]. D'autre part, on a des théorèmes sur la trivialité de certaines classes caractéristiques. En particulier, d'après MM. Seifert, Whitney et Thom, la classe $\overline{W}^N$ est toujours nulle [16], [18], [21]. De plus, si M est orientable, la classe $\overline{W}^N$, qui peut être définie avec les coefficients entiers, est nulle. Cela signifie géométriquement qu'il est possible de définir sur une variété orientable un champ continu de vecteurs normaux non nuls.

II

5. — J'ai beaucoup insisté sur les propriétés topologiques de l'application tangentielle. Il y a des questions plus géométriques qui seraient aussi intéressantes. L'une des plus naturelles est la condition sur l'application $T : M \longrightarrow G(n, N)$ pour qu'elle soit une application tangentielle.

On peut donner immédiatement une condition nécessaire. Soit en effet b un vecteur unitaire fixe. Le produit scalaire $f(x) = bx$, $x \in M$, définit une fonction continue sur M. M étant compacte, cette fonction possède un maximum et un minimum, où on a $bdx = 0$. Cela veut dire que les éléments $T(x)$ correspondants sont situés dans l'hyperplan passant par O et perpendiculaire à b. Par conséquent, pour chaque b il y a au moins deux points de M dont les images par T sont dans l'hyperplan perpendiculaire à b. Pour $n = 1$ cette condition est suffisante pour que l'application T soit l'application tangentielle d'une courbe close. J'ignore si ce résultat s'étend pour n quelconque.

Néanmoins on déduit de cette condition des conséquences intéressantes. Pour simplifier supposons que M soit orientée, de sorte que l'application à considérer soit $\tilde{T} : M \longrightarrow \tilde{G}(n, N)$. Evaluons le volume de l'ensemble des points de l'hypersphère de rayon unité de E^{n+N}, chaque point étant compté un nombre de fois égal au nombre des $\tilde{T}(x)$ contenu dans son hyperplan perpendiculaire. Par la méthode de la géométrie intégrale ce volume peut être exprimé par une formule du type de Crofton. Avec les notations du paragraphe 4, on pose

$$K^*(x) = \frac{c_n}{2\,c_{n+N-1}} \int |\,G(x,\nu)\,|\,d\sigma_{N-1} \geqq 0 \,, \tag{30}$$

où la fonction sous l'intégrale est la valeur absolue de $G(x, \nu)$. Alors on trouve que le volume considéré est égal à

$$\frac{2\,c_{n+N-1}}{c_n} \int_M K^*(x)\,dV \,.$$

Parce que chaque hyperplan contient au moins deux $\tilde{T}(x)$, $x \in M$, ce volume est $\geqq 2\,c_{n+N-1}$, et on a l'inégalité

$$\int_M K^*(x)\,dV \geqq c_n \,, \tag{31}$$

où l'on fait la convention que $c_1 = 2$. Pour une courbe fermée dans l'espace euclidien ordinaire l'intégrale du premier membre de (31) est égale à l'intégrale de la valeur absolue de la courbure de M, divisée par π, et la formule (31) se réduit au théorème bien connu de M. Fenchel. Il est clair que l'invariant $K^*(x)$ dépend de la position de M dans E^{n+N}, et on a $K^*(x) \geqq K(x)$, pour tout $x \in M$.

Dans le cas d'une courbe de l'espace ordinaire ces considérations conduisent au théorème intéressant de MM. Fary et Milnor [9], [12]. Ce théorème se rapporte à une courbe satisfaisant à l'inégalité

$$\int_M K^*(x)\,dV < 2\,c_n \,. \tag{32}$$

Sous l'hypothèse (32) on voit qu'il y a un vecteur unitaire b de sorte que la fonction $f(x) = bx$, $x \in M$, n'a qu'un maximum et un minimum. On voit facilement qu'alors la courbe M est isotope à un cercle et n'est pas un nœud.

Ce résultat peut être étendu au cas général, de la manière suivante[1]: Supposons que la condition (32) soit satisfaite, avec une constante convenable à droite, alors la variété M a ses nombres de Betti modulo 2 nuls

[1] Pour les détails analytiques voir l'*appendice* à la fin de cet article.

pour les dimensions 1, 2, ..., $n-1$. Cela tient au fait que la fonction $f(x) = bx$, $x \in M$, pour un certain b, n'a qu'un maximum et un minimum et ne se réduit pas à une constante. Alors ses nombres de type k, $1 \leqq k \leqq n-1$, au sens de M. MORSE sont tous nuls. L'énoncé est ainsi une conséquence immédiate des inégalités de M. MORSE, qui affirment que le nombre de type k d'une fonction continue sur M est au moins égal au nombre de Betti modulo deux pour la dimension k. Cette généralisation est aussi connue à M. MILNOR.

Il est peut-être justifié d'appeler courbure totale l'intégrale du premier membre de (31), contrairement à l'usage qui remonte à GAUSS. Les résultats ci-dessus montrent que c'est une notion féconde de laquelle on peut faire des applications simples.

7. — La question des implications globales de la métrique riemannienne d'une surface dans l'espace ordinaire a été beaucoup étudiée; deux des problèmes les plus importants sont ceux de réalisation et de rigidité. Quand $n \geqq 3$, la métrique riemannienne a des conséquences très fortes, même localement. Si M est de plus compacte, elle contient un point x_0 à une distance maximum d'un point fixe O de E^{n+N}. L'étude de la géométrie locale en ce point conduit aux résultats dont je vais parler [6].

Appelons d'abord une direction tangentielle direction asymptotique si elle annule toutes les formes Ψ_s:

$$\Psi_s = \sum_{i,j} A_{sij}\, \omega_i\, \omega_j = 0\,. \tag{33}$$

Il est facile de voir qu'au point x_0 il n'y a pas de directions asymptotiques réelles. M. T. OTSUKI a démontré le lemme suivant [1]: Si le second membre de (17) est $\leqq 0$ pour tous les éléments plans déterminés par ξ, η, le système d'équations (33) a des solutions réelles non-triviales ω_i, si $N \leqq n-1$. Ce lemme a été conjecturé par M. KUIPER et moi et démontré dans des cas simples. Il a comme conséquence le théorème géométrique suivant: Si l'espace tangent à chaque point de M contient un espace linéaire à q dimensions tel que la courbure riemannienne soit $\leqq 0$ pour tous ses éléments plans, alors $N \geqq q$. Il est clair que ce théorème n'est pas vrai localement.

[1] [13]. Une autre démonstration a été donnée par M. T. SPRINGER à Leiden, Hollande.

Une autre question de ce genre concerne l'entier $\mu(x)$ tel que $n - \mu(x)$ soit le nombre minimum de formes de Pfaff linéairement indépendantes au moyen desquelles les formes Ω_{ij} dans (14) peuvent être exprimées. Soit $n - \nu(x)$ le nombre minimum de formes de Pfaff linéairement indépendantes au moyen desquelles les formes différentielles ordinaires Ψ_s peuvent être exprimées. M. Kuiper et moi avons démontré les inégalités

$$\nu(x) \leqq \mu(x) \leqq N + \nu(x) . \tag{34}$$

On en déduit la conséquence géométrique suivante: Si $\mu_0 = \inf_{x \varepsilon M} \mu(x)$, alors $N \geqq \mu_0$. En particulier, si la métrique riemannienne induite de M est euclidienne, on a $\mu_0 = n$ et, par suite, $N \geqq n$. Ce résultat est dû à M. Tompkins; il généralise le fait bien connu qu'une surface développable dans l'espace ordinaire n'est pas close.

8. — A côté des invariants arithmétiques introduits ci-dessus, on en a d'autres qui jouent un rôle important dans la géométrie de M dans E^{n+N}. Nous avons vu, dans la formule (13), qu'il y a une forme différentielle quadratique ordinaire (la seconde forme fondamentale) associée à chaque vecteur unitaire normal. Les vecteurs normaux, dont la seconde forme fondamentale est nulle, appartiennent à un sous-espace linéaire de l'espace normal. Son espace perpendiculaire dans l'espace normal de M est appelé le premier espace normal. Sa dimension $p(x)$ est égale au nombre des formes linéairement indépendantes parmi les Ψ_s, d'où $p(x) \leqq n(n+1)/2$.

Un autre invariant arithmétique de M peut être introduit comme il suit. Choisissons les vecteurs e_s dans l'espace normal tels que $e_{n+1}, \ldots, e_{n+p}$ soient dans le premier espace normal. Alors $\Psi_{n+1}, \ldots, \Psi_{n+p}$ sont linéairement indépendantes et $\Psi_{n+p+1}, \ldots, \Psi_{n+N}$ en sont des combinaisons linéaires. On considère les lignes de formes de Pfaff:

$$\omega_{i,n+1}, \ldots, \omega_{i,n+p}, \quad 1 \leqq i \leqq n . \tag{35}$$

Le plus grand entier $\tau(x)$, tel qu'il existe $\tau(x)$ lignes dont les $p\tau(x)$ formes sont linéairement indépendantes s'appelle le type de M au point x. On a évidemment $\tau(x) \leqq [n/p(x)]$, le dernier nombre étant le plus grand entier $\leqq n/p(x)$.

Cela étant, on a le théorème local suivant, qui est dû à M. ALLENDOERFER [1], [4]: Deux variétés isométriques, dont les premiers espaces normaux sont de même dimension, ne diffèrent que par un mouvement (propre ou impropre), si l'une d'entre elles est de type $\geqq 3$.

Ce théorème peut être considéré comme un théorème de rigidité locale. Bien entendu, la condition sur le type est très forte.

III

9. — Pour mieux comprendre la géométrie des sous-variétés, il serait utile d'étudier avec plus de détails le cas d'une surface dans $E^4 (n = N = 2)$. Nous faisons une autre hypothèse simplificatrice en supposant que M est orientée. Alors l'application tangentielle est $\tilde{T}: M \longrightarrow \tilde{G}(2, 2)$. Dans ce cas on peut donner de cette dernière variété une description simple. En effet, soient $p_{\alpha\beta}$, $1 \leqq \alpha, \beta \leqq 4$, les coordonnées plückeriennes dans $G(2, 2)$. Ce sont les coordonnées homogènes assujetties aux conditions

$$p_{\alpha\beta} + p_{\beta\alpha} = 0 \, , \quad p_{12}p_{34} + p_{13}p_{42} + p_{14}p_{23} = 0 \, . \tag{36}$$

Nous les normalisons par la condition

$$\sum_{\alpha, \beta} p^2_{\alpha\beta} = 2 \, . \tag{37}$$

Alors les coordonnées $p_{\alpha\beta}$ satisfaisant aux conditions (36), (37) peuvent être considérées des coordonnées dans $\tilde{G}(2, 2)$, de sorte que les deux plans orientés qui donnent le même plan non orienté aient des coordonnées différant par le signe. Introduisons des coordonnées nouvelles dans $\tilde{G}(2, 2)$ en posant

$$\begin{aligned} x_1 &= p_{12} + p_{34} \, , \quad & x_2 &= p_{23} + p_{14} \, , \quad & x_3 &= p_{31} + p_{24} \, , \\ y_1 &= p_{12} - p_{34} \, , \quad & y_2 &= p_{23} - p_{14} \, , \quad & y_3 &= p_{31} - p_{24} \, . \end{aligned} \tag{38}$$

Avec ces coordonnées x_λ, y_λ, $1 \leqq \lambda \leqq 3$, les conditions (36) et (37) sont équivalentes aux conditions

$$x_1^2 + x_2^2 + x_3^2 = y_1^2 + y_2^2 + y_3^2 = 1 \, . \tag{39}$$

Ceci démontre que la variété $\tilde{G}(2, 2)$ est homéomorphe au produit cartésien de deux sphères S_x et S_y ordinaires. Comme $\tilde{G}(2, 2)$ peut être identifiée avec la variété des droites orientées de l'espace elliptique à trois dimensions, ce fait est à la base d'une représentation de Fubini et Study.

Fixons une orientation de S_x et S_y et désignons par M, S_x, S_y les cycles fondamentaux de ces variétés. Les invariants homologiques qu'on peut déduire de l'homomorphisme $\tilde{T}_*$ sont les entiers d_x, d_y définis par la condition $\tilde{T}_*(M) \sim d_x S_x + d_y S_y$, où $\tilde{T}_*$ désigne l'homomorphisme induit par $\tilde{T}$. On peut démontrer que [7], si les orientations de S_x et S_y sont convenablement choisies, on a $d_x = d_y$ et que la valeur commune est la moitié de la caractéristique d'Euler de M. La démonstration s'appuie sur l'étude de l'homéomorphisme σ introduit dans le n° 5, qui est dans notre cas un homéomorphisme de $\tilde{G}(2, 2)$ en elle-même. Son homomorphisme induit sur les cycles a l'effet de fixer un des cycles S_x et S_y et changer le signe de l'autre. Ce fait et le résultat $\overline{W}^2 = 0$ (à coefficients entiers) conduisent facilement à l'égalité $d_x = d_y$.

Pour exprimer les relations de ces résultats avec les invariants différentiels de M dans E^4, il faut déterminer dans $\tilde{G}(2, 2)$ des formes différentielles extérieures fermées Φ_1, Φ_2 duales aux cycles S_x et S_y, c'est-à-dire telles que

$$\int_{S_\alpha} \Phi_\beta = \delta_\alpha^\beta, \quad 1 \leq \alpha, \beta \leq 2, \quad S_1 = S_x, \; S_2 = S_y, \tag{40}$$

ou δ_α^β est le symbole de Kronecker. Ces formes Φ_1, Φ_2 ne sont pas univoquement déterminées. Cependant on peut démontrer que les choix

$$\begin{aligned} \Phi_1 &= \frac{1}{4\pi}\{\omega_{13} \wedge \omega_{23} + \omega_{14} \wedge \omega_{24} - \omega_{13} \wedge \omega_{14} - \omega_{23} \wedge \omega_{24}\}, \\ \Phi_2 &= \frac{1}{4\pi}\{\omega_{13} \wedge \omega_{23} + \omega_{14} \wedge \omega_{24} + \omega_{13} \wedge \omega_{14} + \omega_{23} \wedge \omega_{24}\}, \end{aligned} \tag{41}$$

satisfont aux conditions (40). On en déduit les formules intégrales suivantes

$$\frac{1}{2\pi}\int_M K\,dV = \chi(M)\,, \tag{42}$$

$$\int_M (A_{311}A_{412} - A_{312}A_{411} + A_{321}A_{422} - A_{322}A_{421})\,dV = 0\,.$$

Ces formules ont été données pour la première fois par M. Blaschke [3].

L'étude de la variété $\tilde{G}(2, 2)$ conduit aussi à un résultat, dû à M. Wu Wen-Tsun, qui a une conséquence géométrique intéressante. C'est le problème de considérer une courbe paramétrique fermée simple dans $G(2, 2)$ et de voir si elle est la projection d'une telle courbe dans $\tilde{G}(2, 2)$. Une telle courbe dans $\tilde{G}(2, 2)$ peut être donnée par $(x(t), y(t))$, $0 \leq t \leq 1$, où $x(t) \in S_x$, $y(t) \in S_y$, et $x(0) = \pm x(1)$, $y(0) = \pm y(1)$, en désignant par $-x(1)$, $-y(1)$ respectivement les points antipodes de $x(1)$, $y(1)$ dans S_x, S_y. M. Wu Wen-Tsun a démontré que si, pour deux valeurs différentes quelconques t', t'' de t, $(t', t'') \neq (0, 1)$, les plans correspondants dans $G(2, 2)$ n'ont que le point O en commun, alors la courbe est la projection d'une courbe fermée simple dans $\tilde{G}(2, 2)$. Interprété dans la géométrie elliptique réglée, cela veut dire qu'une surface réglée dans un espace elliptique à trois dimensions est toujours orientable. Elle est donc homéomorphe à un tore [1].

10. — Je termine cette conférence par quelques questions naturelles:

A) Trouvez des invariants des sous-variétés relatifs à l'homotopie régulière définie dans le nº 1. En particulier, y a-t-il des paires de sous-variétés homéomorphes à deux dimensions dans un espace euclidien à quatre dimensions qui ne sont pas régulièrement homotopes ?

B) Y a-t-il d'autres conditions nécessaires que les conditions déjà connues pour que l'application $T: M \longrightarrow G(n, N)$ soit une application tangentielle ?

C) Dans l'espace euclidien à quatre dimensions y a-t-il une surface compacte à courbure gaussienne toujours négative ?

---

[1] M. H. Hopf m'a fait remarquer que ce théorème est un corollaire d'un théorème plus général, à savoir qu'il n'est pas possible de plonger topologiquement la bouteille de Klein dans l'espace projectif réel à trois dimensions.

Appendice.

Nous nous proposons de donner ici les formules concernant les courbures $K(x)$ et $K^*(x)$ et une démonstration du théorème suivant, énoncé dans le n° 6: L'inégalité (32) entraîne que la variété compacte M a ses nombres de Betti modulo 2 tous nuls pour les dimensions 1, 2, ..., $n - 1$.

Nous utilisons les notations du texte. Désignons par $\nu = \sum_s \nu_s e_s$ un vecteur unitaire normal. Alors la forme (13) est la seconde forme fondamentale de la projection orthogonale $M(\nu)$ de M dans l'espace linéaire à $n + 1$ dimensions déterminé par ν et par le plan tangent à M en x. Par définition, la courbure $G(x, \nu)$ de Gauss-Kronecker de $M(\nu)$ est égale au déterminant:

$$G(x, \nu) = \left| \sum_s \nu_s A_{sij} \right| . \tag{43}$$

Pour calculer l'intégrale dans (23) il faut développer ce déterminant. Le développement sera un polynome homogène de degré n en ν_s, dont un terme général est de la forme

$$\pm \nu_{s_1} \ldots \nu_{s_n} A_{s_1 i_1 j_1} \ldots A_{s_n i_n j_n} = \pm \nu_{t_1}^{\lambda_1} \ldots \nu_{t_k}^{\lambda_k} A_{s_1 i_1 j_1} \ldots A_{s_n i_n j_n} ,$$

où les indices $t_1, \ldots, t_k$ sont tous distincts. Pour qu'un terme donne une valeur non nulle dans l'intégrale (23) il faut que les exposants $\lambda_1, \ldots, \lambda_k$ soient tous pairs. On est conduit ainsi à la démonstration du résultat que l'intégrale (23) est nulle si n est impair et égale à $K(x)$ si n est pair.

Pour exprimer l'élément de volume sur l'hypersphère de rayon unité autour de l'origine O, avec le point courant ν, on prend un repère $a_1, \ldots, a_{n+N}$, dont le dernier vecteur a_{n+N} est identique à ν. Alors l'élément de volume est donné par l'expression

$$\prod_{1 \leqq t \leqq n+N-1} (a_t \, d\nu) = \prod_{1 \leqq t \leqq n+N-1} (a_t \, da_{n+N}) ,$$

où le produit est au sens du produit extérieur. Il est sous-entendu que ce produit est indépendant du choix des $n + N - 1$ premiers vecteurs. Dans le cas présent où v est un vecteur normal en $x \in M$ nous choisissons $a_i = e_i(x)$ et posons

$$a_r = \sum_s u_{rs} e_s(x) ,$$

de sorte que $u_{n+N,s} = v_s$. On trouve alors

$$da_{n+N} = dv = \sum_s dv_s \cdot e_s + \sum_s v_s de_s ,$$

d'où on déduit

$$dv \cdot a_i = \sum_s v_s \omega_{si} = - \sum_{s,j} v_s A_{sij} \omega_j ,$$

$$dv \cdot a_r = \sum_s dv_s u_{rs} + \sum_{s,t} v_s u_{rt} \omega_{st} .$$

Il s'en suit que

$$\prod_{1 \leq t \leq n+N-1} (a_t dv) = \pm \left| \sum_s v_s A_{sij} \right| d\sigma_{N-1} dV .$$

Il s'agit d'intégrer la valeur absolue de cette expression pour tous les vecteurs normaux unitaires en tous les points $x \in M$. Notre discussion dans le n° 6 implique que cette intégrale est $\geqq 2c_{n+N-1}$. Utilisant l'expression (30) pour la courbure totale $K^*(x)$, on obtient l'inégalité (31).

Maintenant supposons que l'inégalité (32) soit valable. Cela implique que les directions auxquelles la fonction coordonnée n'a qu'un maximum et un minimum ont une mesure positive. Il s'ensuit qu'il y a une fonction coordonnée non constante qui n'a qu'un maximum et un minimum. Des inégalités de Morse résulte alors l'énoncé au début de cet appendice.

BIBLIOGRAPHIE

[1] Allendoerfer, C. B., Rigidity for spaces of class greater than one, *Amer. J. Math.* 61, 633-44 (1939).

[2] —— and Weil, A., The Gauss-Bonnet theorem for Riemannian polyhedra, *Trans. Amer. Math. Soc.* 53, 101-129 (1943).

[3] Blaschke, W., Sulla geometria differenziale delle superficie S_2 nello spazio euclideo S_4, *Ann. Math. Pura Appl.* (4), 28, 205-209 (1949).

[4] Chern, S., On a theorem of algebra and its geometrical application, *J. Indian Math. Soc.* 8, 29-36 (1944).

[5] Chern, S., Topics in differential geometry, mimeographed notes, Princeton, 1951.
[6] —— and Kuiper, N. H., Some theorems on the isometric imbedding of compact Riemannian manifolds in Euclidean space, *Annals of Math.* 56, 422-430 (1952).
[7] —— and Spanier, E., A theorem on orientable surfaces in four-dimensional space, *Comm. Math. Helv.* 25, 205-209 (1951).
[8] Ehresmann, C., Sur la topologie de certaines variétés algébriques réelles, *J. Math. Pures Appl.* (9) 16, 69-100 (1937).
[9] Fary, I., Sur la courbure totale d'une courbe gauche faisant un nœud, *Bull. Soc. Math. France* 77, 128-138 (1949).
[10] Fenchel, W., On total curvatures of Riemannian manifolds, *J. London Math. Soc.* 15, 15-22 (1940).
[11] Killing, W., Die nicht-euklidischen Raumformen in analytischer Behandlung, Leipzig 1885.
[12] Milnor, J. W., On the total curvature of knots, *Annals of Math.* 52, 248-257 (1950).
[13] Otsuki, T., On the existence of solutions of a system of quadratic equations and its geometrical application, *Proc. Japan Acad.* 29, 99-100 (1953).
[14] Pontrjagin, L., Characteristic cycles on differentiable manifolds, Rec. Math. (N.S.), 21, 233-284 (1947), *Amer. Math. Soc. Trans.* No. 32.
[15] —— Some topological invariants of closed Riemannian manifolds, Izvestiya Akad. Nauk SSSR, Ser. Math. 13, 125-162 (1949), *Amer. Math. Soc. Trans.* No. 49.
[16] Seifert, H., Algebraische Approximation von Mannigfaltigkeiten, *Math. Z.* 41, 1-17 (1936).
[17] —— und Threlfall, W., Variationsrechnung im Grossen.
[18] Thom, R., Sur les variétés plongées et *i*-carrés, *C. R. Acad. Paris* 230, 507-508 (1950).
[19] Whitney, H., On regular closed curves in the plane, *Comp. Math.* 4, 276-284 (1937).
[20] —— Differentiable manifolds, *Annals of Math.* 37, 645-680 (1936).
[21] —— On the topology of differentiable manifolds, Lectures on topology, 101-141, Univ. of Mich. Press (1941).
[22] Wu Wen-Tsun, Sur les classes caractéristiques des structures fibrées sphériques, *Act. Sci. Indus.* No.1183, Paris (1952).

AN ELEMENTARY PROOF OF THE EXISTENCE OF ISOTHERMAL PARAMETERS ON A SURFACE

SHIING-SHEN CHERN

1. Introduction. Let

$$(1) \qquad ds^2 = E(x, y)dx^2 + 2F(x, y)dxdy + G(x, y)dy^2, \qquad EG - F^2 > 0,\ E > 0,$$

be a positive definite Riemann metric of two dimensions defined in a neighborhood of a surface with the local coordinates x, y. By isothermal parameters we mean local coordinates u, v relative to which the metric takes the form

$$(2) \qquad ds^2 = \lambda(u, v)(du^2 + dv^2), \qquad \lambda(u, v) > 0.$$

In order that isothermal parameters exist it is necessary to impose on the metric some regularity assumptions. In fact, it was shown recently by Hartman and Wintner[1] that it is not sufficient to assume the functions E, F, G to be continuous. So far the weakest conditions under which the isothermal parameters are known to exist were found by Korn and Lichtenstein.[2] To formulate their theorem we recall that a function $f(x, y)$ in a domain D of the (x, y)-plane is said to satisfy a Hölder condition of order λ, $0<\lambda\leq 1$, if the inequality

$$(3) \qquad |f(x, y) - f(x', y')| < Cr^{\lambda}$$

holds for any two points (x, y), (x', y') of D, where C is a constant and r is the Euclidean distance between these two points. With this definition the theorem of Korn-Lichtenstein can be stated as follows:

Received by the editors September 29, 1954 and, in revised form, November 15, 1954.

[1] P. Hartman and A. Wintner, *On the existence of Riemannian manifolds which cannot carry non-constant analytic or harmonic functions in the small*, Amer. J. Math. vol. 75 (1953) pp. 260–276. Also: S. Chern, P. Hartman, and A. Wintner, *On isothermic coordinates*, Comment-Math. Helv. vol. 28 (1954) pp. 301–309.

[2] A. Korn, *Zwei Anwendungen der Methode der sukzessiven Annäherungen*, Schwarz Abhandlungen pp. 215–229; L. Lichtenstein, *Zur Theorie der konformen Abbildung. Konforme Abbildung nichtanalytischer, singularitätenfreier Flächenstücke auf ebene Gebiete*, Bull. Int. de l'Acad. Sci. Cracovie, ser. A (1916) pp. 192–217. Cf. also the paper of C. B. Morrey, *On the solutions of quasi-linear elliptic partial differential equations*, Trans. Amer. Math. Soc. vol. 43 (1938) pp. 126–166, in which the isothermal parameters are shown to exist in a generalized sense under weaker hypotheses.

Added in proof. Weaker conditions were recently found by Hartman and Wintner. Cf. their paper, *On uniform Dini conditions in the theory of linear partial differential equations of elliptic type*, Amer. J. Math. vol. 77 (1955) pp. 329–354.

Suppose, in a domain D of the (x, y)-plane, the functions E, F, G satisfy a Hölder condition of order λ, $0<\lambda<1$. Then every point of D has a neighborhood whose local coordinates are isothermal parameters.

This theorem is rather useful in the global theory of surfaces, when analyticity assumptions are not desirable. We shall present here what seems to be an elementary and rather straightforward proof. Needless to say, the essential ideas of this proof are contained in the works of Korn and Lichtenstein. We believe, however, that some simplifications are achieved by the consistent use of the complex notation.[3]

We first observe that the isothermal parameters are invariant under conformal transformations, that is, multiplications of the Riemann metric (1) by a positive factor. Under a conformal transformation the angle between two vectors remains invariant. If we further assume that the coordinates x, y form a positive system, that is, if we allow only those transformations of local coordinates for which the Jacobian is positive, the angle can be defined, together with its orientation. To express these relations analytically, it will be convenient to introduce complex-valued differential forms, that is, forms $\omega=\alpha+i\beta$, where α, β are real differential forms. We shall write $\bar{\omega}=\alpha-i\beta$.

Since the quadratic differential form in the right-hand side of (1) is positive definite, we can write

$$ds^2 = \theta_1^2 + \theta_2^2, \tag{4}$$

where θ_1, θ_2 are real linear differential forms:

$$\begin{aligned} \theta_1 &= a_1dx + b_1dy, \\ \theta_2 &= a_2dx + b_2dy. \end{aligned} \tag{5}$$

Assuming that $a_1b_2-a_2b_1>0$, the forms θ_1, θ_2 are determined up to a proper orthogonal transformation, that is, one with determinant $+1$. We put

$$\phi = \theta_1 + i\theta_2, \tag{6}$$

so that

$$ds^2 = \phi\bar{\phi}. \tag{7}$$

The form ϕ is then determined up to a complex factor of absolute value 1. A conformal transformation of the Riemann metric is given,

[3] After this paper was submitted for publication, it has come to my attention that a similar proof was given by L. Bers in his mimeographed notes on Riemann surfaces, New York University, 1951–1952.

in terms of the form ϕ, by the multiplication of ϕ by an arbitrary nonzero complex factor. In the recent terminology we say that the complex linear differential form ϕ, determined up to a nonzero complex factor, defines an almost complex structure. As discussed above, such a structure allows the introduction of the notion of oriented angle.

The determination of isothermal parameters u, v is equivalent to that of a complex-valued function $w=u+iv$, such that

$$dw = (1/\rho)\phi. \tag{8}$$

For we have then

$$ds^2 = \phi\bar{\phi} = |\rho|^2 dw d\bar{w} = |\rho|^2(du^2 + dv^2). \tag{9}$$

Conversely, the isothermal parameters u, v determine a function w satisfying (8).

2. **Preliminaries.** We shall first make estimates of certain integrals, which will be needed in the proof.

Let (ξ, η) be any point in the (x, y)-plane, and let

$$r = +((x-\xi)^2 + (y-\eta)^2)^{1/2}. \tag{10}$$

If $g(r)$ is a function of class 1, we have, by exterior differentiation,

$$d\left\{g(r)\,\frac{-(y-\eta)dx + (x-\xi)dy}{r^2}\right\} = \frac{g'(r)}{r}\,dx \wedge dy.$$

It follows by Stokes Theorem that, if D is a domain bounded by a curve C and if $(\xi, \eta) \notin D$, we have

$$\iint_D \frac{g'(r)}{r}\,dxdy = \int_C g(r)\,\frac{-(y-\eta)dx + (x-\xi)dy}{r^2}. \tag{11}$$

This formula remains true, even if $(\xi, \eta) \in D$ (but not on C), provided that the integral in the left-hand side is convergent. The integral

$$\nu = \frac{1}{2\pi}\int_C \frac{-(y-\eta)dx + (x-\xi)dy}{r^2}, \qquad (\xi, \eta) \notin C \tag{12}$$

is an integer and is usually called the order of the point (ξ, η) relative to the curve C.

In particular, if $g(r) = r^\lambda$, $\lambda \neq 0$, formula (11) becomes

$$\lambda \iint_D \frac{r^\lambda dxdy}{r^2} = \int_C r^\lambda\,\frac{-(y-\eta)dx + (x-\xi)dy}{r^2}, \tag{13}$$

and is true either when $(\xi, \eta) \notin D$ or when $\lambda > 0$. It follows that, if the vector joining (ξ, η) to a point $(x, y) \in C$ turns monotonely in the same sense, we have

$$\left| \iint_D \frac{r^\lambda dxdy}{r^2} \right| \leq \frac{2\pi}{\lambda} \Delta^\lambda, \qquad \lambda > 0, \tag{14}$$

$$\left| \iint_D \frac{r^\lambda dxdy}{r^2} \right| \leq \frac{2\pi}{|\lambda|} \delta^\lambda, \qquad \lambda < 0, \tag{15}$$

where Δ and δ denote respectively an upper bound and a lower bound of the distance from (ξ, η) to a point $(x, y) \in C$. We emphasize that formula (15) is valid only under the assumption $(\xi, \eta) \notin D$.

3. **Main lemma.** To simplify our formulas we shall use the complex coordinate $z = x + iy$ in the (x, y)-plane. If $f(x, y)$ is a complex-valued function of class 1, we define $f_z, f_{\bar{z}}$ by the equation

$$df = f_z dz + f_{\bar{z}} d\bar{z} = f_z(dx + idy) + f_{\bar{z}}(dx - idy). \tag{16}$$

They are therefore related to the ordinary partial derivatives f_x, f_y by the equations

$$f_z = (f_x - if_y)/2, \qquad f_{\bar{z}} = (f_x + if_y)/2. \tag{17}$$

We shall write the function as $f(z, \bar{z})$, thus emphasizing the fact that it is in general not analytic in z.

On the other hand, we can define these operators $f_z, f_{\bar{z}}$ directly, without using the partial derivatives f_x, f_y. For instance, we can adopt the definitions:

$$\begin{aligned} f_z &= \lim_{h \to 0} \{ f(z + h/2, \bar{z} + h/2) - f(z, \bar{z}) \\ &\qquad + if(z - hi/2, \bar{z} + hi/2) - if(z, \bar{z}) \}/h, \\ f_{\bar{z}} &= \lim_{h \to 0} \{ f(z + h/2, \bar{z} + h/2) - f(z, \bar{z}) \\ &\qquad + if(z + hi/2, \bar{z} - hi/2) - if(z, \bar{z}) \}/h, \end{aligned} \tag{18}$$

which clearly give (17), when f_x, f_y exist.

LEMMA. *Let D be a disc of radius R about the origin in the $(x, y\text{-})$plane. Let $f(z, \bar{z})$ be a complex-valued continuous function in D, which satisfies the condition*

$$|f(\zeta_1, \bar{\zeta}_1) - f(\zeta_2, \bar{\zeta}_2)| \leq B r_{12}^\lambda, \quad r_{12} = |\zeta_1 - \zeta_2|, \tag{19}$$

for any two points $\zeta_1, \zeta_2 \in D$, where λ and B are constants, $0 < \lambda < 1$. Let the function $F(\zeta, \bar{\zeta})$ be defined by

$$(20)\quad -2\pi i F(\zeta, \bar{\zeta}) = \iint_D \frac{f(z, \bar{z})d\bar{z}dz}{z-\zeta} = 2i \iint_D \frac{f(z, \bar{z})dxdy}{z-\zeta}, \quad \zeta \in D.$$

Then: (1) F_ζ *and* $F_{\bar{\zeta}}$ *exist, and* $F_{\bar{\zeta}} = f(\zeta, \bar{\zeta})$; (2) *If* $|f(z, \bar{z})| \leqq A$, $z \in D$, *the following inequalities are valid*:

$$(21)\quad |F(\zeta, \bar{\zeta})| \leqq 4RA,$$

$$(22)\quad |F_\zeta(\zeta, \bar{\zeta})| \leqq (2^{\lambda+1}/\lambda)R^\lambda B,$$

$$(23)\quad |F(\zeta_1, \bar{\zeta}_1) - F(\zeta_2, \bar{\zeta}_2)| \leqq 2(A + (2^{\lambda+1}/\lambda)R^\lambda B)r_{12},$$

$$(24)\quad |F_\zeta(\zeta_1, \bar{\zeta}_1) - F_\zeta(\zeta_2, \bar{\zeta}_2)| \leqq \mu(\lambda)Br_{12}^{\lambda},$$

where $\mu(\lambda) > 0$ *is independent of* ζ_1, ζ_2.

To prove the lemma we write, according to the second equation of (18),

$$-2\pi i F_{\bar{\zeta}} = \lim_{h \to 0} \frac{1}{2} \iint_D (f(z, \bar{z}) - f(\zeta, \bar{\zeta})) \left\{ \frac{1}{(z-\zeta)(z-\zeta-h/2)} - \frac{1}{(z-\zeta)(z-\zeta-hi/2)} \right\} d\bar{z}dz + f(\zeta, \bar{\zeta}) \frac{\partial}{\partial \bar{\zeta}} \iint_D \frac{d\bar{z}dz}{z-\zeta}.$$

Computation gives

$$(25)\quad \iint_D \frac{d\bar{z}dz}{z-\zeta} = -2\pi i \bar{\zeta}.$$

From these it follows that $F_{\bar{\zeta}} = f(\zeta, \bar{\zeta})$. Similarly, we prove that F_ζ exists and is given by

$$(26)\quad -2\pi i F_\zeta = \iint_D \frac{f(z, \bar{z}) - f(\zeta, \bar{\zeta})}{(z-\zeta)^2} d\bar{z}dz.$$

From (20) we have, by using (14) with $\lambda = 1$,

$$2\pi |F| \leqq 2A \iint_D \frac{dxdy}{r} \leqq 8\pi AR,$$

which gives (21). Similarly, from (26), we have

$$2\pi |F_\zeta| \leqq 2B \iint_D \frac{r^\lambda}{r^2} dxdy \leqq \frac{2^{\lambda+2}}{\lambda} \pi BR^\lambda,$$

which gives (22).

Inequality (23) is trivial. To prove (24) let D_0 be the intersection of D with a circular disc of radius $2r_{12}=2|\zeta_1-\zeta_2|$ about ζ_2. We can write

$$
\begin{aligned}
-2\pi i\{F_\zeta(\zeta_1,\bar\zeta_1)-F_\zeta(\zeta_2,\bar\zeta_2)\}
&= \iint_{D_0} \{f(z,\bar z)-f(\zeta_1,\bar\zeta_1)\}\frac{1}{(z-\zeta_1)^2}\,d\bar z dz \\
&\quad - \iint_{D_0} \{f(z,\bar z)-f(\zeta_2,\bar\zeta_2)\}\frac{1}{(z-\zeta_2)^2}\,d\bar z dz \\
&\quad + \iint_{D-D_0} \{f(z,\bar z)-f(\zeta_1,\bar\zeta_1)\}\frac{1}{(z-\zeta_1)^2}\,d\bar z dz \\
&\quad - \iint_{D-D_0} \{f(z,\bar z)-f(\zeta_1,\bar\zeta_1)\}\frac{1}{(z-\zeta_2)^2}\,d\bar z dz \\
&\quad + \{f(\zeta_2,\bar\zeta_2)-f(\zeta_1,\bar\zeta_1)\}\frac{\partial}{\partial\zeta_2}\iint_{D-D_0}\frac{d\bar z dz}{z-\zeta_2}.
\end{aligned} \tag{27}
$$

We observe that in this sum only the first two integrals are improper integrals, while the last three are proper integrals. To estimate their absolute values we first have

$$
\left|\iint_{D_0}\{f(z,\bar z)-f(\zeta_1,\bar\zeta_1)\}\frac{1}{(z-\zeta_1)^2}\,d\bar z dz\right|
\leq 2B\iint_{D_0}\frac{dxdy}{|z-\zeta_1|^{2-\lambda}}\leq\frac{4\pi B}{\lambda}(3r_{12})^\lambda,
$$

$$
\left|\iint_{D_0}\{f(z,\bar z)-f(\zeta_2,\bar\zeta_2)\}\frac{1}{(z-\zeta_2)^2}\,d\bar z dz\right|
\leq 2B\iint_{D_0}\frac{dxdy}{|z-\zeta_2|^{2-\lambda}}\leq\frac{4\pi B}{\lambda}(2r_{12})^\lambda.
$$

The sum of the third and fourth integrals in (27) is equal to

$$
\begin{aligned}
&\iint_{D-D_0}\{f(z,\bar z)-f(\zeta_1,\bar\zeta_1)\}\int_{\zeta_2}^{\zeta_1} d\,\frac{1}{(z-\zeta)^2}\,d\bar z dz \\
&\quad = 4i\iint_{D-D_0}\int_{\zeta_2}^{\zeta_1}\{f(z,\bar z)-f(\zeta,\bar\zeta)+f(\zeta,\bar\zeta)-f(\zeta_1,\bar\zeta_1)\}\frac{1}{(z-\zeta)^3}\,d\zeta dxdy,
\end{aligned}
$$

where ζ is a point on the segment $\zeta_1\zeta_2$. It follows that its absolute value is

$$\leqq 4r_{12}(F_1 + F_2),$$

where F_1 and F_2 are respectively upper bounds of the integrals

$$\left| \iint\limits_{D-D_0} \frac{f(z, \bar{z}) - f(\zeta, \bar{\zeta})}{(z-\zeta)^3} dxdy \right|,$$

$$\left| \iint\limits_{D-D_0} \{f(\zeta, \bar{\zeta}) - f(\zeta_1, \bar{\zeta}_1)\} \frac{dxdy}{(z-\zeta)^3} \right|$$

along the segment $\zeta_1\zeta_2$. Using (15), we have

$$\left| \iint\limits_{D-D_0} \frac{f(z, \bar{z}) - f(\zeta, \bar{\zeta})}{(z-\zeta)^3} dxdy \right| \leqq B \iint\limits_{D-D_0} \frac{r^{\lambda-1} dxdy}{r^2}$$

$$\leqq \frac{4\pi B}{1-\lambda} \frac{1}{r_{12}^{1-\lambda}},$$

$$\left| \iint\limits_{D-D_0} \frac{f(\zeta, \bar{\zeta}) - f(\zeta_1, \bar{\zeta}_1)}{(z-\zeta)^3} dxdy \right| \leqq B r_{12}^{\lambda} \frac{4\pi}{r_{12}}.$$

The right-hand members of these inequalities can therefore be taken as F_1 and F_2 respectively.

Finally, we have, by (25),

$$\iint\limits_{D-D_0} \frac{d\bar{z}dz}{z-\zeta_2} = \iint\limits_{D} \frac{d\bar{z}dz}{z-\zeta_2} - \iint\limits_{D_0} \frac{d\bar{z}dz}{z-\zeta_2} = -2\pi i \bar{\zeta}_2$$

the integral over D_0 being equal to zero. It follows that the last term of (27) is zero.

Summing up these estimates, we get

$$2\pi \left| F_\zeta(\zeta_1, \bar{\zeta}_1) - F_\zeta(\zeta_2, \bar{\zeta}_2) \right|$$

$$\leqq \frac{4\pi B}{\lambda} (3r_{12})^{\lambda} + \frac{4\pi B}{\lambda} (2r_{12})^{\lambda} + 4\left(\frac{4\pi B}{1-\lambda} + 4\pi B\right) r_{12}^{\lambda}.$$

This gives (24), if we set

$$\mu(\lambda) = \frac{2}{\lambda}(3^{\lambda} + 2^{\lambda}) + 8 \frac{2-\lambda}{1-\lambda}. \tag{28}$$

Thus the proof of the lemma is complete.

4. An existence and uniqueness theorem on an integro-differential equation; application to isothermal parameters.

THEOREM. *In the domain D of the complex z-plane defined by* $|z| \leq R$ (*=positive constant*) *let*

$$Zw = a(z, \bar{z})w_{\bar{z}} + b(z, \bar{z})w_z \tag{29}$$

be a differential operator, whose coefficients $a(z, \bar{z})$, $b(z, \bar{z})$ *satisfy a Hölder condition of order* λ, $0<\lambda<1$, *and vanish at* $z=0$. *Let* $\alpha(z, \bar{z})$ *be a function in D, satisfying a Hölder condition of the same order* λ. *Then the equation*

$$2\pi i w(\zeta, \bar{\zeta}) + \iint_D \frac{(Zw + \alpha w)(z, \bar{z})}{z - \zeta} d\bar{z}dz = \sigma(\zeta) \qquad \zeta \in D, \tag{30}$$

where $\sigma(\zeta)$ *is a complex analytic function, with* $\sigma(0)=0$, *has exactly one solution* $w(z, \bar{z})$, *provided that R is sufficiently small.*

Before proceeding to the proof of this theorem, we shall make some further estimates of integrals, based on the lemma of the last section. The hypothesis of the theorem implies the existence of a number M large enough to fulfill the following inequalities:

$$1 < M, \qquad |\alpha(\zeta, \bar{\zeta})| \leq M, \qquad |\sigma(\zeta)| \leq M, \tag{31}$$

$$|h(\zeta_1, \bar{\zeta}_1) - h(\zeta_2, \bar{\zeta}_2)| \leq M r_{12}^{\lambda}, \tag{32}$$

$$|(Z+\alpha)\sigma(\zeta_1) - (Z+\alpha)\sigma(\zeta_2)| \leq \frac{M^2}{2^\lambda} r_{12}^{\lambda}, \qquad r_{12} = |\zeta_1 - \zeta_2|, \tag{33}$$

where ζ, ζ_1, ζ_2 are any three points of D and where $h(z, \bar{z})$ stands for each of the functions $a(z, \bar{z})$, $b(z, \bar{z})$, $\alpha(z, \bar{z})$, $\sigma(z)$. Since a, b, $(Z+\alpha)\sigma$ all vanish at 0, it follows from the corresponding Hölder inequalities that

$$\begin{gathered} |a(\zeta, \bar{\zeta})| \leq M|\zeta|^\lambda \leq MR^\lambda, \qquad |b(\zeta, \bar{\zeta})| \leq MR^\lambda, \\ |(Z+\alpha)\sigma(\zeta)| \leq M^2R^\lambda. \end{gathered} \tag{34}$$

Consider now the function $F(\zeta, \bar{\zeta})$ defined in (20). Using the notation of the lemma of §3, we have

$$\begin{aligned} &|(Z+\alpha)F(\zeta, \bar{\zeta})| \\ &\quad = |a(\zeta, \bar{\zeta})F_{\bar{\zeta}} + b(\zeta, \bar{\zeta})F_\zeta + \alpha(\zeta, \bar{\zeta})F| \\ &\quad \leq MR^\lambda(A + (2^{\lambda+1}/\lambda)R^\lambda B) + 4MRA \\ &\quad = MR^\lambda\{(1 + 4R^{1-\lambda})A + (2^{\lambda+1}/\lambda)R^\lambda B\}, \end{aligned} \tag{35}$$

$$
\begin{aligned}
&|(Z+\alpha)F(\zeta_1,\bar\zeta_1)-(Z+\alpha)F(\zeta_2,\bar\zeta_2)| \\
&\quad = |a(\zeta_1,\bar\zeta_1)f(\zeta_1,\bar\zeta_1)+b(\zeta_1,\bar\zeta_1)F_{\bar\zeta}(\zeta_1,\bar\zeta_1)+\alpha(\zeta_1,\bar\zeta_1)F(\zeta_1,\bar\zeta_1) \\
&\qquad - a(\zeta_2,\bar\zeta_2)f(\zeta_2,\bar\zeta_2)-b(\zeta_2,\bar\zeta_2)F_{\bar\zeta}(\zeta_2,\bar\zeta_2)-\alpha(\zeta_2,\bar\zeta_2)F(\zeta_2,\bar\zeta_2)| \\
&\quad \leqq Mr_{12}^{\lambda}\{A+(2^{\lambda+1}/\lambda)R^{\lambda}B+4RA\}+MR^{\lambda}(B+\mu(\lambda)B)r_{12}^{\lambda} \\
&\qquad + M(A+(2^{\lambda+1}/\lambda)R^{\lambda}B)\cdot 2r_{12} \\
&\quad \leqq Mr_{12}^{\lambda}\{A+g_1(R)A+g_2(R)B\},
\end{aligned}
\tag{36}
$$

where

$$
\begin{aligned}
g_1(R) &= 4R+2^{2-\lambda}R^{1-\lambda}, \\
g_2(R) &= \{(2^{\lambda+1}/\lambda)+\mu(\lambda)+1\}R^{\lambda}+(8/\lambda)R
\end{aligned}
\tag{37}
$$

are functions of R, which tend to zero with R.

Having the above inequalities, we shall prove the existence of a solution of (30) by successive approximations. For reasons which will be clear later we put the following restrictions on R:

$$
4R^{\lambda} \leqq 1, \qquad 2^{2-\lambda}\frac{\lambda+2}{\lambda}R^{1-\lambda} \leqq 1, \tag{38}
$$

and we choose a constant c, depending only on λ, such that

$$
\begin{aligned}
&\frac{\lambda+2}{\lambda}+\frac{\lambda}{\lambda+2}2^{\lambda} \leqq c, \\
&1+\mu(\lambda)+\frac{3\lambda+2}{\lambda}2^{\lambda} \leqq c.
\end{aligned}
\tag{39}
$$

We now define

$$
\begin{aligned}
2\pi i w_0(\zeta,\bar\zeta) &= \sigma(\zeta), \\
2\pi i w_{n+1}(\zeta,\bar\zeta) &= -\iint_D \frac{(Zw_n+\alpha w_n)(z,\bar z)}{z-\zeta}\,d\bar z dz, \qquad n=0,1,\cdots.
\end{aligned}
\tag{40}
$$

For these functions we shall prove the inequalities:

$$ |w_n(\zeta,\bar\zeta)| \leqq M(cMR^{\lambda})^n, \tag{41} $$

$$ |(Z+\alpha)w_n(\zeta,\bar\zeta)| \leqq M(cMR^{\lambda})^{n+1}, \tag{42} $$

$$ |w_n(\zeta_1,\bar\zeta_1)-w_n(\zeta_2,\bar\zeta_2)| \leqq M(cMR^{\lambda})^n r_{12}^{\lambda}, \tag{43} $$

$$ |(Z+\alpha)w_n(\zeta_1,\bar\zeta_1)-(Z+\alpha)w_n(\zeta_2,\bar\zeta_2)| \leqq (cM^2/2^{\lambda})(cMR^{\lambda})^n r_{12}^{\lambda}. \tag{44} $$

In particular, the last one implies that the function under the integral sign in (40) satisfies a Hölder condition, thus allowing the definition of the next integral.

The inequalities (41)–(44) will be proved by induction on n. For $n=0$, (41) is a consequence of the third inequality of (31), (42) follows from the third inequality of (34), since $c>1$, (43) follows from (32), and (44) from (33).

We now suppose that (41)–(44) are true and proceed to prove the corresponding inequalities for $n+1$. By (21), (23), and the induction hypothesis, we have

$$\begin{aligned} |w_{n+1}(\zeta,\bar\zeta)| &\leqq 4R\cdot M(cMR^\lambda)^{n+1} \leqq M(cMR^\lambda)^{n+1}, \\ |w_{n+1}(\zeta_1,\bar\zeta_1) - w_{n+1}(\zeta_2,\bar\zeta_2)| & \\ &\leqq 2r_{12}\{M(cMR^\lambda)^{n+1} + (2^{\lambda+1}/\lambda)R^\lambda(cM^2/2^\lambda)(cMR^\lambda)^n\} \\ &= M(cMR^\lambda)^{n+1} r_{12}^\lambda(1+2/\lambda)\cdot 2r_{12}^{1-\lambda}. \end{aligned}$$

The last relation gives (43) (for the index $n+1$), on account of the second inequality of (38).

Similarly, we get from (35) and the induction hypothesis,

$$\begin{aligned} |(Z+\alpha)w_{n+1}(\zeta,\bar\zeta)| &\leqq MR^\lambda\{(1+4R^{1-\lambda})cM^2R^\lambda \\ &\quad + (2^{\lambda+1}/\lambda)R^\lambda(cM^2/2^\lambda)\}(cMR^\lambda)^n \\ &= M^2R^\lambda(cMR^\lambda)^{n+1}(1+4R^{1-\lambda}+2/\lambda) \\ &\leqq M(cMR^\lambda)^{n+2}, \end{aligned}$$

on using the first inequality of (39). By (36), we have

$$\begin{aligned} |(Z+\alpha)w_{n+1}(\zeta_1,\bar\zeta_1) - (Z+\alpha)w_{n+1}(\zeta_2,\bar\zeta_2)| & \\ &\leqq Mr_{12}^\lambda\cdot M(cMR^\lambda)^n\{(1+g_1(R))cMR^\lambda + g_2(R)cM/2^\lambda\} \\ &= M^2r_{12}^\lambda(cMR^\lambda)^{n+1}\{1+g_1(R)+(1/(2R)^\lambda)g_2(R)\}. \end{aligned}$$

This gives (44) for $n+1$, if

$$1+g_1(R)+(1/(2R)^\lambda)g_2(R) \leqq c/2^\lambda.$$

But the latter follows from the second inequality of (39). Thus our induction is complete.

It follows that the series

$$\sum_{n=0}^{\infty} w_n(z,\bar z) \tag{45}$$

converges absolutely and uniformly in D if $CMR^\lambda<1$, and defines a function $w(z,\bar z)$, which satisfies (30).

To prove the uniqueness of the solution when R is sufficiently small, let $w'(z, \bar{z})$ be another solution of (30) such that $(Z+\alpha)w'(z, \bar{z})$ satisfies a Hölder condition of order λ. Then the function

$$\tilde{w}(z, \bar{z}) = w(z, \bar{z}) - w'(z, \bar{z})$$

satisfies the equation

$$-2\pi i\tilde{w}(\zeta, \bar{\zeta}) = \iint_D \frac{(Z + \alpha)\tilde{w}(z, \bar{z})}{z - \zeta} d\bar{z}dz.$$

Let A and B be respectively the least upper bounds of

$$\left| (Z + \alpha)\tilde{w}(\zeta, \bar{\zeta}) \right|, \zeta \in D$$

and

$$\left|(Z + \alpha)\tilde{w}(\zeta_1, \bar{\zeta}_1) - (Z + \alpha)\tilde{w}(\zeta_2, \bar{\zeta}_2) \right| / r_{12}^{\lambda}, \qquad \zeta_1, \zeta_2 \in D, \zeta_1 \neq \zeta_2.$$

From (35) and (36) we get

$$A \leqq MR^{\lambda}\left\{(1 + 4R^{1-\lambda})A + (2^{\lambda+1}/\lambda)R^{\lambda}B\right\},$$
$$B \leqq M\left\{A + g_1(R)A + g_2(R)B\right\}.$$

From these we easily conclude that, if R is sufficiently small, $A = 0$. The latter implies that $\tilde{w}(z, \bar{z}) = 0$. Thus the proof of our theorem is complete.

In order to derive from the above existence theorem the theorem of Korn-Lichtenstein we follow the notation of §1. In a neighborhood of the point $z = x + iy = 0$ we suppose the almost complex structure to be given by the complex-valued linear differential form

(46) $$\phi = (1 - a(z, \bar{z}))dz + b(z, \bar{z})d\bar{z},$$

which is determined up to a nonzero factor. By a linear transformation on x, y with constant coefficients and by multiplication of ϕ by a constant factor when necessary, we can suppose that $a(0, 0) = b(0, 0) = 0$. Equation (8) is equivalent to the equations

$$w_{\bar{z}} = b/\rho, \qquad w_z = (1 - a)/\rho.$$

Elimination of ρ gives

$$(1 - a)w_{\bar{z}} - bw_z = 0,$$

or, if we make use of the operator Z in (29),

(47) $$w_{\bar{z}} = Zw.$$

By the above existence theorem, the equation

$$2\pi i w(\zeta, \bar{\zeta}) + \iint_D \frac{Zw(z, \bar{z})}{z - \zeta} d\bar{z}dz = \sigma(\zeta) \qquad \zeta \in D, \sigma(0) = 0, \tag{48}$$

has a solution $w(z, \bar{z})$. If $\sigma(\zeta)$ is not a constant, say $\sigma(\zeta) = \zeta$, this solution has the property that $w_\zeta(0, 0) \neq 0$. Since $Zw(z, \bar{z})$ satisfies a Hölder condition, it follows from our lemma in §3 that $w(z, \bar{z})$ satisfies (47). Thus we have proved the theorem of Korn-Lichtenstein.

REMARKS. 1. As L. Bers observed to me, the same method can be used to establish the existence of a local solution of the equation

$$w_{\bar{z}} = a\bar{w}_{\bar{z}} + bw_z + \alpha w + \beta\bar{w} + \gamma, \tag{49}$$

where $a, b, \alpha, \beta, \gamma$ satisfy a Hölder condition and where $|a| + |b| < 1$.

2. It follows from our proof that the first partial derivatives of the isothermal parameters also satisfy a Hölder condition of order λ.

INSTITUTE FOR ADVANCED STUDY AND
UNIVERSITY OF CHICAGO

Reprinted from
Proc. Amer. Math. Soc. **6** (1955) 771–782

ON SPECIAL W-SURFACES

SHIING-SHEN CHERN

A W-surface is a surface in ordinary Euclidean space for which there is a functional relation between the principal curvatures k_1, k_2:

$$W(k_1, k_2) = 0. \tag{1}$$

We shall be interested in those W-surfaces for which (1) can be written in the form

$$f(H, \mu) = 0, \qquad \mu = H^2 - K, \tag{2}$$

where H and K are the mean curvature and the Gaussian curvature respectively and where f is of class C^1. For such W-surfaces we have

$$\begin{aligned} \partial W/\partial k_1 &= f_H/2 + (k_1 - k_2)f_\mu/2, \\ \partial W/\partial k_2 &= f_H/2 - (k_1 - k_2)f_\mu/2. \end{aligned} \tag{3}$$

It follows that at an umbilic ($k_1 = k_2$) we have

$$\partial W/\partial k_1 \cdot \partial W/\partial k_2 = f_H^2/4 \geqq 0. \tag{4}$$

A W-surface (2) is called special if $f_H \neq 0$ at every umbilic.

Apparently there are very few closed special W-surfaces.[1] The following theorem is due to H. Hopf:[2]

The only closed special analytic W-surfaces of genus zero are spheres.

It was P. Hartman and A. Wintner[3] who succeeded in removing the analyticity assumption in Hopf's Theorem. Their theorem can be stated as follows:

Let S be a closed special W-surface of genus zero, which is C^3-imbedded in Euclidean space. Then S is a sphere.

We shall show that the formalism developed in the preceding paper gives a very simple proof of the theorem of Hartman-Wintner.[4]

Received by the editors November 15, 1954.

[1] The notion of a special W-surface was first introduced by the author in the paper *Some new characterizations of the Euclidean sphere*, Duke Math. J. vol. 12 (1945) pp. 279–290. I take this opportunity to mention that the proof of Theorem 3 in that paper is not valid. So far as I know, the question whether there exist closed surfaces of constant mean curvature and of genus >0 remains unanswered.

[2] H. Hopf, *Über Flächen mit einer Relation zwischen den Hauptkrümmungen*, Mathematische Nachrichten vol. 4 (1950–1951) pp. 232–249.

[3] P. Hartman and A. Wintner, *Umbilical points and W-surfaces*, Amer. J. Math. vol. 76 (1954) pp. 502–508.

[4] The hypotheses of Hartman-Wintner are weakened in one respect in that the function f in (2) is here assumed to be of class C^1, while they supposed f to be of class C^3.

Let x, y be isothermal parameters, and let $z=x+iy$. Let $E(dx^2+dy^2)$, $Ldx^2+2Mdxdy+Ndy^2$ be respectively the first and second fundamental forms of the surface, so that

$$(5)\qquad \begin{aligned} 2EH &= L+N, \\ E^2K &= LN - M^2. \end{aligned}$$

Put

$$(6)\qquad w = (L-N)/2 - iM.$$

Then Codazzi's equations can be written

$$(7)\qquad w_{\bar{z}} = EH_z.$$

Also we have from (5)

$$(8)\qquad \mu = H^2 - K = w\bar{w}/E^2.$$

If the surface is a special W-surface satisfying (2), we have, in a neighborhood of an umbilic,

$$H_z = -(f_\mu/f_H)\mu_z.$$

This means that w satisfies a nonlinear differential equation of the form

$$(9)\qquad w_{\bar{z}} = P(w\bar{w})_z + Qw\bar{w},$$

where

$$(10)\qquad P = -f_\mu/Ef_H, \qquad Q = 2E_z/E^2\cdot f_\mu/f_H.$$

Following the procedure of Hartman-Wintner, the proof of their theorem depends on the lemmas:

LEMMA 1. *Let $w(z,\bar{z})$ be a solution of* (9), *in a sufficiently small neighborhood of $z=0$, at which $w=0$. Then $\lim_{z\to 0} w(z,\bar{z})z^{-k}$ exists if $w=o(|z|^{k-1})$.*

LEMMA 2. *Under the hypotheses of Lemma 1, suppose that $w=o(|z|^{k-1})$ for all k. Then $w(z,\bar{z})\equiv 0$ in a neighborhood of $z=0$.*

From these lemmas we derive immediately the theorem of Hartman-Wintner. In fact, it follows from Lemma 2 that if 0 ($z=0$) does not have a neighborhood which consists entirely of umbilics, there exists an integer k, such that $w=o(|z|^{k-1})$, $w\neq o(|z|^k)$. By Lemma 1, $\lim_{z\to 0} w(z,\bar{z})z^{-k}$ exists and is $\neq 0$. We can therefore write

$$(11)\qquad w(z,\bar{z}) = cz^k + o(|z|^k) \qquad c\neq 0.$$

It follows that the umbilic 0 is isolated and has an index $-k<0$. By well-known arguments this implies the theorem of Hartman-Wintner.

It remains to prove the above lemmas. For this purpose let D be a disc of radius R about 0, and C its boundary circle. There exists a constant $A>0$, such that in D,

$$\left| P(w\bar{w})_z + Qw\bar{w} \right| \leqq A \left| w \right|. \tag{12}$$

Suppose that $w=o(|z|^{k-1})$. Let $\zeta=\xi+i\eta$ be an interior point of D. Then we have, for $\zeta\neq0$,

$$d\left\{\frac{w dz}{z^k(z-\zeta)}\right\} = \frac{P(w\bar{w})_z + Qw\bar{w}}{z^k(z-\zeta)} d\bar{z} \wedge dz. \tag{13}$$

Application of Stokes Theorem gives

$$-2\pi i w(\zeta,\bar{\zeta})\zeta^{-k} + \int_C \frac{w dz}{z^k(z-\zeta)} = \iint_D \frac{P(w\bar{w})_z + Qw\bar{w}}{z^k(z-\zeta)} d\bar{z}dz. \tag{14}$$

It follows that

$$\begin{aligned} 2\pi \left| w(\zeta,\bar{\zeta})\zeta^{-k} \right| &\leqq \int_C \left| \frac{w(z,\bar{z})}{z^k(z-\zeta)} \right| \left| dz \right| \\ &\quad + 2A \iint_D \left| \frac{w(z,\bar{z})}{z^k(z-\zeta)} \right| dxdy. \end{aligned} \tag{15}$$

We multiply this inequality by $d\xi d\eta/|\zeta-z_0|$, $z_0\in D$, and integrate over D. Remembering that

$$\iint_D \frac{dxdy}{|z-\zeta|} < 2R, \tag{16}$$

$$\frac{1}{|(z-\zeta)(z_0-\zeta)|} = \frac{1}{|z-z_0|}\left|\frac{1}{z-\zeta} + \frac{1}{\zeta - z_0}\right|, \tag{17}$$

we get from this integration

$$\begin{aligned} 2\pi \iint_D \left| \frac{w(z,\bar{z})}{z^k(z-\zeta)} \right| dxdy &\leqq 4R \int_C \left| \frac{w(z,\bar{z})}{z^k(z-\zeta)} \right| \left| dz \right| \\ &\quad + 8AR \iint_D \left| \frac{w(z,\bar{z})}{z^k(z-\zeta)} \right| dxdy \end{aligned}$$

or

$$(18)\quad (2\pi - 8AR)\iint_D \left|\frac{w(z, \bar{z})}{z^k(z - \zeta)}\right| dxdy \leqq 4R\int_C \left|\frac{w(z, \bar{z})}{z^k(z - \zeta)}\right| |dz|.$$

We choose R so small that $2\pi - 8AR > 0$. Then

$$\iint_D |w(z, \bar{z})/z^k(z - \zeta)|\, dxdy$$

is bounded as $\zeta \to 0$, and the same is true of $|w(\zeta, \bar{\zeta})\zeta^{-k}|$. It follows that $|(w\bar{w})_{\bar{\zeta}}\zeta^{-k}|$ is bounded and from (14) that $\lim_{\zeta\to 0} w(\zeta, \bar{\zeta})\zeta^{-k}$ exists. This proves Lemma 1.

To prove Lemma 2 we multiply (15) by $d\xi d\eta$ and integrate over D. This gives

$$2\pi \iint_D |w(z, \bar{z})z^{-k}|\, dxdy \leqq 2R\int_C |w(z, \bar{z})z^{-k}|\, |dz| + 4AR\iint_D |wz^{-k}|\, dxdy$$

or

$$(2\pi - 4AR)\iint_D |w(z, \bar{z})z^{-k}|\, dxdy \leqq 2R\int_C |w(z, \bar{z})z^{-k}|\, |dz|.$$

Suppose there exists a z_0 such that $w(z_0, \bar{z}_0) \neq 0$, $|z_0| < R$. Then the left-hand side of the above inequality is $\geqq a|z_0|^{-k}$, and the right-hand side is $\leqq bR^{-k}$, where a and b are positive constants independent of k. The hypothesis of Lemma 2 implies that $|z_0/R|^k \geqq a/b$ for all k, which is a contradiction. It follows that $w(z, \bar{z})$ vanishes identically for $|z| < R$.

INSTITUTE FOR ADVANCED STUDY AND
UNIVERSITY OF CHICAGO

Reprinted from
Proc. Amer. Math. Soc. **6** (1955) 783–786

Reprinted from *Abh. Math. Sem. Univ. Hamburg* **20** (1955) 117–126

On Curvature and Characteristic Classes of a Riemann Manifold

Von Shiing-shen Chern in Chicago

Dedicated to Wilhelm Blaschke on His Seventieth Birthday

1. Introduction

The development in recent years of the theory of fiber bundles in differential geometry has led to relations between the curvature of a compact Riemann manifold and certain global invariants, the so-called characteristic classes. The simplest among these relations is the Gauss-Bonnet formula[1]), which expresses the Euler-Poincaré characteristic of a compact orientable Riemann manifold as an integral involving the components of the Riemann-Christoffel curvature tensor. There are other relations, to be given below, between the Pontrjagin characteristic classes and differential forms constructed from the curvature tensor. The purpose of our paper is twofold: To express these relations in a convenient form and to draw some conclusions from them.

The differential forms which express the Euler-Poincaré characteristic and the Pontrjagin classes bear a close relationship with skew-symmetric determinants and Pfaffian functions. A theorem concerning them, which we will find useful, is given in § 2. In § 3 we shall introduce notions by which our relations can be best formulated. In § 4 we derive some implications of the curvature properties on the characteristic classes. In particular, we give a proof of a theorem of J. Milnor[2]) to the effect that an orientable compact Riemann manifold of four dimensions has positive Euler-Poincaré characteristic, if its sectional curvature is always positive or always negative.

Among the implications of these relations is the fact that they provide a way for effective computation of the characteristic classes. For instance, it will be easy to compute the Pontrjagin classes of the complex projective space, using its elliptic Hermitian metric. Although we will not carry this out, we shall make an application of the knowledge of these classes by deriving the following theorem: The complex projective space P_m of (complex) dimension m cannot be differentiably imbedded in a real Euclidean space of dimension $3m - 1$ or $3m - 2$, according as m is even or odd.

---

[1]) Cf., for instance, Chern, Topics in differential geometry, pp. 38—41, Institute for Advanced Study, Princeton, USA. This monograph will be quoted as TDG.

[2]) I owe this to an oral communication of Milnor.

2. A formula on generalized Pfaffian functions

Let $L_{ik} = -L_{ki}$, $1 \leqq i, k \leqq 2s$, be a set of numbers. We introduce the expressions

$$(1)\quad L(q)=\sum \varepsilon_{i_1 \dots i_{2s}} \varepsilon_{j_1 \dots j_{2s}} L_{i_1 i_2} \dots L_{i_{2q-1} i_{2q}} L_{j_1 j_2} \dots L_{j_{2q-1} j_{2q}} L_{i_{2q+1} j_{2q+1}} \dots L_{i_{2s} j_{2s}},$$
$$0 \leqq q \leqq s,$$

where $\varepsilon_{i_1 \dots i_{2s}}$ is the KRONECKER symbol, being equal to zero when two of the indices are identical and equal to $+1$ or -1, according as $i_1, \dots, i_{2s}$ form an even or odd permutation of $1, \dots, 2s$, and where the summation is extended over all the indices $i_1, \dots, i_{2s}, j_1, \dots, j_{2s}$, running from 1 to $2s$. Clearly we have

$$(2)\quad \begin{aligned} L(0) &= (2s)!\,|L_{ik}|, \\ L(s) &= \{ \sum_{i_1, \dots, i_{2s}} \varepsilon_{i_1 \dots i_{2s}} L_{i_1 i_2} \dots L_{i_{2s-1} i_{2s}} \}^2 . \end{aligned}$$

The expression in the braces of the second formula is usually called a Pfaffian function.

Between the expressions $L(q)$ the following identity is easily established:

$$(3)\quad L(q) = \frac{2q}{2s - 2q + 1} L(q-1).$$

In fact, we introduce the new expression

$$L'(q) = \sum_{\substack{i_1, \dots, i_{2s-1} \\ j_1, \dots, j_{2s-1}, k}} \varepsilon_{i_1 \dots i_{2s-1} k} \varepsilon_{j_1 \dots j_{2s-1} k} L_{i_1 i_2} \dots L_{i_{2q-1} i_{2q}} L_{j_1 j_2} \dots$$
$$\dots L_{j_{2q-1} j_{2q}} L_{i_{2q+1} j_{2q+1}} \dots L_{i_{2s-2} j_{2s-2}} L_{i_{2s-1} k} L_{j_{2s-1} k}, \quad 0 \leqq q \leqq s-1.$$

In the sum for $L(q)$ we let $i_{2q} = j_1, \dots, j_{2s}$ respectively and take the sum. This gives

$$L(q) = 2q L'(q-1).$$

Similarly, letting $i_{2q+1} = j_1, \dots, j_{2s}$ respectively and summing, we get

$$L(q) = (2s - 2q - 1) L'(q).$$

Combination of these two relations gives (3).

By recursion one derives from (3) the relation

$$(4)\quad L(q) = \frac{2^q q!}{(2s - 2q + 1) \dots (2s - 1)} L(0).$$

For $q = s$ this gives the well-known theorem that a skew-symmetric determinant of even order is, up to a numerical factor, the square of a PFAFFIAN function.

3. Formulas for the characteristic classes

Let V be a vector space of dimension n over the real field R, and let V^* be its dual space[3]). There is then a pairing of V and V^* into R, which we denote by $\langle X, X'\rangle \in R$, $X \in V$, $X' \in V^*$. The GRASSMANN algebra $\Lambda(V)$ of V is a graded algebra, admitting a direct sum decomposition

$$\Lambda(V) = \Lambda^0(V) + \Lambda^1(V) + \cdots + \Lambda^n(V), \tag{5}$$

where $\Lambda^r(V)$ is the subspace of all homogeneous elements of $\Lambda(V)$ of degree r. From $\Lambda(V)$ and the GRASSMANN algebra $\Lambda(V^*)$ of V^* we form their tensor product $\Lambda(V) \otimes \Lambda(V^*)$. The latter is generated as a vector space by products of the form $\xi \otimes \xi'$, $\xi \in \Lambda(V)$, $\xi' \in \Lambda(V^*)$. We only remark that, if $\xi' \in \Lambda(V^*)$, $\eta \in \Lambda(V)$, $\xi \in \Lambda(V)$, $\eta' \in \Lambda(V^*)$ then

$$(\xi \otimes \xi') \wedge (\eta \otimes \eta') = (\xi \wedge \eta) \otimes (\xi' \wedge \eta'). \tag{6}$$

Suppose now a scalar product be given in V. We will be interested in the subspace $\Lambda^{2k}(V) \otimes \Lambda^{2k}(V^*)$ of $\Lambda(V) \otimes \Lambda(V^*)$. If e^i, $1 \leqq i \leqq n$, form an orthonormal base of V, the elements $e_{i_1} \wedge e_{i_2} \wedge \cdots \wedge e_{i_{2k}}$, for all combinations $i_1, \ldots, i_{2k}$ of $1, \ldots, n$, constitute an orthonormal base of $\Lambda^{2k}(V)$, and an element of $\Lambda^{2k}(V) \otimes \Lambda^{2k}(V^*)$ can be written in the form[4])

$$A = \sum_{(i_1, \ldots, i_{2k})} (e_{i_1} \wedge \cdots \wedge e_{i_{2k}}) \otimes \xi'_{i_1 \ldots i_{2k}}. \tag{7}$$

We call

$$|A|^2 = \sum_{(i_1, \ldots, i_{2k})} \xi'^2_{i_1 \ldots i_{2k}} \in \Lambda^{4k}(V^*) \tag{8}$$

the square of measure of A; it is clearly independent of the choice of the orthonormal base e_i.

These algebraic notions apply naturally to the space V_x of tangent vectors and the space V_x^* of covectors at a point x of a differentiable manifold M of dimension n (and class C^∞). If $x^1, \ldots, x^n$ form a system of local coordinates at x, we can take as natural bases of V_x the vectors $\frac{\partial}{\partial x^i}$ and those of V^* the linear differential forms dx^k. These bases are dual to each other, that is, $\left\langle \frac{\partial}{\partial x^i}, dx^k \right\rangle = \delta_i^k$.

Let g_{ik} be the fundamental tensor of a positive definite RIEMANN metric on M, and let R^j_{ikl} be its RIEMANN-CHRISTOFFEL tensor. We introduce the expression

$$\Omega = \frac{1}{2} \sum_{i,j,k,l} \left(\frac{\partial}{\partial x^i} \wedge \frac{\partial}{\partial x^j}\right) \otimes R^{ij}{}_{kl} dx^k \wedge dx^l, \tag{9}$$

---

[3]) For notions concerning vector spaces and their multilinear algebra we refer to: N. BOURBAKI, Algèbre multilinéaire, Hermann, Paris.

[4]) In general we shall put under the summation sign the indices over which the sum is taken. When the ranges are not mentioned, they are usually from 1 to n or to $2s$, or otherwise clear by context. Summation over $(i_1, \ldots, i_k)$ means over all such combinations.

where the raising and lowering of indices are understood in the standard fashion, using the fundamental tensor g_{ik} · Ω is a tensor of type (2,2) and assigns to every point $x \in M$ an element of $\Lambda^2(V_x) \otimes \Lambda^2(V_x^*)$. We shall show that the consideration of Ω leads to a convenient formulation of the GAUSS-BONNET formula and its generalizations.

First we wish to remark that Ω can be expressed in terms of other bases in V_x and V_x^*. A convenient choice is to take an orthonormal base in V_x and its dual base in V_x^*. Over a neighborhood we have therefore a family of orthonormal frames $x e_1 \cdots e_n$ and linear differential forms $\omega_1, \ldots, \omega_n$, such that the RIEMANN metric is of the form

$$ds^2 = \omega_1^2 + \cdots + \omega_n^2. \tag{10}$$

The equations of structure of the RIEMANN metric are

$$\begin{aligned} d\omega_i &= \sum_j \omega_j \wedge \omega_{ji}, \quad \omega_{ji} + \omega_{ij} = 0, \\ d\omega_{ij} &= \sum_k \omega_{ik} \wedge \omega_{kj} + \Omega_{ij}, \end{aligned} \tag{11}$$

and our Ω becomes

$$\Omega = \sum_{i,j} (e_i \wedge e_j) \otimes \Omega_{ij}. \tag{12}$$

Suppose $n = 2s$ be even. We find easily

$$\Omega^s = (e_1 \wedge \cdots \wedge e_{2s}) \otimes \sum_{i_1, \ldots, i_{2s}} \varepsilon_{i_1 \cdots i_{2s}} \Omega_{i_1 i_2} \wedge \cdots \wedge \Omega_{i_{2s-1} i_{2s}}.$$

This allows us to formulate the GAUSS-BONNET formula in the following form:

Theorem 1. *Let M be a compact oriented Riemann manifold of even dimension $n = 2s$. Let Δ be the differential form of degree $2s$ defined by the equation*

$$\Omega^s = (e_1 \wedge \cdots \wedge e_{2s}) \otimes \Delta, \tag{13}$$

where $e_1 \wedge \cdots \wedge e_{2s}$ is the generator of $\Lambda^n(V_x)$ determined uniquely by the conditions that it is of measure 1 and that it is coherent with the orientation of V_x. Then the Euler-Poincaré characteristic $\chi(M)$ of M is given by the integral

$$\chi(M) = (-1)^s \frac{1}{2^{2s} \pi^s s!} \int_M \Delta. \tag{14}$$

To describe in a convenient way the differential forms corresponding to the PONTRJAGIN classes we begin by recalling their definition[5]). Let $\tilde{G}(n, N)$ be the GRASSMANN manifold of all oriented n-dimensional linear spaces through a fixed point 0 in a real Euclidean space E^{n+N} of dimension

[5]) L. S. PONTRJAGIN, Caracteristic cycles on differentiable manifolds, Rec. Math. (Mat. Sbornik) N.S. 21, 233—284 (1947); WU WEN-TSUN, Sur les classes caractéristiques des structures fibrées sphériques, Hermann, Paris 1952; also CHERN, TDG.

$n + N$. To integers k, l satisfying $0 \leqq k \leqq n/2$, $0 \leqq l \leqq N/2$ let $L_0^{N+2k-2} = L_0$ and $L_1^{N-2l+2} = L_1$ be fixed linear spaces through 0 of dimensions $N + 2k - 2$ and $N - 2l + 2$ respectively. Denote by $\Sigma(k)$ the subset of $\tilde{G}(n, N)$, whose linear spaces have in common with L_0 a linear space of dimension $\geqq 2k$. Similarly, denote by $\overline{\Sigma}(l)$ the subset of $\tilde{G}(n, N)$, whose linear spaces have in common with L_1 a linear space of dimension $\geqq 2$. It can be proved that both $\Sigma(k)$ and $\overline{\Sigma}(l)$ are orientable pseudomanifolds, of dimensions $nN - 4k$ and $nN - 4l$ respectively. To orient them let $a_1, \ldots, a_{n+N}$ be the coordinate vectors of E^{n+N}, and suppose L_0 be spanned by $a_1, \ldots, a_{2k}, a_{n+1}, \ldots, a_{n+N-2}$. Consider the open subset Σ' of $\Sigma(k)$, whose n-dimensional oriented linear spaces have only the zero vector in common with the N-dimensional space spanned by $a_{n+1}, \ldots, a_{n+N}$, and can be projected orthogonally from the latter into the n-dimensional space spanned by $a_1, \ldots, a_n$, with sense preserved. In an oriented linear space of Σ' there are then n vectors $v_1, \ldots, v_n$, completely determined, whose components are of the form

$$
(15)\quad
\begin{aligned}
v_1 &= (1, 0, \ldots, 0, v_{1,n+1}, \ldots, v_{1,n+N-2}, 0, 0), \\
&\cdots\cdots\cdots\cdots\cdots\cdots\cdots\cdots\cdots \\
v_{2k} &= (0, 0, \ldots, 1, \ldots, 0, v_{2k,n+1}, \ldots, v_{2k,n+N-2}, 0, 0), \\
v_{2k+1} &= (0, 0, \ldots, 0, 1, \ldots, 0, v_{2k+1,n+1}, \ldots, \ldots, v_{2k+1,n+N}), \\
&\cdots\cdots\cdots\cdots\cdots\cdots\cdots\cdots\cdots \\
v_n &= (0, 0, \ldots, 1, v_{n,n+1}, \ldots, v_{n,n+N}).
\end{aligned}
$$

The components v's of the above vectors can be taken as local coordinates in Σ'. When they are arranged in a lexicographic order, by following the rows successively, they define an orientation of Σ'. Since $\Sigma(k)$ is an orientable pseudo-manifold, this defines an orientation of $\Sigma(k)$. In particular, for $k = 0$, our convention orients $\tilde{G}(n, N)$.

Similarly, $\overline{\Sigma}(l)$ can be oriented, but we shall not give the details.

Each of $\Sigma(k)$ and $\overline{\Sigma}(l)$ defines therefore a cycle on $\tilde{G}(n, N)$, with integer or real coefficients. Now suppose our compact manifold M be the base space of a sphere bundle, whose structural group is the rotation group in n variables. If N is sufficiently large, such a sphere bundle can be induced by a continuous mapping $T: M \to \tilde{G}(n, N)$, and T is defined up to a homotopy. By the kth PONTRJAGIN class of the bundle we mean the element $p_k \in H^{4k}(M, R)$[6]) such that, for any $z^{4k} \in H_{4k}(M, R)$, its value is given by

$$
(16)\qquad p_k \cdot z^{4k} = KI(\Sigma(k), T_* z^{4k}),
$$

[6]) If X is a topological space and G an abelian group, $H^r(X, G)$ denotes the r-dimensional cohomology group of X with the coefficient group G and $H_r(X, G)$ denotes the r-dimensional homology group. By R we mean the real field.

where $T_* z^{4k}$ is the image of z^{4k} under T and the right-hand side stands for the intersection number of the two cycles. Similarly, we define the dual PONTRJAGIN class $\overline{p}_l \in H^{4l}(M, R)$ by the relation

$$(17) \qquad -_l \cdot z^{4l} = KI(\overline{\Sigma}(l), T_* z^{4l}), \quad z^{4l} \in H_{4l}(M, R).$$

Unless otherwise specified, we shall mean by the PONTRJAGIN classes of M those arising from its tangent bundle.

Our problem is to give, in the case that M is a RIEMANN manifold, a closed differential form of degree $4k$ which defines the PONTRJAGIN class p_k in the sense of DE RHAM's theorem. The form we found is

$$(18) \qquad \Psi_k = \frac{1}{(2\pi)^{2k}(2k)!} \sum \delta(i_1, \ldots, i_{2k}; j_1, \ldots, j_{2k}) \Omega_{i_1 j_1} \wedge \cdots \wedge \Omega_{i_{2k} j_{2k}}.$$

In this formula $\delta(i_1, \ldots, i_{2k}; j_1, \ldots, j_{2k})$ is zero, except when $j_1, \ldots, j_{2k}$ is a permutation of $i_1, \ldots, i_{2k}$, and is then equal to $+1$ or -1, according as the permutation is even or odd, while the summation is extended over all indices $i_1, \ldots, i_{2k}, j_1, \ldots, j_{2k}$ from 1 to n. The differential form Ψ_k is closed, and defines an element of $H^{4k}(M, R)$ according to DE RHAM's theorem. We have proved that this cohomology class is precisely p_k[7]).

We observe that this relationship can be conveniently expressed in terms of the Ω introduced in (9) and (12). In fact, from (12) we find

$$\Omega^k = \sum_{i_1, \ldots, i_{2k}} (e_{i_1} \wedge \cdots \wedge e_{i_{2k}}) \otimes \Omega_{i_1 i_2} \wedge \cdots \wedge \Omega_{i_{2k-2} i_{2k}}.$$

For a fixed set of integers $i_1, \ldots, i_{2k}$ from 1 to n, the coefficient of $e_{i_1} \wedge \cdots \wedge e_{i_{2k}}$ in Ω^k is equal to

$$\sum_{j_1, \ldots, j_{2k}} \delta(i_1, \ldots, i_{2k}; j_1, \ldots, j_{2k}) \Omega_{j_1 j_2} \wedge \cdots \wedge \Omega_{j_{2k-1} j_{2k}}.$$

By (8), $|\Omega^k|^2$ is the sum of squares of these expressions for all combinations $(i_1, \ldots, i_{2k})$. It follows from (4) that

$$|\Omega^k|^2 = (2^k k!)^2 (2\pi)^{2k} \Psi_k.$$

This gives the following theorem:

Theorem 2. *The differential form $|\Omega^k|^2/(2^k k!)^2(2\pi)^{2k}$ defines the Pontrjagin class p_k in the sense of de Rham's theorem.*

4. Some implications of the curvature on the characteristic classes

The simplest assumption on the curvature is that the sectional curvature be constant. It is well-known that in terms of orthonormal frames this assumption is expressed analytically by the condition

$$(19) \qquad \Omega_{ij} = -K \omega_i \wedge \omega_j,$$

---

[7]) CHERN, TDG, p. 82.

where K is the value of the constant curvature. This condition can be generalized as follows: We put

$$\Omega_{ij} = \tfrac{1}{2} \sum_{k,l} S_{ijkl}\, \omega_k \wedge \omega_l, \quad S_{ijkl} + S_{ijlk} = 0. \tag{20}$$

Then the conditions

$$S_{ijkl} = -K(a_{ik} a_{jl} - a_{il}\, a_{jk}), \tag{21}$$

where $a_{ik} = a_{ki}$, generalize (19). In fact, the latter corresponds to the case $a_{ik} = \delta_{ik}$. We also remark that the K in (21) may not be constant.

When the conditions (21) are fulfilled, we have

$$\Omega^k = K^k (\sum_i e_i \otimes \theta_i)^{2k}, \quad \theta_i = \sum_j a_{ij}\, \omega_j. \tag{22}$$

From Theorems 1 und 2 we derive the theorem:

Theorem 3. *Let M be a compact orientable Riemann manifold, whose curvature satisfies the conditions (21) at every point. Then all its Pontrjagin classes are zero. If it is of even dimension $2s$, and if $K^s |a_{ij}|$ keeps a constant sign, then this sign is the sign of its Euler-Poincaré characteristic. Moreover, under this hypothesis, the Euler-Poincaré characteristic of M is zero, only when $K^s |a_{ij}|$ vanishes identically.*

A similar theorem is valid, when the curvature of the Riemann metric satisfies the following conditions

$$S_{ijkl} = 2 L_{ij} L_{kl} + L_{ik} L_{jl} - L_{il} L_{jk}, \tag{23}$$

where $L_{ik} = -L_{ki}$. In this case we have

$$\Omega = (\sum_{i,j} L_{ij} e_i \wedge e_j) \otimes (\sum_{k,l} L_{kl}\, \omega_k \wedge \omega_l) - (\sum_{i,k} L_{ik} e_i \otimes \omega_k)^2, \tag{24}$$

and we find, when $n = 2s$,

$$\Omega^s = (\sum_{q=0}^{s} (-1)^{s-q} \binom{s}{q} [(\sum_{i,j} L_{ij} e_i \wedge e_j) \otimes (\sum_{k,l} L_{kl}\, \omega_k \wedge \omega_l)]^q (\sum_{i,k} L_{ik} e_i \otimes \omega_k)^{2(s-q)}$$

$$= \left\{ \sum_{q=0}^{s} \binom{s}{q} L(q) \right\} (e_1 \wedge \cdots \wedge e_{2s}) \otimes (\omega_1 \wedge \cdots \wedge \omega_{2s}),$$

where $L(q)$ denotes the expression in (1). From Theorem 1 and formula (4) we derive the theorem:

Theorem 4. *If the curvature of a compact orientable Riemann manifold of even dimension $2s$ has the property (23), its Euler-Poincaré characteristic has the same sign as $(-1)^s$ or is zero. The latter occurs, only when $L(s)$ vanishes identically.*

Perhaps a more natural question is whether the fact that the sectional curvature keeps a constant sign for all plane sections has some implication on the sign of the Euler-Poincaré characteristic. The study of this question would require the choice of frames at a point relative to which the analytical expression of Ω_{ij} takes a simple form. The following method may be useful in this and similar questions:

We recall that if a plane element E is spanned by two vectors with the components X_i. Y_k respectively, its PLÜCKER coordinates are defined to be

$$p_{ik} = X_i Y_k - X_k Y_i. \tag{25}$$

When E is of unit measure, that is, when $\sum_{(ik)} p_{ik}^2 = 1$, the sectional curvature in E is given by

$$K(E) = -\tfrac{1}{4} \sum_{i,\ldots,l} S_{ijkl} p_{ij} p_{kl}. \tag{26}$$

Suppose the plane element $E_0: p_{12} = 1,\ p_{kl} = 0,\ (k, l) \neq (1, 2)$, be the one for which the sectional curvature is a maximum, among all plane elements through the point. In a neighborhood of E_0 we have $p_{12} \neq 0$. For such plane elements a complete set of independent relations between the p_{ik} can be written as

$$p_{12} p_{\alpha\beta} + p_{1\alpha} p_{\beta 2} + p_{1\beta} p_{2\alpha} = 0, \quad 3 \leqq \alpha, \beta \leqq n. \tag{27}$$

To these we add the relation that the measure of the plane element is 1:

$$\sum_{i,k} p_{ik}^2 = 2. \tag{28}$$

Following standard procedure, in order to study a relative extremum of the sectional curvature, we consider the function

$$f(p) = -\tfrac{1}{4} \sum_{i,\ldots,l} S_{ijkl} p_{ij} p_{kl} - \lambda \Big(\sum_{i,k} p_{ik}^2 - 2\Big) - \sum_{\alpha<\beta} 2 \mu_{\alpha\beta} (p_{12} p_{\alpha\beta} + p_{1\alpha} p_{\beta 2} + p_{1\beta} p_{2\alpha}),$$

where λ and $\mu_{\alpha\beta}$ are indeterminates. We find

$$\frac{1}{2} \frac{\partial}{\partial p_{1\alpha}} f(p) = -\frac{1}{4} \sum_{k,l} S_{1\alpha kl} p_{kl} - \lambda p_{1\alpha} - \sum_{\alpha<\beta} \mu_{\alpha\beta} p_{\beta 2}.$$

The condition that these partial derivatives vanish for E_0 gives

$$S_{1\alpha 12} = 0. \tag{29$_1$}$$

By interchanging the indices 1 and 2, we get

$$S_{2\alpha 12} = 0. \tag{29$_2$}$$

We next consider all the plane elements which have a nonzero vector in common with both E_0 and the $(n-2)$-dimensional linear space perpendicular to it. Among these plane elements let E_1 ($p_{13} = 1,\ p_{ik} = 0,\ (i, k) \neq (1, 3)$) be the one for which the sectional curvature attains a maximum. By using the above argument, we get the necessary conditions

$$S_{2313} = 0, \quad S_{1\nu 13} = 0, \quad 4 \leqq \nu \leqq n. \tag{29$_3$}$$

For $n = 4$ our conditions can be summarized in the form

(30) $$S_{1213} = S_{1214} = S_{1223} = S_{1224} = S_{1323} = S_{1314} = 0.$$

It follows that

$$\Omega_{12} \wedge \Omega_{34} + \Omega_{13} \wedge \Omega_{42} + \Omega_{14} \wedge \Omega_{23}$$
$$= (S_{1212}S_{3434} + S^2_{1234} + S_{1313}S_{4242} + S^2_{1342} + S_{1414}S_{2323} + S^2_{1423})\, \omega_1 \wedge \cdots \wedge \omega_4.$$

This expression and the GAUSS-BONNET formula give the following theorem, which was first proved by J. MILNOR:

Theorem 5 (MILNOR). *Let M be a compact orientable Riemann manifold of dimension 4. If its sectional curvatures along perpendicular plane elements always have the same sign, its Euler-Poincaré characteristic is $\geqq 0$. If the sectional curvature is always positive or always negative, the Euler-Poincaré characteristic is positive.*

5. An imbedding theorem

In our definition of the PONTRJAGIN characteristic classes we made use of the mapping $T : M \to \tilde{G}(n, N)$, which induces the bundle. When the bundle in question is the tangent bundle of M, we can define the mapping T in a more geometrical way as follows: Imbed M differentiably in an Euclidean space of dimension $n + N$; to any $x \in M$ define $T(x)$ to be the oriented n-dimensional linear space through a fixed point 0 and parallel to the tangent space to M at x. Since in the definition of the dual PONTRJAGIN class $\bar{p}_l$ the integer l is subject to the condition $2l \leqq N$, it is clear that $N = 2l - 1$ implies $\bar{p}_l = 0$. We get therefore the following criterion for imbedding: *If the Pontrjagin class $\bar{p}_l$ of a compact manifold M is $\neq 0$, M cannot be differentiably imbedded in an Euclidean space of dimension $n + 2l - 1$.*

Similar criteria in terms of the STIEFEL-WHITNEY classes are known[8]). It seems to us that the advantage of this criterion lies in the fact that the classes are with real coefficients and will therefore not leave out certain dimensions because of the arithmetical properties of some integers, although the bound given is not a very sharp one.

We wish to illustrate this criterion by applying it to the complex projective space P_m of (complex) dimension m. If $g \in H^2(P_m, R)$ denotes the cohomology class dual to a hyperplane, then $H^{2k}(P_m, R)$, $k \leqq m$, is generated by g^k (power in the sense of cup product), while all odd-dimensional cohomology groups of P_m are zero. The PONTRJAGIN classes of P_m can be computed by applying our Theorem 2, using the elliptic

[8]) S. CHERN, On the multiplication in the characteristic ring of a sphere bundle, Annals of Math. 49, 362—372 (1948).

Hermitian metric in P_m. As this computation is straightforward and these PONTRJAGIN classes have been determined by other methods, we shall not go into its details. The result can be described by the formula[9])

$$p = \sum_{0 \leq 4k \leq 2m} (-1)^k p_k = (1 - g^2)^{m+1}, \tag{31}$$

where the common expression stands for an element of the cohomology ring of P_m.

To determine the dual PONTRJAGIN classes $\bar{p}_l$ of P_m we make use of the following two facts: 1) Suppose P_m be differentiably imbedded in a real Euclidean space E^{2m+N} of dimension $2m + N$. Let p'_l be the PONTRJAGIN classes of the normal sphere bundle of this imbedding. Then[10]) $p'_l = (-1)^l \bar{p}_l$. 2) Between the PONTRJAGIN classes of the tangent bundle and the normal bundle of this imbedding there is the duality theorem[11]):

$$pp' = 1, \tag{32}$$

where

$$p' = \sum_{0 \leq 4l \leq 2m} (-1)^l p'_l = \sum \bar{p}_l . \tag{33}$$

From (32) and (33) it follows that

$$p\bar{p} = 1, \tag{34}$$

where

$$\bar{p} = \sum_{0 \leq 4l \leq 2m} \bar{p}. \tag{35}$$

Therefore the dual PONTRJAGIN class of P_m is given by

$$\bar{p} = (1 - g^2)^{-m-1}. \tag{36}$$

From this we find

$$\begin{aligned} \bar{p}_l &= \frac{(2l+1)\ldots(3l)}{l!} g^{2l}, \quad \text{if } m = 2l, \\ \bar{p}_l &= \frac{(2l+2)\ldots(3l+1)}{l!} g^{2l}, \quad \text{if } m = 2l+1. \end{aligned} \tag{37}$$

Since these are not zero, we have the theorem:

Theorem 6. *The complex projective space of complex dimension m cannot be differentiably imbedded in a real Euclidean space of dimension $3m - 1$ or $3m - 2$, according as m is even or odd.*

---

[9]) Cf., for instance, F. HIRZEBRUCH, Über die quaternionalen projektiven Räume, Sitzungsberichte der Bayerischen Akad. der Wiss., 1953, p. 305. Our convention on p_k differs by a sign $(-1)^k$ from HIRZEBRUCH'S.

[10]) WU, loc. cit., pp. 30—35.

[11]) CHERN, TDG, pp. 100—104; HIRZEBRUCH, loc. cit., pp. 304—305.

Eingegangen am 21. 3. 1955

Reprinted from *Amer. J. of Math.* **79** (1957) 949–950

A PROOF OF THE UNIQUENESS OF MINKOWSKI'S PROBLEM FOR CONVEX SURFACES.*

By Shiing-shen Chern.

For C'' closed convex surfaces in ordinary Euclidean space E the uniqueness of Minkowski's problem says that the knowledge of the Gaussian curvature (supposed to be strictly positive) as a function of the unit normal vector determines the surface up to a translation. Let S_0 be the unit sphere about the origin 0 in E. Since the normal map onto S_0 of a convex surface M with Gaussian curvature $K > 0$ is one-one, M can be defined by a vector-valued function $x(\xi)$, $\xi \in S_0$.

Let T be the space of all right-handed rectangular frames $0e_1e_2e_3$ about 0. Then T is a circle bundle over S_0, the projection being defined by mapping the frame into the end-point of the last vector e_3. Let $\theta_{ij} = -\theta_{ji} = de_i e_j$. (Throughout this note small Latin indices run from 1 to 3 and small Greek indices from 1 to 2.) Then we have

$$(1) \qquad d\theta_{ij} = \sum_k \theta_{ik} \wedge \theta_{kj}.$$

On the other hand, let F be the space of all right-handed rectangular frames $xf_1f_2f_3$ in E ($\dim F = 6$). If we put

$$(2) \qquad \omega_i = dx \cdot f_i, \qquad \omega_{ij} = -\omega_{ji} = df_i f_j,$$

or

$$(3) \qquad dx = \sum_i \omega_i f_i, \qquad df_i = \sum_j \omega_{ij} f_j,$$

we have

$$(4) \qquad d\omega_i = \sum_j \omega_j \wedge \omega_{ji}, \qquad d\omega_{ij} = \sum_k \omega_{ik} \wedge \omega_{kj}.$$

The mapping $x\colon S_0 \to M$ defined by sending $\xi \in S_0$ into the point $x(\xi) \in M$ at which the unit normal vector is ξ induces a mapping $\tilde{x}\colon T \to F$, which sends the frame $0e_1e_2e_3$ to $x(e_3)e_1e_2e_3$. Let $\tilde{\omega}_i = \tilde{x}^*\omega_i$, $\tilde{\omega}_{ij} = \tilde{x}^*\omega_{ij}$ be the differential forms in T induced by $\tilde{x}$. Then we have $\tilde{\omega}_{ik} = \theta_{ik}$, $\tilde{\omega}_3 = 0$ and we can put

$$(5) \qquad \tilde{\omega}_\alpha = \sum_\beta \lambda_{\alpha\beta}\theta_{\beta 3}, \qquad \lambda_{\alpha\beta} = \lambda_{\beta\alpha}.$$

where $\lambda_{11}\lambda_{22} - \lambda_{12}\lambda_{21} = 1/K(\xi)$.

* Received February 5, 1957.

Suppose there be a second convex surface M' defined by the vector-valued function $x'(\xi)$, with the same Gaussian curvature $K(\xi)$. We denote by dashes notions pertaining to M'. Then we have $\tilde{\omega}'_3 = \tilde{\omega}_3 = 0$, $\tilde{\omega}'_{ij} = \tilde{\omega}_{ij}$, and

$$\tilde{\omega}'_\alpha = \sum_\beta \lambda'_{\alpha\beta}\theta_{\beta 3}, \qquad \lambda'_{\alpha\beta} = \lambda'_{\beta\alpha}. \tag{6}$$

It suffices to prove that $\tilde{\omega}'_\alpha = \tilde{\omega}_\alpha$. For then $\tilde{x}'\tilde{x}^{-1} : \tilde{x}(T) \to \tilde{x}'(T)$ will be a one-one map under which $\omega'_i = (\tilde{x}'\tilde{x}^{-1})^*\omega_i$, $\omega'_{ij} = (\tilde{x}'\tilde{x}^{-1})^*\omega_{ij}$. By a well-known theorem on moving frames (E. Cartan, La théorie des groupes finis et continus . . . Paris, 1937), $\tilde{x}'\tilde{x}^{-1}$ is the identity, if it is the identity for one frame of $\tilde{x}(T)$. But this can be achieved by a translation.

To complete the proof we derive from (3), (5), (6) the formula

$$\begin{aligned} d(x, x', dx') &= p'(\tilde{\omega}_2 \wedge \tilde{\omega}'_1 - \tilde{\omega}_1 \wedge \tilde{\omega}'_2) + 2p\tilde{\omega}'_1 \wedge \tilde{\omega}'_2 \\ &= \{p' \begin{vmatrix} \lambda'_{11} - \lambda_{11} & \lambda'_{12} - \lambda_{12} \\ \lambda'_{21} - \lambda_{21} & \lambda'_{22} - \lambda_{22} \end{vmatrix} + \frac{2}{K}(p - p')\}\theta_{13} \wedge \theta_{23} \end{aligned}$$

where $p(\xi)$ is the distance from 0 to the tangent plane at $x(\xi)$. By Stokes Theorem we get

$$\int\int_{S_0} \{p' \begin{vmatrix} \lambda'_{11} - \lambda_{11} & \lambda'_{12} - \lambda_{12} \\ \lambda'_{21} - \lambda_{21} & \lambda'_{22} - \lambda_{22} \end{vmatrix} + \frac{2p}{K} - \frac{2p'}{K}\}\theta_{13} \wedge \theta_{23} = 0.$$

Since $(\lambda_{\alpha\beta})$ and $(\lambda'_{\alpha\beta})$ are positive definite symmetric matrices with the same determinant, it is well-known that

$$\begin{vmatrix} \lambda'_{11} - \lambda_{11} & \lambda'_{12} - \lambda_{12} \\ \lambda'_{21} - \lambda_{21} & \lambda'_{22} - \lambda_{22} \end{vmatrix} \leqq 0,$$

and that the equality sign holds only when $\lambda'_{\alpha\beta} = \lambda_{\alpha\beta}$. It follows that

$$\int\int_{S_0} \frac{1}{K}(p - p')\theta_{13} \wedge \theta_{23} \leqq 0$$

($p' < 0$, by choosing 0 to be in the interior of M'). But the relationship between M and M' is symmetrical, so that the relation still holds, when p and p' are interchanged. Hence the above integral is zero, and we have

$$\int\int_{S_0} p' \begin{vmatrix} \lambda'_{11} - \lambda_{11} & \lambda'_{12} - \lambda_{12} \\ \lambda'_{21} - \lambda_{21} & \lambda'_{22} - \lambda_{22} \end{vmatrix} \theta_{13} \wedge \theta_{23} = 0.$$

Here the integrand keeps a constant sign; the relation is possible, only when the integrand vanishes identically, which in turn implies $\lambda'_{\alpha\beta} = \lambda_{\alpha\beta}$. This completes our proof.

UNIVERSITY OF CHICAGO.

ON THE TOTAL CURVATURE OF IMMERSED MANIFOLDS, II

Shiing-shen Chern and Richard K. Lashof

Let M^n be a compact differentiable manifold of dimension n, and let

$$x: M^n \to E^{n+N}$$

be a differentiable mapping of M^n into a Euclidean space of dimension $n + N$ with the property that the functional matrix is everywhere of rank n. Then M^n is said to be immersed in E^{n+N}. If x is one-one, it is said to be imbedded in E^{n+N}. To each unit normal vector $\nu(p)$ of an immersed manifold M^n at $p \in M$, we draw through the origin O of E^{n+N} the unit vector parallel to it. This defines a mapping, to be called $\tilde{\nu}$, of the normal sphere bundle B_ν of M^n into the unit hypersphere S_0 about O. In a previous paper [1; this paper will be referred to as TCI], we studied the volume of the image of $\tilde{\nu}$ and called it the *total curvature* of M^n. It will be advantageous to normalize this volume by dividing it by the volume c_{n+N-1} of S_0, c_{n+N-1} being of course an absolute constant. Throughout this paper, we will understand by the total curvature of M^n the normalized one. Then, if $E^{n+N} \subset E^{n+N'}$ $(N < N')$, the total curvature $T(M^n)$ of M^n remains the same, whether M^n is considered as a submanifold of E^{n+N} or of $E^{n+N'}$ (Lemma 1, Section 1). One of the theorems we proved in TCI states that $T(M^n) \geq 2$. We shall show below (Section 1) that the same argument can be used to establish the following more general theorem.

THEOREM 1. *Let* M^n *be a compact differentiable manifold immersed in* E^{n+N}, *and let* β_i $(0 \leq i \leq n)$ *be its ith Betti number relative to a coefficient field. Then the total curvature* $T(M^n)$ *of* M^n *satisfies the inequality*

$$T(M^n) \geq \beta(M^n), \tag{1}$$

where $\beta(M^n) = \Sigma_{i=0}^{n} \beta_i$ *is the sum of the Betti numbers of* M^n.

The right-hand side of (1) depends on the coefficient field. For the real field, the lower bound in (1) cannot always be attained. In fact, we have the following theorem.

THEOREM 2. *If the equality sign holds in* (1) *with the real field as coefficient field, then* M^n *has no torsion.*

For a compact differentiable manifold M^n given abstractly, the total curvature $T(M^n)$ or $T_x(M^n)$ is a function of the immersion $x: M^n \to E^{n+N}$ (N arbitrary). Obviously, the number $q(M^n) = \inf_x T_x(M^n)$ is a global invariant of M^n itself. Theorem 1 says that $q(M^n) \geq \beta(M^n)$. In this connection, there is another invariant $s(M^n)$ of M^n, namely the minimum number of cells in a cell complex covering M^n. Clearly, we have $s(M^n) \geq \beta(M^n)$. If M^2 is a compact orientable surface of genus g, it is easy to see that

$$q(M^2) = s(M^2) = \beta(M^2) = 2 + 2g.$$

Received October 18, 1957.

Work done while S. S. Chern was under a contract with the National Science Foundation.

Generally, it can be shown that $q(M^n)$ is an integer; but the proof will not be included in this paper. It seems likely that $q(M^n) = s(M^n)$.

Another problem in this order of ideas is the characterization of the immersions of M^n by which the minimal total curvature of M^n is realized, that is, for which $T(M^n) = q(M^n)$. If M^n is homeomorphic to an n-sphere, it is a consequence of Theorem 3 of TCI that such an immersion is characterized by the property that M^n is imbedded as a convex hypersurface in a linear space of dimension $n+1$. The general problem can therefore be regarded as a natural generalization of the theory of convex hypersurfaces in Euclidean space. When M^n is immersed as a hypersurface, that is, when $N = 1$, the Gauss-Kronecker curvature $K(p)$ $(p \in M^n)$, a local invariant of M^n, plays an important role in our problem. It is defined only up to sign when n is odd. The answer to our problem is most complete in the case of compact surfaces imbedded in ordinary Euclidean space $(n = 2, N = 1)$:

THEOREM 3. *A compact orientable surface of genus* g *is imbedded in the three-dimensional Euclidean space with total curvature* $2g + 2$, *if and only if the surface lies at one side of the tangent plane at every point of positive Gaussian curvature.*

For oriented compact hypersurfaces $(N = 1)$ with Gauss-Kronecker curvature $K(p) \geq 0$ for all $p \in M^n$, we have the following theorem.

THEOREM 4. *A compact orientable surface immersed in three-dimensional Euclidean space with Gaussian curvature* ≥ 0 *is imbedded and convex. There are examples of nonconvex compact orientable hypersurfaces, of dimension* ≥ 3, *whose Gauss-Kronecker curvature is everywhere* ≥ 0.

The main point of this theorem is that $K(p)$ is assumed merely to be ≥ 0, and not strictly > 0. In the latter case, a well-known argument due to Hadamard shows that M^n is imbedded as a convex hypersurface. Theorem 4 implies that a conjecture made by us in TCI (p. 318) is true for $n = 2$ and false for $n \geq 3$.

1. TOTAL CURVATURE AND THE SUM OF BETTI NUMBERS

LEMMA 1. *Let* $x: M^n \to E^{n+N}$ *be an immersion of a compact differentiable manifold of dimension* n *in* E^{n+N}, *given by*

$$x: p \to x(p) = (x^1(p), \cdots, x^{n+N}(p)) \qquad (p \in M^n).$$

Let $x': M^n \to E^{n+N'}$ $(N < N')$, *be the immersion defined by*

$$x'(p) = (x^1(p), \cdots, x^{n+N}(p), 0, \cdots, 0).$$

Then the immersed mainfolds $x(M^n)$ *and* $x'(M^n)$ *have the same total curvature.*

The lemma is intuitively obvious. For if B'_ν is the normal sphere bundle, and $\tilde{\nu}': B' \to S^{n+N'-1}$ the corresponding normal map of the immerson x', then clearly $\tilde{\nu}'$ is the $(N' - N)$-fold suspension of $\tilde{\nu}$ on each fiber. Since $S^{n+N'-1}$ is the $(N' - N)$-fold suspension of S^{n+N-1}, it follows that the ratio of the area covered by $\tilde{\nu}'$ on $S^{n+N'-1}$ to the area covered by $\tilde{\nu}$ on S^{n+N-1} is the same as the ratio of the areas of $S^{n+N'-1}$ and S^{n+N-1}. In spite of this short argument, we give a more analytical proof as follows:

It suffices to prove the lemma for the case $N' - N = 1$. The general case will then follow by induction on the difference $N' - N$.

We follow the notation of TCI, and consider the bundle B of all frames

$$x(p)e_1 \cdots e_{n+N} \qquad (p \in M^n)$$

such that $e_1, \cdots, e_n$ are tangent vectors and $e_{n+1}, \cdots, e_{n+N}$ are normal vectors at $x(p)$, If we put

$$\omega_{n+N,A} = de_{n+N} \cdot e_A \qquad (1 \leq A \leq n+N),$$

then the total curvature is, according to our definition,

$$T_x(M^n) = \frac{1}{c_{n+N-1}} \int_{B_\nu} \omega_{n+N,1} \wedge \cdots \wedge \omega_{n+N,n+N-1} ,$$

where the integral is taken in the measure-theoretic sense. It is to be pointed out that, as stated in the Introduction, we have inserted the factor $1/c_{n+N-1}$ to normalize the total curvature.

Let a be one of the two unit vectors perpendicular to E in E^{n+N-1}. A unit normal vector at $x'(p)$ can be written uniquely in the form

$$e'_{n+N+1} = (\cos\theta)\, e_{n+N} + (\sin\theta)\, a \qquad \left(-\frac{\pi}{2} < \theta \leq \frac{\pi}{2}\right),$$

where e_{n+N} is the unit vector in the direction of its projection in E^{n+N}. Let

$$e'_{n+N} = (\sin\theta) e_{n+N} - (\cos\theta)\, a, \qquad e'_s = e_s \qquad (1 \leq s \leq n+N-1)$$

and

$$\phi_{n+N+1,A} = de'_{n+N+1} \cdot e'_A .$$

Then the total curvature of the immersed manifold $x'(M^n)$ is equal to

$$T_{x'}(M^n) = \frac{1}{c_{n+N}} \int_{B'_\nu} \phi_{n+N+1,1} \wedge \cdots \wedge \phi_{n+N+1,n+N} .$$

Now we have

$$de'_{n+N+1} = (\cos\theta) de_{n+N} + \{-(\sin\theta) e_{n+N} + (\cos\theta) a\} d\theta = (\cos\theta) de_{n+N} - e'_{n+N} d\theta .$$

Since

$$de_{n+N} \cdot e'_{n+N} = -(\cos\theta)(de_{n+N} \cdot a) = (\cos\theta)(e_{n+N} \cdot da) = 0,$$

we find that

$$\phi_{n+N+1,n+N} = de'_{n+N+1} \cdot e'_{n+N} = -d\theta .$$

Also

$$\phi_{n+N+1,s} = (\cos\theta)(de_{n+N} \cdot e_s) = (\cos\theta) \omega_{n+N,s} .$$

It follows that

$$T_{x'}(M^n) = \frac{c_{n+N-1}}{c_{n+N}} \int_{-\pi/2}^{\pi/2} |\cos^{n+N-1}\theta|\, d\theta \cdot T_x(M^n) = \frac{2c_{n+N-1}}{c_{n+N}} \int_0^{\pi/2} \cos^{n+N-1}\theta\, d\theta \cdot T_x(M^n).$$

That $T_{x'}(M^n) = T_x(M^n)$ is then a consequence of the following well-known formulas:

$$c_k = \frac{2\left[\Gamma\left(\frac{1}{2}\right)\right]^{k+1}}{\Gamma\left(\frac{k+1}{2}\right)}, \qquad \int_0^{\pi/2} \cos^k\theta\, d\theta = \frac{\Gamma\left(\frac{1}{2}\right)\Gamma\left(\frac{k+1}{2}\right)}{2\Gamma\left(\frac{k+2}{2}\right)}.$$

This completes the proof of the lemma.

Since the total curvature is clearly invariant under motions in space, Lemma 1 implies that the total curvature of $x(M^n)$ in E^{n+N} remains unchanged if $x(M^n)$ is considered as a submanifold of a high-dimensional Euclidean space which contains E^{n+N} as a linear subspace.

We wish now to give a proof of Theorem 1. As in the proof of Theorem 1 of TCI, we consider the map $\tilde{\nu}: B_\nu \to S_0^{n+N-1}$ defined by assigning to each unit normal vector the end-point of the unit vector through the origin parallel to it. The total curvature of M^n is by definition the volume of the image of B_ν under $\tilde{\nu}$. The singular points of $\tilde{\nu}$, that is, the points where the functional determinant of $\tilde{\nu}$ is zero, are exactly the points where the quadratic differential form $\nu \cdot d^2x = -d\nu \cdot dx$ is of rank $< n$. By Sard's theorem, their image on S_0^{n+N-1} has measure zero. Hence, for almost all unit vectors ν, the function $\nu \cdot x(p)$ on M^n, with ν fixed, has only nondegenerate critical points. By Morse's inequalities, the total number of critical points is $\geq \Sigma\beta_i(M^n) = \beta(M^n)$. Now the image of B_ν under $\tilde{\nu}$ is the same as the set of points $\nu \in S_0^{n+N-1}$, each counted a number of times equal to the number of critical points of the function $\nu \cdot x(p)$ on M^n. It follows that the measure of the image is $\geq c_{n+N-1}\beta(M^n)$, and hence that the total curvature of M^n is $\geq \beta(M^n)$.

2. IMMERSIONS WITH MINIMAL TOTAL CURVATURE

Proof of Theorem 2.† This theorem follows immediately from Theorem 1. In fact, let $\beta_i(M^n, F)$ be the ith Betti number of M^n with the coefficient field F, and let $\beta(M^n, F)$ be the sum of these Betti numbers. If R denotes the real field and Z_p the field mod p (p a prime), we have

†We are indebted to the referee for this elementary argument. Our original proof makes use of results of R. Thom [3] and of Eilenberg and Shiffman [2, p. 53] concerning the cell decomposition of a manifold on the basis of a real-valued function on it. The result of Eilenberg and Shiffman can be stated as follows: If a compact differentiable manifold M has a differentiable function on it with k nondegenerate critical points of indices $i_1, \cdots i_k$, respectively, then M is of the same homotopy type as a cell complex consisting of k cells of dimensions $i_1, \cdots, i_k$, respectively. Theorem 2 follows immediately from this, because there is a coordinate function with $\beta(M^n,R)$ nondegenerate critical points. The theorem of Eilenberg and Shiffman also gives more information on manifolds satisfying the hypothesis of Theorem 2. For instance, it follows easily that the fundamental group of M^n is isomorphic to its first homology group.

$$\beta_i(M^n, R) \leq \beta_i(M^n, Z_p) \quad (0 \leq i \leq n).$$

Now by hypothesis, $T(M^n) = \beta(M^n, R)$, and by Theorem 1, $T(M^n) \geq \beta(M^n, Z_p)$, so that $\beta(M^n, R) \geq \beta(M^n, Z_p)$. In view of the inequalities above, this is possible only when $\beta_i(M^n, R) = \beta_i(M^n, Z_p)$, which means that M^n has no torsion.

From now on we study the particular case of hypersurfaces $(N = 1)$. Here the most important local invariant is the Gauss-Kronecker curvature $K(p)$ $(p \in M^n)$. If $K(p) \neq 0$, the principal curvatures of M^n at p are all different from zero. In this case, we call the signature of M^n at p the nonnegative integer which is the excess of the number of principal curvatures of one sign over that of the opposite sign. A part of Theorem 3 is true for n dimensions, and we state it as a lemma:

LEMMA 2. *Let* $x: M^n \to E^{n+1}$ *be an immersion of a compact manifold as a hypersurface in Euclidean space with total curvature equal to the sum* $\beta(M^n)$ *of the Betti numbers of* M^n *relative to the coefficient field* mod 2. *Then every point* $p \in M^n$ *with* $K(p) > 0$ *and signature* n *lies on the outside of* M^n; *that is,* $x(M^n)$ *lies on one side of the tangent hyperplane at* $x(p)$.

To prove the lemma, we suppose the contrary, namely, that there is a point p with $K(p) > 0$ and signature n such that $x(M^n)$ lies on both sides of the tangent hyperplane at $x(p)$. Then there is a neighborhood U of p whose points have the same property. We orient U by choosing a field of unit normal vectors. The normal map $\nu: U \to S_0^n$ is then defined. Since $K(p) > 0$ $(p \in U)$, ν can be supposed to be one-one (with the choice of a smaller neighborhood, if necessary). The image $\nu(U)$ is therefore of positive measure on S_0^n. It follows that there exists a set F of points of positive measure in $\nu(U)$ such that the function $\nu \cdot x(p)$ on M^n $(\nu \in F)$ has only nondegenerate critical points, and such that there is a tangent hyperplane to $x(M^n)$ perpendicular to ν which divides $x(M^n)$ and which is tangent to $x(M^n)$ at a point of signature n. The latter is a critical point of index 0 or n for the function $\nu \cdot x(p)$, and it is neither a maximum nor a minimum. On the other hand, since the total curvature is equal to $\beta(M^n)$, and since $\nu \cdot x(p)$, for almost all $\nu \in S_0^n$, has at least β_i critical points of index i for each dimension i, the number of critical points of $\nu \cdot x(p)$ of indices 0 and n must be 1 $(= \beta_0 = \beta_n)$, except for a set of points $\nu \in S_0^n$ of measure zero. These critical points are obviously the maximum and the minimum of the function $\nu \cdot x(p)$. Thus we arrive at a contradiction, and the lemma is proved.

We proceed to give a proof of Theorem 3. Half of the theorem follows from Lemma 2, because a point $p \in M^2$ with $K(p) > 0$ has the signature 2, the two principal curvatures being either both positive or both negative.

Suppose now that the surface M^2 is imbedded in E^3 in such a way that it lies on one side of the tangent plane at every point of positive Gaussian curvature. Suppose also that its total curvature is $> 2g + 2$. Then there exists a set of points $\nu \in S_0^2$ of positive measure such that the function $\nu \cdot x(p)$ on M^2 has only nondegenerate critical points, whose number exceeds $2g + 2$. Let m_i $(0 \leq i \leq 2)$ be the number of critical points of index i of this function. Then we have by hypothesis

$$m_0 + m_1 + m_2 > 2g + 2,$$

and by Morse's relation, $m_0 - m_1 + m_2 = 2 - 2g$. Combination of these two relations gives $m_0 + m_2 > 2$. It follows that there are at least three distinct points of positive Gaussian curvature on M^2, whose tangent planes are perpendicular to ν. According to our hypothesis, two of these three tangent planes must coincide, and $x(M^2)$ is contained between the two tangent planes and is tangent to one of them, say π, in two distinct points. Since x is an imbedding, it is geometrically clear that we can rotate

π slightly so that the new plane is again tangent to $x(M^2)$ at a point of positive Gaussian curvature and divides $x(M^2)$. This contradiction proves Theorem 3.

Remark. Examples can easily be given to show that Theorem 3 is not true if x is an immersion.

3. HYPERSURFACES WITH NONNEGATIVE GAUSS-KRONECKER CURVATURE

As remarked before, an immersed compact orientable hypersurface with $K(p) > 0$ is convex. Its total curvature is equal to 2. We will show that the class of immersed compact hypersurfaces with $K(p) \geq 0$ is much wider.

LEMMA 3. *Let* $x: M^n \to E^{n+1}$ *be an immersion such that (1)* n *is even; (2)* M^n *is compact and orientable; (3)* $K(p) \geq 0$ $(p \in M^n)$. *Then* M^n *has no torsion, the odd-dimensional Betti numbers of* M^n *are zero, and its total curvature is equal to* $\beta(M^n)$.

As usual, let $\nu \in S_0^n$ be a unit vector such that the function $\nu \cdot x(p)$ on M^n has only nondegenerate critical points. The second-order terms in the expansion of the function at a critical point are given by $\nu \cdot d^2x = -d\nu \cdot dx$, which is the second fundamental form of the hypersurface. Since the critical points are nondegenerate, the Gauss-Kronecker curvature is > 0 at these points, and the numbers of negative principal curvatures and hence positive principal curvatures are both even. This means that the critical points of $\nu \cdot x(p)$ are of even indices. By the Theorem of Eilenberg and Shiffman stated in the footnote of Section 2, the manifold M^n is of the same homotopy type as a cell complex which consists only of even-dimensional cells. Hence the odd-dimensional Betti numbers of M^n are zero, and M^n has no torsion.

The degree of the normal map ν is equal to one-half of the Euler-Poincaré characteristic of M^n, which is in this case equal to $\beta(M^n)/2$. Since the image under ν of the set of points with $K(p) = 0$ is of measure zero, and since $K(p) > 0$ otherwise, the number of times which almost every point of S_0^n is covered by ν is $\beta(M^n)/2$. It follows that the total curvature of M^n is $\beta(M^n)$, because at every point of M^n there are two unit normal vectors, one being the negative of the other.

Remark. If besides the hypotheses of Lemma 3 we further suppose that the signature of M^n at p is equal to n at all points where $K(p) > 0$, then it follows that the Euler-Poincaré characteristic of M^n is $\beta(M^n)/2 = 1$. By Theorem 4 of TCI, we conclude that M^n is imbedded as a convex hypersurface.

Proof of Theorem 4. The first statement on compact orientable surfaces follows immediately from Lemma 3 and Theorem 4 of TCI. To prove the second statement, it suffices to exhibit some examples of hypersurfaces.

First let n be odd. In E^{n+1} with the coordinates $x^1, \cdots, x^{n+1}$, we consider the hypersurface with the equation

$$(r - 2)^2 + (x^{n+1})^2 = 1,$$

where

$$r^2 = (x^1)^2 + \cdots + (x^n)^2 \qquad (r \geq 0).$$

This hypersurface is obtained by rotating a unit circle about the x^{n+1}-axis, and is hence homeomorphic to the Cartesian product $S^1 \times S^{n-1}$. Its equation can also be written

$$x^{n+1} = \varepsilon\,\phi(r),$$

where $\varepsilon = \pm 1$ and

$$\phi(r) = +(1 - (r-2)^2)^{1/2}.$$

Then we have

$$dx = (dx^1, \cdots, dx^n, \varepsilon\, d\phi), \qquad d^2x = (0, \cdots, 0, \varepsilon\, d^2\phi),$$

$$\nu = \frac{\varepsilon}{+(1 + \phi_1^2 + \cdots + \phi_n^2)^{1/2}} (+\phi_1, \cdots, +\phi_n, -1),$$

where

$$\phi_i = \frac{\partial \phi}{\partial x^i}, \quad \phi_{ij} = \frac{\partial^2 \phi}{\partial x^i \partial x^j} \quad (1 \le i, j \le n).$$

This determination of ν is inward. It follows that

$$\nu \cdot d^2x = \frac{-1}{(1 + \phi_1^2 + \cdots + \phi_n^2)^{1/2}} d^2\phi = \frac{-1}{(1 + \phi_1^2 + \cdots\ \phi_n^2)^{1/2}} \sum_{i,j} \phi_{i,j}\, dx^i dx^j.$$

If ϕ' and ϕ'' denote the first and second derivatives of ϕ with respect to r, we have

$$\phi(r) = (-r^2 + 4r - 3)^{1/2}, \quad \phi'(r) = \frac{2-r}{\phi(r)}, \quad \phi''(r) = -\frac{1}{\phi(r)^3}\{\phi(r)^2 + (2-r)^2\},$$

and

$$\phi_i = \phi' \frac{x^i}{r}, \quad \phi_{ij} = \phi'' \frac{x^i x^j}{r^2} + \frac{\phi'}{r^3}(\delta^{ij} r^2 - x^i x^j).$$

The Gauss-Kronecker curvature $K(p)$ is equal to the determinant of the second fundamental form divided by the determinant of the first fundamental form. Since the latter is positive, the sign of $K(p)$ is the same as that of $-\det(\phi_{ij})$. Since our hypersurface is a hypersurface of revolution, it suffices to consider those of its points in the (x^n, x^{n+1})-plane for which $x^1 = \cdots x^{n-1} = 0$. At such a point we have

$$-\det(\phi_{ij}) = -\frac{(\phi')^{n-1}}{r^{n-1}} \phi'' \ge 0.$$

The example for n even $(n \ge 4)$ is similar. It is a hypersurface obtained by rotating a two-dimensional sphere about a two-dimensional coordinate plane, and it has the equation

$$(r-2)^2 + (x^n)^2 + (x^{n+1})^2 = 1, \quad r^2 = (x^1)^2 + \cdots + (x^{n-1})^2 \quad (r \ge 0),$$

or

$$x^{n+1} = \varepsilon \psi(x^n, r) \quad (\varepsilon^2 = 1),$$

where

$$\psi(x^n, r) = \{1 - (x^n)^2 - (r - 2)^2\}^{1/2} \geq 0.$$

As in the preceding example, K(p) has the same sign as $\det(\psi_{ij})$, where $\psi_{ij} = \frac{\partial^2 \psi}{\partial x^i \partial x^j}$. It is a straightforward computation to show that $\det(\psi_{ij}) \geq 0$ or $K(p) \geq 0$; we omit the details.

The following corollaries are obvious.

COROLLARY 1. *If a compact manifold* M *can be imbedded in* E^n, *then* $M \times S^n$ *can be imbedded in* E^{n+N}.

COROLLARY 2. *The product of spheres* $S^{n_1} \times \cdots \times S^{n_r}$ *can be imbedded in* $E^{n_1 + \cdots + n_r + 1}$ *with minimal total curvature* 2^r.

REFERENCES

1. S. S. Chern and R. K. Lashof, *On the total curvature of immersed manifolds*, Amer. J. Math. 79 (1957), 306-318.

2. M. Shiffman, *Notes on topology (critical point theory)*, Mimeographed, Stanford University, Spring 1950.

3. R. Thom, *Sur une partition en cellules associée à une fonction sur une variété*, C. R. Acad. Sci. Paris 228 (1949), 973-975.

University of Chicago

Reprinted from
Michigan Math. J. **5** (1958) 5–12

Differential Geometry and Integral Geometry

By Shiing-shen Chern

Integral geometry, started by the English geometer M. W. Crofton, has received recently important developments through the works of W. Blaschke, L. A. Santalò, and others. Generally speaking, its principal aim is to study the relations between the measures which can be attached to a given variety. It is my purpose in the present paper to discuss the services it can render to some problems in differential geometry.

1. Measure of the Spherical Image of a Closed Submanifold in Euclidean Space

A submanifold of dimension n in an Euclidean space E^{n+N} of dimension $n + N$ is given by an abstract differentiable manifold M^n of dimension n and a differentiable map x: $M^n \to E^{n+N}$, whose Jacobian matrix has everywhere the rank n. We say that M^n is imbedded, if the map x is one-one, i.e. if $x(M^n)$ does not intersect itself. All the unit normal vectors of $x(M^n)$ form a bundle of spheres of dimension $N - 1$ over M^n and constitute a manifold B_v of dimension $n + N - 1$. If O is a fixed point of E^{n+N} and S_0 the unit hypersphere with origin O, there is a mapping T: $B_v \to S_0$ which maps a unit normal vector of M^n into the end-point of the unit vector through O and parallel to it. T is a generalization of the normal mapping of Gauss in the theory of surfaces.

Suppose from now on that M^n is compact. Then B_v is also compact $T(B_v)$ divided by the volume of S_0 itself, each point of $T(B_v)$ counted a and we define as the *total curvature* of $x(M^n)$ the volume of the image number of times equal to the number of points mapped into it. This total curvature we will denote by $T_x(M^n)$. It is in a sense a measure of the curvedness of the submanifold. For a closed space curve, for instance, its total curvature is, up to a constant factor, the integral of the absolute value of the curvature.

Concerning the total curvature Lashof and I[3,4] proved the following theorems:

(1) The total curvature $T_x(M^n)$ is greater than or equal to the sum of Betti numbers of M^n relative to an arbitrary coefficient field. As a corollary it follows that $T_x(M^n) \geq 2$, a result which can be derived directly by an elementary argument on the maxima and minima of the co-ordinate functions on M^n.

(2) If $T_x(M^n) < 3$, M^n is homeomorphic to a sphere. (This result was proved also by Milnor[8].)

(3) $x(M^n)$ is a convex hypersurface imbedded in a subspace of dimension $n + 1$, if and only if $T_x(M^n) = 2$.

A basic reason for these theorems is the existence of the large number of co-ordinate functions on M^n. Morse's critical point theory then furnishes one of the essential tools in the proofs.

For a differentiable manifold abstractly given, one is led to study the immersions for which $T_x(M^n)$ is as small as possible. Two questions naturally arise: (a) What is the minimum value of $T_x(M^n)$, expressed in terms of M^n itself only, for all possible immersions x? (*b*) To characterize the immersions x for which the total curvature attains this minimum value. Theorem 3 answers these questions for the case when M^n is homeomorphic to a sphere.

For general compact manifolds very little is known about the two questions. Theorem 1 implies that $T(M^n) = \min T_x(M^n)$ is greater than or equal to the sum of Betti numbers mod 2 of M^n. There are sufficient indications to support the conjecture that $T(M^n)$ is equal to the minimum number of cells by which M^n can be subdivided into a cell complex, but the truth of this remains undecided.

As for Question (*b*), a conjecture of N. H. Kuiper says that if M^n is immersed in E^{n+N} with the minimum total curvature $T(M^n)$, then $N \leq \frac{1}{2}n(n+1)$. Some necessary conditions are known when M^n is a hypersurface ($N = 1$), and has minimum total curvature.

A simple case is when M^n is a closed space curve ($n = 1, N = 2$). Then Theorem 2 can be sharpened to the following form (Fary[5] and Milnor[8]): A closed space curve with total curvature < 4 is unknotted. The theorem thus gives a simple necessary condition for a knot in space.

Another application is the following consequence of the above Theorem 3: a closed surface immersed in ordinary Euclidean space with Gaussian curvature $K \geq 0$ is an imbedded convex surface. Under the stronger assumption that $K > 0$ the conclusion follows from a well-know argument of Hadamard. It may be of interest to remark that a similar statement is not valid in higher dimensions; there are examples of non-convex closed hypersurfaces in an Euclidean space of four or higher dimensions whose Gauss–Kronecker curvature is everywhere non-negative.

2. Measure of the Image of a Complex Analytic Mapping

Entirely analogous to the theory of submanifolds in Euclidean space is that of complex analytic submanifolds in a complex projective space. Let M_n be a complex manifold of (complex) dimension n and $Z: M_n \to P_{n+N}$ be a complex analytic mapping of M_n into the complex projective space P_{n+N} of dimension $n + N$. The study of such mappings includes as particular cases various classical theories. In fact, if M_n is compact, $Z(M_n)$ is an algebraic variety. If M_n is the complex Euclidean line E_1 (or the Gaussian plane, as is commonly called), the complex analytic mapping $Z: E_1 \to P_{1+N}$ defines a meromorphic curve in the sense of H. Weyl, J. Weyl and Ahlfors. In particular, the notion of the complex analytic mapping $Z: E_1 \to P_1$ is identical with that of a meromorphic function defined in the Gaussian plane.

Starting with the classical theorem of Picard, a main problem in such investigations is the determination of the maximum size of the set of linear spaces of dimension N, which will be disjoint with the image $Z(M_n)$. For meromorphic curves a satisfactory solution is provided by the following theorem of E. Borel: Suppose that the meromorphic curve is non-degenerate (i.e. that it does not lie in a hyperplane of P_{1+N}). Given $N + 3$ hyperplanes in general position, the image $Z(E_1)$ meets one of them. Obviously this theorem contains as a particular case the theorem of Picard than an entire function in the Gaussian plane omits at most one value.

That the theory is mainly geometrical can be justified by the following generalization of Borel's theorem, which follows easily from results of Ahlfors: Let $Z: E_1 \to P_{1+N}$ be a non-degenerate meromorphic curve. Given $\binom{N+2}{k+1} + 1$ linear spaces of dimension $N - k$ in general position, $0 \le k \le N$, one of them must meet an osculating linear space of dimension k of the curve.

In the establishment of these and related results, integral geometry plays a role on at least two occasions. Although the theorems relate only to the incidence of the curve with the linear subspaces, it is necessary to use the elliptic Hermitian metric in P_{n+N}. Then, for compact M_n, $Z(M_n)$ has a finite volume and this volume is, up to a numerical factor, equal to the order of the algebraic variety. This identification of volume and order, of sufficient interest in the compact case, will be of paramount importance in the case when M_n is non-compact. For then the notion of order does not exist, while the volume does. As it turns out, the volume does fulfil many of the functions of the order.

Since a non-compact manifold will be exhausted by a sequence of expanding polyhedra with boundaries, we are led to the study of a complex analytic mapping $Z: M_n \to P_{n+N}$, where M_n is compact and is with or without boundary. The first problem is the following: Given a generic linear space L of complementary dimension N, to determine the difference between the number of points of intersection of L and $Z(M_n)$, each counted with its proper multiplicity, and the volume of $Z(M_n)$. This problem was solved by Levine[7], who expressed the difference as an integral over the boundary ∂M_n of M_n. His result can be stated as follows:

Let $Z = (z_0, z_1, \ldots, z_{n+N}) \neq 0$ be a homogeneous co-ordinate vector of P_{n+N}, so that Z and $\lambda Z = (\lambda z_0, \lambda z_1, \ldots, \lambda z_{n+N})$, where λ is a non-zero complex number, define the same point. For Z and $W = (w_0, w_1, \ldots, w_{n+N})$ we introduce the Hermitian scalar product

$$(Z, W) = \overline{(W, Z)} = \sum_{k=0}^{n+N} z_k \overline{w}_k. \tag{1}$$

The linear space L of dimension N can be defined by the equations

$$l_i \equiv (Z, A_i) = 0 \qquad (1 \le i \le n), \tag{2}$$

where we suppose $(A_i, A_j) = \delta_{ij}$, $1 \le i, j \le n$. Then, for $\xi \in M_n$, the function

$$u(\xi, L) = \frac{\|L\|}{|z|} \le 1 \qquad (|z| = +(Z, Z)^{1/2}, \|L\| = +(l_1 \bar{l}_1 + \cdots + l_n \bar{l}_n)^{1/2}), \tag{3}$$

where $Z = Z(\zeta)$ is a homogeneous co-ordinate vector of the image point of ζ, is a well-defined real-valued function in M_n, and vanishes, if and only if $Z(\zeta) \in L$. Similarly, we define the exterior differential forms

$$\left.\begin{aligned}\Phi &= \frac{i}{\pi} d'd'' \log \|L\|,\\ \Psi &= \frac{i}{\pi} d'd'' \log |Z|,\end{aligned}\right\} \tag{4}$$

and

$$\Lambda = \frac{1}{2\pi i}(d' - d'')\log u \wedge \sum_{0 \leq k \leq n-1} \Phi^k \wedge \Psi^{n-1-k}. \tag{5}$$

Then we have the formula

$$v(M_n) - n(M_n, L) = -\int_{\partial M_n} \Lambda, \tag{6}$$

where $n(M_n, L)$ is the number of points common to L and $Z(M_n)$, counted with their multiplicities, and $v(M_n)$ is the volume of $Z(M_n)$, suitably normalized. It follows in particular that, if M_n is without boundary and if $Z(M_n)$ is non-degenerate, so that $v(M_n) \neq 0$, then $Z(M_n)$ meets every linear space of dimension N in P_{n+N}.

Perhaps the first example of a non-compact complex manifold is the complex Euclidean space E_n of dimension n. Let $\zeta_1, \ldots, \zeta_n$ be the co-ordinates in E_n. We will exhaust E_n by the domains $M(r)$:

$$\zeta_1\bar{\zeta}_1 + \cdots + \zeta_n\bar{\zeta}_n \leq r^2, \tag{7}$$

as $r \to \infty$. This seems to be the most natural exhaustion, because if we compatify E_n by adding a hyperplane π at infinity, the complement of $M(r)$ in E_n will form a tubular neighborhood about π.

Consider first the classical case of a meromorphic function $Z: E_1 \to P_1$. Let $v(r)$ be the volume of the image of $M(r)$. For a generic point $L \in P_1$ let $n(r, L)$ be the number of times L is covered by $Z(M(r))$. Then (6) can be written

$$v(r) - n(r, L) = -\frac{1}{2\pi}\int_0^{2\pi} r\frac{\partial \log u}{\partial r}\, d\theta \qquad (\xi_1 = re^{i\theta}). \tag{8}$$

This induces us to put

$$T(r) = \int_{r_0}^{r} \frac{v(t)\, dt}{t}, \qquad N(r, L) = \int_{r_0}^{r} \frac{n(t, L)\, dt}{t} (r_0 > 0). \tag{9}$$

By integrating (8) with respect to r, we get

$$T(r) - N(r, L) = -\frac{1}{2\pi}\int_0^{2\pi} \log u\, d\theta \Big|_{r_0}^{r}. \tag{10}$$

This is the so-called first main theorem in the theory of meromorphic functions. Our introduction of the order function $T(r)$ is exactly the way it was introduced by Shimizu-Ahlfors.

Since the first main theorem involves a generic point L of P_1, it is natural to integrate it over P_1. If we perform the integration with the invariant density dL, we shall get a formula of the Crofton type

$$T(r) = \int_{L \in P_1} N(r, L)\, dL, \tag{11}$$

which implies that the average of the right-hand side of (10) is zero. On the other hand, we derive from the first main theorem the fundamental inequality

$$T(r) - N(r, L) > \text{const.} \tag{12}$$

If we integrate this inequality over a non-invariant density, it is easy to get the theorem that the complement of the image set $Z(E_1)$ in P_1 has measure zero. An idea initiated by F. Nevanlinna and simplified by Ahlfors[1] consists in the use of a density with singularities. It is the integration of (12) relative to such a density that leads to a proof of the Picard–Borel theorem.

In the case of complex analytic mappings $Z: E_2 \to P_2$ there are known examples which show that the complement of the image $Z(E_2)$ may contain open subsets of P_2. We shall give a brief discussion of the proper restrictions on the mapping Z in order that general statements can be made. In fact, the first main theorem on complex analytic functions has the following generalization:

Let $v(r)$ be the volume of the image of $M(r)$ and, for a generic point $L \in P_2$ let $n(r, L)$ be the number of times L is covered by $Z(M(r))$. Let

$$T(r) = \int_{r_0}^{r} \frac{v(t)\, dt}{t^3}, \qquad N(r, L) = \int_{r_0}^{r} \frac{n(t, L)\, dt}{t^3} \qquad (r_0 > 0). \tag{13}$$

Then we have the inequality

$$T(r) - N(r, L) > \text{const} - S(r, L), \tag{14}$$

where

$$\left.\begin{aligned} S(r, L) &= -\frac{2}{\pi^2}\int_{r_0}^{r} \frac{dt}{t} \int (v_{11} + v_{22}) \log u\, dV \geq 0, \\ v_{kk} &= \frac{\partial^2}{\partial \zeta_k \partial \xi} \log (|Z| \cdot \|L\|) \qquad (k = 1, 2). \end{aligned}\right\} \tag{15}$$

the integration being over the unit hypersphere in E_n. It is clear from this inequality that in order to have a statement on the image set $Z(E_n)$ we must have $T(r) \to \infty$ as $r \to \infty$. The latter is automatic in the 1-dimensional case, but is an additional assumption in the 2-dimensional case. In fact, the well-known examples of Fatou-Bieberbach do not have this property. If, moreover,

$$S(r, L) = o(T(r)), \tag{16}$$

then we have the theorem that $Z(E_2)$ omits at most a set of measure zero.

The assumption (16) is unsatisfactory in the sense that it involves the generic point $L \in P_2$. The expression for $S(r, L)$ suggests that a 'mixed order function' should be introduced. In fact, let Ω and Ω_0 be the associated two-forms of P_2 and E_2 respectively. Then

$$\int_{M(r)} Z^*(\Omega) \wedge \Omega_0 = v_1(r), \tag{17}$$

where $Z^*(\Omega)$ is the inverse image of Ω under the mapping Z, is a mixed volume of the domain $M(r)$. Put

$$S(r) = \int_{r_0}^{r} \frac{v_1(t)\,dt}{t^3}. \tag{18}$$

It is conceivable that condition (16) can be replaced by a condition on the relative growth of $T(r)$ and $S(r)$.

It seems to me that these problems on complex analytic mappings deserve much further study.

3. Integral Formulae and Rigidity Theorems

I believe my discussion of relations between differential and integral geometry will leave a big gap, if I do not touch on the role that integral formulae play in the proofs of rigidity or uniqueness theorems. Perhaps the most well-known example of such considerations is Herglotz's proof of the uniqueness of Weyl's problem. In spite of these important applications it should be of independent interest to derive integral formulae for compact immersed submanifolds for their owns sake. A little analytic manipulation shows that there are few such formulae, unless the latter are allowed to involve other geometrical elements in the space, such as fixed points, fixed linear subspaces, fixed directions, etc. The reason is simple: For an immersed submanifold $x\colon M^n \to E^{n+N}$, the co-ordinate vector $x(p)$, $p \in M^n$, depends on the choice of the origin.

The simplest case is that of a strictly convex hypersurface $x\colon M^n \to E^{n+1}$. Naturally we orient it so that the Gauss-Kronecker curvature is everywhere > 0. Since the normal mapping of the hypersurface $\Sigma = x(M^n)$ into the unit hypersphere S_0 about the origin is one–one and has a non-zero Jacobian everywhere, the hypersurface can be defined by $x\colon S_0 \to \Sigma \subset E^{n+1}$, where x maps a point ξ of S_0 into the point of Σ having ξ as the unit normal vector.

To get rigidity theorems suppose $x'\colon S_0 \to \Sigma'$ is a second strictly convex hypersurface. It is then possible to write down a number of globally defined exterior differential forms on S_0. For our purpose we shall restrict ourselves to the following:

$$\left.\begin{aligned} A_{rs} &= (x, \xi, d\xi, \ldots, d\xi, dx, \ldots, dx, dx', \ldots, dx'),\\ A'_{rs} &= (x', \xi, d\xi, \ldots, d\xi, dx, \ldots, dx, dx', \ldots, dx'). \end{aligned}\right\} \tag{19}$$

Each of these expressions is a determinant of order $n + 1$, whose rows are the components of the respective vectors or vector-valued differential forms, with the convention that in the expansion of the determinant the multiplication of differential forms is in the sense of exterior multiplication. The subscript r refers to the number of entries dx and the subscript s that of the entries dx'. Since A_{rs} and A'_{rs} are globally defined on S_0, their integrals over S_0 are zero.

The integral formulae so obtained can be expressed in a more geometrical form as follows: Let $\mathrm{III} = d\xi^2$ be the fundamental form of S_0, and let

$$\mathrm{II} = -dx\,d\xi, \qquad \mathrm{II}' = -dx'\,d\xi \tag{20}$$

be the second fundamental forms of Σ, Σ' respectively. Let $\Delta(y, y')$ be the determinant the ordinary quadratic differential form $y\text{II} + y'\text{II}' + \text{III}$ relative to a local co-ordinate system, so that $\Delta(y, y')/\Delta(0, 0)$ is independent of the choice of the local co-ordinate system. Let

$$\frac{\Delta(y, y')}{\Delta(0, 0)} = \sum_{0 \leq r+s \leq n} \frac{n!}{r!s!(n-r-s)!} y^r y'^s P_{rs}, \tag{21}$$

where P_{rs}, P'_{rs} are mixed invariants of Σ, Σ'. In particular, P_{l0}, P_{0l} are, up to numerical factors, the lth elementary symmetric functions of the principal radii of curvature of Σ, Σ' respectively. Then our integral formulae can be written

$$\int_{S_0} (pP_{rs} - P_{r+1,s})dV = 0, \qquad \int_{S_0} (p'P_{rs} - P_{r,s+1})\, dV = 0, \tag{22}$$

where dV is the volume elements of S_0 and p, p' are the support functions of Σ, Σ' respectively. An important consequence of (22) consists of the formulae

$$\int pP_{0l}\, dV = \int p'P_{1,l-1}\, dV, \qquad \int pP_{l-1,1}\, dV = \int p'P_{l0}\, dV \qquad (l \geq 1), \tag{23}$$

which give

$$2\int p(P_{0l} - P_{l-1,1})\, dV = \int \{p'(P_{1,l-1} - P_{l0}) - p(P_{l-1,1} - P_{0l})\}\, dV. \tag{24}$$

It is important to observe that the right-hand side of (24) is antisymmetric in the hypersurfaces Σ, Σ'.

Formula (24) reduces to a purely algebraic problem the proof of the following uniqueness theorem of Minkowski, A. D. Alexandroff[2], and Fenchel and Jessen[6]: If two closed strictly convex hypersurfaces are such that at points with parallel normals, the lth (for fixed $l \geq 2$) elementary symmetric functions of the principal radii of curvature have the same value, then they differ from each other by a translation.

The theorem is also true for $l = 1$, but it will have a different (and simpler) proof.

The algebraic lemma needed has been communicated to me by L. Gårding, as a consequence of his work on hyperbolic polynomials. It can be stated as follows:

Let (λ_{ik}) be an $n \times n$ asymmetric matrix, and let

$$\det(\delta_{ik} + y\lambda_{ik}) = \sum_{0 \leq r \leq n} P_r(\lambda)y^r. \tag{25}$$

Let $P_r(\lambda^{(1)}, \ldots, \lambda^{(r)})$ be the completely polarized form of $P_r(\lambda)$, so that $P_r(\underbrace{\lambda, \ldots, \lambda}_{r}) = P_r(\lambda)$. Then, for $r \geq 2$ and for positive definite matrices $(\lambda_{ik}^{(1)}), \ldots, (\lambda_{ik}^{(r)})$, the following inequality is valid:

$$P_r(\lambda^{(1)}, \ldots, \lambda^{(r)}) \geq P_r(\lambda^{(1)})^{1/r} \ldots P_r(\lambda^{(r)})^{1/r}. \tag{26}$$

Equality sign holds, if and only if the r matrices are pairwise proportional.

The uniqueness theorem then follows immediately from the lemma and the integral formula (24). For the hypothesis says that $P_{0l} = P_{l0}$. From (26) it follows that $P_{l-1,1} - P_{0l} \geq 0$. By (24) this is possible only when $P_{l-1,1} - P_{0l} = 0$. Again by the lemma it follows that the second fundamental forms of the hypersurfaces are equal.

So far as I am aware, it is not known whether a similar uniqueness theorem is valid, if the lth elementary symmetric function of the principal curvatures is prescribed as a function of the normal vector. Alexandroff proved that a closed convex surface in ordinary Euclidian space is defined up to a translation, if its mean curvature is a given function of the normal. His proof made use of a maximum principle. It would be interesting if this theorem can be proved by using integral formulas.

RFERENCES

[1] Ahlfors, L. The theory of meromorphic vurves. *Acta Soc. Sci. fenn.*, A, 3, 1–31 (1941).

[2] Alexandroff, A. D. Zur Theorie der gemischten Volumina von konvexen Körpern. (Russian). *Rec. Math.* (Série nouvelle), 2, 947–972, 1205–1238; 3, 27–46, 227–251 (1937–38).

[3] Cherns, S. and Lashof, R. K. On the total curvature of immersed manifolds *Amer. J. Math.* 79, 302–318 (1957).

[4] Chern, S. and Lashof, R. K. On the total curvature of immersed manifolds. II. To appear in *Michigan Math. J.*

[5] Fary, I. Sur la courbure totale d'une courbe gauche faisant un noeud. *Bull. Soc. Math. Fr.* 77, 128–138 (1949).

[6] Fenchel, W. and Jessen, B. Mengenfunktionen und konvexe Körper. *K. danske vidensk. Selsk.* (*Math.-fysiske Medd.*). 16, 1–31 (1938).

[7] Levine, Harold I. Contributions to the theory of analytic mappings of complex manifolds into projective space. University of Chicago thesis, 1958.

[8] Milnor, J. W. On the total curvature of knots. *Ann. Math.* 52. 248–257 (1950).

Reset from
Proc. Int. Congr. Math. Edinburgh (1958) 441–449

CHERN, SHIING-SHEN, and CHUAN-CHIH HSIUNG
Math. Annalen 149, 278—285 (1963)

On the Isometry of Compact Submanifolds in Euclidean Space*

By

SHIING-SHEN CHERN and CHUAN-CHIH HSIUNG in Berkeley, California

Introduction

The question as to the conditions on an isometry between compact hypersurfaces in euclidean space in order that it be a congruence has been the subject of extensive researches in differential geometry. In this paper we will prove a theorem which states that a volume-preserving diffeomorphism between two compact submanifolds in euclidean space satisfying certain conditions (see § 2) is an isometry. Our main tool consists of some integral formulas. The latter should be of independent interest.

§ 1. Preliminaries on Algebra

Let V be a real vector space of dimension n, and let G be a bilinear real-valued function over $V \times V$. G is completely determined by the values $g_{ik} = G(e_i, e_k)$, $1 \leq i, k \leq n$, where e_i form a basis of V. Under a change of basis $e_i \to e_i^* = \sum_k t_i^k e_k$, the matrix $\|g_{ik}\|$ is changed to $T\|g_{ik}\|^t T$, where $T = \|t_i^k\|$ and ${}^tT$ is the transpose of T.

Let H be another bilinear real-valued function over $V \times V$, and let $h_{ik} = H(e_i, e_k)$. Consider the determinant

$$\det(g_{ik} + h_{ik}\lambda) = \det(g_{ik}) + n\lambda P(g_{ik}, h_{ik}) + \cdots + \lambda^n \det(h_{ik}) \,. \tag{1}$$

Since it will be multiplied by $(\det T)^2$ under a change of basis, the ratio of any two coefficients in the polynomial (1) in λ is independent of the choice of basis. In particular, if G is non-singular, i.e., if $\det(g_{ik}) \neq 0$, the quotient

$$H_G = P(g_{ik}, h_{ik})/\det(g_{ik}) \tag{2}$$

depends only on G and H. For example, if $g_{ik} = \delta_{ik}$ $(=1,\ i = k;\ = 0,\ i \neq k)$, then $H_G = \sum_i h_{ii}/n$.

This construction of H_G is linear in H. Hence, if H is a bilinear function over $V \times V$ with values in a vector space W, the above construction can be generalized and H_G is an element in W. H_G can be called the contraction of H relative to G. In the language of tensors we can express our construction by saying that H_G is a vector in W constructed from a covariant tensor H of

*) Work done under National Science Foundation Research Grant NSF G-19137.

order two with values in W relative to a non-singular covariant tensor G of order two.

In what follows we will need an inequality, due to LARS GÅRDING [2], on H_G. It can be stated as follows:

Let G and H be symmetric positive definite bilinear real-valued functions over $V \times V$. Let

$$g = \det(g_{ik}), \quad h = \det(h_{ik}). \tag{3}$$

Then

$$H_G \geqq (h/g)^{1/n}, \tag{4}$$

and the equality sign holds if and only if $h_{ik} = \varrho g_{ik}$, for a certain ϱ.

§ 2. Preliminaries on Differential Geometry and Statement of Theorem

Let M and M^* be two C^2-riemannian manifolds of the same dimension n, and let $f: M \to M^*$ be a C^2-mapping. In M there are then two connections: the Levi-Civita connection defined by its riemannian metric and the connection induced by the mapping f from the Levi-Civita connection of M^*. Their difference is a tensor field Δ of contravariant order 1 and covariant order 2. If G denotes the fundamental tensor field of M, the construction in § 1 gives a vector field Δ_G. We will call f an *almost isometry*, if $\Delta_G = 0$.

Next consider an immersion $x: M \to E$, i.e., a C^2-mapping x of M into an euclidean space E of dimension $n + m$, such that the induced linear mapping on the tangent spaces is univalent everywhere. Then $x(p)$, $p \in M$, is a vector in E and will be called the position vector of the submanifold. We will denote by a parenthesis the scalar product of two vectors in E. If ν is a normal vector at $x(p)$, then $(\nu, d^2x(p))$, where d^2x is the second differential in the ordinary sense, is the second fundamental form relative to ν. Since the latter is a symmetric covariant tensor of order two, we can form its contraction $(\nu, d^2x(p))_G$ relative to the fundamental tensor G of the induced riemannian metric on M. There is therefore exactly one normal vector field H over $x(M)$ defined by

$$(\nu, d^2x(p))_G = (\nu, H). \tag{5}$$

This normal vector field H is called the *mean curvature vector*. It generalizes the notion of mean curvature in ordinary surface theory. A submanifold is called *minimal*, if the mean curvature vector is identically zero.

Consider now two immersions x, x^* of M into E, and a diffeomorphism f as given by the commutative diagram

$$\begin{array}{ccc} M & \xrightarrow{\;x\;} & x(M) \subset E \\ & \searrow^{x^*} & \downarrow f \\ & & x^*(M) \subset E. \end{array}$$

The mapping f is called an isometry, if $(dx^*(p), dx^*(p)) = (dx(p), dx(p))$, i.e., if it maps the induced riemannian metric of the one into that of the other.

It is called volume-preserving, if it maps the volume element of one into that of the other. As a consequence of this definition a volume-preserving diffeomorphism exists only if M is oriented and the diffeomorphism is then orientation-preserving. We will denote by G, G^* respectively the fundamental tensors of the two induced metrics.

Our main theorem is the following:

Let $x, x^* : M \to E$ be two immersed compact submanifolds and let $f : x(M) \to \to x^*(M)$ be a volume-preserving diffeomorphism. Suppose that f has the properties:

1) *It is an almost isometry relative to G^*, i.e. $\Delta_{G^*} = 0$;*

2) $(x^\perp(p),\ d^2x(p))_G \leqq (x^\perp(p), d^2x(p))_{G^*}$, *where $x^\perp(p)$ is the orthogonal projection of the position vector $x(p)$ in the normal space to $x(M)$ at $x(p)$.*

Then f is an isometry.

§ 3. Integral Formulas

Over the abstract manifold M there are now two induced riemannian metrics with the same volume element, namely $(dx(p), dx(p))$ and

$$(dx^*(p), dx^*(p)) = (d(f\circ x)(p), d(f\circ x)(p)) .$$

The notion of frames $e_1, \ldots, e_n$ having measure 1 and coherently oriented with that of M has thus a sense in both metrics. At a point $p \in M$ any such frame can be obtained from a fixed one by a linear transformation of determinant 1. Since the induced linear map x_* on tangent spaces is univalent, we identify e_i with $x_*(e_i)$. Let $\omega^1, \ldots, \omega^n$ be the coframe dual to $e_1, \ldots, e_n$, so that the volume element of M is

$$dV = \omega^1 \wedge \cdots \wedge \omega^n . \tag{6}$$

Let $e_{n+1}, \ldots, e_{n+m}$ be a frame in the normal bundle at $x(p)$. We introduce the matrices

$$\omega = \|\omega^1, \ldots, \omega^n\|,$$
$$e = \begin{Vmatrix} e_1 \\ \vdots \\ e_n \end{Vmatrix}, \quad a = \begin{Vmatrix} e_{n+1} \\ \vdots \\ e_{n+m} \end{Vmatrix}. \tag{7}$$

Then we have the matrix equations

$$dx = \omega e, \quad de = \Omega e + \theta a , \tag{8}$$

where

$$\Omega = \|\omega_i^k\| ,$$
$$\theta = \|\omega_i^{n+s}\| , \qquad 1 \leqq i, k \leqq n, 1 \leqq s \leqq m , \tag{9}$$

ω_i^k, ω_i^{n+s} being linear differential forms in the bundle induced over M from the principal bundle of E. Exterior differentiation of the first equation of (8) gives

$$d\omega = \omega \wedge \Omega, \quad \omega \wedge \theta = 0 . \tag{10}$$

Before proceeding, we shall give the geometrical meaning of these equations. It is well-known that the matrix Ω gives the connection form of the induced riemannian metric by x. Written explicitly, the second equation of (10) gives

$$\sum_i \omega^i \wedge \omega_i^{n+s} = 0 . \tag{10a}$$

This allows us to put

(11) $$\omega_i^{n+s} = \sum_j A_{ij}^{n+s} \omega^j ,$$

where A_{ij}^{n+s} is symmetric in the lower indices:

(12) $$A_{ij}^{n+s} = A_{ji}^{n+s} .$$

These quantities are the ones which enter into the second fundamental form. In fact, ν being a normal vector at $x(p)$, we find

(13) $$(\nu, d^2 x) = \sum_{s,i,j} (\nu, e_{n+s}) A_{ij}^{n+s} \omega^i \omega^j ,$$

where the quadratic differential form is in the sense of the symmetric algebra, multiplication being commutative. We put

(14) $$(x, e_{n+s}) = y_{n+s} , \qquad 1 \leq s \leq m ,$$

and introduce the matrix

(15) $$h = \|y_{n+s}\| = (x, {}^t a)^{1)} .$$

Then we have the one-rowed matrix

(16) $$h\, {}^t\theta = \left\| \sum_{s,j} y_{n+s} A_{ij}^{n+s} \omega^j \right\| .$$

On the other hand, we have

(17) $$x^\perp = \sum_s y_{n+s} e_{n+s} .$$

Thus we find the following explicit expression for the quadratic differential form in the second condition of our theorem:

(18) $$Q = (x^\perp, d^2 x) = \sum_{s,i,j} y_{n+s} A_{ij}^{n+s} \omega^i \omega^j .$$

The notation Q will hereafter be used as an abbreviation of this expression.

Furthermore the condition that the frames are of measure 1 is

$$(e_1 \wedge \cdots \wedge e_n, e_1 \wedge \cdots \wedge e_n) = 1 ,$$

where the left-hand side is the scalar product of multivectors defined by the scalar product in E. Differentiating this equation and using the second equation of (8), we get

(19) $$\operatorname{Tr} \Omega = \sum_i \omega_i^i = 0 .$$

To derive our integral formula we have to introduce exterior differential forms globally defined over M. Their study will be facilitated by the adoption of notations developed by H. FLANDERS [1]. This is to consider tensor products of multivectors and exterior differential forms. Differentiation of multivectors will be taken in the sense of (8), while differentiation of exterior differential forms will be exterior differentiation. Multiplication of matrices will be by

[1]) For an $(m \times n)$-matrix $a = \|a_{\lambda j}\|$ and an $(n \times p)$-matrix $b = \|b_{jk}\|$, whose elements are vectors in E, we will use the notation (a, b) to denote the matrix of real numbers given by

$$(a, b) = \left\| \sum_j (a_{ij}, b_{jk}) \right\| .$$

the usual row-by-column law. With this understanding we introduce the following matrices:

(20) $$G = (e, {}^t e) = {}^t G = \|g_{ij}\| ,$$

(21) $$G^* = (f_*(e), f_*({}^t e)) = {}^t G^* = \|g^*_{ij}\| ,$$

(22) $$\Lambda = \|\omega_1, \ldots, \omega_n\| = \omega G, \quad \Lambda^* = \|\omega^*_1, \ldots, \omega^*_n\| = \omega G^* ,$$

(23) $$u = \Lambda e, \quad u^* = \Lambda^* e ,$$

(24) $$Y = (x, {}^t e) = \|y_1, \ldots, y_n\| ,$$

(25) $$r = Y e ,$$

(26) $$z = r u^{*n-1} = \varphi e_1 \wedge \cdots \wedge e_n .$$

The diffeomorphism f is an isometry, if and only if $G = G^*$.

Clearly the form φ is an exterior differential form of degree $n-1$ globally defined over M. Our integral formula in question will be

(27) $$\int_M d\varphi = 0 ,$$

M being supposed to be compact. It is therefore our problem to calculate $d\varphi$ or dz.

We have

(28) $$dY = (dx, {}^t e) + (x, d\,{}^t e) = \Lambda + Y\,{}^t\Omega + h\,{}^t\theta ,$$

where h is defined in (15). It follows that

(29) $$dr = \{\Lambda + Y(\Omega + {}^t\Omega) + h\,{}^t\theta\} e + Y\theta a .$$

Now we have

(30) $$dG = (de, {}^t e) + (e, d\,{}^t e) = \Omega G + G\,{}^t\Omega ,$$

and

(31) $$d\Lambda = (d\omega) G - \omega\, dG = -\Lambda\,{}^t\Omega .$$

For the riemannian metric induced by x^*, we have

(32) $$d\Lambda^* = -\Lambda^*\,{}^t\Omega^* ,$$

whence

(33) $$du^* = -\Lambda^*({}^t\Omega^* + \Omega) e - \Lambda^*\theta a .$$

By noting that

(34) $$d(u^{*n-1}) = (n-1)(du^*) u^{*n-2} ,$$

we get

(35) $$dz = \{\Lambda + Y({}^t\Omega + \Omega) + h\,{}^t\theta\} e u^{*n-1} - (n-1) r \Lambda^*({}^t\Omega^* + \Omega) e u^{*n-2} + \\ + Y\theta a u^{*n-1} - (n-1) r \Lambda^* \theta a u^{*n-2} .$$

Since the frames are of measure 1, $d(e_1 \wedge \cdots \wedge e_n)$ contains no term in $e_1 \wedge \cdots \wedge e_n$. Equating the terms in $e_1 \wedge \cdots \wedge e_n$ on both sides of (35), we get

(36) $$(d\varphi) e_1 \wedge \cdots \wedge e_n = \Lambda e u^{*n-1} + h\,{}^t\theta e u^{*n-1} + \\ + Y\{({}^t\Omega + \Omega) e u^* - (n-1) e \Lambda^*({}^t\Omega^* + \Omega) e\} u^{*n-2} .$$

Lemma. *The following identity is true*:

(37) $$\Omega e u^{n-1} - (n-1) e \Lambda \Omega e u^{n-2} = 0 .$$

To prove this we observe that the left-hand side is a one-columned matrix. There is no loss of generality in proving only that its bottom element is zero. Assuming the following ranges of indices:

$$1 \leqq i, j \leqq n, \quad 1 \leqq \alpha, \beta \leqq n-1 ,$$

we can write the bottom element in $\Omega e u^{n-1}$ as

$$\begin{aligned}\sum_i \omega_n^i e_i u^{n-1} &= \Big(\sum_\alpha \omega_n^\alpha e_\alpha + \omega_n^n e_n\Big)\Big(\sum_\beta \omega_\beta e_\beta + \omega_n e_n\Big)^{n-1} \\ &= \Big(\sum_\alpha \omega_n^\alpha e_\alpha + \omega_n^n e_n\Big)\Big\{\Big(\sum_\beta \omega_\beta e_\beta\Big)^{n-1} + (n-1)\Big(\sum_\beta \omega_\beta e_\beta\Big)^{n-2} \omega_n e_n\Big\} \\ &= e_n \Big\{\omega_n^n\Big(\sum_\beta \omega_\beta e_\beta\Big) + (n-1)\omega_n \sum_\alpha \omega_n^\alpha e_\alpha\Big\} u^{n-2} \\ &= e_n \Big\{-\sum_\alpha \omega_\alpha^\alpha\Big(\sum_\beta \omega_\beta e_\beta\Big) + (n-1)\omega_n \sum_\alpha \omega_n^\alpha e_\alpha\Big\} u^{n-2} .\end{aligned}$$

Now consider the expression

$$e_n \sum_{\alpha,\beta} \omega_\beta \omega_\beta^\alpha e_\alpha u^{n-2} .$$

Its non-zero summands are those for which $\beta = \alpha$. It is also clear that $e_n \omega_\alpha e_\alpha u^{n-2}$ is independent of α. Hence

$$e_n \sum_{\alpha,\beta} \omega_\beta \omega_\beta^\alpha e_\alpha u^{n-2} = e_n \sum_\alpha \omega_\alpha \omega_\alpha^\alpha e_\alpha u^{n-2} = -\frac{1}{n-1} e_n \sum_\alpha \omega_\alpha^\alpha u^{n-1} .$$

It follows that

$$\sum_i \omega_n^i e_i u^{n-1} = (n-1) e_n \sum_{i,\alpha} \omega_i \omega_i^\alpha e_\alpha u^{n-2} .$$

But this is the bottom element of the second term in (37). Thus the lemma is proved.

Since the identity in the lemma is a purely algebraic identity, it remains true when Ω, Λ, and u are replaced by ${}^t\Omega^*$, Λ^*, and u^* respectively.

Using (37), we can simplify (36) to the form

(38) $$(d\varphi) e_1 \wedge \cdots \wedge e_n = \Lambda e u^{*n-1} + h\,{}^t\theta e u^{*n-1} + Y({}^t\Omega - {}^t\Omega^*) e u^{*n-1} .$$

The integral formula (27) says that, for a pair of compact immersed submanifolds under a volume-preserving diffeomorphism, the integral over M of the coefficient of $e_1 \wedge \cdots \wedge e_n$ of the right-hand side of (38) is zero.

§ 4. Proof of the Theorem

To apply the integral formula of the last section, it remains to examine its integrand more closely. We observe that it involves terms of the form $\pi e u^{*n-1}$, where

(39) $$\pi = \|\pi_1, \ldots, \pi_n\|$$

is a one-rowed matrix of linear differential forms. By expanding, we get

$$\pi e u^{*n-1} = \Big(\sum_i \pi_i e_i\Big)\Big(\sum_j \omega_j^* e_j\Big)^{n-1} = (\Sigma\, \varepsilon_{i_1 \dots i_n} \pi_{i_1} \wedge \omega_{i_2}^* \wedge \cdots \wedge \omega_{i_n}^*)\,(e_1 \wedge \cdots \wedge e_n)\,,$$

where $\varepsilon_{i_1 \dots i_n}$ is equal to $+1$ or -1 according as $i_1, \dots, i_n$ constitute an even or odd permutation of $1, \dots, n$, and is otherwise equal to zero, and where the summation is extended over all $i_1, \dots, i_n$ from 1 to n. Putting

$$\pi_i = \sum_j h_{ij} \omega^j \tag{40}$$

and remembering that

$$\omega_i^* = \sum_j g_{ij} \omega^j\,,$$

we get

$$\pi e u^{*n-1} = P(g_{ij}^*, h_{ij})\,(e_1 \wedge \cdots \wedge e_n)\, dV\,, \tag{41}$$

where $P(g_{ij}^*, h_{ij})$ is the expression introduced in (1).

Of the terms in (38) we have therefore

$$\Lambda e u^{*n-1} = G_{G^*}(e_1 \wedge \cdots \wedge e_n)\, dV\,. \tag{42}$$

By (16) and (18) we have

$$h^t \theta e u^{*n-1} = Q_{G^*}(e_1 \wedge \cdots \wedge e_n)\, dV\,. \tag{43}$$

The third term at the right-hand side of (38) is zero, if the diffeomorphism f is an almost isometry relative to G^*, i.e., if condition 1) of the theorem is satisfied. Hence for a pair of compact immersed submanifolds under a volume-preserving almost isometry, we have the integral formula

$$\int_M (G_{G^*} + Q_{G^*})\, dV = 0\,. \tag{44}$$

This formula remains valid when the submanifolds are identical. In this case we have $G_G = 1$ by definition, so that we have

$$v(M) = -\int_M Q_G\, dV\,, \tag{45}$$

where $v(M)$ is the total volume of M. Formula (45) is a generalization of a classical formula of Minkowski on convex bodies.

Taking the difference of (44) and (45), we get

$$\int_M \{(G_{G^*} - 1) + (Q_{G^*} - Q_G)\}\, dV = 0\,.$$

By GARDING's inequality (4) we have

$$G_{G^*} - 1 \geqq 0\,,$$

while by condition 2) of our theorem, we have

$$Q_{G^*} \geqq Q_G\,.$$

It follows that $G_{G^*} - 1$ is identically zero. Again by GÅRDING's theorem this is possible only when $G = G^*$. Hence the theorem is proved.

Formula (45) gives the following corollary, which may be of interest to note:

Corollary. *A complete minimal submanifold is non-compact.*

Bibliography

[1] FLANDERS, H.: Development of an extended exterior differential calculus. Trans. Am. Math. Soc. **75**, 311—326 (1953).

[2] GÅRDING, L.: An inequality for hyperbolic polynomials. J. Math. Mech. **8**, 957—965 (1959).

(Received June 1, 1962)

HERMITIAN VECTOR BUNDLES AND THE EQUIDISTRIBUTION OF THE ZEROES OF THEIR HOLOMORPHIC SECTIONS

BY

RAOUL BOTT and S. S. CHERN

Harvard University, Cambridge, Mass., U.S.A., and University of California, Berkeley, Calif., U.S.A. (1)

1. Introduction

At present a great deal is known about the value distribution of systems of meromorphic functions on an open Riemann surface. One has the beautiful results of Picard, E. Borel, Nevanlinna, Ahlfors, H. and J. Weyl and many others to point to. (See [1], [2].) The aim of this paper is to make the initial step towards an n-dimensional analogue of this theory.

A natural general setting for the value distribution theory is the following one. We consider a complex n-manifold X and a holomorphic vector bundle E over X whose fiber dimension equals the dimension of X and wish to study the zero-sets of holomorphic sections of E.

When X is compact (and without boundary) then it is well-known that if the zeroes of any continuous section are counted properly then the algebraic sum of these zero-points is independent of the section and is given by the integral of the nth Chern (2) class of E over X: Thus we have

$$\text{Number of zeroes of } s = \int_X c_n(E), \tag{1.1}$$

and this formula is especially meaningful for a holomorphic section because the indexes of all the isolated zeroes of such a section are necessarily positive.

The central question of the value distribution theory is to describe the behavior of the zeroes of holomorphic sections when X is not compact. (For continuous sections there

---

(1) This work was partially supported by a grant from the National Science Foundation. The second author was a professor of the Miller Institute at the University of California (Berkeley) and received partial support from the Office of Naval Research.

(2) With misgivings on the part of the second author we have adopted a terminology now commonly used.

are no restrictions in that case, for instance there is always a section which does not vanish at all!)

The main results, all concerned with the case $\dim X = 1$, then take the following form. One considers a finite-dimensional "sufficiently ample" subspace V of the space of all holomorphic sections of E and shows that under suitable convexity conditions on E and X "most" of the sections in V vanish the "same number of times". Depending on how "most" and "same number of times" are defined one gets results of various degrees of delicacy and difficulty. For example, the classical Picard theorem asserts that when X is the Gauss-plane, so that E may be taken as the trivial line bundle $\mathbf{C}$, and $\dim V = 2$, then at most 2 sections of V in general position can fail to vanish on X. The Borel generalization of this theorem asserts that when $\dim V = n$, then at most n sections in V, in general position, can fail to vanish. Here, as throughout, the term general position is used in the following sense: A set of n elements $v_1, \ldots, v_n$ of a vector-space V is called *in general position*, if any subset of k elements span a k-dimensional subspace of V, for $k = 1, \ldots, \dim V$.

In the Nevanlinna theory one again deals with $X = \mathbf{C}$, $\dim V = 2$, but now a deficiency index $\delta(s)$ is defined for every $s \in V - 0$, which measures the extent to which s behaves unlike the generic section in V. In particular δ has the properties $\delta(\lambda s) = \delta(s)$, if $\lambda \in \mathbf{C} - 0$; $0 \leqslant \delta(s) \leqslant 1$; and finally: $\delta(s) = 1$ if s does not vanish on X. The "first main theorem" may then be interpreted as asserting that δ considered as a function on the projective space $P_1(V)$ of lines in V, is equal to 0 almost everywhere. Thus "most" sections in the measure sense behave the same way.

The second main theorem yields the much stronger inequality:

$$\sum_{i=1}^{m} \delta(s_i) \leqslant 2 \tag{1.2}$$

valid for any system of sections $s_i \in V$ in general position. The Ahlfors generalization deals with the case $\dim V = n$ and again proves among other things that $\delta(s) = 0$ nearly everywhere, and that now the inequality

$$\sum_{i=1}^{m} \delta(s_i) \leqslant n$$

is valid for any system of $s_i \in V$, which are in general position.

Usually these results are stated in terms of maps of X into the Riemann-sphere, (i.e., meromorphic functions) for the Picard and Nevanlinna theory, while the Borel and Ahlfors generalizations deal with maps of X into complex projective spaces of higher dimensions. The transition to our formulation is quite trivial. Indeed consider the evaluation map: $e_x : V \to E_x$ which attaches to each section in V, the value of s at x. By definition, a space of sections V will be "sufficiently ample" if and only if:

α) $e_x: V \to E_x$ is onto for each $x \in X$.

β) V contains a section which vanishes to the first order at some point of X.

Now let $k(x)$ be the kernel of e_x. This is then a subspace of a fixed dimension $m = \dim V - \dim E_x$, in V, so that the assignment $x \to k(x)$ defines a map $e_V: X \to P_m(V)$ of X into the Grassmannian of m-dimensional subspaces in V.

Now for each $s \in V$, let $z(s)$ be the subvariety of $P_m(V)$ consisting of those subspaces which contain s. Then, for $s \neq 0$, $z(s)$ has codimension n in $P_m(V)$, and it is clear that the zeroes of s on X correspond precisely to the intersections of $e_V(X)$ with $z(s)$ in $P_m(V)$.

In particular, when $\dim E_x$ is 1, $P_m(V)$ is just a projective space, and $z(s)$ is a hyperplane, so that we may reformulate our statements in the terms of the number of hyperplanes which the image of X avoids.

Conversely, starting with a map $e: X \to P_m(V)$, one may pull back the *quotient bundle* of $P_m(V)$ (see the end of Section 6) to obtain a bundle E over X, together with a finite dimensional subspace, V, of sections of E, for which $e_V = e$. Indeed, let $K \subset X \times V$, consist of the subset (x, v) for which $v \in e(x)$. Then K is a sub-bundle of the trivial bundle $X \times V$, and the corresponding quotient bundle, $X \times V/K$ is the desired bundle E. The constant sections of $X \times V$ over X, then go over into the desired subspace, V, of sections of E. Thus these two points of view are completely equivalent.

The aim of this paper is to discuss the n-dimensional case and we are able to push to an analogue of the first main theorem. Thus we obtain the weak equidistribution in the measure sense only. On the other hand this generalization is not quite immediate and in fact depends on a formula in the theory of characteristic classes, which seems to us of independent interest. To formulate this result we need to recall two facts: Namely 1) That the complex structure on X induces a natural "twisted boundary operator", d^c, on the real differential forms, $A(X)$, of X, and 2) That a given Hermitian structure on E determines definite representatives, $c_k(E) \in A(X)$, $k = 1, \ldots, n$, of the Chern classes of E. With this understood, we consider a given Hermitian, complex n-bundle E, over X and its Chern form $c_n(E) \in A(X)$. Also let $B^*(E) = \{e \in E \mid 0 < |e| < 1\}$ be the subset of vectors in e which are of length greater than 0 and length less than 1, and set $\pi: B^*(E) \to X$ equal to the natural projection. Then our first and principal result is expressed by the theorem:

THEOREM I. *There exists a real valued form ϱ on $B^*(E)$ which is of type $(n-1, n-1)$ and for which*

$$\pi^* c_n(E) = \frac{dd^c}{4\pi} \varrho. \tag{1.3}$$

Further if E is non-negative then ϱ may be chosen to be non-negative.

Remark that $B^*(E)$ has the homotopy type of the unit sphere bundle $S(E)$, of E, and it is of course well known that $c_n(E)$, when lifted to $S(E)$, becomes a boundary. Hence Theorem I refines this result for the complex analytic model $B^*(X)$ of $S(E)$.

The method which leads to Theorem I also yields the following auxillary result.

PROPOSITION 1.4. *Let E be a complex analytic bundle and let $c(E)$ and $c'(E)$ be the Chern forms of E relative to two different Hermitian structures. Then $c(E)-c'(E)=dd^c\cdot\lambda$ for some λ.*

In other words, if we define $\hat{H}^k(X)$ by:

$$\hat{H}^k(X)=A^{k,k}(X)\cap \operatorname{Ker}(d)/dd^cA^{k-1,k-1}(X)$$

then the class in $\hat{H}^*(X)=\sum\hat{H}^k(X)$ of the Chern form $c(E)$, of E relative to some Hermitian structure on E, is independent of that Hermitian structure, so that we may define a "*refined* Chern class" $\hat{c}(E)\in\hat{H}^*(X)$. (Cf. Section 3 for definition of $A^{k,k}(X)$.)

In fact, Theorem I will follow directly from the following Whitney type duality theorem concerning these refined Chern classes:

PROPOSITION 1.5. *Let $0\to E'\to E\to E''\to 0$ be an exact sequence of holomorphic vector-bundles over X. Then their refined Chern classes satisfy the duality formula:*

$$\hat{c}(E')\cdot\hat{c}(E'')=\hat{c}(E).$$

The formula (1.2) is very pertinant for the whole Nevanlinna theory; for instance in the one-dimensional case, ϱ is just a real valued function on $B^*(E)$, and is seen to be minus the logarithmic "height" function:

$$\varrho(e)=-\log|(e)|^2,\quad e\in B^*(E).$$

Indeed one may roughly express the situation by saying that the first "main inequality" of the Nevanlinna theory is just a *twice integrated* version of (1.3).

The plan of the paper is as follows: In Section 2 we review the theory of characteristic classes as found in [3], [5]. We then go on to refine this theory for complex analytic Hermitian bundles in Sections 3 to 5. Section 6 is devoted to a proof of the generalized Gauss-Bonnet theorem which fits into the context of this paper. In Section 7 we define the order function, while in Section 8 we formulate and start to prove the equidistribution theorem. Sections 9 and 10 then complete this proof. Our final section brings a leisurely account of the classical Nevanlinna theorem. This Section 11 is included primarily to show how much more will have to be done before an n-dimensional analogue of this delicate theorem is established.

2. Curvature and characteristic classes

In this section E will denote a C^∞-bundle over the C^∞ manifold X. We write $T=T(X)$ for the cotangent bundle of X, and $A(X)=\sum A^i(X)$ for the graded ring of C^∞ complex valued differential forms on X. The differential operator on $A(X)$ is denoted by d. More generally we write $A(X; E)$ for the differential forms on X with values in E. Thus if $\Gamma(E)$ denotes the C^∞ sections of E, then $A(X; E)=A(X)\otimes_{A^0(X)}\Gamma(E)$.

The natural pairing from $\Gamma(E)\otimes_{A^0(X)}\Gamma(F)$ to $\Gamma(E\otimes F)$ (1) will often be written simply as multiplication.

Our aim here is to give an elementary and essentially selfcontained review of the geometric theory of characteristic classes, as developed by Chern and Weil. More precisely, we will describe how the curvature of a connection on the vector bundle E can be used to construct closed differential forms on X whose cohomology classes are independent of the connection chosen and therefore furnish topological invariants of the bundle E. Of the many definitions of a connection we will use the differential operator one. It leads to the simplest local formulae. We will also thereby avoid the possibly less elementary concept of principal bundles. For a more general account of this theory see [3], [4], [5].

Definition 2.1. *A connection on E is a differential operator $D:\Gamma(E)\to\Gamma(T^*\otimes E)$ which is a derivation in the sense that for any $f\in A^0(X)$:*

$$D(fs)=df\cdot s+f\cdot Ds,\quad s\in\Gamma(E). \tag{2.2}$$

Remarks. In general a differential operator from $\Gamma(E)$ to $\Gamma(F)$ is just a C-linear map which decreases supports. If such an operator is also $A^0(X)$ linear, then it is induced by a linear map from E to F, i.e., by a section of Hom (E, F). Thus if D_1 and D_2 are connections then D_1-D_2 is induced by an element of

$$\Gamma\,\mathrm{Hom}\,(E^*, T^*\otimes E)=A^1(X;\mathrm{Hom}\,(E,E)).$$

Suppose now that E is equipped with a definite connection D. One may then construct the Chern form of E relative to the connection D in the following manner.

Let $s=\{s_i\}$, $i=1, ..., n$ be a set of sections (2) of $E|U$ where U is open in X, such that the values $\{s_i(x)\}$ form a base for each E_x, with $x\in U$. (Such a set s will be called a *frame* of E over U.) In view of (2.2) a formula of the type:

$$Ds_i=\sum_j\theta_{ij}s_j,\quad \theta_{ij}\in A^1(U), \tag{2.3}$$

(1) The tensor product is over C unless otherwise indicated.
(2) We will be dealing with smooth sections throughout.

must then exist and serves to define a matrix of 1-forms on $U:\theta(s; D)=\|\theta_{ij}\|$—the so-called connection matrix relative to the frame s.

In terms of $\theta(s, D)$ one now defines a matrix $K(s, D)=\|K_{ij}\|$ of 2-forms on U by the formula: $K_{ij}=d\theta_{ij}-\sum_\alpha \theta_{i\alpha}\wedge\theta_{\alpha j}$. In matrix notation:

$$K(s, D)=d\theta(s, D)-\theta(s, D)\wedge\theta(s, D). \tag{2.4}$$

This is the curvature matrix of D relative to the frame s. Because even forms commute with one another it makes sense to take the determinant of the matrix $1+iK(s, D)/2\pi$ and so to obtain an element $\det\{1+iK(s, D)/2\pi\}\in A(U)$.

A priori, this form depends on the frame s. However as we will show in a moment, $\det\{1+iK(s, D)/2\pi\}$ is actually independent of the frame s, and therefore defines a global form, the Chern form of E relative to D, $c(E, D)$ in $A(X)$. More precisely $c(E, D)$ is defined as follows: We cover X by $\{U_\alpha\}$ which admit frames s^α over U_α, and then set $c(E, D)\,|\,U_\alpha=\det\{1+iK(s^\alpha, D)/2\pi\}$. On the overlap these definitions agree because of the asserted independence of our form on the frame s.

Consider then two frames s and s' over U. Then there exist elements $A_{ij}\in A^0(U)$ such that $s_i'=\sum_j A_{ij}s_j$ and in matrix notation we write simply $s'=As$. From (2.2) it follows that $Ds'=\{dA+A\theta(s, D)\}s$. Further, by definition, $Ds'=\theta(s', D)s'$. Hence the connection matrices are related by

$$dA+A\theta(s, D)=\theta(s', D)A, \quad s'=As, \tag{2.5}$$

from which one directly derives the important formula:

$$AK(s, D)=K(s', D)A, \quad s'=As. \tag{2.6}$$

This transformation law of the curvature matrix, together with the invariance of the determinant under conjugation now immediately implies the desired independence of our form $\det\{1+iK(s, D)/2\pi\}$ on s.

Thus we now have defined $c(E, D)$ explicitly and our next aim is to show that $c(E, D)$ is closed and its cohomology class independent of D. For this purpose it is expedient to analyse the above construction a little more carefully, and then to generalize the whole situation.

Note first of all that the transformation law (2.6) is characteristic of the elements of $A(X; \operatorname{Hom}(E, E))$. Indeed if $\xi\in A^p(X; \operatorname{Hom}(E, E))$ and if s is a frame for E over U, then ξ determines a matrix of p-forms $\xi(s)=\|\xi(s)_{ij}\|$ by the formula:

$$\sum_j \xi(s)_{ij}s_j=\xi\cdot s_i, \quad s=\{s_i\}, \tag{2.7}$$

and under the substitution $s'=As$, these matrices transform by the law $\xi(s')A=A\xi(s)$.

The converse is equally true so that in particular the curvature matrix $K(s, D)$ represents a definite element $K[E, D] \in A^2(X; \mathrm{Hom}(E, E))$.

Next we observe that the "determinant construction" really becomes more understandable when formulated in this manner.

We let M_n denote the vector-space of $n \times n$ matrices over $\mathbf{C}$. A k-linear function φ on M_n will be called *invariant* if for all $y \in GL(n, \mathbf{C})$:

$$\varphi(x_1, \dots, x_k) = \varphi(yx_1y^{-1}, yx_2y^{-1}, \dots, yx_ky^{-1}), \quad x_i \in M_n. \tag{2.8}$$

The vector-space of all k-linear invariant forms shall be denoted by $I_k(M_n)$. Now given $\varphi \in I_k(M_n)$ and $U \subset X$, we extend φ to a k-linear map denoted by φ_U—from $M_n \otimes A(U)$ to $A(U)$ by setting:

$$\varphi_U(x_1w_1, x_2w_2, \dots, x_kw_k) = \varphi(x_1, \dots, x_k)w_1 \wedge w_2 \wedge \dots \wedge w_k, \quad x_i \in M_n, \; w_i \in A(U). \tag{2.9}$$

With this understood consider k elements $\xi_i \in A(X; \mathrm{Hom}(E, E))$ and let $\varphi \in I_k(M_n)$. It is then clear that there is a well-defined form $\varphi(\xi_1, \dots, \xi_k) \in A(X)$, which has the local description:

Given a frame s over U, then

$$\varphi(\xi_1, \dots, \xi_k) \mid U = \varphi_U\{\xi_1(s), \dots, \xi_k(s)\} \tag{2.10}$$

where the $\xi_i(s)$ are the matrices of ξ_i relative to s and hence elements of $A(U) \otimes M_n$.

We will abbreviate $\varphi(\xi, \xi, \dots, \xi)$ i.e., the case all ξ_i equal, to $\varphi((\xi))$. Now given a connection D on E, and a $\varphi \in I_k(M_n)$ we have well-determined forms $\varphi((K[E, D]))$ and $\varphi((1 + iK[E, D]/2\pi))$ in $A(X)$, and our Chern form is clearly of the latter type. Indeed we need only take for φ the n-multilinear form det on M_n obtained by polarizing the polynomial function $x \to \det x$ on M_n, to describe the Chern form in the present frame work:

$$c(E, D) = \det((1 + iK[E, D]/2\pi)). \tag{2.11}$$

It is now also an easy matter to construct elements $\varphi_k \in I_k(M_n)$ so that

$$c(E, D) = \sum \varphi_k((K[E, D])).$$

In short, the two properties of $c(E, D)$ which we are after will follow from the more conceptual assertion that for any $\varphi \in I_k(M_n)$, the form $\varphi((K[E, D]))$ is closed and its homology class independent of D.

We will now derive both these properties from the invariance identity (2.8). Note first that differentiation with respect to y leads to the identity

$$\sum_{-1}^{k} \varphi(x_1, \dots, [x_i, y], \dots x_k) = 0, \quad x_i, y \in M_n \tag{2.12}$$

and conversely—because $GL(n, \mathbf{C})$ is connected—(2.12) implies (2.8). This identity now generalizes in a straight forward manner to the extension of φ_U and takes the following form in matrix notation. An element $x^p \in M_n \otimes A^p(U)$ (called of deg p) is represented by a matrix of p-*forms*. Matrix multiplication therefore gives rise to a pairing $x \otimes y \to x \wedge y$, of elements of deg p and deg q to elements of deg $(p+q)$. In terms of this multiplication one now defines the bracket $[x^p, y^q]$ by the usual formula for graded Lie-algebras:

$$[x^p, y^q] = x \wedge y - (-1)^{pq} y \wedge x. \tag{2.13}$$

In this terminology the following invariance law for any $\varphi \in I_k(M_n)$ follows directly from (2.9) and (2.10) and (2.12):

$$\sum (-1)^{qf(\alpha)} \varphi_U(x_1, \ldots, [x_\alpha, y], \ldots, x_k) = 0 \tag{2.14}$$

whenever the x_α and y are homogeneous elements with $q = \deg y$, and $f(\alpha) = \sum_{\beta > \alpha} \deg x_\beta$.

From the derivation property of d it follows further that, with the x_α as above:

$$d\,\varphi_U(x_1, \ldots, x_k) = \sum (-1)^{g(\alpha)} \varphi_U(x_1, \ldots, dx_\alpha, \ldots, x_k) \tag{2.15}$$

where now $g(\alpha) = \sum_{\beta < \alpha} \deg x_\beta$.

PROPOSITION 2.16. *Let D be a connection for E over X, and let φ be an invariant form on M_n. Then $\varphi((K[E, D]))$ is closed.*

Proof. This is a local matter. Hence it is sufficient to show that if s is a frame over U, then $\varphi_U((K))$, $K = K(s, D)$, is closed on U. From (2.14) we obtain

$$d\varphi_U((K)) = \sum_{i=1}^{k} \varphi_U(K, \ldots, dK, \ldots, K).$$

On the other hand the definition, (2.4), of $K(s, D)$ immediately implies the "Bianchi-identity":

$$dK(s, D) = -[K(s, D), \theta(s, D)]. \tag{2.17}$$

Substituting (2.17) and applying (2.14) now yields the desired result.

PROPOSITION 2.18. *Let D_t be a smooth one parameter family of connections on E. Then the function $s \to dD_t s/dt$, $s \in \Gamma(E)$, is $A^0(X)$-linear and hence determines an element $\dot{D}_t \in A^1(X; \mathrm{Hom}\,(E, E))$.*

Further if $\varphi \in I_k(M_n)$, then

$$\Big|_a^b \varphi((K[E, D_t])) = d \int_a^b \varphi'((K[E, D_t]; \dot{D}_t))\, dt \tag{2.19}$$

where $\varphi'((\xi; \eta))$ stands for $\sum_\alpha \varphi(\xi, \ldots, \eta_\alpha, \ldots, \xi)$. (1)

(1) $\big|_a^b f(t)$ stands for $f(b) - f(a)$.

Proof. We have $$\frac{d}{dt} D_t f s = \frac{d}{dt}\{f D_t s + df s\} = f \frac{d}{dt} D_t s.$$

This proves the first part, and so in particular that $\varphi'((K[E, D_t], \dot{D}_t))$ is a well defined form. For the rest we may work locally relative to a frame s over U. It is then easy to see that the matrix of $\dot{D}_t$ relative to s is simply the t-derivative (denoted by a dot) of the connection from $\theta = \theta(s, D_t)$. Thus $\dot{D}_t(s) = \dot{\theta}$. Hence (2.19) will follow from the identity

$$\dot{\varphi}((K)) \,|\, U = d \sum_i \varphi_U(K, \ldots, K, \underset{(i)}{\dot{\theta}}, K, \ldots, K); \quad K = K(s, D_t). \tag{2.20}$$

Consider now the right hand side (=R.H.S.) of this expression. By (2.15) and (2.17) we obtain:

$$\begin{aligned} \text{R.H.S} = & -\sum_i \sum_{\alpha<i} \varphi_U(K; \ldots; \underset{(\alpha)}{[K, \theta]}; \ldots; \underset{i}{\dot{\theta}}; \ldots, K) \\ & + \sum_i \varphi_U(K; \ldots, K; \underset{(i)}{d\dot{\theta}}; K, \ldots, K) \\ & + \sum_i \sum_{\alpha>i} \varphi_U(K, \ldots, \underset{(i)}{\dot{\theta}}, \ldots, \underset{\alpha}{[K, \theta]}, K). \end{aligned}$$

Using (2.14) this simplifies to:

$$\text{R.H.S.} = \sum_i \varphi_U(K, \ldots, K, \underset{(i)}{d\dot{\theta} - [\dot{\theta}, \theta]}, K, \ldots, K).$$

Finally we recall that $K = d\theta - \theta \wedge \theta$. Hence $\dot{K} = d\dot{\theta} - [\dot{\theta}, \theta]$. Thus the R.H.S. takes the form:

$$\sum_{j=1}^{k} \varphi_U(K, \ldots, \underset{(j)}{\dot{K}}, \ldots, K)$$

which manifestly is just the left hand side $\dot{\varphi}((K))$. Q.E.D.

COROLLARY 2.21. *The cohomology class of $\varphi((K[E, D]))$ is independent of the connection on E.*

Indeed if D_1 and D_0 are two connections on E then for each t, $D_t = tD_1 + (1-t)D_0$ is again a connection on E. Hence (2.19) implies the corollary.

Remarks. This concludes our elementary and therefore necessarily rather pedestrian account of the theory of characteristic classes for vector-bundles. A slightly more conceptual path to the same results might run along these lines.

One first notes the following general properties:

(2.22). A pairing of bundles from $E \otimes F$ to G induces a pairing from

$$A^p(X; E) \otimes A^q(X; F)$$

to $A^{p+q}(X; G)$ by combining the above pairing with exterior multiplication. All pairings of this type will be written as a multiplication, i.e., denoted by a dot.

(2.23). If D is a connection for E, then the dual bundle E^* has a unique connection —also denoted by D—which satisfies the equation

$$d\langle s, s^*\rangle = \langle Ds, s^*\rangle + \langle s, Ds^*\rangle, \quad s\in\Gamma(E),\ s^*\in\Gamma(E^*).$$

(2.24). If D_i are connections on E_i, $i=1, 2$ then the formula

$$D(s_1\otimes s_2) = D_1 s_1\cdot s_2 + s_1\cdot D_2 s_2$$

defines a connection on $E_1\otimes E_2$, $s_i\in\Gamma(E_i)$.

(2.25). The connection D on E extends uniquely to an antiderivation of the $A(X)$ module $A(X; E)$, i.e., so as to satisfy the law:

$$D(\theta\cdot s) = d\theta\cdot s + (-1)^p\theta\cdot Ds, \quad \theta\in A^p(X),\ s\in\Gamma(E).$$

Now, with these trivialities out of the way, one may argue as follows. First one shows that there is a unique element $K[E, D]\in A^2(X; \mathrm{Hom}\,(E, E))$ such that

$$D^2 s = K[E, D]\cdot s \quad \text{for any } s\in\Gamma(E).$$

One next observes (as we did) that $\varphi\in I_k(M_n)$ defines a definite homomorphism

$$\varphi: \mathrm{Hom}\,(E, E)^{(k)}\to 1,$$

of the kth tensor power of $\mathrm{Hom}(E, E)$ into the trivial bundle, and so induces a map

$$\varphi: A(X, \mathrm{Hom}\,(E, E)^{(k)})\to A(X).$$

Now, our earlier $\varphi((K[E, D]))$ is defined simply as $\varphi\{K[E, D]^{(k)}\}$.

One next shows that extension to $A(X; \mathrm{Hom}\,(E, E)^{(k)})$ of the connection which D induced on $\mathrm{Hom}(E, E)^k$, by (2.23) and (2.24) is compatible, with φ. That is,

$$d\varphi = \varphi\cdot D.$$

Then the proof of Proposition 2.21 follows directly from this compatibility and the *Bianchi-identity*: $DK=0$.

3. Hermitian vector-bundles

Let E be a vector-bundle over X. Then a real-valued function $N: E\to\mathbf{R}$ is said to define a hermitian structure on E—or more briefly to be a *norm* for E—if the restriction of N to any fiber is a Hermitian norm on that fiber. Thus for each $x\in X$, the expression:

$$\tfrac{1}{2}\{N(u+v)-N(u)-N(v)\}+i\tfrac{1}{2}\{N(u+iv)-N(u)-N(v)\},\quad u,\, v\in E_x$$

is to define a positive definite Hermitian form on E_x. This form will generally be denoted by $\langle u, v\rangle_x$, or simply $\langle u, v\rangle$, and upon occasion by $\langle u, v\rangle_N$. We of course have $N(u)=\langle u,u\rangle$.

When a complex analytic bundle is endowed with a norm, then the inter-play between these two structures gives rise to several interesting phenomena which will be reviewed in this section.

Recall first of all that on a complex manifold the complex valued differential forms $A(X)$ split into a direct sum $\sum A^{p,q}(X)$ where $A^{p,q}(X)$ is generated over $A^0(X)$ by forms of the type $df_1\wedge\ldots\wedge df_p\wedge d\bar f_{p+1}\wedge\ldots\wedge d\bar f_{p+q}$, the f_i being local holomorphic functions on X. As a consequence d splits into $d'+d''$ where

$$d':A^{p,q}\to A^{p+1,q}\quad\text{and}\quad d'':A^{p,q}\to A^{p,q+1}.$$

These two halves of d are then related by:

$$d'^2=d''^2=0,\quad d'd''+d''d'=0. \tag{3.1}$$

If E is a vector-bundle over X, then this decomposition of $A(X)$ induces a corresponding decomposition of $A(X;E)$ into $A^{p,q}(X,E)=A^{p,q}(X)\otimes_{A^0(X)}\Gamma(E)$, and hence any connection D on E, splits canonically into the sum of

$$D':\Gamma(E)\to A^{1,0}(X;E)\quad\text{and}\quad D'':\Gamma(E)\to A^{0,1}(X;E).$$

With these preliminaries out of the way we come to the first consequence of the simultaneous existence of a holomorphic and Hermitian structure on E.

PROPOSITION 3.2. *Let N be a Hermitian norm on the analytic bundle E. Then N induces a* canonical connection $D=D(N)$ *on E which is characterized by the two conditions:*

(3.3) *D preserves the norm N.*

(3.4) *If s is a holomorphic section of $E\,|\,U$ then $D''s=0$ on U.*

The first condition is expressed by the formula:

$$d\langle s, s'\rangle=\langle Ds, s'\rangle+\langle s, Ds'\rangle,\quad s,\, s'\in\Gamma(E), \tag{3.5}$$

where we defined $\langle s, s'\rangle$ as the function $\langle s, s'\rangle(x)=\langle s(x), s'(x)\rangle_x$ and we have in general set $\langle s\otimes\theta, s'\otimes\theta'\rangle$ equal to $\theta\wedge\bar\theta'\cdot\langle s, s'\rangle$.

The proof of Proposition 3.2 is straightforward. If s is a frame for E, over U, we write $N(s)$ for the matrix of functions:

$$N(s)=\|\langle s_j, s_k\rangle\|. \tag{3.6}$$

6 – 652932 *Acta mathematica* 114. Imprimé le 11 août 1965.

This is the *norm* of the frame. Now, let s be a holomorphic frame over U, and let θ be a prospective connection matrix for E, relative to s. Then (3.5) applied to all the pairs $\langle s_j, s_k\rangle$ implies the relation

$$\theta N + N\bar{\theta}^t = dN, \quad N = N(s). \tag{3.7}$$

Hence if θ is to satisfy this condition, and is also to be of type (1, 0) so as to satisfy (3.4), then we must have:

$$\theta = d'N \cdot N^{-1} \quad \text{on } U. \tag{3.8}$$

Thus there is at most one connection with the properties (3.5) and (3.6).

Conversely, let $s = \{s_i\}$ be a *holomorphic* frame over U, and set $N(s) = \|\langle s_i, s_j\rangle\|$ as before. Then the formula:

$$Ds_i = \sum (d'N \cdot N^{-1})_{ij} s_j, \quad N = N(s), \tag{3.9}$$

defines a connection over U, which is seen to be independent of the *holomorphic* frame s chosen, and hence induces a global connection $D(N)$ on E which manifestly satisfies the condition of our proposition. The independence of D on s is proved as follows:

Let $s_1 = As$ be another *holomorphic* frame, over U. Then $dA = d'A$, because A is a holomorphic matrix. Further $N_1 = N(s_1) = AN\bar{A}^t$. Hence

$$d'N_1 \cdot N_1^{-1} = dA \cdot A^{-1} + Ad'N \cdot N^{-1}A^{-1}$$

which shows that the matrices $\theta(s, N) = d'N \cdot N^{-1}$, $N = N(s)$ transform like the connection matrix of a connection.

COROLLARY 3.10. *Let E be a holomorphic bundle with Hermitian norm N, and let θ, K denote the connection and curvature matrices of $D(N)$ relative to a holomorphic frame over U. Then on U one has:*

$$\theta \text{ is of type } (1, 0), \text{ and } d'\theta = \theta \wedge \theta. \tag{3.11}$$

$$K = d''\theta, \text{ whence } K \text{ is of type } (1, 1) \text{ and } d'' \cdot K = 0. \tag{3.12}$$

$$d'K = -[K, \theta]. \tag{3.13}$$

Proof. The first line follows directly from $\theta = d'N \cdot N^{-1}$, where $N = N(s)$ is the norm of s. Indeed $d'\theta = d'd'N \cdot N^{-1} - d'N \cdot d'N^{-1}$ and $d'N^{-1} = -N^{-1} \cdot d'N \cdot N^{-1}$. The others are even more straightforward. Note that because K is of type (1, 1), *the characteristic classes of the connection $D(N)$ always are of type (p, p).*

These formulae become especially simple when E is a line bundle. Then a holomorphic frame is simply a nonvanishing holomorphic section s, so that, relative to s, $\theta = d' \log N(s)$ and $K = d''d' \log N(s)$. Thus in particular, if E admits a global nonvanishing holomorphic section s, then

$$c_1(E) = \frac{i}{2\pi} d''d' \log N(s). \tag{3.14}$$

The next proposition is a refinement of the earlier homotopy formula (2.19). For simplicity, we will abbreviate $K[E, D(N)]$ to $K[E, N]$.

PROPOSITION 3.15. *Consider a smooth family of norms N_t, on the holomorphic bundle E. Then the function $(s, s') \to d\langle s, s'\rangle_{N_t}/dt$ is Hermitian linear over $A^0(X)$ and hence determines a section $L_t \in \Gamma$ Hom (E, E), by the formula*

$$\langle L_t \cdot s, s'\rangle_{Nt} = \frac{d}{dt}\langle s, s'\rangle_{Nt}, \quad s, s' \in \Gamma(E).$$

If φ is any invariant form in $I_k(M_n)$; $n = \dim E$, then

$$\Big|_a^b \varphi((K[E, N_t])) = d''d' \int_a^b \varphi'((K[E, N_t]; L_t))\, dt, \tag{3.16}$$

where, as before, $\varphi'((\xi; \eta)) = \Sigma_i \varphi(\xi, \ldots, \xi, \underset{(i)}{\eta}, \xi, \ldots, \xi)$.

Proof. We have
$$\frac{d}{dt}\langle fs, f's'\rangle = f\bar{f}' \cdot \frac{d}{dt}\langle s, s'\rangle;$$

so that L is well defined. Hence $\varphi'((K[E, N_t]; L_t))$ is a global form and it suffices to check the formula

$$\frac{d}{dt}\varphi((K[E, N_t])) = d''d'\varphi'((K[E, N_t]; L_t)) \tag{3.17}$$

locally. We therefore choose a *holomorphic* frame $s = \{s_i\}$ over U, and set $N = N_t(s)$, $K = K(s, D(N_t))$, $\theta = \theta(s, D(N_t))$. Then the matrix of L relative to s, is easily computed to be $\dot{N}N^{-1}$, the dot denoting the t-derivative as before. Let us denote this matrix by L also. Finally, we will abbreviate $\varphi'((K[E, N_t]; L_t)) \,|\, U$ to $\square$. Then

$$d'\square = \sum_{j \neq i} \varphi(K, \ldots, \underset{(i)}{-[K, \theta]}, \ldots, \underset{(j)}{L}, \ldots, K) + \sum_j \varphi(K, \ldots, \underset{(j)}{d'L}, \ldots, K). \tag{3.18}$$

Applying the invariance identity one obtains

$$d'\square = \sum_j \varphi(K, \ldots, \underset{(j)}{d'L + [L, \theta]}, \ldots, K). \tag{3.19}$$

Finally, one now computes directly from $L = \dot{N}N^{-1}$ and $\theta = d'N \cdot N^{-1}$, that

$$\dot{\theta} = d'L + [L, \theta]. \tag{3.20}$$

Hence $d'\square$ is the form $\varphi'((K[E, D(N_t)], \dot{D}(N_t)))$ of Proposition 2.18 so that (2.19) implies (3.16). Q.E.D.

This proposition now directly proves Proposition 1.2 of the Introduction. Indeed if N_1 and N_2 are two Hermitian norms on E, then $N_t=(1-t)N_1+tN_2$ defines a smooth family between these two Hermitian norms, so that the formula of the proposition becomes a special case of (3.16).

As another direct application we have:

COROLLARY 3.21. *Suppose E is an n-dimensional complex vector bundle over X, with Hermitian norm N. Suppose also that E admits n holomorphic sections which span the fiber at each point. Then the refined Chern classes $\hat{c}_i(E)$ are zero for $i>0$ so that:*

$$\hat{c}(E)=1. \tag{3.22}$$

Proof. Let s be the global frame determined by the sections in question, and define a Hermitian norm on E, by setting $N_1(s)=\text{identity}$. For this norm $\theta(s)$, $K(s)$ and hence $c_i(E, N_1)$, $i>0$ clearly vanish. Q.E.D.

Remark. The deformation $D_t=D(N_t)$ induced by the variation of N_t is not the linear one encountered earlier. Rather, D_t satisfies the differential condition:

$$\dot{D}_t(s)=d'L_t(s)+[L_t(s), \theta(s, D_t)], \quad s \text{ any frame.} \tag{3.23}$$

In other words $\dot{D}_t$ is the D'-derivative of L_t, and it is clear that much of the foregoing depends on just the existence of some $L_t \in \Gamma \operatorname{Hom}(E, E)$ for which (3.23) is valid.

In the remainder of this section we will formulate a generalization of (3.16) along these lines.

DEFINITION 3.24. *A connection D on the holomorphic bundle E over X, is called of type* (1, 1) *if:*

(3.25) *For any holomorphic section s of $E|U$, $D''s=0$.*

(3.26) *The curvature matrix $K(s, D)$ of D relative to a frame s over U, are of type* (1, 1), *i.e.,*

$$K[E, D] \in A^{1,1}(X; \operatorname{Hom}(E, E)).$$

This is then clearly an extension of the class of connections induced by Hermitian norms on E.

Next consider a family of connections D_t of type (1, 1). Such a family will be called bounded by $L_t \in A^0(X; \operatorname{Hom}(E, E))$ if the relation (3.23) holds between

$$\dot{D}_t \in A^1(X; \operatorname{Hom}(E, E))$$

and L_t. Note that the elements of $\Gamma \operatorname{Hom}(E, E)$ may be thought of as defining degree zero differential operators on $A(X; E)$ *and* on $A(X; \mathrm{Hom}(E, E))$, the latter action being induced by the composition of endomorphisms. With this understood, the bounding condition (3.23) may quite equivalently be expressed by:

$$[D_t', L_t]s = \dot{D}_t s \quad \text{for } s \in \Gamma(E). \tag{3.27}$$

In any case it is now easy to check that our earlier argument leading to (3.16) also proves the following more general homotopy lemma.

PROPOSITION 3.28. *Let D_t be a smooth family of connections of type* (1, 1) *on the holomorphic bundle E. Suppose further that D_t is bounded by $L_t \in A^0(X; \mathrm{Hom}(E, E))$. Then for any $\varphi \in I_k(M_n)$, $n = \dim E$, we have the relation:*

$$\Big|_a^b \varphi((K[E, D_t])) = d''d' \int_a^b \varphi'((K[E, D_t]; L_t))\, dt. \tag{3.29}$$

We note in conclusion that if D_t is related to $L_t \in \Gamma\{\mathrm{Hom}(E, E)\}$ by (3.27), and if D_0 is of type (1, 1), then D_t will be of type (1, 1) for all t. Indeed, D_t is of type (1, 1) if and only if $\theta = \theta(s, D_t)$ satisfies the two conditions: $\theta_{ij} \in A^{1,0}$, $d'\theta = \theta \wedge \theta$, whenever s is a holomorphis frame.

Differentiating these conditions with respect to time one obtains:

$$\dot{\theta}_{ij} \in A^{1,0}, \quad d'\dot{\theta} = [\dot{\theta}, \theta]. \tag{3.30}$$

Now if (3.23) holds then—setting $L_t(s)$ equal to L—we have $d'L + [L, \theta] = \dot{\theta}$. It follows that $d'\dot{\theta} = [d'L, \theta] + [L, d'\theta]$ which by resubstituting (3.23) leads to $d'\dot{\theta} = [\dot{\theta}, \theta]$. Thus (3.23) together with $\theta_{ij} \in A^{1,0}$ imply the differentiated identities. Q.E.D.

4. The duality formula

Consider an exact sequence of holomorphic vector bundles:

$$0 \to E_I \to E \to E_{II} \to 0 \tag{4.1}$$

over the base manifold X. We wish to prove the duality formula: $\hat{c}(E) = \hat{c}(E_I)\hat{c}(E_{II})$ for the refined Chern classes of these bundles.

For this purpose consider a norm N on E. Such a norm then induces norms N_I on E_I and N_{II} on E_{II} in a natural manner: The restriction of N to E_I defines N_I, and the restriction of N to the orthocomplement of E_I—denoted by $E_I^\perp$—determines N_{II}, via the C^∞ isomorphism of E_{II} and $E_I^\perp$ induced by (4.1).

Thus (4.1) gives rise to three Chern forms in $A(X)$: $c(E) = c(E, N)$, and $c(E_i, N_i)$,

$i = I, II$; and the duality formula will be established once we prove the following proposition.

PROPOSITION 4.2. *There exists a form ξ in $A(X)$ such that*

$$c(E) - c(E_I) \cdot c(E_{II}) = d''d'\xi. \tag{4.3}$$

The proof of (4.3) is based upon a specific deformation of the canonical connection $D = D(N)$. To describe this deformation we need certain preliminaries concerning the geometric implications of the exact sequence (4.1).

First we introduce the *orthogonal* projections

$$P_i : E \to E_i, \quad i = I, II, ^{(1)} \tag{4.2}$$

which this situation naturally defines. These are then elements of Γ Hom (E, E) and therefore—interpreted as degree zero operators, they lead to a decomposition of $D = D(N)$ into four parts:

$$D = \sum_{ij} P_i D P_j, \quad ij = I, II.$$

LEMMA 4.3. *In the decomposition just introduced, $P_i D P_i$ induces the connection $D(N_i)$ on E_i, while $P_i D P_j$, $i \neq j$, are degree zero operators of type* (1, 0) *and* (0, 1) *respectively:*

$$P_{II} D'' P_I = 0, \quad P_I D' P_{II} = 0. \tag{4.4}$$

Proof. We first show that $P_i D P_j$, $i \neq j$ is $A^0(X)$ linear. Consider then $P_i D P_j(fs)$. Using the derivation properly of D we get $P_i D P_j(fs) = df \cdot P_i P_j s + f P_i D P_j s$. Hence as $P_i P_j = 0$ if $i \neq j$, it follows that $P_i D P_j$ is a degree zero operator. We next show that $P_{II} D'' P_I = 0$. This is clearly also a degree zero operator. Hence it is sufficient to show that given $e \in E_x$, there exists *some* section s of E near x, such that $s(x) = e$ and $P_{II} D'' P_I s = 0$ at x. Now because E_I is a *holomorphic* sub-bundle of E we may choose a holomorphic section s_1 near x such that the two conditions, $s_1(x) = P_I e$; $P_I s_1 = s_1$ (i.e., $s_1 \in \Gamma(E_I)$) hold near x. We may also choose a C^∞ section s_2 of $E_I^\dagger$ which satisfies the two conditions $s_2(x) = P_{II} e$; $P_{II} s_2 = s_2$ near x. Now setting $s = s_1 + s_2$, we clearly have $s(x) = e$. Further $P_I D'' P_I(s_1 + s_2) = 0$ near x because $P_I s_2 = 0$ and $D'' s_1 = 0$ there.

The companion statement of (4.4) now follows directly from the fact that D preserves the inner product: We have $d\langle P_I s, P_{II} s'\rangle = 0$ whence

$$\langle P_{II} D P_I s, s'\rangle = -\langle s, P_I D P_{II} s'\rangle \tag{4.5}$$

so that in particular $P_{II} D'' P_I = 0$ implies $P_I D' P_{II} = 0$.

---

(1) Here as in what follows we use the natural projection $E_I^\dagger \to E_{II}$ to identify these two bundles.

It is now quite straightforward to check that the $P_i DP_i$ interpreted as differential operators on E_i, satisfy the two conditions which characterize $D(N_i)$. We note that one precisely needs (4.4) to show that the connection induced by $P_i DP_i$ on E_i is of type (1, 0) —that is satisfies (3.4).

The deformation which we need for the duality theorem is now defined by:

$$D_t = D + (e^t - 1)\delta, \quad \delta = P_{II} D P_I. \tag{4.6}$$

By our lemma δ is a degree zero operator and hence D_t a connection for every $t \in \mathbf{R}$. We have further:

LEMMA 4.7. *The family D_t defined by (4.6) is "bounded" by the element*

$$P_I \in \Gamma \operatorname{Hom}(E, E).$$

Proof. We have to show that $[D_t', P_I] = \dot{D}_t$, or in other words, that

$$[D', P_I] + (e^t - 1)[\delta, P_I] = e^t \delta,$$

where $[A, B]$ stands for the commutator $AB - BA$. Now it is clear that $[\delta, P_I] = \delta$. Hence we just have to show that $P_{II} D' P_I \cdot P_I - P_I \cdot P_{II} D' P_I = P_{II} D P_I$ and that follows directly from (4.4). Q.E.D.

We next investigate the curvature form $K[E, D_t]$, and its decompositions according to the P_i. Using some obvious identifications the formulae take the following form:

LEMMA 4.8. *Let $P_i K[E, D_t] P_j$ be denoted by $K_{ji}[E, D_t]$. Then,*

$$K_{I,I}[E, D_t] = K[E_I, D_I] - e^t \delta^* \wedge \delta, \tag{4.9}$$

$$K_{II,II}[E, D_t] = K[E_{II}, D_{II}] - e^t \delta \wedge \delta^*, \tag{4.10}$$

$$K_{I,II}[E, D_t] = e^t K_{I,II}[E, D_0]; \quad K_{II,I}[E, D_t] = K_{II,I}[E, D_0], \tag{4.11}$$

where δ^* denotes the adjoint of the form $\delta \in A^1(X; \operatorname{Hom}(E, E))$, and hence by (4.5) represents the operator $-P_I D P_{II}$.

These formulae clearly show the pertinence of our deformation D_t to the problem at hand. When $t = -\infty$, we see that $K_{I,II} = 0$ and the $K_{i,i}[E, D_t]$ reduces to $K[E_i, D_i]$. As the Chern form $c(E, D_t)$ is defined by $\det((1 + i/2\pi K[E, D_t]))$ it follows directly that

$$\lim_{t \to -\infty} c(E, D_t) = c(E_I, D_I) \cdot c(E_{II}, D_{II}). \tag{4.12}$$

The proof of Lemma 4.8 is quite straightforward. If one interprets K as the operator D^2, the terms can be just read off. Alternately one may choose a frame $s = (u, v)$ which is naturally suited to the problem—namely of the following type: 1) The frame s is unitary, and its first k components, u, span E_I. (Hence the remaining components—v—span $E_I^\perp$.) For such a frame $\theta(s, D_t)$ breaks into blocks $\theta_{ij}(s, D_t)$—corresponding to the operators $P_j DP_i$—and then (4.8) follows directly from (2.4).

We have now nearly completed our argument. Indeed, in view of (4.7) and the general homotopy lemma, one has the formula:

$$\Big|_t^0 c(E, D_t) = d''d' \int_t^0 \det{}'((1+\varkappa K;\varkappa P_I))\,dt, \qquad K=K[E,D_t], \varkappa = i/2\pi \tag{4.13}$$

valid for all $t \in \mathbf{R}$.(1) Hence if we could simply put $t=-\infty$, in (4.13), we would be done. However the integral will not converge in general. In fact, it follows from (4.8) that

$$K[E, D_t] = K[E, D_{-\infty}] + e^t \square, \tag{4.14}$$

where $\square \in A^2(X; \operatorname{Hom}(E, E))$ is independent of t. Therefore:

$$\det{}'((1+\varkappa K[E,D_t];\varkappa P_I)) = \sum_{\alpha=0}^{n-1} a_\alpha e^{\alpha t}, \quad a_\alpha \in A(X) \tag{4.15}$$

with
$$a_0 = \det{}'\,((1+\varkappa K[E, D_{-\infty}]; \varkappa P_I)). \tag{4.16}$$

Hence the integral will converge only if $a_0=0$, and that is generally not the case. Note however that by (4.7) a_0 can be re-expressed as

$$\det{}'\,((1+\varkappa K[E_I, D_I]; \varkappa 1))\cdot \det\,((\varkappa K[E_{II}, D_{II}])).$$

It then follows easily that a_0 is a linear combination of the Chern forms $c_\alpha(E_I, N_I)$ multiplied by $c_{n-k}(E_{II}, N_{II})$:

$$a_0 = \left\{\sum_{\alpha=0}^{k} (n-k)\, c_\alpha(E_I)\right\} c_{n-k}(E_{II}). \tag{4.17}$$

In any case a_0 will be a *closed* form.

Hence (4.13) will remain valid even when a_0 is deleted from under the integral sign. But once this is done one may clearly integrate and pass to the limit $t=-\infty$, in (4.13) to obtain:

$$c(E) - c(E_I)\cdot c(E_{II}) = d''\, d' \left\{\sum_{\alpha=1}^{n} \alpha^{-1}\cdot a_\alpha\right\}; \quad a_\alpha \text{ as in (4.15).} \tag{4.18}$$

This then completes the proof of the duality formula, and also gives us the explicit form $\sum_{\alpha=1}^{n} \alpha^{-1}\cdot a_\alpha$ for ξ. In the proof of Theorem I we need to compute the highest component of ξ for the case, $\dim E_I = 1$. Thus we want to expand $\varkappa^n \det((K[E, D_{-\infty}] + e^t \square; P))$

(1) det′ is the function denoted by φ', with $\varphi = \det$, in Section 3; see also (4.19) below.

under that assumption. Again using (4.7) we see that because $\dim E_I = 1$ this expression reduces to $\varkappa^n \det((K[E_{II}, N_{II}] + e^t \Box_{II}))$ where $\Box_{II} = K_{II,II}[E, D] - K[E_{II}, N_{II}]$. Hence if we define $\det^\alpha((\xi;\eta))$ by the identity

$$\det((\xi + \lambda\eta)) = \sum \lambda^\alpha \det{}^\alpha((\xi;\eta)), \tag{4.19}$$

then the coefficients a_α which we are after, are given by $\varkappa^n \det{}^\alpha(K[E_{II}, N_{II}]; \Box_{II})$. We record this fact for later reference:

PROPOSITION 4.20. *Let $0 \to E_I \to E \to E_{II} \to 0$ be an exact sequence of holomorphic vector bundles, and let $c(E)$, and $c(E_i)$, $i = I, II$ denote the Chern forms induced by a norm N on E. Then if $\dim E_I = 1$,*

$$c_n(E) - c_1(E_I) \cdot c_{n-1}(E_{II}) = \varkappa d'' d' \{ \sum_{\alpha > 0} \alpha^{-1} \det{}^\alpha((\Omega[E_{II}]; -\Delta_{II}[E])) \} \tag{4.21}$$

where $\Omega[E_{II}] = \varkappa K[E_{II}, N_{II}]$, $\Omega_{II}[E] = \varkappa K_{II,II}[E; N]$, and $-\Delta_{II} = \Omega_{II}[E] - \Omega[E_{II}]$. Hence if E_I admits a nonvanishing holomorphic section s, then by 3.15,

$$c_n(E) = \varkappa d'' d' \{ \log N(s) \cdot c_{n-1}(E_{II}) + \sum_{\alpha > 0} \alpha^{-1} \cdot \det{}^\alpha((\Omega[E_{II}], -\Delta_{II}[E])) \}. \tag{4.22}$$

Note that aside from the positivity assertion, (4.22) proves Theorem I. Indeed consider the projection $\pi_1 : E \to X$. The identity map of E into itself, then induces a holomorphic section s of $\pi_1^*(E)$ over E, which vanishes only on the zero section $X \subset E$. Hence if $\pi : B^*(E) \to X$ is the restriction of π_1 to the subset $B^*(E) = \{ e \in E \mid 0 < N(e) < 1 \}$ of E then the section s of $\pi^* E = \pi_1^*(E) \mid B^*(E)$ does not vanish. We may therefore apply (4.22) to $\pi^*(E)$ and so obtain a formula of the type $c_n\{\pi^*(E)\} = \varkappa d'' d' \xi$. Now by the obvious functoriality of the Chern forms relative to holomorphic isomorphisms, $c_n\{\pi^*(E)\} = \pi^* c_n(E)$, so that we are done. In the next section we will discuss the positivity of the ξ given by (4.22). Let us close this one with a direct consequence of the duality formula which generalizes (4.22) but in a less specific fashion.

COROLLARY. *Let E be a holomorphic bundle over E, which admits k linearly independent* holomorphic *sections. Then*

$$\hat{c}_i(E) = 0 \quad \text{for } i > n - k.$$

Proof. Let E_I be the bundle spanned by the sections. Then by (3.22) $\hat{c}(E_I) = 1$. Hence by the duality formula $\hat{c}(E) = \hat{c}(E/E_I)$ and the bundle E/E_I has dimension $(n-k)$. Q.E.D.

5. Remarks on Positivity. The proof of Theorem I completed

As we have seen (4.22) already proves the combinatorial aspects of Theorem I and it remains only to discuss the "sign" of the ξ there constructed.

We recall first of all that $A^{p,p}(X)$ contains a well determined convex cone of positive ($\geqslant 0$) elements. By definition a form Ω is in the cone—noted by $\Omega \geqslant 0$—if and only if there exist forms $\theta_\alpha \in A^{p,0}(X)$ such that

$$\Omega = i^{p^2} \sum_\alpha \theta_\alpha \wedge \bar{\theta}_\alpha.$$

We may extend this definition to matrix valued forms in the following fashion:

DEFINITION 5.1. *Let Ω be an $n \times n$ matrix of forms of type (p, p). Then Ω is positive, if there exist $n \times m$ matrices N_r, of type $(p, 0)$ such that*

$$\Omega = i^{p^2} \sum_r N^r \wedge \bar{N}^t_r. \tag{5.2}$$

Note that if A is any nonsingular $n \times n$ matrix of functions, then Ω is positive with $A\Omega\bar{A}^t$. This enables us to define positivity in $A^{p,p}(X; \mathrm{Hom}\,(E, E))$ for any bundle E with a Hermitian structure. Namely if $\xi \in A^{p,p}(X; \mathrm{Hom}\,(E, E))$ we define $\xi \geqslant 0$ to mean that the *matrix* of ξ relative to a *unitary frame* s, be positive: $\xi(s) \geqslant 0$. As unitary frames are related by unitary transformations—for which therefore $\bar{A}^t = A^{-1}$,—this concept is thereby well defined.

Hence in particular, if E is a holomorphic bundle with Hermitian norm N – then it makes sense to ask whether the "*real curvature form*" $\Omega[E, N] = \varkappa K[E; N]$ is positive or not.

To simplify the notation we will, in the sequel, call such a holomorphic bundle with a given Hermitian norm simply a Hermitian bundle, denote it by a single letter E, and write $K[E]$, $\Omega[E]$, $c(E)$ etc., instead of $K[E; N]$, $\Omega[E, N]$, $c(E)$ etc. Such a bundle is called positive if $\Omega[E] \geqslant 0$.

That these notions of positivity on the form and the vector-bundle level are compatible follows readily from the following lemma:

LEMMA 5.3. *Let E be a Hermitian bundle of dimension n, and let ξ_r be positive elements of $A^{p,p}(X; \mathrm{Hom}\,(E, E))$; $r = 1, \ldots, n$. Then if p is odd* $\det(\xi_1, \ldots, \xi_n) \in A^{pn,pn}(X)$ *is positive.*

Proof. We may find forms $N^r_{i\alpha} \in A^{p,0}(U)$, so that with respect to some unitary frame s over U,

$$\xi_r(s) = i^{p^2} \sum_{\alpha_r} N^r_{i\alpha_r} \bar{N}^r_{j\alpha_r}, \quad 1 \leqslant \alpha_r \leqslant \beta_r.$$

Hence $\det(\xi_1, \ldots, \xi_n)$ is given by the sum:

$$i^{p^2 \cdot n}(n!^{-1}) \sum_{\sigma,\tau,\alpha} (-1)^{\sigma} N^{\tau(1)}_{1\alpha_{\tau(1)}} \bar{N}^{\tau(1)}_{\sigma(1)\alpha_{\tau(1)}} \dots N^{\tau(n)}_{n\,\alpha_{\tau(n)}} \bar{N}^{\tau(n)}_{\sigma(n)\alpha_{\tau(n)}}$$

where σ and τ vary over the group of permutations and $\underline{\alpha}$ denotes independent summation over the α_i's.

If we now take all the barred terms to the right and reorder them in ascending order according to their first lower index then, because p is odd $(-1)^{\sigma}$ cancels out and this expression is seen to take the form:

$$(n!)^{-1} i^{p^2 n} \cdot i^{p^2(n^2-n)} \sum_{\lambda,\tau,\alpha} N^{\tau(1)}_{1\cdot} \dots N^{\tau(1)}_{n\cdot} \cdot \bar{N}^{\lambda(1)}_{1\cdot} \dots \bar{N}^{\lambda(n)}_{1\cdot}$$

where $\lambda = \tau \circ \sigma^{-1}$ and we have denoted the appropriate α-index by a dot. Hence

$$\det(\xi_1, \dots, \xi_n) = (n!)^{-1} i^{(pn)^2} \sum_{\alpha} \theta_\alpha \wedge \bar{\theta}_\alpha$$

where $\theta_\alpha = \sum_\lambda N^{\lambda(1)}_{1\cdot} N^{\lambda(2)}_{2\cdot} \dots N^{\lambda(n)}_{n\cdot}$ and therefore clearly $\geqslant 0$. Q.E.D.

As an immediate corollary we have:

(5.4) The forms $\det^{\alpha}((\xi;\eta))$ are positive if $\xi, \eta \in A^{1,1}(X; \operatorname{Hom}(\underset{\cdot}{E}, E))$ are positive. Further if $\xi \geqslant \xi' \geqslant 0$, $\eta \geqslant \eta' \geqslant 0$ then $\det^{\alpha}((\xi, \eta)) \geqslant \det^{\alpha}((\xi'; \eta'))$.

Indeed, $\det^{\alpha}(\xi, \eta)$ is just the sum $\sum \det(\xi_1, \dots, \xi_n)$ where α of the ξ_i are set equal to η and the remaining ones are equal to ξ. In particular then, we have:

$$\Omega(E) \geqslant 0 \Rightarrow c(E) \geqslant 0. \tag{5.5}$$

When applied to the exact sequence of Hermitian vector bundles (4.1):

$$0 \to E_I \to E \to E_{II} \to 0$$

our lemma yields the following inequalities: In the notation of that section, define $\Omega_i[E]$, $i = I, II$, to be the form $\varkappa P_i K[E] P_i$ *interpreted* as a section of $A^2(X; \operatorname{Hom}(E_i, E_i))$. Then in view of (4.4) and (4.8) we immediately obtain the inequalities:

$$\Delta_I(E) = \Omega[E_I] - \Omega_I[E] \leqslant 0 \tag{5.6}$$

$$\Delta_{II}(E) = \Omega[E_{II}] - \Omega_{II}[E] \geqslant 0. \tag{5.7}$$

Put differently, sub-bundles are less positive and quotient bundles more positive, than the bundle itself.

We next return to the formula (4.22):

$$c_n(E) = \varkappa d''d'\{\log N(s) \cdot c_{n-1}(E_{II}) + \xi\}$$

where

$$\xi = \sum_{\alpha > 0} \alpha^{-1} \det^{\alpha}((\Omega[E_{II}]; \Omega_{II}[E] - \Omega[E_{II}])).$$

Assume now that $\Omega[E]\geqslant 0$. Then $\Omega_{II}[E]\geqslant 0$ and hence by (5.6) $\Omega[E_{II}]\geqslant\Delta_{II}[E]\geqslant 0$. From this it follows that $c_{n-1}(E_{II})\geqslant 0$. The form ξ can be written as

$$\sum_{\alpha>0}\alpha^{-1}(-1)^{\alpha}\det{}^{\alpha}\{\Omega[E_{II}];\Delta_{II}(E)\}.$$

Hence ξ is an alternating sum of positive terms and therefore neither positive or negative. However, this is not serious. In fact we can add to ξ a closed form ξ_0 so as to make $\xi+\xi_0\geqslant 0$ and $\xi-\xi_0\leqslant 0$. This is done as follows: Let

$$\xi_0=\sum_{\alpha>0}\alpha^{-1}\det{}^{\alpha}((\Omega[E_{II}];\Omega[E_{II}])). \tag{5.8}$$

Then by the definition of $\det^{\alpha}$, $\xi_0=\sum\alpha^{-1}\binom{n}{\alpha}c_{n-1}(E_{II})$, and hence is a closed form. Further note that in view of (5.4), we have

$$\det{}^{\alpha}((\Omega[E_{II}];\Omega[E_{II}]))\geqslant\det{}^{\alpha}((\Omega[E_{II}];\Delta_{II}(E))),$$

and so our assertion concerning ξ_0 is correct.

We next replace ξ by $\xi-\xi_0$ in (4.1) and use the definition

$$d^c=i(d''-d'). \tag{5.9}$$

The formula (4.7) then takes the form:

$$c_n(E)=\frac{dd^c}{4\pi}\{\log N^{-1}(s)\,c_{n-1}(E_{II})+(\xi_0-\xi)\} \tag{5.10}$$

with the bracketed term $\geqslant 0$ wherever $\log N^{-1}(s)>0$, i.e., wherever $N(s)<1$. Applied to $B^*(E)$, this formula therefore precisely proves Theorem I.

6. The relative Gauss Bonnet theorem

We already remarked in the introduction that the first main inequality of the Nevanlinna theory may be thought of as a twice integrated version of the formula (1.2) in Theorem I. The first integral of (1.2) leads to the generalized theorem of Gauss-Bonnet (for the complex case) and so serves to give a geometric interpretation of the Chern classes $c_i(E)$.

In this section we will, for the sake of completeness, briefly derive this development. The situation we wish to study is the following one: let E be a holomorphic n-bundle with a Hermitian norm N, over the complex n-manifold X with boundary $\partial X=Y$, and assume that s_y is a nonvanishing section of E over Y. The question now arises when s_y may be extended to all of X without vanishing, and Theorem I, in the explicit form given by (4.22) may be interpreted as giving an answer to this question.

Indeed, let $E_0 \subset E$, be the subset $\{e \mid N(e) > 0\}$ complementary to the zero-section in E, and let $\pi_0 : E_0 \to E$ be the projection. As we already remarked, the identity inclusion $E_0 \to E$ then induces a nonvanishing section s_I of $\pi_0^{-1}(E)$ over E_0, so that the formula (5.10) gives rise to a definite form ϱ over E_0, for which

$$\frac{dd^c}{4\pi}\varrho = \pi_0^* c_n(E).$$

At this stage we will actually only need the form

$$\eta(E) = \frac{d^c}{4\pi}\varrho,$$

for which we therefore clearly have the identity

$$\pi_0^* c_n(E) = d\eta(E). \tag{6.1}$$

In terms of this form, the answer to our question is given by the following proposition.

PROPOSITION 6.2. *The section s_y of $E_0 \mid Y$ may be extended to all of E_0 if and only if*

$$\int_X c_n(E) - \int_X s_y^* \eta(E) = 0.$$

The proof of this proposition follows directly from quite elementary obstruction theory, once it is established that the expression $\nu(X; Y; s_y) = \int_X c_n(E) - \int_Y s_y^* \eta(E)$, always measures the number of times any extension of s_y to X has to vanish. To be more precise we need to recall the topological definition of the order of vanishing of a section s of E at a point p which is an isolated zero of s. This is an integer, denoted by zero$(s; p)$, which is defined as follows:

Let B_ε be a disc of radius $\varepsilon > 0$ about p, relative to local coordinates centered at p. Also, let $\varphi_p : E \mid B_\delta \to E_p$, be a trivialization of E over B_δ, i.e., a retraction of $E \mid B_\delta$ onto E_p which is an isomorphism on each fiber. For small enough ε_0, the map $\varphi_p \circ s$ then maps ∂B_ε into $E_p - 0$, for all $0 < \varepsilon < \varepsilon_0 \leqslant \delta$. The degree of this map is by definition the number zero$(s; p)$:

$$\text{zero}(s; p) = \deg(\varphi_p \circ s) \quad \text{on } \partial B_\varepsilon. \tag{6.3}$$

(The orientation of B_ε is taken to be the one given by positive forms; similarly we orient the unit disc of E_p by the positive forms. There are canonical induced orientations on ∂B_ε, and the unit sphere, $S(E_p)$, in E_p. Using the retraction $E_p - 0 \to S(E_p)$ this class serves to orient $E_p - 0$, so that $\deg(\varphi_p \circ s)$ is well-defined.)

With this understood, (6.2) becomes an easy consequence of the following theorem.

PROPOSITION 6.4. *Let s be a smooth section of E with the following properties:*

α) $s \neq 0$ *on* $\partial X = Y$

β) s *has isolated zeroes only.*

Under these circumstances one has the formula:

$$\sum \operatorname{zero}(p; s) = \int_X c_n(E) - \int (s \mid Y)^* \eta(E) \tag{6.5}$$

where p ranges over the set of zeroes p_i, $i = 1, \ldots, m$ of s.

Proof. We first derive (6.5) from the following proposition:

PROPOSITION 6.6. *Let $j_p : S(E_p) \to E_0$ be the inclusion of the unit sphere of E_p, into E_0. Then,*

$$j_p^* \eta(E) = -\textit{the orientation class of } S(E_p). \tag{6.7}$$

Granted (6.7) we proceed as follows.

Let X_ε be obtained from X by removing the interiors of little discs B_ε^i of radius ε about p_i from X. Then there is a $\delta > 0$ so that s will not vanish on X_ε for $0 < \varepsilon \leqslant \delta$. Also choose trivializations φ_i of $E \mid B_\delta^i$. We then have

$$\int_{X_\varepsilon} c_n(E) = \int_{X_\varepsilon} s^* \pi_0^* c_n(E)$$

because s is a section of E_0 over X_ε. Now by Stokes formula it follows that

$$\int_{X_\varepsilon} c_n(E) = \int_{\partial X_\varepsilon} \eta(E) = \int_Y s^* \eta(E) - \sum \int_{\partial B_i^\varepsilon} s^* \eta(E). \tag{6.8}$$

Using the φ_i it is now clear that $\int_{\partial B_i^\varepsilon} s^* \eta(E)$ is approximately $\int_{\partial B_i^\varepsilon} \varphi_i^* j_{p_i}^* \eta(E)$ when ε is small. Hence by (6.7) $-\int_{\partial B_i^\varepsilon} s^* \eta(E) \to \operatorname{zero}\ (s, p_i)$. Thus (6.8) goes over directly into (6.5) as $\varepsilon \to 0$. Q.E.D.

It is Proposition 6.6 which therefore lies at the center of these formulae. To prove it one may explicitly integrate the form described by (5.10). Alternatively one may apply the argument we just gave in reverse, to a situation where (6.5) can be established by some other means. We will follow the second alternative because many of the concepts which are needed for this special example will also be used later. Note finally, that because of the functorial definition of $\eta(E)$ $j_p^* \eta(E)$ is a *well determined form on S_{2n-1} modulo only unitary transformations of that sphere.* In short, to prove (6.7) it will be sufficient to find an example of a Hermitian bundle E over the complex manifold X with $\partial X = \emptyset$ together with a section s of E, such that:

α) s has a single isolated zero at $p \in X$, with zero$(s, p) = 1$

β) $\int_X c_n(E) = 1$.

As we will now show, an example of this type is furnished by the complex projective space and the "quotient bundle" over it.

Let then V be a fixed complex vector space, of dim $(n+1)$, and let $P_1(V)$ be the projective space of 1-dimensional subspaces of V. (Note that $\dim_{\mathbf{C}} P_1(V) = n$.)

Over $P_1(V)$ we have the *canonical exact sequence*

$$0 \to S_1(V) \to T_1(V) \to Q_1(V) \to 0 \tag{6.9}$$

of holomorphic vector-bundles, defined in the following "tautologous" manner.

(6.10) $T_1(V)$ is the product bundle $P_1(V) \times V$ over $P_1(V)$

(6.11) $S_1(V)$ is the subset of $T(V)$ consisting of pairs (l, v)—where $l \in P_1(V)$ is a line in V, and $v \in V$—for which $v \in l$.

(6.12) $Q_1(V)$ is the quotient bundle $T_1(V)/S_1(V)$.

The bundle $Q_1(V)$ over $P_1(V)$ is the one we called the quotient bundle of the projective space $P_1(V)$. Note that each $v \in V$, determines a holomorphic section s_v of $Q_1(V)$, defined by the projection of the constant section: $x \to (x, v)$, $x \in P_1(V)$, of $T_1(V)$ into $Q_1(V)$. Clearly, if $v \neq 0$ then s_v vanishes at only one point $l \in P_1(V)$ namely at the subspace, $[v]$, generated by v. Further it is not hard to see that zero$[s_v, [v]] = 1$. Thus α) is satisfied.

To check the condition β) we need a hermitian structure on $T_1(V)$ which we of course take to be the trivial one induced by a hermitian structure on V. Thus the curvature form of $T_1(V)$ is equal to zero. Hence by the inequality (5.7) we note that $\Omega\{Q_1(V)\} \geqslant 0$. Thus in any case $\int_{Q_1(V)} c_n\{Q_1(V)\} \geqslant 0$. Hence it will be sufficient to show that $c_n\{Q_1(V)\}$ is an orientation class for $P_1(V)$ to establish β).

Consider the case $n = 1$, first. Let v_1, v_2 be an orthonormal base for V and let $z \to [v_1 + zv_2]$ be a local parameter near $[v_1]$. Also let s_1 be the section s_{v_1}, which is therefore holomorphic and $\neq 0$ on $P_1(V) - [v_1]$. Hence on this set

$$c_1\{Q(V)\} = \frac{dd^c}{4\pi} ln |s_1|^{-1},$$

where $|s_1|(l)$ is the norm of the section at l. Thus near $[v_1]$ we have,

$$|s_1|^2([v_1 + zv_2]) = |z|^2/(1 + |z|^2).$$

It follows, again by Stokes, that if $B_\varepsilon = (|z| < \varepsilon)$ then

$$\int_{P_1(V)} c_1\{Q(V)\} = \lim_{\varepsilon\to 0} -\int_{\partial B_\varepsilon} \frac{d^c}{4\pi}[ln|z|^2 + ln(1+|z|^2)].$$

Clearly the second term tends to zero, while the first tends to $+1$, as is seen directly if we write $z=re^{i\theta}$, $\log z = \log r + i\theta$ and recall that $d^c = id'' - id'$. Thus β) is true for $n=1$.

To get β) in general one may use the Whitney duality formula. In the present instance this formula yields:

$$c\{S_1(V)\}\cdot c\{Q_1(V)\} = c\{T_1(V)\} = 1.$$

Thus $c_n\{Q(V)\} = [-c_1\{S_1(V)\}]^n$. For $n=1$, this implies that $c_1\{S_1(V_2)\}$ is an orientation class of $P_1(V_2)$. Now under the inclusion $V_2 \to V$, $S_1(V)$ clearly restricts to $S_1(V_2)$. Hence $c_1 S_1(V)\}$ restricts to an orientation class of $P_1(V_2)$. But then $c_1\{S_1(V)\}$ must generate $H^2(X; Z)$, $X = P_1(V)$ and hence $(-1)^n c_1\{S_1(V)\}$ must be an orientation class for $P_1(V)$ in general. Q.E.D.

An important corollary of (6.5) is the following interpretation of $c_n(E)$:

COROLLARY 6.13. *Let E be a holomorphic n-bundle over the complex n-manifold X, and let $s: X \to E$ be a smooth section of E, which is $\neq 0$ on ∂X, and which is transversal to the zero section of X in E. Then* zero(s) *has a natural structure of a C^∞ manifold of real codimension $2n$ in X, and the proper orientation class of* zero(s) *is the Poincaré dual of $c_n(E)$.*

Proof. Let γ be a smooth singular n-cycle in the interior of X, which is transversal to zero(s), i.e., every singular simplex σ which intersects zero(s), meets it in an isolated interior point. Just as in the proof of (6.4) one now concludes from (6.7) that

$$\int_\sigma c_n(E) = \text{intersection } (\sigma, \text{zero}(s)) + \int_{\partial\sigma} s^*\eta(E).$$

Hence summing over σ in γ, we obtain:

$$\int_\gamma c_n(E) = \text{intersection } (\text{zero}(s), \gamma). \quad \text{Q.E.D.}$$

Remark I. It is of course artificial to bring in any assumption of complex analyticity when dealing with the Gauss-Bonnet theorem, and one could modify this account by defining η directly on any smooth hermitian bundle. However as we are primarily interested in the complex analytic case here and the more general approach would have taken us even further afield, we only discussed that case. In the next integration the analytic structure plays a vital role.

Remark II. There are two quite straightforward generalizations of the exact sequence

$$0 \to S_1(V) \to T_1(V) \to Q_1(V) \to 0$$

over $P_1(V)$, for which we will have use later on.

Namely, if $P_n(V)$ denotes the Grassmanian of n-dimensional subspaces of V, we have the corresponding sequence

$$0 \to S_n(V) \to T_n(V) \to Q_n(V) \to 0$$

over $P_n(V)$, with $T_n(V) = P_n(V) \times V$, and $S_n(V)$ being the subset of pairs (A, v) with $v \in A$.

Finally this construction makes sense when V is replaced by a vector bundle E over X. That is, one defines $P_n(E)$ as the pairs (A, x) consisting of a point $x \in X$, and an n-dimensional subspace A in E_x. One lets $T_n(E)$ be the bundle induced from E over $P_n(E)$ by the projection $P_n(E) \to X$, and then obtains an exact sequence

$$0 \to S_n(E) \to T_n(E) \to Q_n(E) \to 0 \quad \text{over } P_n(E)$$

where $S_n(E)$ consists of the triples (A, x, e) with $e \subset A$.

7. The second integration; definition of the order function

We are now in a position to discuss the generalized first inequality of the Nevanlinna theory. Just as in Section 6, we will be dealing with a holomorphic hermitian vector bundle E over the complex manifold X, however instead of assuming that X is compact we assume only that X admits a "concave exhaustion" f. By definition, such an exhaustion is a smooth real valued function, f, on X such that

(7.1) f maps X onto $\mathbf{R}^+$

(7.2) f is proper, that is, $f^{-1}(K)$ is compact whenever K is.

(7.3) The (1, 1) form $dd^c f$ is $\leqslant 0$ for large values of f.

With respect to such an exhaustion of X, one defines the *order-function* of E, by the formula

$$T(r) = \int_{-\infty}^{r} \left\{ \int_{X_r} c_n(E) \right\} dr; \quad X_r = \{x \mid f(x) \leqslant r\}. \tag{7.4}$$

The behavior of $T(r)$ as $r \to +\infty$ is then to be thought of as the analogue of $\int_X c_n(E)$ in the compact case.

One next defines a corresponding *order function* for the number of zeroes of a section s on E which is assumed to have only isolated zeroes, by the formula

$$N(r, s) = \int_{-\infty}^{r} \operatorname{zero}(s, X_r)\, dr \tag{7.5}$$

where $\operatorname{zero}(s, X_r) = \sum \operatorname{zero}(s, p)$, p ranging over the zeroes of s interior to X_r.

7 – 652932 *Acta mathematica* 114. Imprimé le 11 août 1965.

We note that if the integral along the boundary of X_r could be disregarded, the formula (6.5) would imply that $N(r, s) = T(r)$. This is of course false in general, however we do have the following estimate of this error term under certain circumstances.

FIRST MAIN THEOREM. *Let E be a positive Hermitian bundle over X where X has a concave exhaustion f. Let s be a holomorphic section of E with isolated zeroes, and let $N(r, s)$ be the order function of these zeroes. Then*

$$N(r, s) < T(r) + \text{constant} \tag{7.6}$$

where $T(r)$ is the order function of E.

In particular if $c_n(E) > 0$ at some point of X, then $\overline{\lim}\,\{N(r, s)/T(r)\} \leqslant 1$. Hence the deficiency measure of s, defined by: $\delta(s) = 1 - \lim\,\{N(r, s)/T(r)\}$ satisfies the inequality

$$0 \leqslant \delta(s) \leqslant 1. \tag{7.7}$$

Proof. Let $\Gamma \subset X \times \mathbf{R}$ be the graph of f, and let W be the region in $X \times \mathbf{R}$, which is "above" Γ and "below" the slice $X \times r$:

$$W = \{(x, t) \mid f(x) \leqslant t \leqslant r;\ x \in X,\ t \in \mathbf{R}\}.$$

The natural projection $W \to X_r$ will be denoted by σ.

It is then clear that
$$T(r) = \int_W \sigma^* c_n(E)\, dt$$

with W the orientation induced by the product orientation on $X \times \mathbf{R}$, and dt the volume element on $\mathbf{R}$.

Suppose now that $s \neq 0$ on X_r. Because $|s| < 1$ we may think of s as a section of $B^*(E)$ so that on X_r

$$c_n(E) = \frac{1}{4\pi} s^* dd^c \varrho$$

where $\varrho = \varrho(E)$ is the form given by Theorem I on $B^*(E)$.

We may therefore write $\sigma^*(c_n(E) \wedge dt)$ as $d\{\sigma^* s^* d^c \varrho \wedge dt\}/4\pi$ and apply Stokes' formula to obtain:

$$T(r) = \frac{1}{4\pi} \int_{\partial W} \sigma^* s^* d^c \varrho \wedge dt. \tag{7.8}$$

Now the boundary of W clearly falls into the top-face $X_r \times r$, and the bottom face Γ_r, which is the graph of $f \mid X_r$:

$$\partial W = (X_r \times r) \cup \Gamma_r.$$

Further, the integrand in (7.8) clearly restricts to zero on the top-face, as dt does. Hence, keeping track of the orientation we obtain $-1/4\pi\int_{\Gamma_r}\sigma^*s^*d^c\varrho\wedge dt$ for this integral, so that identifying Γ_r with X_r one obtains:

$$T(r)=\frac{1}{4\pi}\int_{X_r}-s^*d^c\varrho\wedge df\cdot \tag{7.9}$$

We next use the fact that s is holomorphic. This implies that $s^*d^c\varrho=d^cs^*\varrho$ and, furthermore, that $s^*\varrho\in A^{n-1,n-1}(X)$.

Now a direct verification shows that the following identity is valid:

Proposition 7.10. *If X is an n-dimensional complex manifold, and $f\in A^0(X)$, $\lambda\in A^{n-1,n-1}(X)$, then*

$$df\wedge d^c\lambda=d(d^cf\lambda)-\lambda dd^cf. \tag{7.11}$$

When this identity is substituted into (7.9) and the Stokes formula is used once more in the first term we obtain the relation:

$$T(r)=\frac{1}{4\pi}\int_{\partial X_r}d^cf\cdot\lambda-\frac{1}{4\pi}\int_{X_r}\lambda dd^cf,\quad \lambda=s^*\varrho(E) \tag{7.12}$$

and this is the basic integral relation which lies behind the first main theorem when s does not vanish on X_r.

In the case when s vanishes at isolated points p_i, $i=1,\ldots,m$, in X_r let X_r^ε be obtained from X_r by deleting ε discs $D_i(\varepsilon)$ about the p_i, and let $W(\varepsilon)$ be W with the solid cylinders $C_i(\varepsilon)$ above these discs removed. Now

$$T(r)=\frac{1}{4\pi}\lim_{\varepsilon\to 0}\int_{W(\varepsilon)}\sigma^*s^*dd^c\varrho\wedge dt,$$

and on $W(\varepsilon)$ we may apply our earlier argument. However this time $\partial W(\varepsilon)$ also contains the boundaries of the cylinders $C_i(\varepsilon)$, contributing the extra term

$$\frac{1}{4\pi}\sum\int_{\partial C_i(\varepsilon)}\sigma^*s^*d^c\varrho\wedge dt,$$

which by (6.5) is easily seen to tend to $N(r,s)$ as $\varepsilon\to 0$. Hence (7.12) is modified to:

$$T(r)-N(r,s)=\frac{1}{4\pi}\lim_{\varepsilon\to 0}\left[\int_{\partial X_r^\varepsilon}d^cf\lambda-\int_{X_r^\varepsilon}\lambda dd^cf\right];\quad \lambda=s^*\varrho(E). \tag{7.13}$$

We next apply the following lemma which will be proved later by an estimate.

LEMMA. *In the situation just described;*

(7.14) $\lim_{\varepsilon\to 0}\int_{\partial D(\varepsilon)} |d^c f\wedge\lambda| = 0, \quad \lambda = s^*\varrho(E).$

(7.15) *The form $\lambda dd^c f$ is absolutely integrable on X_r.*

(7.16) *The form $d^c f\wedge\lambda$ is absolutely integrable on ∂X_r.*

In view of this good state of affairs we may pass to the limit in (7.13) to obtain the fundamental integral formula:

$$T(r) - N(r,s) = \frac{1}{4\pi}\int_{\partial X_r} d^c f\wedge\lambda - \frac{1}{4\pi}\int_{X_r}\lambda dd^c f, \quad \lambda = s^*\varrho(E). \tag{7.17}$$

The inequality of the first main theorem now follows directly. Indeed, by Theorem I, $\lambda \geqslant 0$ on X_r. By assumption $dd^c f \leqslant 0$ on the complement of some X_{r_0}. Hence $-\lambda dd^c f \geqslant 0$ there, and so the second term on the right hand side is greater than some constant.

The term $\int_{\partial X_r} d^c f\wedge\lambda$ is actually *non-negative*, for the following reason. Recall first that X_r was oriented by the positive (n,n)-forms on X. Recall also that the orientation induced by the Stokes formula $\int_{X_r} d\omega = \int_{\partial X_r} i^*\omega$, on ∂X_r is characterized by the condition:

A real $(2n-1)$ form ξ on X restricts to a positive form on ∂X_r, relative to the induced orientation, if and only if $df\wedge\xi$ is positive on X, near X_r.

Hence the sign of $\int_{\partial X_r} d^c f\wedge\lambda$ is determined by the sign of $df\wedge df^c\wedge\lambda$ on X_r. But if f is any real valued function, then $df\wedge d^c f\wedge\lambda$ is also positive. Q.E.D.

The inequality now follows as we have proved that $T(r) - N(r,s) > \text{constant}$.

Proof of the lemma. We need to estimate the form $\lambda = s^*\varrho$ near an isolated singularity, p, of s. For this purpose choose a holomorphic trivialization $\varphi: E\to E_p$, of E near p. Then $s^*\varrho$ will be close to $(\varphi\circ s)^* j_p^*\varrho$ near p, so that it is sufficient to study this form near p. Our first task is therefore to describe $j_p^*\varrho$.

Let $\pi: E_p\to p$, and set $\boldsymbol{E} = \pi^{-1}(E_p)$ be the induced bundle over E_p. The identity map $E_p\to E_p$, then defines a section s of $\boldsymbol{E}$, which does not vanish on $E_{p,0} = E_p - 0$, and so generates a sub-bundle $\boldsymbol{E}_I$ of $\boldsymbol{E}$ there. Let $j_p: E_{p,0}\to \boldsymbol{E}$ be the inclusion. The form $j_p^*\varrho$ is then made out of the curvature forms of $\boldsymbol{E}_I$ and $\boldsymbol{E}_{II} = \boldsymbol{E}/\boldsymbol{E}_I$, according to the prescription (5.10). Now as $\boldsymbol{E}$ is clearly the trivial Hermitian bundle over $E_p - 0$, the curvature of $\boldsymbol{E}$ vanishes identically. Hence $K(\boldsymbol{E}_{II})$ has the form $\delta\wedge\delta^*$, where δ is the degree zero operator $P_{II}DP_I$ of Section 4, and may be computed explicitly. Indeed let u_α, $\alpha = 1, \ldots, n$, be an orthonormal frame for E_p, and let z_α be the corresponding local coordinates on E_p so that

$$\sum z_\alpha(q)\, u_\alpha = q, \quad q\in E_p,$$

and let $r(q)=(\sum|z_\alpha(q)|^2)^{\frac{1}{2}}$. If we interpret the u_α as the constant sections of E then the identity section s is given by $s(q)=\sum z_\alpha(q)\,u_\alpha$, and so $Ds(q)=\sum dz_\alpha u_\alpha$. It follows that at a point q, with $z_1(q)=r(q)$, $z_\beta(q)=0$, $\beta=2,\dots,n$, the curvature matrix relative to the frame of E_{II} determined by the u_β, $\beta=2,\dots,n$, is simply given by

$$\frac{1}{r^2}dz_\alpha\wedge d\bar z_\beta,\quad \alpha,\beta=2,\dots,n. \tag{7.18}$$

In particular then,

$$c_{n-1}(E_{II})=\frac{\text{const}}{r^{2(n-1)}}dz_2\wedge d\bar z_2\wedge\dots\wedge dz_n\wedge d\bar z_n.$$

With the aid of (7.18) one may estimate all the terms of (5.10) and so conclude that:

$$j_p^*\varrho=r^{-2(n-1)}\log r\cdot\omega_1+\sum_{i\geq 2}r^{-2n-i}\,\omega_i \tag{7.19}$$

where ω_i is bounded on all of $E_{p,0}$.

The lemma now follows easily from (7.19).

Assume first that s is transversal to the zero section at p. Then the Jacobian of $\varphi\circ s$ is not zero at p. For our convergence questions $\varphi\circ s$ may therefore be replaced by the identity map. Now let $D(\varepsilon)$ be the ball of radius ε about 0 in $\mathbf{C}_n$. Then if ϱ is of the type given by (7.19) we clearly have

$$\int_{\partial D(\varepsilon)}\varrho\wedge\theta\to 0,\quad\text{and}\quad\int_{D(\varepsilon)}\varrho\wedge\theta\wedge\bar\varphi\to 0$$

for any bounded 1-forms θ and φ because the volume of the sphere of radius r is of the order r^{2n-1} and so dominates $r^{2(n-1)}\log r$. The lemma therefore is clear in that case. For a general *isolated* zero of s, there exist arbitrarily small perturbations of s with only a finite number of nondegenerate zeroes near p. Hence our lemma also holds in that case.

8. Equidistribution in measure

In this section we derive the generalized first equidistibution theorem from the first main theorem with the aid of two essentially known but hard to refer to propositions which are then taken up in later sections.

We start with a statement of the theorem we are after:

EQUIDISTRIBUTION THEOREM. *Let E be a complex vector bundle of fiber-dimension n, over the complex connected manifold X, and let $V\subset\Gamma(E)$ be a finite dimensional space of holomorphic sections of E. Assume further that,*

(8.1) *X admits a concave exhaustion f, in the sense of Section 7.*

(8.2) *V is sufficiently ample in the sense that:*

α) *The map $s \to s(x)$, maps V onto E_x for each $x \in X$.*

β) *There is some $s \in V$, and some $x_0 \in X$, so that $s: X \to E$, is transversal to the zero-section of E at x_0.*

Under these circumstances nearly every section in V vanishes the same number of times. Precisely, a hermitian structure on V defines a hermitian structure on E, and hence a deficiency measure $\delta(s)$ on the generic sections of V. The assertion is that except for a set of measure 0, $\delta(s) = 0$.

Proof. We first of all remark more explicitly on how the hermitian structure on V defines $T(r)$, $N(r, s)$ etc.

For this purpose let $m = \dim V - n$, and consider the exact sequence

$$0 \to S_m(V) \to T_m(V) \to Q_m(V) \to 0 \quad \text{over } P_m(V).^{(1)} \tag{8.3}$$

By (8.2) part α, the map $\varepsilon_x : V \to E_x$ which sends s into $s(X)$ is onto. Hence, k_x, the kernel of ε_x has dim m. Now it is clear from (8.3) that the induced map $e_V : X \to P_m(V)$, defined by $x \to k_x$, determines an isomorphism of $Q_m(V)$ with E: That is

$$e_V^{-1}\{Q_m(V)\} \simeq E. \tag{8.4}$$

A hermitian structure on V induces one on $T_m(V)$ and hence on $Q_m(V)$ and $S_m(V)$ and hence by (8.4) also on E.

Note further that $Q_m(V)$ is positive in this structure as $T_m(V)$ clearly has zero curvature and "quotient bundles are always more positive" (see Section 4). Hence E is also positive. Finally, the "height of a section s" in V at any point $x \in X$ is clearly bounded by the length of s "qua element" in V. In short we may, after possibly multiplying s by a suitable constant, not only apply the notions of Section 7 to E, but we also obtain the inequality of the first main theorem:

$$N(r, s) < T(r) + \text{constant}$$

valid for sections with isolated signularities.

Now condition β, of (8.2) is seen to imply by an explicit check, that $e_V^* c_n\{Q_n(V)\}$ is *strictly* positive near x_0 (see remark at end of Section 9). Hence $T(r) \to +\infty$, so that (8.4) implies the inequality:

$$0 \leqslant \overline{\lim} \frac{N(r, s)}{T(r)} \leqslant 1. \tag{8.5}$$

We now need the following two propositions:

PROPOSITION 8.6. *Under the assumption* (8.2) *nearly all $s \subset V$, have only isolated zeroes. In fact nearly all sections $s \in V$ are* transversal *to the zero section of E.*

(1) See the remark at the end of Section 6.

Proposition 8.7. *Under the assumptions* (8.2) *we have the equality*

$$\int_{P_1(V)} N(r, s)\,\omega = T(r), \quad s \in [s] \in P_1(V) \tag{8.8}$$

where ω is the volume on $P_1(V)$ invariant under the group of isometries of V and normalized by $\int_{P_1(V)}\omega = 1$; while $N(r, s)$ is the order function interpreted as a function on $P_1(V)$.

The equidistribution theorem: $\delta(s) = 0$ almost everywhere now follows directly. Indeed by (8.5) $0 \leqslant \delta(s) \leqslant 1$. Hence $\int_{P_1(V)} \delta(s)\omega \geqslant 0$. On the other hand by (8.7) we have

$$\overline{\lim} \int_{P_1(V)} \{N(r, s)/T(r)\}\,\omega = 1, \quad \text{whence} \quad \int_{P_1(V)} \overline{\lim}\, \{N(r, s)/T(r)\}\,\omega \geqslant 1,$$

and so finally
$$\int_{P_1(V)} \delta(s)\,\omega \geqslant 0. \quad \text{Q.E.D.}$$

9. The proof of Proposition 8.6

This assertion is clearly a variant of Bertini's theorem, and is proved along the same lines. Briefly the argument runs as follows.

Let $K = e_V^{-1}\{S_m(V)\}$, and consider the associated projective bundle $P_1(K)$ over X. (See the remark at the end of Section 6 for these concepts.) There is a natural imbedding of $P_1(K)$ in $X \times P_1(V)$ as the subset:

$$P_1(K) = \{(x, l) \text{ with } l \subset k_x\}$$

and we let $\pi : P_1(K) \to P_1(V)$ be the projection on the second factor. Next let $\Sigma \subset P_1(K)$ be the subset of those pairs (x, l) for which l is generated by a section $s : X \to E$, which is not transversal to the zero section at x.

This is the singular set in $P_1(K)$, and it is clear that the complement in $P_1(V)$ of the image of Σ under π consists of transversal sections. Now $\dim P_1(K) = \dim P_1(V)$ because $\dim X$ equals the fiber-dimension of E. Hence if we can show that the codimension of Σ in $P_1(K)$ is $\geqslant 1$, then Σ and $\pi(\Sigma)$ will have measure zero and the proposition will be established.

Our aim is therefore to show that Σ is the zero set of a not identically zero section of a certain line-bundle over $P_1(K)$. To see this, remark first that if $s \in k_x$ and if $U \in X_x$ is a tangent vector to X at x, then the derivative of s in the direction U, is a well determined element $U \cdot s$ of E_x. (Recall that $s \in k_x \Leftrightarrow s(x) = 0$; to differentiate general sections one of course needs a connection, however at the zeroes of s all connections on E define the same derivative.)

This operation therefore leads to a map

$$J^t: k_x \to \mathrm{Hom}(X_x, E_x)$$

and it is easy to see that $\Sigma = \{(x, l) \in P_1(K) \mid \det J^t(s) = 0, [s] = l\}$.

Now when lifted to $P_1(K)$, $\det J^t$ may be interpreted as a section of the line bundle

$$L = \mathrm{Hom}\,\{S(K), \mathrm{Hom}\,(\Lambda^n T, \Lambda^n E\}$$

where Λ^n denotes the nth exterior power and T denotes the tangent bundle of X lifted to $P_1(K)$. Thus Σ = zero set of $\det J^t \in \Gamma(L)$. On the other hand condition (8.2) β) demands precisely, that $\det J^t$ be non-zero at some point of $P_1(K)$. Because X is connected it follows that $\operatorname{codim} \Sigma = 1$. Q.E.D.

Remark. The transpose of J^t is given by

$$J: X_x \to \mathrm{Hom}\,(k_x, E_x)$$

and may be identified with the Jacobian of e_V at x. Thus condition (8.2) β) implies that e_V is an immersion near x_0. From this it follows easily that $c_n(E) > 0$ near x_0.

10. Some remarks on integral geometry. The proof of Proposition 8.7

Suppose $\pi: Y \to X$ is a smooth fibering of compact manifolds with oriented fiber F. In that situation there is a well-defined operation

$$\pi_*: A^k(Y) \to A^{k-f}(X), \quad f = \dim F$$

called integration over the fiber, which "realizes" the adjoint of π^* in the sense that if X and Y are oriented compatibly then for any $\varphi \in A(X)$, $\psi \in A(Y)$ we have the identity:

$$\int_Y \psi \pi^* \varphi = \int_X (\pi_* \psi) \cdot \varphi. \tag{10.1}$$

The existence of π_* on the "form level", suggests the following definition.

DEFINITION 10.2. *Let $\varphi \in A^k(X)$. By an integral representation of φ we mean a triple, (Y, Z, ω) where $Y \xrightarrow{\pi} X$ is an oriented fibering over X, with projection π, and $\omega \in A^m(Z)$ is a volume element* (1) *on the oriented m-manifold Z, together with a map $\sigma: Y \to Z$, such that*

$$\varphi = \pi_* \circ \sigma^* \omega. \tag{10.3}$$

In general the question whether a given closed form φ on X admits an integral repre-

(1) Volume element means a nonvanishing form of top dimension, in the orientation class.

sentation seems quite difficult. Certainly φ must have integral periods and there are most probably much more subtle conditions which also have to be satisfied. For our purposes it will however be sufficient to show that the characteristic class $c_n(E)$ of a hermitian bundle which is ample in the sense of (9.2) α) *always* has an *integral representation.* Note that if φ has an integral representation, then any pull-back $f^*\varphi$ also has an integral representation. Hence it will be sufficient to get a representation theorem for $c_n[Q_n(V)]$ over $P_n(V)$.

In the next proposition we describe a quite general representation theorem for the Grassmann-varieties $P_n(V)$. We will first simplify the notation as follows: V will denote a fixed hermitian vector space of dimension d; and we write simply P_n, Q_n etc., for $P_n(V)$, $Q_n(V)$ etc. The bundle Q_n is always considered in the *hermitian structure induced on Q_n by the trivial structure on T_n*; so that the Chern forms $c(Q_n)$ are well-defined.

Now let $0<n<m<d$ be two integers and define $P_{n,m}=P_{n,m}(V)$ as the "flagmanifold" of pairs $(A_n\subset B_m)$ of subspaces of dimension n and m respectively in V. Let $P_{n,m}\xrightarrow{\sigma}P_n$, and $P_{n,m}\xrightarrow{\pi}P_m$ be the natural projections, $\sigma(A,B)=A$; $\pi(A,B)=B$, and consider the diagram:

$$\begin{array}{ccc} P_{n,m} & \xrightarrow{\sigma} & P_n \\ {\scriptstyle\pi}\downarrow & & \\ P_m & & \end{array} \tag{10.4}$$

We then have the following proposition.

Representation theorem. *In the diagram* (10.4) *one has the relation:*

$$c_{d-m}\{Q_m\}=\pi_*\cdot\sigma^*c_{d-n}\{Q_n\}. \tag{10.5}$$

Proof. One may of course compute everything explicitly in these examples and so verify (10.5). There is also a much simpler global proof based on the corollary (6.13). The argument runs as follows.

Consider the action of the group of isometries of V, say $I(V)$, on P_n. From the rather canonical definition of the bundles Q_n it is then not hard to see that their Chern forms $c\{Q_n\}$, are *invariant* under $I(V)$.

It is also easy to see that P_n is a symmetric space of $I(V)$ whence every real cohomology class of P_n contains a *single* invariant differential form. We may therefore prove (10.5) by checking it on the cohomology level. Alternately we may pass to homology by Poincare duality. Then σ^* corresponds simply to inverse image of the dual cycle to $c_{v-n}\{Q_n\}$ and π_* corresponds to projection.

Now let v be a non-zero element of V, and consider the section s_v it determines in Q_n over P_n. This is clearly a transversal section and we have: $\text{zero}(s_v) = \{A_n \supset v\}$. Similarly v determines the section s_v' of Q_m over P_m, and, $\text{zero}(s_v') = (B_m \supset v)$. Thus we get the formula:

$$\text{zero}(s_v') = \pi \circ \sigma^{-1} \circ \text{zero}(s_v). \tag{10.6}$$

Finally, by Proposition 6.13 these zero-sets are duals of the corresponding Chern classes, so that (10.6) proves (10.5) on the homology level. (Because our sections are holomorphic there is no problem with orientations.)

We discuss next, the geometric implications of an integral representation. Consider then the diagram:

$$\begin{array}{ccc} Y & \xrightarrow{\sigma} & Z \\ \downarrow{\scriptstyle\pi} & & \\ X & & \end{array} \tag{10.7}$$

with $\varphi = \pi_* \circ \sigma^* \omega$. If z is not in the critical set σ (a point p is critical if at some point of $\sigma^{-1}(p)$, the differential $d\sigma$ is not onto its tangent space Z_p). Then $\sigma^{-1}(z)$ is a well-defined oriented manifold, so that the pair $(\sigma^{-1}(z); \pi)$ determines a smooth oriented singular submanifold X which we denote by $c(z)$. As already remarked, each $c(z)$ represents the homology class dual to φ. However, we have more than that; the family $c(z)$ determines not only the dual homology class of φ but also the value of φ on any singular k-submanifolds $f: K \to X$.

PROPOSITION 10.8. *Let the k-form $\varphi \in A^k(X)$ have an integral representation ω on Z, in the sense of* (10.1). *Then for any compact singular submanifold $f: K \to X$ of dimension k; one has:*

(10.9) *The intersection $n(K, c(z))$ of K with $c(z)$ is well-defined except for a set of measure zero in Z.*

$$\int_K f^* \varphi = \int_Z n(K, c(z))\, \omega. \tag{10.10}$$

Outline of Proof. Let $Y(f) = f^{-1}(Y)$ be the bundle induced by Y over K, under f, and let $f': Y(f) \to Y$ be the bundle map covering f. If $\pi_K: Y(f) \to K$ is the projection we have:

$$\int_K f^* \varphi = \int_K \pi_{K*} f'^* \circ \sigma^* \cdot \omega = \int_{Y(f)} (\sigma \circ f')^* \omega \tag{10.11}$$

where the first step follows from the identity $f^* \circ \pi^* = \pi_K^* \circ f'^*$ and the second one from the adjoint property (10.1) of π_*.

One next considers the map $\lambda = \sigma \circ f': Y(f) \to Z$. A count of dimension shows that dim $Y(f) = \dim Z$. Hence on the complement of the critical value set of λ the degree of λ at z, is well-determined and computes the algebraic number of sheets with which some vicinity of z is covered. Thus

$$\int_{Y(f)} \lambda^* \omega = \int_Z \deg(\lambda; z)\, \omega.$$

Finally consider the points of $\lambda^{-1}(z)$, with z a regular (i.e., not critical) value of λ. We see first of all that these points correspond precisely to the intersections of $c(z)$ with $f(K)$ and furthermore that all these intersections are transversal so that the intersection numbers are well-defined and their algebraic sum, is precisely $\deg(\lambda, z)$. The theorem now follows from the fact that the critical set of smooth maps have measure zero.

The proposition (8.7) which motivated this excursion is a direct consequence of the formulae (10.5) and (10.10). Indeed, let $r = d - n$, and consider the exact sequence

$$0 \to S_r(V) \to T_r(V) \to Q_r(V) \to 0 \quad \text{over } P_r(V).$$

Let $e_V : X \to P_r(V)$ be the evaluation map, so that $e_V^{-1}[Q_r(V)] = E$. We now apply (10.5) with $n = 1$, and $m = r$. Thus the diagram we need is

$$\begin{array}{ccc} P_{1,r} & \longrightarrow & P_1(V) \\ \downarrow & & \\ P_r(V) & & \end{array}$$

Applying (10.5) one obtains $c_n[Q_r] = \pi_* \circ \varphi^* c_{d-1}\{Q_1\}$ and it is clear that $c_{d-1}(Q_1)$ is a volume of measure 1 on $P_r(V)$—because s_v vanishes at a single point for instance! Now one applies (10.8) with X_r replacing K, and e_V replacing f, to obtain:

$$\int_X c_n(E) = \int_{P_1(V)} n(r, s)\, \omega$$

with $n(r, s)$ simply the number of zeroes of $s \in V$ on X_r (s a generic section). Integrating with respect to r, we conclude that

$$T(r) = \int_{P_1(V)} N(r, s)\, \omega. \quad \text{Q.E.D.}$$

11. The Nevanlinna Theorem

We conclude this paper with a short account of the classical Nevanlinna theorem. In particular we would like to show that the second main theorem of the Nevanlinna theory is also a consequence of the integral formula which yields the first main inequality.

We therefore specialize all our constructions as follows:

(11.1) For X we take the plane $\mathbf{C}$, with an exhaustion, $f(z)$, for which $f(z)=\log|z|$ when $|z|\geqslant 1$.

(11.2) E is the trivial line bundle over X, so that $\Gamma(E)$ is the space of holomorphic functions on $\mathbf{C}$.

(11.3) $V\subset\Gamma(E)$ is a 2-dimensional sufficiently ample subspace of $\Gamma(E)$, i.e., one generated by two functions s_1 and $s_2\in\Gamma(E)$ which are not proportional, and which do not have any common zeroes.

Remarks. α) The assumption that E is the trivial bundle is really no restriction as all holomorphic bundles over $\mathbf{C}$ are known to be trivial.

β) The function $\log|z|$, is harmonic for large $|z|$ and therefore has the property $dd^cf=0$. Hence f does define a "concave" exhaustion. On the other hand, the function $|z|^2$ would *not* do, because

$$dd^c|z|^2=2idz\wedge d\bar{z}\geqslant 0.$$

γ) By (11.3) every $s\in V$ is of the form as_1+bs_2. Hence the zeroes of s correspond to the points where $r(z)=s_1(z)/s_2(z)=-b/a$. In short we are dealing precisely with the value distribution of the meromorphic function $r(z)$.

The refinement of the equidistribution theory which is possible in this situation is in the first place a consequence of the fact that our exhaustion function is harmonic, so that $dd^cf=0$ for large values of $|z|$. It follows that (7.16) specializes to the formula:

$$T(r)-N(r,s)=\frac{1}{4\pi}\int_{\partial X_r}\log\{1/|s|^2\}\cdot d^cf+\text{const.} \tag{11.4}$$

Indeed, in the case of line bundles, the form ϱ of Theorem I may simply be taken to be $\log(1/N(s))$. The formula (11.4) is furthermore valid for all $s\in V-0$, because all these sections vanish at isolated points.

Consider then a set of q sections $s_i\in V$, *no two of which are dependent*. Our aim is the Nevanlinna inequality:

$$\sum_{i=1}^{q}\delta(s_i)\leqslant 2 \tag{11.5}$$

where

$$\delta(s_i)=1-\overline{\lim}\,\frac{N(r,s_i)}{T(r)}.$$

For this purpose observe first of all that a repeated application of (11.4) yields,

$$q\,T(r)-\sum N(r,s_i)=\frac{1}{4\pi}\int_{\partial X_r}\log\mu^2\cdot d^cf+\text{const} \tag{11.6}$$

with
$$\mu^2 = \{\prod_{1\leq i_q} |s_i|^2\}^{-1}. \tag{11.7}$$

Hence if $M(r)$ denotes the term $1/4\pi \int_{\partial X_r} \log \mu^2 d^c f$, then we need to prove the inequality:
$$\underline{\lim} \frac{M(r)}{T(r)} \leq 2 \tag{11.8}$$
to establish (11.5).

The estimation of $M(r)$ proceeds by first *correcting* $M(r)$ for the *singular* points of our evaluation map
$$e_V : X \to P_1(V). \tag{11.9}$$

Let $\tau(X)$ and $\tau[P_1(V)]$ be the respective holomorphic tangent bundles of X and $P_1(V)$, so that de_V becomes a section of the line bundle:
$$\text{Hom}\,[\tau(X), e_V^{-1} \cdot \tau[P_1(V)]]. \tag{11.10}$$

We consider the global section $\partial/\partial z$ of X, and set $t = de_V(\partial/\partial z) = de_V/dz$. Then t is a holomorphic section of $L = e_V^{-1}\tau[P_1(V)]$ and the singular points of e_V are precisely the zeroes of t.

Now the fixed Hermitian structure on V which underlies all our constructions, induces a Hermitian structure on $\tau\{P_1(V)\}$ through the well known isomorphism:
$$\tau[P_n(V)] = \text{Hom}\,[S_n(V), Q_n(V)]. \tag{11.11}$$

One may therefore apply the first main integral formula to the section, t, of $L = e_V^{-1} \circ \tau[P_1(V)]$, and so obtains:
$$\int_{-\infty}^{r} dr \int_{X_r} c_1(L) - N(r, t) = \frac{1}{4\pi} \int_{\partial X_r} \log 1/|t|^2 \cdot d^c f + \text{const}. \tag{11.12}$$

Once one identifies the first integral in (11.12) with $2T(r)$, this formula becomes the so-called *second* main theorem of the Nevanlinna theory. Actually, that identification follows directly from the following quite general proposition:

PROPOSITION 11.13. *In the natural Hermitian structure on $\tau\{P_n(V)\}$ induced by* (11.11), *one has the identity of Chern forms:*
$$c_1\{\tau[P_n(V)]\} = \dim V \cdot c_1[Q_n(V)]. \tag{11.14}$$

We bring only the proof for $n = 1$, the general case being similar but involving some complicated identities about the determinant. Let then n be 1, and consider the sequence (see Section 6) of bundles:
$$0 \to S_1 \to T_1 \to Q_1 \to 0 \tag{11.15}$$

over $P_1(V)$. Taking $\mathrm{Hom}(S, \cdot)$ of this sequence we get

$$0 \to 1 \to \mathrm{Hom}(S_1, T_1) \to \tau\{P_1(V)\} \to 0. \tag{11.16}$$

Now we again use the Whitney formula on the homology level, and invariance under the isometries of V to deduce a relation on the form-level. Namely, from (11.6)

$$c_1\{\mathrm{Hom}(S_1, T_1)\} = c_1\{\tau[P_1(V)]\},$$

while from (11.15), $c_1(S_1) = -c_1(Q_1)$ and hence $c_1\{\mathrm{Hom}(S_1, T_1)\} = \dim V \cdot c_1(Q_1)$ as was to be shown.

The formula (11.12) may therefore be used to give the following estimate:

$$2T(r) \geqslant \frac{1}{4\pi}\int_{\partial X_r} \log 1/|t|^2 d^c f + \text{const} \tag{11.17}$$

and this is now precisely the second fundamental inequality of the subject.

The proof of (11.8) now proceeds as follows:

Choose $0 < \lambda < 1$ and write:

$$\lambda M(r) = \frac{1}{4\pi}\int_{\partial X_r} \log \{\mu^{2\lambda}|t|^2\} d^c f + \frac{1}{4\pi}\int_{\partial X_r} \log \{1/|t|^2\} d^c f.$$

Then if the two expressions on the right are denoted by $A(r)$ and $B(r)$ respectively, we get

$$\lambda \varliminf M(r)/T(r) \leqslant \varliminf A(r)/T(r) + \varlimsup B(r)/T(r),$$

whence by the second main theorem,

$$\lambda \varliminf M(r)/T(r) \leqslant \varliminf A(r)/T(r) + 2.$$

Hence if it can be shown that $\varliminf A(r)/T(r) = 0$ for every $0 < \lambda < 1$ we will be done.

For this estimate, one first uses the concavity of the logarithm: Namely, if $z = re^{i\theta}$ are the usual coordinates in C, then for r large enough; $r \geqslant r_0$; we have $f = lnr$, whence $d^c f = d\theta$ and $df = dr/r$. In particular, $\frac{1}{2}\pi \int_{\partial X_c} d^c f = 1$. It follows from the concavity of log that for $c > lnr_0$

$$A(c) \leqslant \log \frac{1}{2\pi}\int_{\partial X_c} |u|^{2\lambda}|t|^2 d\theta$$

or equivalently:

$$e^{A(c)} \leqslant \frac{1}{2\pi}\int_{\partial X_c} u^{2\lambda}|t|^2 d\theta. \tag{11.18}$$

This last relation is now exploited to construct an integral inequality which in some sense relates $e^{A(c)}$ with $T(c)$. For this purpose one needs the following identity, which relates $|t|^2$ to the Chern-form $c_1(E)$ which occurs in the definition of $T(c)$. The relation in question is the following one:

$$c_1(E) = \frac{1}{\pi} \cdot |t|^2 r\, dr\, d\theta. \tag{11.19}$$

To see this formula, consider a point $x \in C$, and choose a section s_1 in V, with $|s_1| = 1$, and $s_1(x) = 0$. Thus s_1 spans k_x and is of unit length. Choose s_2 to be orthogonal to s_1 and also of unit length. Then near x there is a well determined holomorphic function $\alpha(z)$ such that $s_1(z) - \alpha(z) s_2(z)$ generates k_1 for z near x. It follows immediately from the formulae of Section 5 that in terms of this α,

$$c_1(E) = \frac{i}{2\pi} \cdot \left| \frac{d\alpha}{dz} \right|^2 dz \wedge d\bar{z}. \tag{11.20}$$

Finally noting that $|d\alpha/dz|^2 = |t|^2$, and that $i\, dz \wedge d\bar{z} = 2r\, dr\, d\theta$ one obtains (11.19).

Integrating with respect to c, one now deduces from (11.18) that

$$\int_{c_0}^{c} e^{[2c + A(c)]} dc \leqslant \frac{1}{2} \int_{X_c - X_{c_0}} c_1(E) \mu^{2\lambda}$$

and so finally that

$$\int_{c_0}^{c} dc \int_{c_0}^{c} e^{[2c + A(c)]} dc \leqslant \frac{1}{2} \int_{c_0}^{c} dc \int_{X_c - X_{c_0}} c_1(E) \mu^{2\lambda}. \tag{11.21}$$

The concluding steps of the proof are now expressed by the following two lemmas:

Lemma 11.22. *In the notation used above,*

$$\int_{c_0}^{c} dc \int_{X_c - X_{c_0}} c_1(E) \mu^{2\lambda} < K_1 T(c) + K_2$$

where the K_i are constants.

Lemma 11.23. *The integral inequality*

$$\int_{c_0}^{c} dc \int_{c_0}^{c} e^{[2c + A(c)]} dc \leqslant K_1 T(c) + K_2$$

implies that the inequality

$$2c + A(c) \leqslant k^2 \log [K_1\{T(c)\} + K_2], \quad k > 1,$$

hold for arbitrarily large values of c.

We can clearly conclude from (11.21) and these two lemmas that $\underline{\lim}\, A(c)/T(c) = 0$, so that the Nevanlinna theorem is a direct consequence of (11.22) and (11.23).

Both Lemmas are well-known, see for instance [2]. The first one follows from an integral geometry argument, while the second one is a purely real variable inequality.

References

This paper links classical function theory with differential geometry; it is difficult to give an adequate bibliography satisfactory to readers from both fields. We will restrict ourselves in giving some standard literature from which other references can be found:

[1]. Classical value distribution theory:

NEVANLINNA, R., *Eindeutige analytische Funktionen*. Berlin, 1936.

[2]. Holomorphic curves in projective space:

AHLFORS, L., The theory of meromorphic curves. *Acta Soc. Sci. Fenn.*, Ser. A. 3, no. 4 (1941).

WEYL, H., *Meromorphic Functions and Analytic Curves*. Princeton, 1943.

[3]. Differential geometry of connections:

KOBAYASHI, S. & NOMIZU, K., *Foundations of differential geometry*. Interscience, 1963.

[4]. Characteristic classes:

HIRZEBRUCH, F., *Neue topologische Methoden in der algebraischen Geometrie*. Springer 1962.

[5]. Curvature of connections and characteristic classes:

CHERN, S., Differential geometry of fiber bundles. *Proc. International Congress* 1950, 2 (1952) 397–411.

GRIFFITHS, P. A., On a theorem of Chern. *Illinois J. Math.*, 6, 468–479 (1962).

Received August 17, 1964

Reprinted from
Acta. Math. **114** (1965) 71–112

List of Ph.D. Theses Written Under the Supervision of S.S. Chern

I. *At the University of Chicago*

1. Nomizu, Katsumi, Invariant affine connections on homogeneous spaces. June 1953
2. Auslander, Louis, Contribution to the curvature theory of Finsler spaces. June 1954
3. Liao, San Dao, On the theory of obstructions of fiber bundles. March 1955
4. Spanier, Jerome, Contributions to the theory of almost complex manifolds. September 1955
5. Rodrigues, Alexandre, Characteristic classes of homogeneous spaces. March 1957
6. Hertzig, David, On simple algebraic groups. August 1957
7. Levine, Harold I., Contributions to the theory of analytic maps of complex manifolds into projective space. August 1957
8. Suzuki, Haruo, On the realization of Stiefel–Whitney characteristic classes by submanifolds. August 1957
9. Wolf, Joseph Albert, On the manifolds covered by a given compact, connected Riemannian homogeneous manifold. December 1959
10. Petridis, Nicholas C., Quasiconformal mapping and pseudo-meromorphic curves. June 1961.

II. *At the University of California at Berkeley*

1. Pohl, William Francis, Differential geometry of higher order. September 1961
2. Do Carmo, Manfredo Perdigao, The cohomology ring of certain Kählerian manifolds. January 1963
3. Amaral, Leo Huet. Hypersurfaces in non-Euclidean spaces. June 1964
4. Banchoff, Thomas Francis, Tightly embedded two-dimensional polyhedral manifolds. June 1964
5. Garland, Howard, On the cohomology of lattices in Lie groups. June 1964.
6. Gardner, Robert Brown, Differential geometric methods in partial differential equations. June 1965
7. Smoke, William, Differential operators on homogeneous spaces. June 1965
8. Weinstein, Alan David, The cut locus and conjugate locus of a Riemannian manifold. March 1967

9. Shiffman, Bernard, On the removal of singularities in several complex variables. June 1968
10. Reilly, Robert, The Gauss map in the study of submanifolds of spheres. September 1968
11 Wolf, R., Some integral formulas related to the volume of tubes. September 1968
12 Eisenman, D., Intrinsic measures on complex manifolds and holomorphic mappings. June 1969
13 Jordan, Steve, Some invariants for complex manifolds, September 1970
14. Leung, Dominic, Deformations of integrals of exterior differential systems. September 1970
15. Lai, Hon-Fei, Characteristic classes of real manifolds immersed in complex manifolds. June 1971
16. Yau, Shing Tung, On the fundamental group of compact manifolds of non-positive curvature. June 1971
17. Barbosa, Lucas, On the minimal immersions of S^2 in S^{2m}. December 1972
18. Bleecker, David, Contributions to the theory of surfaces. June 1973
19. Millson, John James, Chern–Simons invariants of constant curvature manifolds, December 1973
20. Simoes, Plinio, A class of minimal cones in $\mathbb{R}^n$, $n \leq 8$, that minimize area. December 1973
21. Cheng, Shiu-Yuen, Spectrum of the Laplacian and its applications to differential geometry. June 1974
22. Donnelly, Harold, Chern–Simons invariants of reductive homogeneous spaces. June 1974
23. Webster, Sidney Martin, Real hypersurfaces in complex spaces. June 1975
24. Sung, C. H., Contributions to holomorphic curves in complex manifolds. December 1975
25. Dunham, Douglas, Holomorphic and meromorphic vector fields on compact hermitian symmetric spaces. 1976
26. Faran, James, Segre families and real hypersurfaces. June 1978
27. Li, Peter, Eigenvalues of the Laplacian on a Riemannian manifold. June 1979
28. Shifrin, Ted, Kinematic formula in complex integral geometry. June 1979
29 Smith, Stuart Preston, Contributions to the eigenvalue problem for the Laplacian. December 1979
30. Wang, Ai-Nung, Contributions to Differential Geometry. June 1981
31. Wolfson, Jon, Minimal Surfaces in Complex Manifolds. June 1982

Permissions

Springer-Verlag would like to thank the original publishers of Chern's papers for granting permissions to reprint specific papers in his collection. The following list contains the credit lines for those articles.

[1] Reprinted from *Science Reports Tsing Hua University* **1,** © 1932 by Tsing Hua University.

[3] Reprinted from *Tohoku Mathematical Journal* **40,** © 1935 by Tohoku University.

[6] Reprinted from *Comptes Rendus de l'Academie des Sciences de Paris* **204,** © 1937 by l'Academie des Sciences.

[8] Reprinted from *Annals of Mathematics* **39,** © 1938 by Princeton University Press.

[9] Reprinted from *Journal University Yunnan* **1,** © 1938 by University of Yunnan.

[10] Reprinted from *Bull. Sci. Math.* **63,** © 1939 by Editions Bordas, Dunod, Gauthier-Villars.

[12] Reprinted from *Science Reports Tsing Hua University* **4,** © 1940 by Tsing Hua University.

[14] Reprinted from *Comptes Rendus de l'Academie des Sciences de Paris* **210,** © 1940 by l'Academie des Sciences.

[15] Reprinted from *Boll. Un. Mat. Ital.* **2,** © 1940 by Matematica Italiana.

[17] Reprinted from *Acta Pontif. Acad. Sci.* **5,** © 1941 by Acta Pontificia Academia Scientiarvm.

[20] Reprinted from *Annals of Mathematics* **43,** © 1942 by Princeton University Press.

[21] Reprinted from *Academia Sinica Science Record,* **1,** © 1942 by Institute of Mathematics, Taiwan, ROC.

[22] Reprinted from *Proceedings National Academy of Sciences USA* **29,** 1943.

[24] Reprinted from *Proceedings National Academy of Sciences USA* **30,** 1944.

[26] Reprinted from *Proceedings National Academy of Sciences USA* **30,** 1944.

[29] Reprinted from *Duke Mathematical Journal* **12,** pages 279–290, © 1945 by Duke University Press.

[32] Reprinted from *Bulletin American Mathematical Society* **52,** © 1946 by American Mathematical Society.

[34] Reprinted from *Science Reports Tsing Hua University* **4,** © 1947 by Tsing Hua University.

[40] Reprinted from *Academia Sinica Science Record* **2,** © 1948 by Institute of Mathematics, Taiwan, ROC.

[42] Reprinted from *Science Reports Tsing Hua University* **5,** © 1948 by Tsing Hua University.

[43] Reprinted from *Transactions American Mathematical Society* **67,** © 1949 by American Mathematical Society.

[45] Reprinted from *Proceedings National Academy of Sciences USA* **36,** 1950.

[46] Reprinted from *Proc. Int. Congr. Math.* **II,** © 1950 by American Mathematical Society.

[48] Reprinted from *American Journal of Mathematics* **74,** © 1952 by The Johns Hopkins University Press.

[49] Reprinted from *Bulletin American Mathematical Society* **58,** © 1952 by American Mathematical Society.

[50] Reprinted from *Annals of Mathematics* **56,** © 1952 by Princeton University Press.

[53] Reprinted from *Duke Mathematical Journal* **10,** © 1953 by Duke University Press.

[56] Reprinted from *l'Ens. Math.* **40,** © 1955 by Institut de Mathematiques.

[57] Reprinted from *Proceedings American Mathematical Society* **6,** © 1955 by American Mathematical Society.

[58] Reprinted from *Proceedings American Mathematical Society* **6,** © 1955 by American Mathematical Society.

[59] Reprinted from *Abh. Math. Sem. Hamburg* **20,** © 1955 by Universität Hamburg.

[64] Reprinted from *American Journal of Mathematics* **79,** © 1957 by The Johns Hopkins University Press.

[66] Reprinted from *Michigan Math. Journal* **5,** © 1958 by University of Michigan Press.

[67] Reprinted from *Proc, Int. Congs. Math. Edinburgh,* © 1958 by Cambridge University Press.

[80] Reprinted from *Acta Mathematica* **114,** © 1965 by Institut Mittag-Leffler.

The following paper was originally published by Springer-Verlag Heidelberg.

[77] Reprinted from *Mathematische Annalen* **149,** © 1963.